G.D. Sharma

The Alaskan Shelf

Hydrographic, Sedimentary, and Geochemical Environment

With 345 Illustrations

Springer-Verlag
New York Heidelberg Berlin

G.D. Sharma
Alaska Sea Grant Program
University of Alaska
Fairbanks, Alaska 99701
USA

Library of Congress Cataloging in Publication Data. Sharma, Ghanshyam Datt, 1931- The Alaskan shelf. Includes bibliographies and indexes. 1. Continental shelf—Alaska. 2. Oceanography—Alaska. 3. Marine sediments—Alaska. I. Title. GC85.2.A4S5 557.98 79-10454

Printed in the United States of America.

9 8 7 6 5 4 3 2 1

ISBN 0-387-90397-6 Springer-Verlag New York Heidelberg Berlin
ISBN 3-540-90397-6 Springer-Verlag Berlin Heidelberg New York

To the memory of my father
Ram Chandra

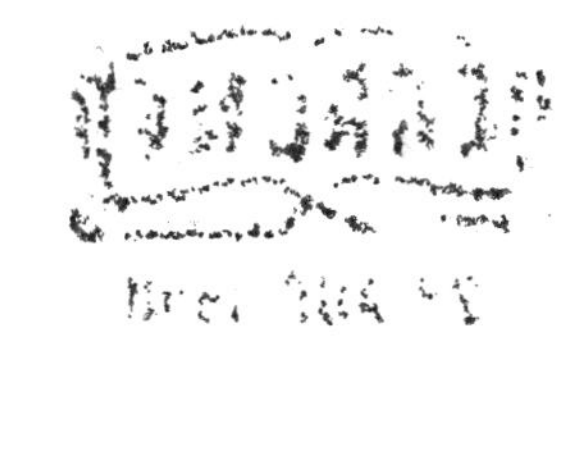

Preface

From a general point of view the importance of striving to minimize environmental disturbances on the continental shelf cannot be overemphasized. Coastal areas are sites of population centers, navigation and recreation activities, and resource development, all of which contribute to environmental stress on the shelf. Proper management of the shelf for optimum use requires a thorough understanding of shelf processes. Complex problems, such as the influence of hydrodynamics on sediment dispersal, element differentiation and migration, physiochemical changes at the sediment water interface, the relationship between the pollutants and sediments, and the type of substrate with regard to benthic community and/or man-made structures require a multidisciplinary approach to their solution. The present study interrelates meteorologic, hydrographic, sedimentologic, and geochemical parameters to define specific environments on the Alaskan Shelf. These observations are then related to geologic principles in an effort to elucidate the sedimentary processes and elemental migration on the shelf.

Attempts have also been made to relate the sediment texture to the geochemistry of the sediments. Obviously the chemistry is complicated as a result of biogenic contributions and variable provenance; however, to some extent elemental differentiation accompanies textural differentiation in sediments.

The distribution of elements in various phases of crustal (source) rocks is generally interpreted on the basis of crystallographic concepts, especially the concept of isomorphism. This is the basis of our conception of distribution and abundance of the elements and is a starting point for the hypothesis of the origin of elements. Studies of geochemical balance in different processes, such as weathering of crystalline rocks and sedimentation, have provided us with information on the mechanism of separation and concentration of the elements. The understanding of the sedimentary rocks and their elemental distribution, however, requires the understanding of the processes of their formation as well as of aquatic chemical processes. This concerns the migration of the chemical elements, in various specific forms (i.e., ions, complexes, atoms), and the development of a physiochemical theory related to marine geologic processes. Emphasis is placed on this theme throughout this treatise. The textural, mineralogic, and geochemical parameters are extensively used to identify the sources and pathways of sediments on the shelf. The interrelations between the texture and geochemistry of shelf sediments reveal the effects of marine transport on the mineral and elemental differentiation.

There are undoubtedly shortcomings in this book, particularly in regard to station coverage, and perhaps a need for some additional analyses on sediments. Although it is always highly desirable to have additional data, the logistic problems and the harsh climatic conditions place formidable limitations on the collection of comprehensive data.

G. D. Sharma
Professor of Marine Science
Alaska Sea Grant Program
University of Alaska
Fairbanks, Alaska

Acknowledgment

This volume represents the work achieved over a number of years during which many students and assistants provided their talents in the laboratory as well as onboard ship. Special thanks are extended to Miss Shalini Sharma and Mr. David Burbank who helped generously during collection and analysis of samples, data processing, and drafting.

I am especially grateful to Dr. Joe S. Creager for providing sediment samples from some regions of the Chukchi and Bering seas and for his generous cooperation and assistance extended to me during my stay at the University of Washington.

The computer analysis was performed under the supervision of Mr. Ivan Frohne, Dr. Dan Hawkins, and Dr. James Kelley, and their expert assistance is much appreciated.

The book was reviewed by Dr. Tom Mowatt, Prof. Al Belon, Dr. D. M. Hopkins, and Prof. G. M. Friedman. These reviews deserve many thanks for their cooperation and quality efforts.

Finally, I wish to thank my wife, Norma, for typing the original draft and for her patience.

Contents

CHAPTER 1
Regional Setting

INTRODUCTION

Alaska is a subcontinental and peninsular landmass with an unusually long coastline. This extended shoreline, bordered on the south by an active and young mountainous belt, provides an extensive interplay between the land and adjacent marine environments. The general coastline of Alaska (general coastline is a term to describe the coast; it includes bays, but crosses narrow inlets and river mouths) is 10,685 km (6,640 miles*) long, representing 54% of the total 19,928 km (12,383 miles) general coastline of the United States. The tidal shoreline of Alaska (intersection of high-tide water with the shore) is much longer and is about 76,120 km (47,300 miles) long, representing about 53% of the total 142,637 km (88,633 miles) of the tidal shoreline of the United States.

The great areal coverage of the shallow shelf off Alaska is noteworthy. The continental shelves adjacent to Alaska are the southeastern Alaska Shelf, the Gulf of Alaska, the Bering Sea, the Chukchi Sea, and the Beaufort Sea. Together they cover an area of 2,149,690 km^2 (830,000 sq miles), about 74% of the total 2,900,785 km^2 (1,120,000 sq miles) of the United States continental shelf. Some parts of the shelf form a broad transition zone between continental crust and oceanic mass, while other parts of the shelf display a distinct boundary. The shelf is susceptible to seismicity, destructive earthquakes, volcanism, and has major petroleum and mineral resources. It is estimated that more than 50 billion barrels of oil are trapped in the various basins of the Alaskan Shelf. No systematic attempt has been made so far to assess the availability of other minerals. The biological resources harvested annually constitute one-fifth of the total of the nation's commercial fisheries. The increased fishing in past decades and recent development in Alaska have resulted in increasing utilization of the waters around the state. Currently, there is concern over the use and abuse of many regions along the Alaskan coast, in particular the oil impact regions.

From the academic standpoint, the Alaskan Shelf is well suited to various geological studies. In southeastern Alaska, the shelf is a part of or lies adjacent to an active tectonic belt and receives freshly eroded sediments. The study of early changes in texture and mineralogy of these sediments during marine transport

---

*Marine linear measurements are, conventionally, in terms of nautical miles nm; on land, however, measurements are in terms of statute miles.

should be of value. Farther north, the tectonic belt merges into a volcanic arc bordered by a trench. The trench is a classic subduction zone, where the Pacific Plate underthrusts the Aleutian Chain. Such a setting is an ideal place to investigate the redistribution of elements in sedimentary rocks at depths and metallogenesis along a subduction zone.

The large shallow shelves in the Bering and Chukchi seas are unique. These shelves contain complex records of the major transgressions during Quaternary time. Extremely smooth featureless shelves are also an ideal natural laboratory for scrutiny of sediment transport by various processes prevalent in subarctic and arctic climates. The northern Alaskan Shelf is dominated by ice during most of the year, and therefore ice-related sedimentary marine processes can be studied in this region.

Significant variability in sediment character is observed throughout the shelf. The vast latitudinal expanse of the Alaskan Shelf provides an opportunity to compare differences in textural and compositional parameters of the sediments formed in temperate and arctic environments. Changes in sediment texture, mineralogy and geochemistry are commonly attributed to source, weathering, and regional setting (tectonic and hydrographic), and related sedimentary processes. The early stages of differentiation at the source which provides detritus for the shelf are related to the initial composition of the source rock as well as to the weathering agents. There are resultant differences in mineralogy, manifested in their geochemistry, in these products of weathering. Sediment texture distribution is mostly related to the regional setting and the sedimentary processes. This suggests that an approach attempting to elucidate relationships among processes, latitude and corresponding sediment textures ought to be of some significance.

In addition to the enormous length and breadth of the Alaskan Shelf, the environment prevailing over it is unique in many respects. The river runoff and coastal sediment discharge to the shelf have seasonal extremes. In southeast Alaska, fluctuation in discharge is primarily controlled by variations in glacial melts, whereas along the arctic shores it is dependent on the variable ice breakup periods. Mountain- and glacial-spawned winds are an active agent of sediment erosion and transport in the southeastern Alaska region. Volcanic and seismic activities influence the sedimentary character of the central Alaskan Shelf. The annual cycle of coastal thawing and sea ice movement has the main effect on sedimentary processes in the high latitudes. Obviously the vast latitudinal expanse of the Alaskan Shelf encompasses a variety of meteorologic and oceanographic regimes. Each of these has a characteristic sedimentary environment (Fig. 1-1).

With these factors in mind, it was considered appropriate to describe the Alaskan Shelf in terms of various geographic sections, so that each section would exemplify a dominant sedimentary process and a related sedimentary environment on the shelf (Fig. 1-2). The entire coastal region, including the shelf, therefore is divided and described as follows:

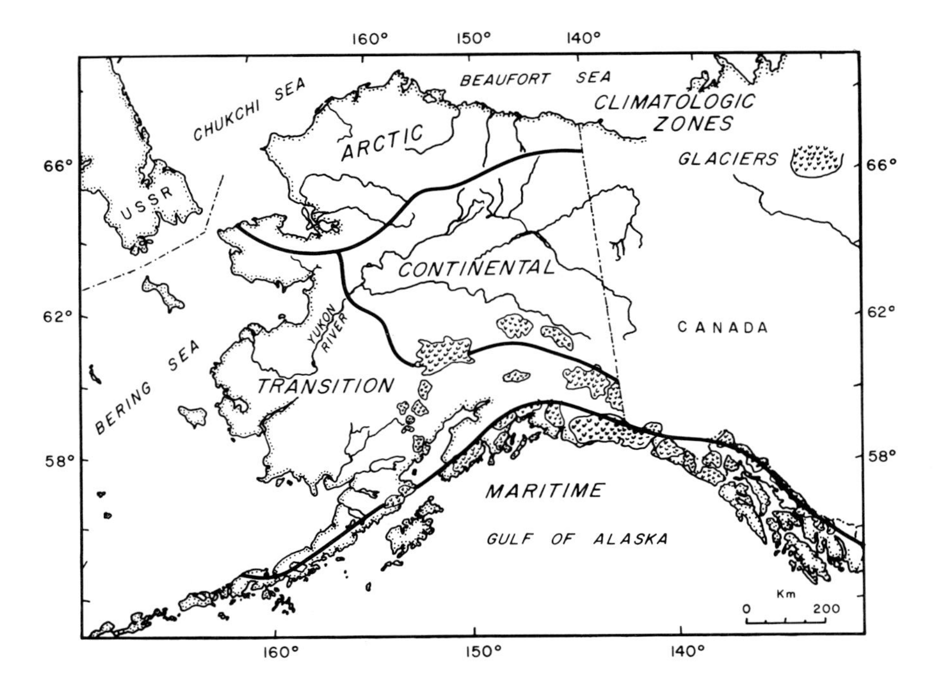

Figure 1-1. Major climatic zones in Alaska.

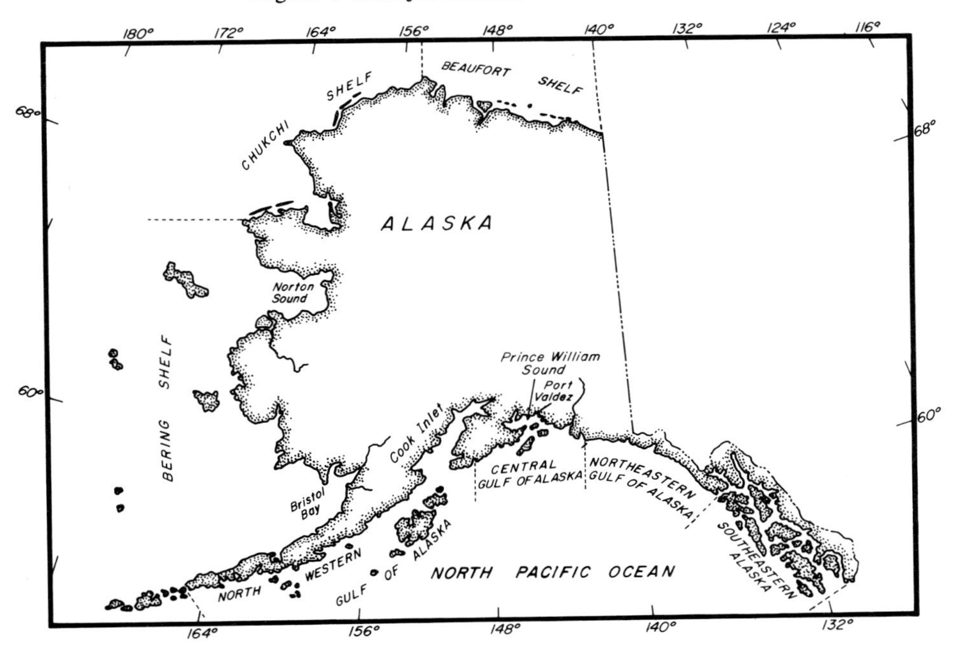

Figure 1-2. Major divisions of Alaskan Shelf.

1. Southeastern Alaska: Alaskan Southeast Panhandle and the Inside Passage.
2. Northeastern Gulf of Alaska: The region between Cross Sound and Kayak Island.
3. Central Gulf of Alaska: The complex and arcuate region between Kayak Island and Resurrection Bay.
4. Prince William Sound: A semi-enclosed basin landward of the Central Gulf of Alaska.
5. Port Valdez: The deep fjordal basin at the northwest periphery of Prince William Sound.
6. Northwestern Gulf of Alaska: The region between Resurrection Bay and Unimak Pass, including the shelf surrounding Kodiak Island.
7. Cook Inlet: A large high-latitude tidal estuary.
8. Bering Sea: A large region between Alaska and Siberia and between the Alaska Peninsula and the Bering Strait.
9. Chukchi Sea: The region between the Bering Strait, Point Barrow and the Northwest Coast of the Chukotka Peninsula of Siberia.
10. Beaufort Sea: The region between Point Barrow and Demarcation Point (Canadian border).

These are only broad subdivisions of the Alaskan coastal region. There are numerous marine-land-atmospheric systems which are unique to gulfs, bays, sounds, inlets, and estuaries within each subdivision. It is beyond the scope of this book to describe the complex systems prevailing in each of these water bodies. Nevertheless, the general sedimentary processes and resultant sedimentary products prevailing in such water bodies have been exemplified by describing a representative area in considerable detail.

SOUTHEASTERN ALASKA

The southeastern Alaska region is dominated by the Coast Mountains, part of the Pacific Mountain System. The southern part of the region consists of two mountain ranges separated by the Kupreanof Lowlands. Short distances (20-50 km) from the shoreline, these mountains reach altitudes between 2,500 and 3,000 m. The rugged and precipitous coast has a complex structure with a web of intersecting faults and folds. The entire coast is cut by channels and straits, which isolate numerous small as well as large islands; the region is commonly termed the Alexander Archipelago. The indented shoreline contains numerous fjords, inlets, bays, and estuaries. The shoreline is estimated at 50,000 km, which constitutes approximately 63% of the total Alaskan shoreline.

The region has been repeatedly glaciated, and is the site of contemporary glacial activity as well. There are numerous permanent ice fields, from which glaciers descend to the lowlands, sometimes directly into the sea as the tidal glaciers. For the most part, the region is characterized by fjords, often with one or more sills; estuaries with sedimentary deltas; and channels. The tidal glaciers, glacial meltwater streams, and rivers entering fjords and inlets directly all influence the marine environment. In fjords, the hydrography is controlled by the sill depth at the entrance and by the tidal flux. Exchange of water between a fjord and the sea occurs only to sill depth. Freshwater flow and tidal oscillation are often sufficient to set up a circulation which permits year-round mixing in the basin. The resulting sediment distribution is intricately related to the movement of water in fjords and estuaries.

NORTHEASTERN GULF OF ALASKA

The physiography of the northeastern Gulf of Alaska is dominated by the narrow (up to 25 km) coastal plain and by the adjacent rugged, massive St. Elias, Chugach, and Fairweather Mountain Ranges which rise to an altitude of 6,500 m. These mountains are primarily eroded by glaciers. Some of the largest known piedmont glaciers, Malaspina and Bering glaciers, occur in this region. The glaciers generally descend from higher elevation to the coastal plain, where they form numerous glacial-melt streams. These swift, short, and turbid streams build outwash deltas and carry glacier discharge to the shelf.

This region, except for a few inlets, has a relatively smooth shoreline. The largest of the inlets is Yakutat Bay; the other two and smaller indentations are Lituya and Icy bays.

CENTRAL GULF OF ALASKA

Landward, the central Gulf of Alaska region is bordered by the extremely rugged, arcuated Chugach-Kenai Range. From the shore, the mountains rise very rapidly to an elevation of about 4,000 m. The central, northernmost part is presently covered by ice and has been extensively glaciated. Generally, the coast is deeply indented by long northeast-trend fjords and inlets which are often separated by narrow ridges. These ridges commonly rise to about 1,500 m above sea level. The tidal shoreline of the region is estimated at 9,000 km, about 12% of the total Alaskan tidal shoreline.

The eastern part of the Gulf receives the discharge of the Copper River, a major river which roughly bisects the Chugach Mountains and receives its drainage from the north. The river has built an unusual delta near its mouth and a string of barrier islands offshore.

The entire region is the "burial place" for southwesterly storms, which approach from the north-central Pacific and Aleutian areas. The region is ravaged by high surface winds, violent sea states and unpredictable currents. The winds reach velocities of 100 to 150 km/hr. The open shoreline, where most of this energy dissipates, consists of narrow, high-energy beaches.

PRINCE WILLIAM SOUND

Prince William Sound, a large, crescent-shaped, semi-enclosed basin, is separated from the gulf by a chain of northeast trending islands. Restricted exchange of water between the open ocean and the sound occurs through various shallow openings between islands. The shoreline within the sound is typical of a glaciated region and is characterized by glacial fjords and other indentations. A few glaciers are tidal, of which the best known and most frequently visited is Columbia Glacier. The coastal topography is steep and rugged, but a few outwash deltas are built near the stream and river mouths.

PORT VALDEZ

Port Valdez is the deepest reentrant in the Chugach Mountains bordering the northeastern corner of Prince William Sound. It is a typical fjord with steep sides, but with an unusually flat floor. The fjord receives the discharge of several rivers near its head. The rivers have built an impressive delta at the head of the fjord. Near its entrance, Port Valdez has two sills.

NORTHWESTERN GULF OF ALASKA

The northwestern Gulf of Alaska region extends from Resurrection Bay to Unimak Pass, including Kodiak, Afognak and numerous other small islands. It encompasses the southwestern portion of the Kenai Mountains and the south side of the Alaska Peninsula. This tidal shoreline is estimated at 13,000 km, which is approximately 18% of the Alaskan total. The hinterland rises abruptly from the shoreline and effectively traps moisture of the northward moving humid North Pacific airmass.

The Kenai Mountains are rugged and have a typical glaciated topography; the shoreline is indented with numerous fjords, inlets, and sounds. The northeast trending structure is dissected by numerous northwest oriented valleys.

The Kodiak-Afognak islands group is also a northeast trending structure, and is extensively dissected by northwest trending valleys. The mountain elevation reaches about 1,500 m. The coastline is rugged and irregular and consists of many fjords and islands. The drainage basins are small, with short and swift streams.

The Alaska Peninsula is approximately 800 km long and 30 to 100 km wide. It is dotted with numerous volcanoes, dormant as well as active. The topography is rugged and glaciated, and the shoreline is indented with numerous fjords and inlets. Drainage is divided into small areas, giving rise to small streams that are short but swift.

COOK INLET

Cook Inlet is a large tectonic downwarp which forms an elongated arm of the sea. It receives drainage from the Chugach-Kenai Mountains to the east, the Alaska Range and Talkeetna Mountains to the north, and the Aleutian Range to the northwest. This drainage area is rather large and thus the inlet receives a significant amount of freshwater inflow. Major rivers draining into Cook Inlet are the Knik, Matanuska, Susitna, and Drift.

Whereas the shoreline of the lower third of the inlet, near the entrance, is rugged and glaciated, the upper two-thirds are bordered by extensive muddy tidal flats. The tidal shoreline of Cook Inlet is approximately 650 km, less than 1% of the Alaskan total shoreline.

A unique feature of this high-latitude estuary is the unusual amplification of tides near its head. The main diurnal tidal range varies from 4.1 m near the entrance to 8.9 m at Anchorage and the extreme tidal height causes swift currents and considerable turbulence, particularly in the region of the Forelands where current velocities reach about 200 cm/sec.

Upper Cook Inlet is covered with ice from December through March. Currents continually move ice to the south, where it breaks up into small pieces and rapidly dissipates. The southward moving ice significantly affects the hydrographic and sedimentary regimes in the inlet.

BERING SEA

In the Bering Sea the shelf extends from the northern shores of the Alaska Peninsula to the Bering Strait. Diversity in physiography is characteristic. Among the major physiographic features are the Alaska Peninsula to the south, the Kuskokwim Mountains adjacent to Bristol Bay, a large complex of deltas between the Kuskokwim Mountains and the Seward Peninsula, and the Seward and Chukotka peninsulas forming the Bering Strait. The extremely smooth shelf has three large and a few small islands. There are three broad regions: Bristol Bay, which is the shelf area lying between the Alaska Peninsula and Kuskokwim Mountains; the shelf lying adjacent to the Yukon-Kuskokwim Delta; and the region between the Yukon Delta and Seward Peninsula.

Bristol Bay, a large bight, is bordered by the Alaska Peninsula to the south and the Kuskokwim Mountains to the north. Its eastern and northeastern shores are surrounded by the littoral lowlands. The Alaska Peninsula, to the south, rises gradually to altitudes of up to about 1,250 m. The peninsula is also dotted with volcanic cones which rise to elevations of between 1,500 and 3,000 m. Extensive glaciation occurs and it is drained to the north through the lowlands by streams with braided channels. The coastal lowlands contain numerous small and few large lakes, while the upper reaches of the lowlands generally consist of moraines and outwash, often covered with eolian sand and silt. The northern part of

Bristol Bay is dominated by the southeastern flank of the Kuskokwim Mountains, generally known as the Ahklum Mountains. These steep and rugged mountains are heavily glaciated.

Alaska's two largest rivers, the Yukon and the Kuskokwim, form a huge delta. Between the Kuskokwim Mountains and the Seward Peninsula these rivers have formed an extensive triangular delta of subcontinental magnitude, covered with marshland, lakes, ponds, and meandering streams. Presently the Yukon River is actively prograding the delta in the central and northern Bering Sea. On the other hand, the Kuskokwim River, in contrast, flows south and forms a large estuary in Bristol Bay. The deltas of these two rivers are separated by a topographic high, part of which forms Nunivak and Nelson islands.

The Alaskan portion of the northern Bering Sea region is bordered by Seward Peninsula to the north and the Norton Sound coastal area to the east. Seward Peninsula consists ob broad convex hills and uplands and the region has been extensively glaciated in the past. Much of the peninsula is drained by two rivers; the Koyuk River flows eastward and south into Norton Bay, and the Kuzitrin River flows westward into the Imuruk Basin-Port Clarence region.

The Nulato Hills border on eastern Norton Sound. To the south the sound is bordered by a small strip of coastal lowland and the Yukon Delta.

Although the Bering Shelf constitutes a large portion of the Alaskan Shelf the tidal shoreland is relatively small, and tidal shoreline in the eastern Bering Sea is estimated at 2,900 km, which is about 4% of the tidal shoreline of Alaska.

CHUKCHI SEA

The Chukchi Sea region is bordered by the Seward and Chukotka peninsulas in the south and the Arctic Ocean to the north. Dominating major coastal physiography is the western flank of the Brooks Range, a major east-west mountain range which cuts across northern Alaska. This range forms coastal promontories along the eastern margin of the Chukchi Sea. For the most part, however, the eastern shoreline lies along the coastal plain, which was submerged until recently (circa 125,000 years ago). In some areas this coastal plain forms a narrow belt; in other areas it extends deep into the hinterland.

A major feature of the large shallow epicontinental shelf is the large embayment in the southeastern corner, Kotzebue Sound. This shallow embayment receives the discharges of the Kobuk and Noatak rivers and is being rapidly filled with sediments.

The eastern nearshore zone of the Chukchi Sea, except for the sea cliffs Cape Thompson and Cape Lisburne, is universally characterized by coastal lagoons and barrier islands. These islands are formed by sediment drifts and longshore currents.

North of Point Hope the coast is aligned along structural lineaments. The coastline between Point Hope and Cape Lisburne is formed by the structural block of the Brooks Range, with north-south lineaments, whereas the coastline between Point Lay and Point Barrow has three 90-km-long linear coastal stretches which are successively offset to the east by approximately 25 km. Linear trends of the coast are apparently associated with the major 35° structural lineaments.

BEAUFORT SEA

The Beaufort Sea region lies along the northern coast of Alaska and extends from Point Barrow to Demarcation Point (Canadian border). Landward, the region is bordered by a wide coastal plain. All the major rivers of arctic Alaska rise along the 1,000-km-long Brooks Range, flow across the tundra hills and coastal plains, and debouch along the shelf. This coastal plain is literally covered with lakes, ponds and meandering streams.

The coastal zone is irregular and is characterized by a combination of arcuate barrier islands separated by broad expanses of open water and backed by lagoons, numerous large deltas, embayments, and tundra shoreline. Its shoreline is estimated at about 800 km or less than 1% of the tidal shoreline of Alaska. Four major barrier islands and the general tundra coastline are aligned to major lineament trending 312°. Because of prevailing northeasterly winds and waves there is a westerly migration of bars and spits.

Rivers draining into the Beaufort Sea commonly build deltas, the largest of which is formed by the Colville River. Such river mouths are bell shaped and have well-developed offshore bars. Landward, the deltas have braided and meandering channels, abandoned channel scars, and channel lakes.

Shallow lagoons formed between the offshore islands and the mainland are generally narrow and elongated. The southern periphery of these lagoons is characterized by crenulated shorelines and bluffs. These shore bluffs attain elevations of about 6 m.

Polar pack and nearshore fast ice affect the coastline and the shallow shelf. In some cases offshore barrier islands and shorefast ice protect the coast from the erosive polar pack ice.

SEDIMENT PARAMETERS

Textural

Introduction. Numerous sediment properties have been used to characterize contemporary marine sediments. One property widely used and universally accepted is the particle size distribution, employing various statistical measures. Alternately, the particle size distribution can also be expressed on a triangular plot of gravel, sand, silt, and clay percents. Recent attempts to apply these parameters to determining the source and the sedimentary environments of deposition have shown some inadequacies. Therefore, sediment grain size distribution, when used to delineate the source and sedimentary environments, essentially requires supplementary information concerning the sediments as well as the water carrying them. Almost all the conventional sediment parameters have been used throughout this book to characterize the sediments of the Alaskan Shelf.

Mean Size. Sediment mean size on the Alaskan Shelf is primarily a function of mixing of subequal proportion of the two main modes; silt-clay and gravelly, silty, sand. The silt-clay mode mostly represents sediments carried in suspension, whereas the latter mode is transported near the bottom as traction. The movement of coarse material as traction develops a progression of various modes of decreasing medium size in the offshore direction. For instance, in Bristol Bay, where the high-energy environment inhibits deposition of fine fractions and sediment deposition is a result of traction (essentially bed load), a succession of sediments with seaward decreasing mean size becomes apparent. In most instances, however, the coarser fraction invariably is mixed with the finer fraction and the net result is deposition of bimodal sediments.

Additional complexity in the lateral distribution of the mean size parameter is introduced by the exposed glacial relicts and accumulation of biogeneous sediments on the shelf. These anomalous regions may sometimes lie farther offshore and thus do not fit in the normal scheme of water depth-mean size relationship commonly observed in the adjacent region. Furthermore, in numerous cases these exposed relicts serve as a source of sediments for the surrounding region. Finally, the mean size distribution may be influenced by extreme fluctuations in both intensity and nature of the transporting agents. These include storms, terrestrial and aqueous slides, and ice rafting.

Generally, there is a marked relationship between the sediment mean size and water depth: mean size decreases with increasing water depth. The interesting feature of the variance of mean size with water depth is the interrelation between the steepness of slope, wave energy and the unimodality of sediments. This is particularly prevalent in the nearshore areas.

Because of the mean size-depth relationship, except for the exposed relicts, topographic highs are well identified by the sediment mean size. Contours of mean size generally conform with bathymetry to a depth of effective sorting by wave action. Beyond that depth the mean size contours hardly conform to the isobaths. This transition suggests change in either depositional environment or transporting agent. This generally is the silt-clay environment.

Last, the mean size distribution on the shelf may be influenced by the bottom currents. Currents on the various parts of the shelf have been measured and are quite competent to carry sand particles. Generally, the more intense and hydrodynamic the regime, the larger the mean size of the sedimentary zone. Alternate seaward decrease and increase in sediment mean size observed on the shelf can be understood if it is borne in mind that the sediments in some areas are primarily transported by currents along the shore and are partly carried offshore. Offshore shelf currents flowing parallel to the shore usually develop such patterns.

Sorting. Sorting and mean size in sediments are so interdependent that sometimes mean size distribution also reflects sorting. The best sorting occurs in sands deposited on the midshelf; the sorting progressively deteriorates offshore and onshore from midshelf. This decrease in sorting results from mixing of sand with the coarse mode landward and silt-clay mode seaward.

Sorting in sediments is also affected by the tidal, longshore, and permanent currents, which tend to inhibit deposition of fine fractions. Winnowing of finer fractions from the topographic high by currents is also quite common.

Poor sorting in coarse sediments from offshore can be readily used to identify relict sediments of glacial origin. Poor sorting in sediments may also be attributed to the biogenic contributions.

Skewness. Skewness is a good indicator of the modal nature of sediments. Unimodal distributions are symmetrical, as are equal mixtures of two modes. The mixtures of unequal amounts of modes cause extreme values of skewness. Depending upon the dominance of coarse or fine modes the skewness value will be positive or negative. An excellent example of skewness as a function of biomodality is displayed by sediments from Bristol Bay. The unimodal sediments from the midshelf display a near-symmetrical distribution, but seaward the finer mode increases and sediments become positively skewed. Similarly, the coarser fraction increases shoreward and sediments become negatively skewed.

Winnowing of fine sediments from topographic highs and areas with bottom currents is also reflected by skewness. The changes in skewness values are, however, generally minor.

Kurtosis. Kurtosis, like skewness, is a valuable indicator of the modal relationship in sediments. The nearly unimodal sediments are sharp peaked (leptokurtic), while the sediments with subequal proportions of two modes become progressively mesokurtic, and sediments with equal amounts of the modes are flat and broad peaked (platykurtic).

Variations in kurtosis generally show similarlity in the sorting of the sediments. Although skewness and kurtosis both express the modal character of the sediments, the patterns are often dissimilar. This is primarily due to the extreme sensitivity of the kurtosis measure. Kurtosis varies from extremely leptokurtic to extremely platykurtic.

Triangular Plot. The triangular plot with gravel-sand, silt, and clay as corners, provides a convenient means of describing gross texture of the sediments. One plot for each region is included in each chapter and the comparison of plots should provide an easy and convenient reference for regional textural variations.

It may be noted that invariably gravel and sand are taken as one unit. These diagrams generally indicate the mixing of major modes.

Summary. The sediment texture can be adequately described in terms of the major modes, gravel, sand, silt, and clay, and their distribution is the combined result of sorting by waves and currents, and source character. Sediment mean size and sorting delineate sediment trends in terms of their movement and distinguish some hydrodynamic conditions as well. Skewness and kurtosis, on the other hand, are effective indicators of the modal character of the sediments. All these parameters combined are helpful in characterizing the sediments, their source, and their migration in a particular environment. Simultaneously they also reveal the genesis of sediment particles and their deposition on the shelf.

Chemical Analyses

Each sediment sample was dried and powdered. A portion of the powdered sample was weighed and ignited in the muffle furnace for 15 minutes at 780°C, and loss of weight was determined. Another portion of the powdered sample was placed in the oven set at 60°C for 24 hours for drying. One-half gram of the dried sample was put in a platinum crucible, covered with perchloric acid, and placed in an electric furnace for fuming. The crucible was removed from the furnace and, upon cooling, approximately 3-5 ml of concentrated hydrofloric acid was added. The crucible with the sample and HF was then placed on a water bath for 24 hours. At the end of the HF digestion, the sample was dried and moistened with a few drops of perchloric acid and once again placed in the electric furnace for fuming. The dried sample was put into solution in 50 ml of 30% hydrochloric acid for elemental determination.

The cold digestion was carried out in batches of ten samples. Each batch consisted of three standard samples AGV, BCR, and PCC, five sediment samples and two duplicates. Three standard and two duplicate samples provided control over the digestion process. In case the chemical components of any of the standard samples varied more than 3% of those reported, the entire batch was repeated for digestion. With variations greater than 3% in the components of the duplicate sample, the batch was also discarded. The elemental distribution in samples was determined using routine atomic spectrophotometric analytical techniques. Special dilutions and methods were used for a few of the elements. All determinations were made against the standards (AGV, BCR, and PCC) digested during each batch.

CHAPTER 2
Southeastern Alaska

INTRODUCTION

Southeastern Alaska lies in a tectonic belt with structural trends oriented parallel to the Pacific Coast, as noticed by Suess (1892). These trends are now explained by the theory of plate tectonics, which postulates lateral movement of lithospheric plates. The region lies along the actively moving oceanic plate and has structural trends parallel to the contact. Mountainous belts formed along the edge of the continental plate have extensive strike-slip faults and folds. Tectonically mobile hinterland is fronted by a narrow continental shelf. Orogenic activity is high and the region is known for volcanic and earthquake activity. Consequently, the steep mobile hinterland provides a large input of sediments and marine sedimentation is accelerated by frequent earthquakes and related tsunamis.

Two events which have led to the present major features of the continental shelf and the contemporary sedimentation in this region are:

1. Uplift of a north-northwest oriented Paleozoic eugeosyncline during the Cretaceous-Tertiary.
2. Glaciation and sea level changes during the Pleistocene and perhaps Mio-Pliocene.

The major morphologic features have their origin in the uplift with its associated intense metamorphism and deformation. The surficial morphology, on the other hand, was controlled by Pleistocene glaciation and Holocene changes in sea levels.

GEOLOGY

Pre-Cambrian and Cambrian rocks are not exposed in Southeast Alaska; therefore, the early history of the region remains obscured. However, the absence of these sequences may indicate that the region was not part of the crust prior to the Cambrian. From the Early Paleozoic through the Late Mesozoic, the north-northwest oriented eugeosyncline intermittently was filled primarily with graywacke, pelite and volcanic deposits. During the Late Mesozoic and the Tertiary, the region was uplifted, deformed, and metamorphosed. The Tertiary Orogeny was accompanied by the emplacement of granitic masses, in particular the Coast

Range composite batholith. Concurrently strike-slip faulting and extensive folding along a north and northwest structural trend occurred.

The intrusive rocks, mainly coarse-grained granodiorite, form the great masses of the Coast Range bordering the mainland and occupy wide areas in the central portions of many of the islands. The mode of occurrence is at many places related to geologic structure, and their longitudinal axes and lines of contact are usually parallel with the strike of the bedded rocks. The occurrence of these intrusive rocks is found from south of the Canadian border to the northern end of Lynn Canal, with their western boundary paralleling the main coastwise channels. Bordering this intrusive core of the Coast Range is a band of closely folded crystalline metamorphosed Paleozoic strata. Low topographic areas or arms of the sea occupy the crest or trough lines of anticlinorial or synclinorial structures.

The Tertiary Uplift was followed by deposition of continental sandstones, conglomerates and volcanics. The Late Tertiary deposits consist of sandstones, siltstone and volcanics.

Extensive faulting has formed a maze of islands, inlets, bays, and water passages (Fig. 2-1) throughout the region, and appears to be part of one of the major northwestern Pacific Ocean rim faults (Ovenshine and Brew, 1972). One of the most extensive and evident faults, Chatham Strait, is about 425 km long and forms a link between the Shakwak Fault to the north and the Fairweather-Queen Charlotte Fault to the south. This strait is an elongated fjord eroded along the fault zone and has a width between 5 and 10 km and depths between 330 and 900 m; it cuts across the general trend of the mountain range and bedrock structure.

During the Pleistocene, the geomorphic development of the region was strongly influenced by glaciation and sea level changes. Glaciation has been the most significant factor in modifying the land form. Many glaciers have scoured the area, the greater number of which are presently receding. Past glaciation by ice sheets and alpine glaciers has had marked effects. The courses of many present and past glaciers have been partly determined by faults and shear zones in this region. Preglacial drainage lines were widened and deepened to form U-shaped valleys and fjords, and the lower mountain peaks and ridges were rounded.

Presently the landscape is characterized by steep topography, many streams, heavy precipitation, numerous alpine glaciers and many snow and ice fields. Mountains are separated by steep-sided glacial valleys and fjords. Deep sounds and long arms of the sea extend inland, and many small inlets and bays form an extremely irregular coastline. The shoreline characteristics are affected largely by the composition and structure of the rocks. Intrusive rocks tend to form resistant headlands; limestone tends to form a block shoreline where it is exposed to the wave action of the open sea, or irregular shorelines where it is protected. The dangerous offshore reefs and rugged coastlines of southeast Alaska are formed by widespread volcanic rocks, often associated with thinly interbedded rocks of many different lithologies.

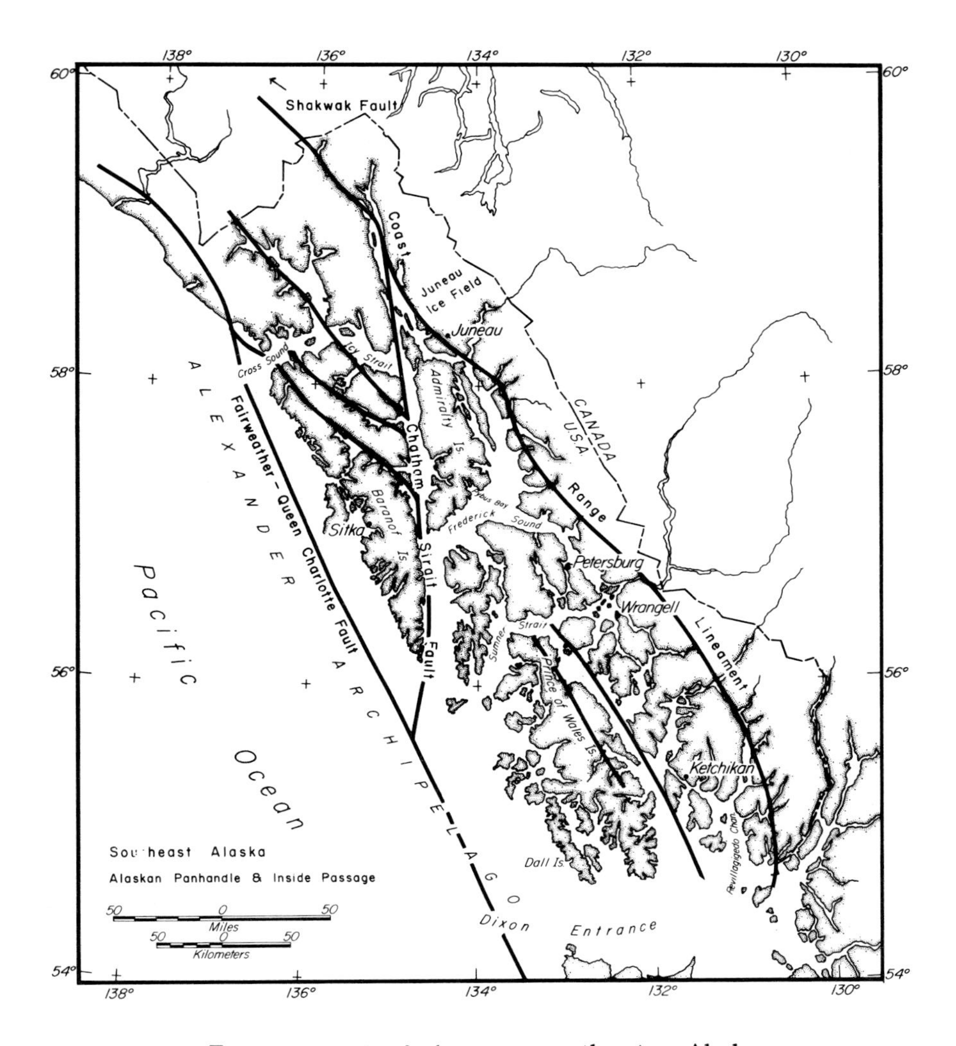

Figure 2-1. Major fault system, southeastern Alaska.

The southeastern Alaska Panhandle is a primary or youthful coast. Its shoreline character was produced primarily by diastrophism (faults and folds), subaerial erosion, and subsequent drowning as a result of the rise in sea level accompanying deglaciation. The submergence of glacial troughs is indicated by:

a. their longitudinal profiles, which are not graded but frequently show basins and reverse slopes, and
b. hanging valleys which enter from the sides.

Bedrock exposures in southeast Alaska are mostly Paleozoic sedimentary and volcanic rocks. A smaller part of the bedrock is comprised of a belt of Mesozoic rocks as well as scattered remnants of the Cenozoic system. Igneous dikes, sills, stocks and small batholithic masses cut through the Paleozoic and Mesozoic sediments.

The lithology is widely varied both laterally and vertically. Paleozoic rocks typically include thinly interbedded graywacke, tuff, conglomerate, lava flow, slate, tuffaceous or arkosic sandstone, limestone, dolomite and chert in varying combinations. Volcanic rocks are abundant, as are thick sections of marine limestones, within the Paleozoic sequence. Most volcanic sediments of the Paleozoic and the Mesozoic eras are closely folded and metamorphosed to greenstone green schist; the Tertiary rocks are relatively flat lying and unaltered.

Intrusive rocks are widely distributed and are generally believed to be genetically related to the granitic-dioritic Coast Range batholiths complex. The metamorphosed rocks are siliceous mica schists, feldspathic schists, and chloritic schists and occasional beds of crystalline limestones. These schists are cut by a network of pegmatite dikes and quartz veinlets. Close to intrusive masses, some of the sediments have been metamorphosed to gneiss, often grading to schists and slates away from core.

BATHYMETRY

Southeast Alaska has many inlets, straits, canals, arms, and bays (Fig. 2-2). These terms are informally applied as place names and do not have any consistent geomorphologic or hydrographic significance. The entire coast of southeast Alaska is indented with numerous long, narrow, fjord-type inlets, bays, estuaries, and passages. Longitudinal profiles of the bottom morphology of some inlets have a single sill at the mouth, whereas others have several sills along the longitudinal axes. The sill depths in these inlets are tabulated in order from the mouth (entrance) to the head of each inlet (Table 2-1). The listing of each parameter which basically defines a configuration of the major basins is an attempt to provide a description, although limited, of the sedimentologic and hydrographic conditions prevailing in each area.

The sills composed of bedrock are generally formed as a result of variance in bedrock lithology, structure or both. In other cases sills may be end moraines deposited during the Pleistocene ice retreat. Subbottom seismic data suggest that most sills have bedrock cores, even if the bedrock is covered by unconsolidated materials. Such bedrock cover tends to be oriented parallel to regional structures. The coarse glacial sediments observed on some sills are an enhancement of the original bedrock sills. It appears that during the ice retreat the glacier's terminus tended to remain stationary for some period when it reached a topographic high. The prolonged ablation at the glacier face near the topographic high resulted in sediment accumulation, which tended to cover the underlying, structurally aligned bedrock sills.

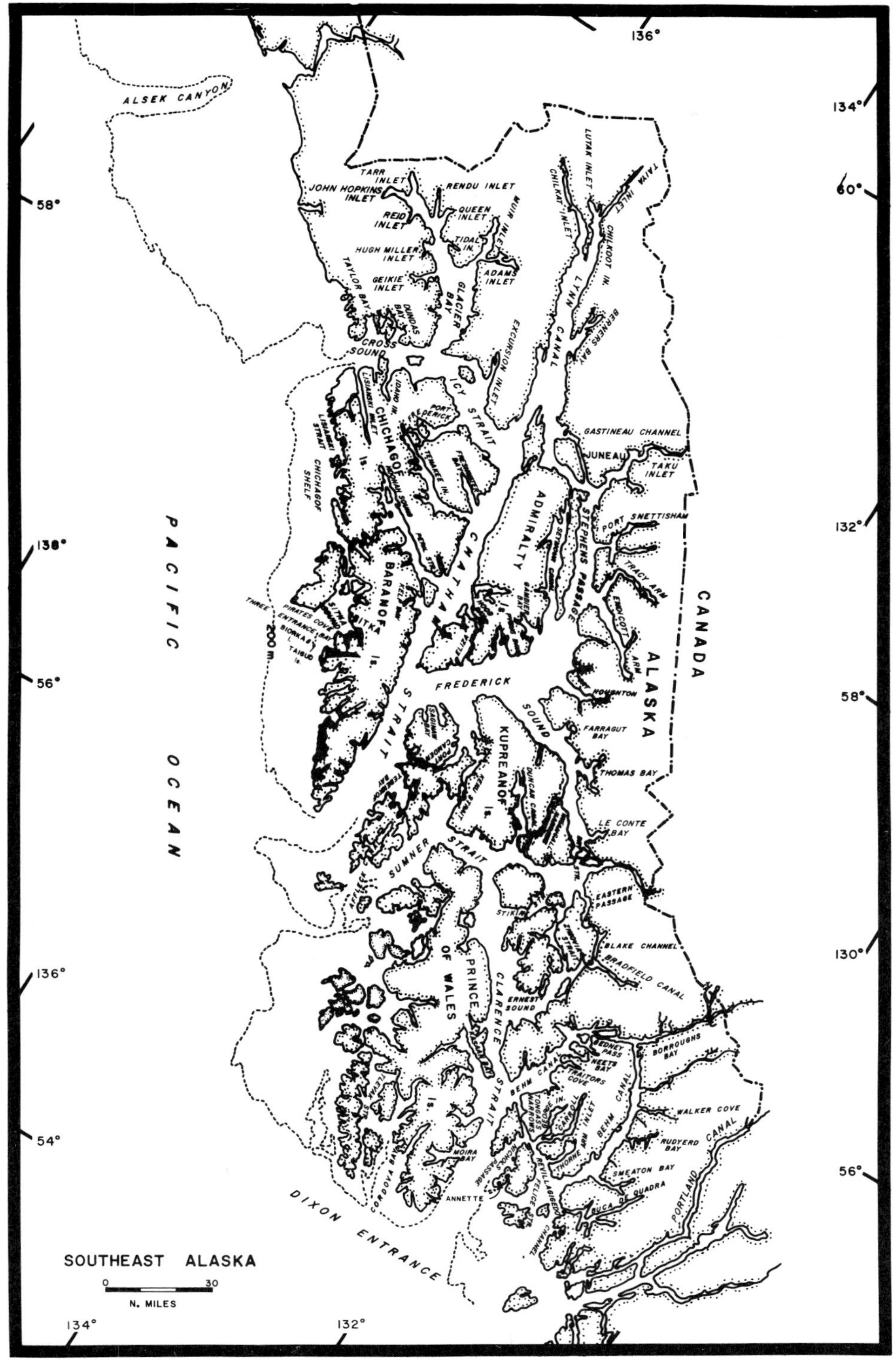

Figure 2-2. Index map, southeastern Alaska.

TABLE 2-1

Names, Locations and Some Characteristics of the Major Physiographic Features in Southeast Alaska
(J. B. Matthews, unpublished data)

Name	Location Latitude Longitude		Length (km)	Mean Width (km)	Surface Area (km^2)	Maximum Depth (m)	Sill Depth Mouth → Head	Number of Basins
Adams Inlet	58°54.9′ 135°45.6′	58°49.8′ 136° 4.5′	19.1	1.9	10.34	73		1
Affleck Canal	56°19.3′ 133°59.5′	56° 0.1′ 134° 8.0′	30.2	3.7	93.76	181	133	1
Behm Canal	56° 0.5′ 130°52.0′	55° 8.0′ 131°57.9′	185.2	6.1	961.53	621	208 281 356 64 336	4
Berners Bay	58°50.7′ 134°54.2′	58°39.4′ 135° 1.0′	16.7	5.2	63.52	228	228	1
Blake Channel	56°21.2′ 131°54.0′	56°12.3′ 132° 0.8′	16.3	1.5	1694	188	47 51	2
Buca De Quadra	55°20.9′ 130°28.0′	55° 3.8′ 131° 1.0′	56.5	1.7	84.86	393	89 84	2
Bradfield Canal	56°13.8′ 131°31.8′	56°10.9′ 131°53.9′	25.2	2.2	38.54	287	287 84	2
Burroughs Bay	56° 3.0′ 131° 5.7′	56°59.1′ 131°13.7′	9.6	2.4	20.45	349	349	1

TABLE 2-1 (Continued)

Name	Location Latitude Longitude		Length (km)	Mean Width (km)	Surface Area (km²)	Maximum Depth (m)	Sill Depth Mouth → Head			Number of Basins
Carroll Inlet	55°38.5′ 131°17.2′	55°18.4′ 131°29.0′	40.7	1.7	61.45	292	245	51		2
Chatham Strait	58°11.5′ 134°13.6′	55°54.4′ 135° 3.3′	240.8	15.2	3354.69	835	512	640	493	3
Charpentier Inlet	58°45.2′ 136°27.2′	58°40.1′ 136°33.3′	10.7	1.3	9.39	157	91			1
Chilkat Inlet	58°13.7′ 135°18.3′	59° 4.2′ 135°31.6′	18.9	4.1	75.09	171	171			1
Chilkoot Inlet	58°17.3′ 135°12.0′	59° 3.9′ 135°26.7′	24.6	4.1	94.10	234	234	76		2
Clarence Strait	56°21.4′ 131°20.7′	54°41.7′ 133°10.5′	201.5	13.0	2451.05	634	398	384	32	2
Cordova Bay	55° 8.3′ 132°17.5′	54°39.8′ 132°42.4′	52.8	10.2	692.63	480	197			1
Cross Sound	58°23.1′ 135°53.7′	58° 1.5′ 136°46.5′	59.3	14.4	514.25	347	234 65	228	128	3
Dry Strait	56°38.6′ 132°29.7′	56°35.1′ 132°36.7′	7.0	2.6	27.34	1	1			1

TABLE 2-1 (Continued)

Name	Location Latitude Longitude	Length (km)	Mean Width (km)	Surface Area (km^2)	Maximum Depth (m)	Sill Depth Mouth → Head	Number of Basins
Duncan Canal	56°51.6′ 56°31.2′ 133° 0.3′ 133°21.3′	44.4	3.3	128.68	73	73	1
Dundas Bay	58°27.3′ 58°14.7′ 136°16.3′ 136°35.3′	23.0	3.1	44.77	274	274 18	2
Eastern Passage	56°31.7′ 56°20.4′ 132° 0.0′ 132°23.0′	31.7	2.8	69.81	140	45 9 73	2
Eliza Harbor	57°15.3′ 57° 9.1′ 134°13.7′ 134°18.5′	12.6	1.1	10.17	135	9	1
Endicott Arm	57°45.8′ 57°30.5′ 132°59.0′ 133°34.8′	45.2	3.3	136.49	367	32 153	2
Ernest Sound	56°13.6′ 55°43.8′ 131°53.1′ 132°22.0′	79.6	6.9	415.69	528		
Excursion Inlet	58°30.6′ 58°19.5′ 135°23.7′ 135°31.3′	18.7	1.9	31.53	151	140 113 111 12	4
Farragut Bay	57°11.2′ 57° 5.5′ 133° 7.6′ 133°17.7′	9.6	3.1	53.38	113	109 53	2
Felice Strait	55°10.1′ 54°56.2′ 131°12.4′ 131°36.0′	28.7	7.0	262.89	131	93	1

TABLE 2-1 (Continued)

Name	Location Latitude Longitude		Length (km)	Mean Width (km)	Surface Area (km^2)	Maximum Depth (m)	Sill Depth Mouth → Head			Number of Basins
Frederick Sound	57°18.5′ 132°31.5′	56°38.4′ 134°36.8′	151.3	13.0	1592.16	466				5
Freshwater Bay	57°57.4′ 134°56.1′	57°48.4′ 135°12.4′	21.7	2.8	54.88	230				1
Gambier Bay	57°32.2′ 133°51.4′	57°23.0′ 134° 6.8′	17.6	4.6	69.81	193	193	54		2
Gastineau Channel	58°22.1′ 134°13.0′	58°11.7′ 134°38.0′	29.6	1.9	38.02	64	64	36	26	2
Gedney Pass	55°51.7′ 131°32.8′	55°49.8′ 131°43.4′	10.2	2.2	24.70	371	129	65		1
Geikie Inlet	58°41.3′ 136°19.0′	58°35.1′ 136°31.7′	14.8	2.2	34.17	155	140			1
George Inlet	55°31.2′ 131°25.2′	55°18.9′ 131°32.6′	25.4	2.2	41.55	407	245	135		2
Glacier Bay	58°57.3′ 135°50.0′	58°22.0′ 136°56.0′	88.5	11.1	764.76	512	69	201	204	3
Hassler Pass	55°54.2′ 131°33.4′	55°51.1′ 131°37.0′	5.7	1.5	8.33	197	69	65		1

TABLE 2-1 (Continued)

Name	Location Latitude Longitude		Length (km)	Mean Width (km)	Surface Area (km^2)	Maximum Depth (m)	Sill Depth Mouth → Head			Number of Basins
Holkham Bay	57°48.0′ 133°32.2′	57°42.6′ 133°41.9′	8.3	7.8	46.35	195				1
Hood Bay	57°28.5′ 134°19.0′	57°20.2′ 134°36.5′	21.1	2.6	49.39	89	78	27	38	3
Hoonah Sound	57°46.6′ 135°25.5′	57°35.1′ 135°50.8′	29.6	5.6	114.63	202	299	43		2
Hugh Miller Inlet	58°50.5′ 135°26.3′	58°44.0′ 136°39.2′	17.0	2.2	37.53	98	21	42		2
Icy Strait	58°25.9′ 134°56.7′	58° 2.2′ 135°55.0′	59.6	13.9	379.74	543	65	140		2
Idaho Inlet	58°13.5′ 136° 5.5′	58° 3.8′ 136°16.9′	20.7	2.8	38.02	73	42			1
Johns Hopkins Inlet	58°55.7′ 136°55.5′	58°49.1′ 137° 8.6′	16.7	2.4	34.17	448	219			1
Kasaan Bay	55°37.7′ 132° 9.5′	55°21.9′ 132°34.7′	35.9	5.2	158.02	276	153			1
Keku Strait	57° 1.5′ 133°38.8′	56°24.6′ 134°15.5′	75.9	6.9	651.39	164	73	12		3

TABLE 2-1 (Continued)

Name	Location Latitude Longitude		Length (km)	Mean Width (km)	Surface Area (km²)	Maximum Depth (m)	Sill Depth Mouth → Head			Number of Basins
Kelp Bay	57°19.8′ 134°48.8′	57°15.0′ 134°57.2′	9.4	5.2	50.00	336	195	20		2
King Salmon Bay	58° 6.1′ 134° 9.0′	57°57.0′ 134°20.7′	17.2	3.0	40.43	43	38	7		2
Le Conte Bay	56°49.1′ 132°25.3′	56°43.5′ 132°34.3′	13.3	1.9	21.36	137	18			1
Lisianski Inlet	58° 7.3′ 136° 1.6′	57°50.6′ 136°29.5′	39.8	1.7	47.15	223	195	117		2
Lisianski Strait	58° 1.0′ 136°20.5′	57°50.2′ 136°27.3′	19.6	0.9	19.24	268	5	179	9	3
Lutak Inlet	59°19.6′ 135°24.0′	59°16.3′ 135°33.2′	8.9	1.9	15.39	126	126			1
Lynn Canal	59° 5.1′ 134°47.0′	58° 5.8′ 135°24.2′	107.0	10.6	912.97	755	543	457		2
Middle Arm (Kelp Arm)	57°21.2′ 134°55.2′	57°19.5′ 135° 4.8′	9.3	1.1	9.39	177	140			1
Moira Sound	55° 6.5′ 131°58.6′	54°59.3′ 132°10.5′	16.1	4.1	60.96	345	115	120		2

TABLE 2-1 (Continued)

Name	Location Latitude Longitude		Length (km)	Mean Width (km)	Surface Area (km^2)	Maximum Depth (m)	Sill Depth Mouth → Head		Number of Basins
Muir Inlet	59° 2.8′ 136° 1.0′	58°44.7′ 136°11.2′	33.9	3.5	104.93	318	62		1
Neets Bay	55°49.0′ 131°29.0′	55°45.1′ 131°42.2′	14.8	3.3	31.96	277	277	157	2
Nichols Passage	55°17.4′ 131°34.0′	55° 2.2′ 131°44.7′	32.7	5.6	186.68	347	193	54	1
Peril Strait	57°36.9′ 134°49.9′	57°21.7′ 135°42.7′	70.9	3.1	256.54	316			6
Portage Arm (Kelp Bay)	57°23.8′ 134°53.1′	57°19.7′ 135° 0.8′	10.4	0.7	7.24	96	96	1	1
Port Althorp	58°11.6′ 136°15.9′	58° 6.0′ 136°21.9′	9.8	2.2	16.60	160	129		1
Port Camden	56°53.0′ 133°51.5′	56°37.7′ 134° 3.4′	26.5	3.3	71.36	67	62		1
Port Frederick	58°10.3′ 135°24.0′	57°56.7′ 135°49.8′	33.3	3.7	94.82	166	74	80	2
Port Houghton	57°23.2′ 133° 3.2′	57°17.2′ 133°31.2′	29.6	3.7	91.95	234	182	10	2

TABLE 2-1 (Continued)

Name	Location Latitude Longitude		Length (km)	Mean Width (km)	Surface Area (km^2)	Maximum Depth (m)	Sill Depth Mouth → Head			Number of Basins
Port Snettisham	58° 8.5′ 133°40.7′	57°55.8′ 133°53.7′	23.1	2.4	79.17	259	259	142		2
Pybus Bay	57°23.8′ 133°58.0′	57°12.5′ 134° 9.2′	17.2	4.6	83.46	113	85			1
Queen Inlet	58°58.3′ 136°30.0′	58°51.7′ 136°35.0′	12.4	2.4	27.42	109	95			1
Reid Inlet	58°52.9′ 136°48.0′	58°50.7′ 136°50.7′	3.9	1.5	5.77	42	42			1
Rendu Inlet	59° 1.7′ 136°35.0′	58°53.3′ 136°43.7′	17.2	1.7	26.71	182	128			1
Revillagigedo Channel	55°19.5′ 130°54.0′	54°46.9′ 131°36.8′	68.2	8.0	637.01	426	254	78	118	2
Rudyerd Bay	55°39.6′ 130°38.2′	55°32.5′ 130°52.5′	22.2	0.9	24.06	303	56	65	29	3
Saginaw Bay	56°56.0′ 134° 7.1′	56°50.4′ 134°17.3′	13.1	3.3	35.44	122	122			1
Seymoor Canal	57°59.8′ 133°48.7′	57°27.8′ 134°21.0′	62.0	5.6	344.25	283	248	104	31	3

TABLE 2-1 (Continued)

Name	Location Latitude Longitude		Length (km)	Mean Width (km)	Surface Area (km^2)	Maximum Depth (m)	Sill Depth Mouth → Head			Number of Basins
Smeaton Bay	55°23.8′ 130°36.2′	55°17.2′ 130°54.5′	14.8	1.5	28.80	281	148			1
South Arm (Kelp Arm)	57°19.5′ 134°56.7′	57°15.2′ 135° 1.0′	8.3	1.1	7.52	135	36			1
Stephens Passage	58°36.5′ 133°28.2′	57°11.2′ 133°57.5′	175.6	11.1	1502.59	420	283	186	53	3
Stikine Strait	56°27.3′ 132°20.2′	56°10.3′ 132°51.5′	31.5	6.5	145.59	415	325			1
Summer Strait	56°36.8′ 132°21.4′	55°50.9′ 134° 8.0′	146.9	14.8	1746.08	543	296	175	316	6
Taiya Inlet	59°30.5′ 135°20.0′	59°16.2′ 135°24.0′	25.6	2.0	38.25	438	126			1
Taku Inlet	58°26.4′ 133°57.8′	58°10.6′ 134° 9.0′	29.6	4.1	51.49	206	206			1
Tarr Inlet	59° 4.6′ 136°51.1′	58°56.2′ 137° 4.8′	17.8	3.3	51.00	384	384			1
Taylor Bay	58°22.1′ 136°28.6′	58°14.6′ 136°39.2′	11.5	5.0	34.89	171	171			1

TABLE 2-1 (Continued)

Name	Location Latitude Longitude		Length (km)	Mean Width (km)	Surface Area (km^2)	Maximum Depth (m)	Sill Depth Mouth → Head			Number of Basins
Tebenkof Bay	56°33.7′ 134° 5.2′	56°24.7′ 134°18.9′	14.8	6.9	159.23	210	64			1
Tenakee Inlet	58° 0.1′ 134°56.3′	57°42.3′ 135°52.7′	64.1	3.1	220.42	316	316	74	43	3
Thomas Bay	57° 6.8′ 132°46.5′	56°57.7′ 133° 0.0′	18.0	3.1	70.58	254	14			1
Thorne Arm	55°25.3′ 131°11.3′	55°14.7′ 131°21.2′	21.3	3.3	59.10	318	146			1
Tidal Inlet	58°50.1′ 136°16.0′	58°48.9′ 136°25.0′	8.7	1.3	8.67	228	54			1
Tlevak Strait	55°16.2′ 132°40.3′	54°54.3′ 133° 7.3′	47.2	8.3	317.51	442	269			1
Tracy Arm	57°55.7′ 133°10.5′	57°46.2′ 133°39.5′	42.6	1.9	75.09	374	21	164	58	4
Traitors Cove	55°44.2′ 131°31.5′	55°41.5′ 131°42.0′	12.6	1.1	10.91	140	64	9	16	3
Tongass Narrows	55°27.2′ 131°32.8′	55°17.3′ 131°51.0′	24.1	2.2	40.37	160	160	29	27	3

TABLE 2-1 (Continued)

Name	Location Latitude Longitude		Length (km)	Mean Width (km)	Surface Area (km^2)	Maximum Depth (m)	Sill Depth Mouth → Head		Number of Basins
Wachusetts Inlet	58°57.2′ 136° 7.5′	58°55.7′ 136°13.0′	4.8	1.1	4.82	84	84		1
Walker Cove	55°45.9′ 130°42.1′	55°42.2′ 130°54.1′	14.4	1.1	12.26	299	12	160	2
Wrangell Narrows	56°50.2′ 132°54.0′	56°30.5′ 133° 0.7′	37.6	1.3	41.93	40	40	16	1
Zimovia Strait	56°22.4′ 132° 4.3′	56° 6.3′ 132°25.8′	30.6	2.8	77.10	162	162	14	2

Basins are generally formed by steeply sloping subareal and submarine bedrock walls and are generally filled with sediments. The sediment-filled basins are relatively smooth and flat in the center. Near the steep walls some small mounds and hill-like structures interrupt the otherwise smooth, flat floors and most likely are formed by landslides or subaqueous slides.

HYDROLOGY

Because the climate in southeast Alaska is alpine maritime, the surface water flow is a function of summer precipitation and the thermal regime which controls storage of water as snow and glaciers. The average annual precipitation is greater than 3,000 mm, with an average potential annual evaporation of less than 500 mm, and solar radiation between 0 and 100 kcal/cm^2. In brief, the weather is wet but with mild temperatures.

The summer weather is influenced by the high-pressure system prevailing over the North Pacific. This gives rise to clockwise rotating descending air masses. In this case, the potential for holding water vapors by air masses is increased by two processes. First, it is moving from north to south or, in other words, toward a warmer region and, second, it picks up more and more influence from the land as it descends. This results in clearer, dryer weather conditions. In southeast Alaska, however, the warmed air parcels begins to feel the cooling effects of the marine environment and thus begin to lose some of their moisture.

The transition from a high-pressure to a winter low-pressure system over the North Pacific each fall leads to considerable precipitation. During the transition, the prevailing winds move air masses from the west to the southwest. Thus the air mass in southeast Alaska during the fall arrives directly from the oceans and is laden with moisture.

In winter, the weather in southeast Alaska is influenced by very large low-pressure system in the region of the Aleutian Islands, the "Aleutian Low." This rising air mass moves in counterclockwise rotation and approaches southeast Alaska from a southerly direction. Precipitation occurs when the warm, moisture-laden oceanic air mass rises and is chilled by the coastal terrain.

This annual cycle of meteorologic conditions sets up a fairly consistent precipitation pattern that prevails throughout southeast Alaska, although there is some regional variation in precipitation. At Juneau and Annette, Alaska, maximum precipitation occurs during the spring and fall seasons, with minimum precipitation during summer (Pickard, 1967). The freshwater discharge monitored as river runoff clearly reflects the influence of seasonal precipitation. Snow and glacier melt in some areas, however, determine the stream runoff. The amount of runoff also depends on the catchment basin: The larger the drainage basin, the larger the river output (Hood and Wallen, 1971).

The freshwater output pattern of drainage basins in southeast Alaska can be classified into two basic categories: bimodal and unimodal. The bimodal discharge pattern is triggered by melting of winter snow during spring and is a function of radiation input, which corresponds to the rise of the mean temperature above freezing. During summer, the discharge is low because of low precipitation and increased transpiration and evaporation. During late summer and fall, the discharge increases rapidly and reflects storm activity and an increase in precipitation. The unimodal discharge is stored runoff and is generally due to seasonal glacial ablation. The drainage basins which are influenced by glaciers and ice fields show a unimodal discharge pattern. The discharge begins to increase with increased radiation input and reaches its maximum in summer.

Factors controlling the type of discharge pattern, whether unimodal or bimodal, in southeastern Alaska become apparent from their geographic distribution. Rivers and streams of the coastal islands invariably have a bimodal discharge pattern because they lack large glaciers in their drainage basins. The unimodal discharge pattern, on the other hand, is generally found on the mainland. It is interesting to note that the peak discharge of the unimodal pattern occurs between the peak flows of the bimodal discharge pattern. Peak discharges of the bimodal pattern occur during spring and fall, while the unimodal pattern peak discharge occurs during summer.

HYDROGRAPHY

The oceanographic regimes in southeast Alaska can be classified as inlets, fjords, bays, estuaries, channels, passages and straits. The inlets, fjords, bays and estuaries characteristically have a prism of fresh water on the surface near the head, but an essentially marine environment at the mouth. Channel passages and straits, on the other hand, have a marine environment with different and variable oceanographic characteristics at each end.

Because sediment transport and deposition is primarily controlled by water movement, it is essential that the general water circulation of the water bodies be understood. This should include mechanisms of water mass formation, exchange between water masses, and the distribution and circulation of water properties in a basin.

Most waters in southeast Alaska are semi-enclosed bodies having free connections with the open sea. This description fits well with the classical estuary as defined by Pritchard (1952,1967). Water properties and circulation in an estuary are primarily controlled by the land drainage, the bottom morphology, and the characteristics of the adjoining water with which it exchanges its water. The water bodies in southeast Alaska, however, differ from classical estuaries in a unique way. An estuary has free access to the adjoining sea but many of the water bodies of southeast Alaska are partly isolated by sills or narrows, so that identifiable sedimentary as well as hydrographic basins are formed.

The sum of precipitation and runoff greatly exceeds evaporation in southeast Alaska. Glacially formed estuaries or fjords are like drowned river mouths with a distinct unimodal discharge pattern at the heads of the inlet. The excess of freshwater input over evaporation and the glaciation and drowning of valleys should make these inlets and fjords positive estuaries (Pritchard, 1952). In positive estuaries there is a net dilution inside the basin which causes an outward flow at the surface. An intrusion of subsurface marine water toward or into the basin must occur to maintain a horizontal density equilibrium. In southeast Alaska, however, there are many inlets which have no vertical stratification and so are in effect negative estuaries. However, the formation of negative or positive water structures in southeast Alaska waterways is complex and is not solely a function of geomorphology and season. The seasonal dilution results in a density structure and an advective flow due to surface input. In a complicated morphologic and structural setting, such as in southeast Alaska, the drainage system of an inlet is usually subdivided into various smaller subflows which enter the estuary along its longitudinal axis. In many fjords, therefore, the discharge pattern tends to be unimodal near the glaciated head, with a bimodal pattern near the mouth. More than one flow entering into fjords and inlets also tends to form localized dilution and circulation patterns. Besides these surface flows, the complex dilution within the basin may be induced by the hydrographic parameters in the adjoining basin.

Although these water bodies can be broadly categorized in the general scheme of estuarine classification, it must be pointed out that each inlet develops an individual water structure and circulation pattern. The water structure primarily results from a unimodal, bimodal or mixed hydrology; the geomorphology of the basin; the location of diluting sources; and the hydrography of the adjacent basin. These water structures vary seasonally and annually. Water circulation in Alaska inlets is also controlled by numerous factors which vary seasonally and annually. These factors are precipitation and runoff, tides, surface air temperatures, ice, and mixing with adjacent water. The combined effects of these factors set up a water circulation pattern which is singular to each inlet.

As discussed earlier, the excess of precipitation and runoff over evaporation is the prime reason for circulation in an inlet in summer. During maximum runoff the waters in the inlet are highly stratified with respect to both salinity and temperature. An addition of water at the surface results in the development of a strong thermocline and halocline at depths varying from 5 to 30 m. Once the pycnocline is formed, the main driving force for the inlet circulation is the inflow of fresh water from glaciers, rivers and peripheral streams. The accumulation of fresh water toward the head of the inlet results in a seaward slope, which induces a seaward movement of the brackish water. The seaward movement of brackish water causes entrainment of more saline water from the underlying water mass. This forms an estuarine circulation of the inlet. Sometimes the stratification includes two thermoclines near the surface in the vicinity of the freshwater source.

During winter the runoff is minimal. The cooling of surface water and wind mixing generally lower the thermocline and halocline and sometimes ultimately destroy the stratification. The result of an extended winter season is often ice sheets which extend several miles from the head of some northern inlets.

The stratified waters above the pycnocline in Alaskan inlets generally are warm and brackish (8-10°C and 5-6‰ salinity). A uniform increase in density with depth, indicated by σt, suggests general stability of the water structure in the inlets. Surface and bottom waters are generally well oxygenated. The dissolved oxygen ranges from 4-10 mℓ/ℓ in surface waters to 2-10 mℓ/ℓ in bottom waters. The pH is greatly influenced by the extent of mixing between fresh and sea waters, and variations between 7.1 and 8.4 have been recorded.

The circulation in Alaskan inlets is very complex. A relatively simple estuarine circulation may be set up in an inlet with uncumbered geometry and sufficient runoff. In general the location and height of the sill(s) control the circulation regime of the inlet. The presence of a sill in most inlets usually restricts the outflow to a thin surface layer, which is balanced by an inflow of saline water immediately underneath it. Conservative properties of the external water mass at a sill depth govern the corresponding properties of the water mass within the inlet.

Circulation in inlets with relatively small runoff is strongly influenced by tides. In large inlets the runoff is generally less than 10% of the tidal prism and therefore water movement due to tides is much larger than that due to net outflow. The tidal currents and resulting circulation are usually complex, arising from the inlet morphology, coriolis force and local inputs. The surface currents intermittently are controlled by winds.

The vertical advective mixing of the water mass, from depth in an inlet, is dependent upon the type of drainage basin which provides discharge to the inlet. The vertical mixing can be classified in three broad categories–glacial, low watershed runoff, and high watershed runoff–and they correspond to the extent of total runoff contributed by spring melt and precipitation.

The surface circulation in south Alaska inlets during winter is strongly influenced by tidal mixing and becomes progressively dominated by runoff during late summer in unimodal drainage inlets. Even during maximum surface flow, the tidal prism, which oscillates at intermediate or near-sill depth, is an important component of water movement in the inlet and often sets up a three-layered system, contrary to the two-layered system of conventional estuarine circulation (Fig. 2-3).

SEDIMENTS

Sediment distribution on the southeastern Alaskan Shelf will be described from two broad regions: the open shelf and the Inside Passage. The detritus sediment pattern in several bays is discussed along with the sediments in the larger water bodies in the Inside Passage. Each area chosen for description typifies numerous similar water bodies in southern Alaska. The sediment distribution in each area is

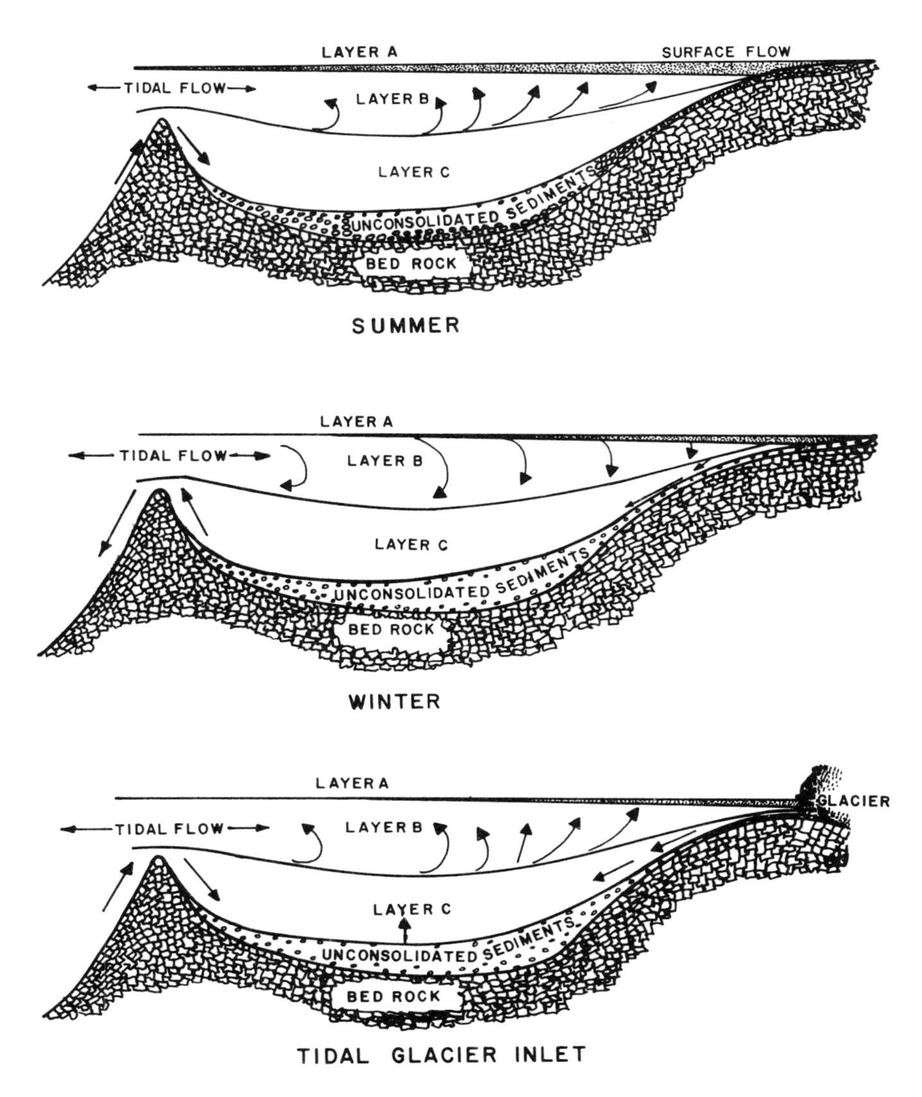

Figure 2-3. Idealized water circulation in southeast Alaskan fjord with seasonal (summer and winter) stream discharge and in fjord with glacier discharge.

therefore characteristic of other water bodies with similar settings. The biogenic sediments, common to the entire Inside Passage, are described separately.

The open shelf adjoining the southeastern Alaskan Panhandle has not been studied in detail. Bottom sediments and a few cores were obtained from the Cross Sound during 1966. Sediments from inlets of the Inside Passage have been studied by Hoskin and Nelson (1969 and 1971), Hoskin and Burrell (1972), and

Sharma (1970a, 1970b). Geochemical and mineralogic studies have been conducted by Kunze *et al.* (1968), O'Brien and Burrell (1970), and Sharma (1970a, 1970b).

Detrital Sediments

Open Shelf. The open shelf in southeast Alaska has been glaciated recently and is mostly covered by sediments deposited during glaciation, which have been only slightly reworked during the Holocene transgression. Some areas of the open shelf, however, are covered with modern sediments; they are generally adjacent to glaciers, river mouths, or channels. For example, a large fan of modern sediments, consisting predominantly of silt and clay, is formed at the junction of Icy Strait and Cross Sound. Cores up to a few meters in length from this fan consistently are composed of gray sediments of uniform textures.

The region between Dixon Entrance and Cross Sound also has not been investigated in detail. Most probably, the entire shelf of this region is mantled with glacial relict sediments. Near the coast, these are probably covered with modern sediments.

Sediment texture on the open shelf varies significantly and may include mixture of gravel, sand, silt and clay. When relict sediments are partly mixed with modern sediments, the predominant mode generally lies in the silt and clay fraction. Coarse gravel, often found offshore, is probably the winnowed remnant of relict glacial sediments.

Surface sediments from shallow waters and beaches in Sitka Sound have been studied by Hoskin and Nelson (1969 and 1971). These sediments are bi- and polymodal, coarse-skewed, gravelly, carbonate sands. Major components of the sand are fragmented skeletons of barnacles, echinoids, byrozoans, and mollusks, and silicate rock fragments. Carbonate sediments were also observed on the shelf west of Chichagof Island and Symonds Bay at Biorka Island. The modern carbonates deposited are mostly of low Mg-calcite with lesser amounts of aragonite and high Mg-calcite.

The Inside Passage. Modern sediment accumulation in the Inside Passage of southeast Alaska is dominated by silt and clay with variable amounts of sand and occasional gravel. Sand and occasional gravel occur on the upper slopes of troughs and near source areas. Where bottom topography is rugged, a mixture of gravel, sand and silt occurs on the banks, whereas silts and clays occur in the intervening basins. The sediments are light to dark greenish gray. The average clay content varies from 20 to 60% by weight.

The sediments from southeast Alaska show striking uniformity in mineralogy of the sand-silt fraction and clay-sized fraction deposited over large areas (Sharma, 1970a, 1970b). The mineralogy of the sediments is characteristic of the provenance and its consistency in sediments of southeast Alaska is attributed to:

1. uniform lithology of exposed rocks over large areas.
2. physical grinding of rock by ice, with minimal chemical weathering,
3. mixing and homogenization of sediments during transport and deposition by glaciers.

The lack of chemical weathering in the source rock region is best illustrated by the mineralogy of the clay-sized fraction of sediments (see Fig. 2-22). X-ray diffraction studies seem to support findings that much of the chloritic material in the sediments closely resembles the chlorite found in adjacent bedrock source terrains. Little evidence of appreciable weathering or diagenetic change have been observed (Kunze *et al.*, 1968; O'Brien and Burrell, 1970). Electron photomicrographs of clay aggregates also show the dominant effects of mechanical grinding and hence substantiate the x-ray interpretations.

Variations in bulk mineralogy of the sediments, however, have been observed and are direct results of varying admixture of the silt and clay fractions. Sandy and silty sediments are primarily feldspar and quartz, whereas clay-sized fractions are mainly chloritic and micaeous illitic material (referred to as "illite" in this treatise). The distribution of minerals in various sand- and silt-sized fractions from Taku Glacier and adjacent marine environments is shown in Figure 2-7. The mineral distribution in both environments is similar except for a variable content of hornblende.

Tenakee Inlet. Tenakee Inlet is a long, narrow inlet on the northeastern part of Chichagof Island (58°01′-57°42.3′N and 134°56.3′-135°52.7′W; Figs. 2-2 and 2-4). The inlet is 64 km long and has a mean width of 3.1 km, a surface area of approximately 220 km^2, and three basins separated by two sills (Fig. 2-4). A shallow basin with a maximum depth of 78 m is enclosed in the northwestern part of the inlet by a broad sill which rises to a depth of 43 m. A second sill at a depth of 74 m separates the central basin, with a maximum depth of 113 m, from the southeastern basin. The bottom of the southeastern basin slopes gradually to 316 m where it merges with the Chatham Strait.

Tenakee Inlet is a northwest-southwest oriented indentation running parallel to Icy Strait, Lisianski Inlet, Hoonah Sound and Peril Strait, apparently as part of a major fault system. The outcrops in the vicinity of the inlet are of Silurian age, consisting of thick-bedded limestones interbedded with coarse conglomerates and sandstones.

There are no major rivers draining into the inlet but numerous small streams enter at the head and at various bays along the southern shore. In spite of limited surface flow, the waters in the inlet are stratified during the summer months.

The bottom surficial sediments in Tenakee Inlet are poorly to very poorly sorted sandy, clayey silts. The sediments are nearly symmetrical to negatively skewed and leptokurtic in the inlet but become positively skewed and platykurtic near the entrance; the sediment mean size becomes finer from the head to the entrance of the inlet. In essence, then, the sediments are typical of fjordal silt.

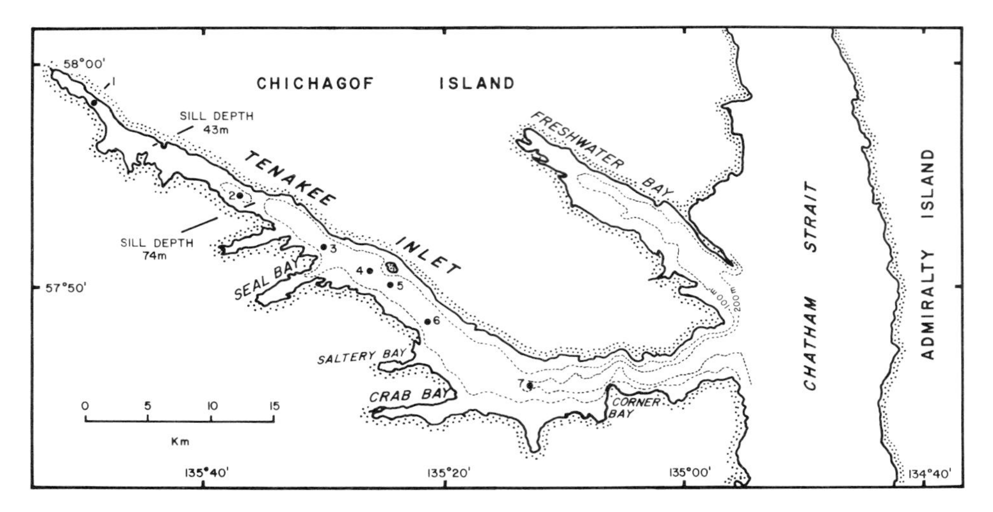

Figure 2-4. Index map showing core locations in Tenakee Inlet, Chatham Strait.

There are no major sediment input areas, only numerous small cirques which are drained by small but fast streams and creeks. Larger streams generally enter bays and deposit their loads as modest deltas. The general decrease in sediment size along the length of the inlet suggests that sediment may move from the head to the entrance. This is a rather long and narrow inlet without significant water input and definable sediment source. Sediment deposition in such an inlet is particularly important because it may reveal transport mechanisms in the absence of major water flow. To investigate sediment transport seven cores were retrieved from the inlet. One core was retrieved from the deepest part of each of the two small basins near the head, and five cores were obtained at increasing depths in the southeastern basin of the inlet (Fig. 2-4).

The sediments in cores from Tenakee Inlet vary significantly (Fig. 2-5). Mean size variations are primarily controlled by sand content: increase in sand results in coarser mean size. Significant increase of sand in cores occurs at specific horizons. Sandy horizons A, B and C in Figure 2-5 can be easily correlated in cores 1 and 2 from the northwestern and central basins, respectively. The A horizon, however, becomes less distinct in cores 3 and 4, while horizon B retains its integrity. Because of the few size analyses for core 5 it is not possible to locate the A and B horizons, but the two lower sand maxima, C and D, are distinct. Core 7, near the entrance, shows all four sandy horizons.

A close relationship of mean size distribution and the occurrence of sandy silts in cores from the two basins suggests three distinct episodes of sedimentation in these basins. Cores 3, 4, 5, and 6 do not show such relationship with depth, but this may be due to their location in relation to the main channel through which sediments were transported. Cores 3 and 4 are located on the periphery of a saucer-type basin. In particular, cores 4 and 5 are on opposite sides of a ridge

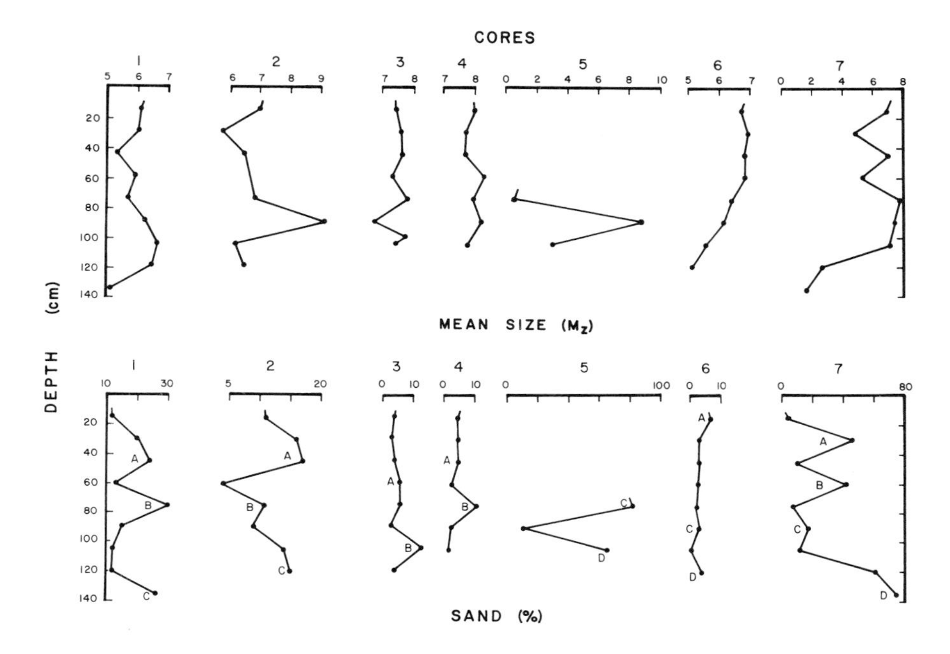

Figure 2-5. Sediment variations with depth in cores from Tenakee Inlet.

that forms a constriction in the inlet; part of this ridge is exposed as an island. This constriction causes a jet flow, and the result may be significant turbulence near core 5. Core 6 is located a little off the main channel and, therefore, does not contain horizons A and B. Near the entrance of the inlet, core 7 distinctly shows four sandy horizons, which are probably related to others in the inlet.

A cursory evaluation of the sandy horizon might suggest that it may have originated as a slump in the vicinity of cores 4 and 5 and moved toward the head and the entrance. A closer look at the mean size distribution, however, quite clearly refutes such an origin. The mean size in all sandy horizons in cores 1, 2, 3, and 4 decreases successively, suggesting a source near the head. Similar lateral grading of sediment in each horizon occurs in cores 6 and 7. The overall coarseness of sediments in these cores is perhaps caused by the constriction northwest of core 5. Three distinct episodes of sedimentation in cores 1 and 2 and possibly in cores 4 and 5 are apparent. Core 7 near the entrance displays four sedimentary episodes. It also appears that sediments have travelled from one basin to another, by spilling over the sills, and have preserved the characteristic sedimentary parameters of each episode. Fining upward of sediment mean size suggests that sandy sediments may have been deposited as turbidities. Mass sediment movement may have originated as landslides or submarine slides, perhaps triggered by earthquakes.

Taku Inlet. Taku Inlet is a fjord-like estuary which lies 20 km east of Juneau (58°26.4 -58°10.6′N and 133°57.8′-134°9.0′W; Figs. 2-2 and 2-6). It is a part of the extensive inland waterways that include Chatham Strait and Stephans Passage. To the north of the inlet lies the extensive Juneau ice field which feeds numerous glaciers, two of which terminate near the periphery of the inlet.

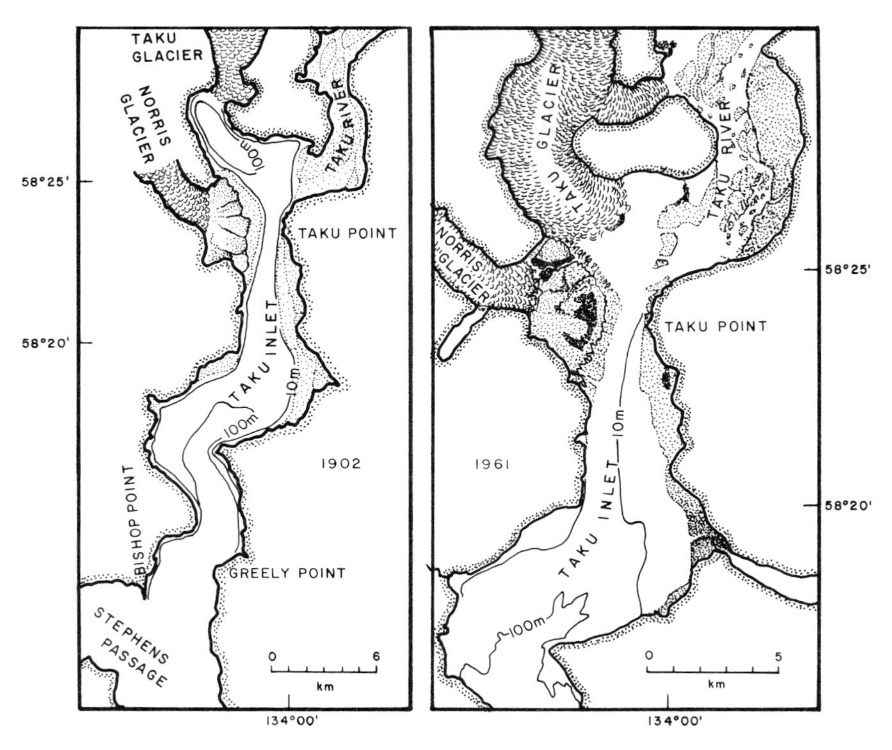

Figure 2-6. Charts of Taku Inlet showing positions of glacier termini.

The Taku River enters at the head of the inlet, and along the northwest side numerous streams originate in meltwater lakes or at glacier termini. Taku River is one of the larger rivers in southeast Alaska and has a mean flow of 262 m^3/sec (nine-year average). From the mouth of the river the inlet extends 29.6 km southward, where it merges with Stephans Passage (Fig. 2-6). The inlet has a mean width of 4.1 km and a surface area of 51.5 km^2. Surrounding headlands are mostly low-lying terminal moraines, including in particular the 20.7-km^2 outwash fan of Norris Glacier. End moraines, which are generally exposed at low tide, separate the termini of Taku and Norris glaciers from the inlet.

Fresh water forms a surface prism near the head of the inlet, which becomes progressively more saline toward the mouth. Temperatures and salinities during July 1964 varied from 11°C and 0.0‰ near the head to 4-5°C and 33.5‰

near the mouth of the inlet. The water is stratified, and a strong thermocline and halocline are formed at a depth varying between 15 and 30 m. Its freshwater prism is maintained by the underlying saline water which has increasing density with depth.

Tides are primary control on the circulation in the inlet. The tidal range increases from about 1 m at the entrance to 5 m at the northern end. These tides are of mixed types, and tidal currents are quite strong.

Taku and Norris glaciers and their sediments have been the subject of many recent investigations. Although the glaciers are quite close to each other, in recent history Norris Glacier has receded somewhat, while Taku Glacier has actively advanced. The changes in the positions of both glacier termini between 1961 and the close of the last century are shown in Figure 2-6. Norris Glacier receded about 5 km from the eastern edge of the inlet to its present position between 1750 and 1890 (Lawrence, 1950) and deposited extensive outwash. These sediments have been described by Slatt and Hoskin (1968).

Recently, advance of Taku Glacier has been dramatic. A Coast and Geodetic Survey chart prepared during 1888 to 1902 shows the Taku Glacier terminus 5 km northeast of its present position. At that time the glacier was tidal and entered a northwest extension of the inlet that has since been filled in by sediments. This extension apparently had two basins, separated by a shallow sill near Taku Point. According to the 1888-1902 bathymetric chart, the northwestern basin was 104 m deep. During the early 1930s the glacier was still tidal. However, at the end of the last decade, a push moraine appeared above tide level and shortly after became a permanent land feature. Based on the 1961 hydrographic charts shown in Figure 2-6, Jordan (1962) estimated that over 370 million cubic meters of sediments were deposited in the upper inlet between 1890 and 1960. The source of these sediments was most likely Taku Glacier, but the Taku River might have contributed significantly.

In spite of the large sediment volume deposited in the northwest extension described above, the bathymetry of the portion which comprised the present inlet has not changed significantly since 1888-1902. A narrow channel shown on this early chart, runs north-south and slightly west of mid-inlet. Comparison of 1961 and 1888-1902 bathymetry charts shows that there has been no sediment filling of the main part of the inlet during this century.

Bottom sediments in Taku Inlet are mostly sand at shallow depths, with clayey-sandy silt in deeper waters near the mouth. Sands are moderately well sorted to well sorted whereas silts are poorly sorted.

Mineral distribution in various sand and silt size fractions from Taku Inlet was described by Sharma (1970a) and is reproduced here in Figure 2-7. Clay mineralogy of 2 μm fractions of the bottom and suspended sediment has been reported by Kunze *et al.* (1968). The clay minerals are mainly primary silicate minerals which are a product of physical abrasion by ice. The major components of clay size fractions are "illite" (trioctahedral micas) and iron-rich chlorite; amphibole is a minor component.

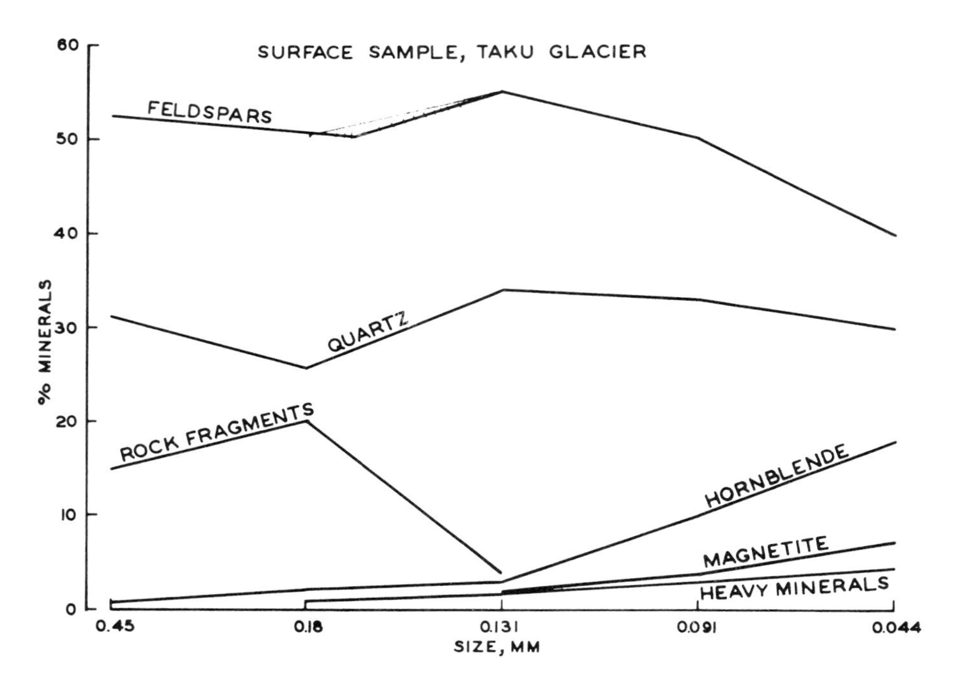

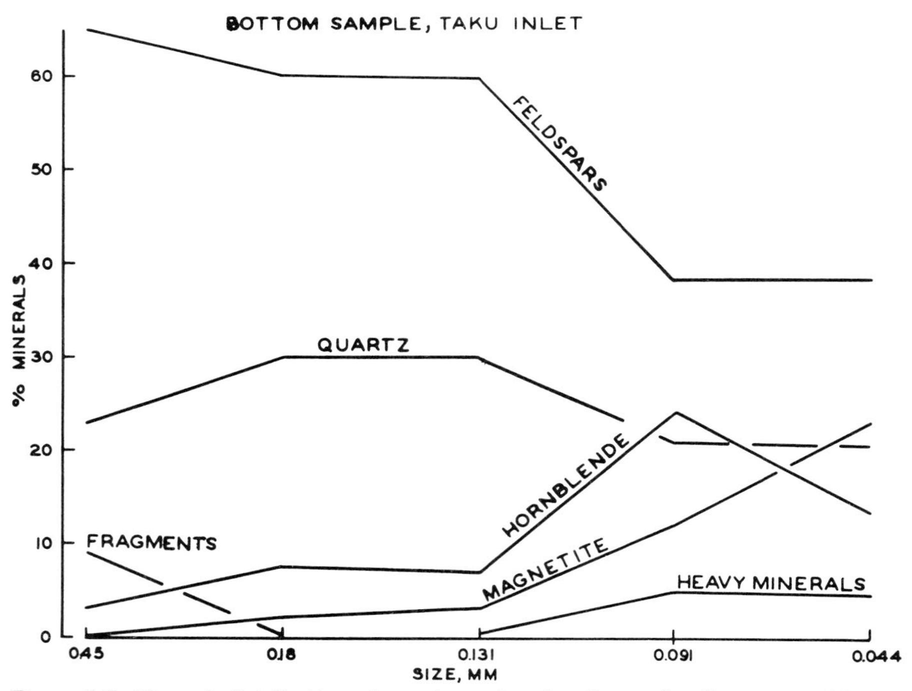

Figure 2-7. Mineral distributions in various size fractions of sediment resulting from glacier transport (Taku Glacier) and marine transport (Taku Inlet).

Suspended sediments are derived from the subglacial streams and from the Taku River, which discharges near the head of the inlet. These waters are very turbid, and a suspended load in excess of 10,000 mg/ℓ is not uncommon during the summer months. Sediments in suspension generally consist of fine and coarse silts with some clay. Most sediments are carried seaward down the inlet in near-surface waters; however, subaqueous sediment transport has been postulated by Wright (1971).

Subsurface sediments near the mouth of Taku Inlet were studied for the U.S. Army Corps of Engineers by the U.S. Geological Survey (unpublished preliminary report to U.S. Corps of Engineers by the Pacific Coast Marine Geology Unit, U.S. Geological Survey, Menlo Park, California, February 1966). Reconnaissance seismic reflection data indicated that the inlet is typical fjordal basin with steeply inclined walls and is filled with horizontal to slightly inclined unconsolidated sediments. The thickness of unconsolidated sediments near the mouth of the inlet approaches 110 m. Subbottom profiles show subsurface continuation of the fjord walls exposed above the sea level. The steep bedrock ridges of the subbottom show irregular and disturbed contacts with overlying unconsolidated sediments.

Based on the disturbed and distorted nature of the sedimentary fill, the report concluded that structural movement of the underlying bedrock is at least partially responsible for the deformation of sediments. The movement appeared to be a relative dipolar coupling along the axis of the inlet.

Gastineau Channel. Gastineau Channel lies between 58°11′and 58°22′N latitude and 134°13′and 134°38′W longitude. The channel is 29.6 km long, with a mean width of 1.9 km and an area of 38 km^2. It follows a northwest-southeast fault line depression which runs between Douglas Island and the mainland (Fig. 2-2). The channel is bordered on both sides by rugged topography, with mountains rising abruptly from the water's edge, and elevations of 800-1200 m are common within a kilometer or so of the shore. The rugged mountain peaks, some exceeding 1750 m, have been glaciated during the Pleistocene epoch. At present, part of the Juneau ice field is drained into the channel. The steep topography gives rise to numerous short streams, which enter Gastineau Channel from both sides. All of these streams have very small catchment areas, and none exceeds 5 km in length. The valleys are U-shaped, with morainal deposits near the cirques which occupy the headward portion.

The regional geology of the channel area has been described by A. C. Spencer (1906). He distinguished five northwest-southeast oriented rock types: schists, carbonaceous slates, greenstones with slate, black slate, and greenstones. The Coast Range Batholith, to the northeast, is bordered along its southwest margin by a belt of schist up to 3.0 km wide, and a 1.5-km-wide band of black carbonaceous slates. Continuing southwest of the edge of Gastineau Channel are bedded greenstones with slates. Douglas Island along Gastineau Channel consists of a metamorphic-intrusive igneous complex about 3 km wide. This includes black slate and greenstones intruded by diorite, and a dike of diorite-porphyry along the shore near the lower end of the island. Thick-bedded greenstones form the rest of Douglas Island, including the shore of Stephans Passage.

Exposed rocks adjacent to Gastineau Channel are of Paleozoic and early Mesozoic ages. Glaciomarine deposits generally cover lower slopes, alluvium and talus occur in some creek beds, and deltas have been formed where streams enter the channel. Gold and Sheep creeks, for example, have developed significant deltas, but spits, lagoons, or offshore bars are not present.

Two basins are the main features of bottom topography along the channel axis. The near-shore bathymetry reflects subaqueous deltas deposited by major creeks. The smaller basin (approx. 2 km long with a maximum depth of 38 m) lies between Juneau and Douglas and is separated by a bedrock sill from the larger basin (8 km in length and up to 50 m deep) to the southeast. This divide is located near Douglas and has a sill depth of about 16 m. The larger basin has a shallow saucer shape and is partly restricted to the southeast by a deep sill where it joins Stephans Passage. This broad, rounded sill is located approximately 2 km from the entrance and rises to a minimum depth of 35 m. From the entrance sill the bottom slopes gradually to a depth of 65 m in Stephans Passage.

In Figures 2-8 through 2-11, the areal distribution of various textural parameters of the sediments in Gastineau Channel are shown. Sediments are sandy and clayey silts in the basin, with sand and gravelly sand in shallow areas near shore and around sills. Sediments vary from poorly sorted to very poorly sorted. Sorting improves, although it is still poor, near the creek deltas and the deeper parts of the channel. Most sediments are positively skewed and leptokurtic.

Variations in textural parameters are controlled by the water depth and hydrodynamics in the channel. In general, sediment size decreases with increasing water depth in Gastineau Channel. The texture is affected by the tidal flow, which forms a multilayered complex current pattern in the channel. Water currents, at various depths in the channel as measured by the Federal Water Pollution Control Board, show a three-layered flow scheme, consisting of surface inflow, intermediate but shallow outflow, and some inflow at depth. The sorting is indicative of water flushing at moderate speed, as suggested by the coarse grain size.

Gastineau Channel is a typically U-shaped trough filled with glacial sediments. The sediments reach a maximum thickness of 125 m in the central part of the larger basin (Thomas P. Ross, Jr., Senior Marine Geologist, Global Marine, Inc.: personal communication). The sediments thin out toward the periphery of the elongated basin as the bedrock rises to form sills. The structural sill near Douglas Island, which separates two basins, is devoid of contemporary sediments. The absence of sediments may be due to swift tidal currents, which prevent sedimentation.

Berners Bay. Berners Bay is a rectangular re-entrant on the eastern margin of Lynn Canal; it lies between 58°40′ and 58°51′N latitude and 134°54′ and

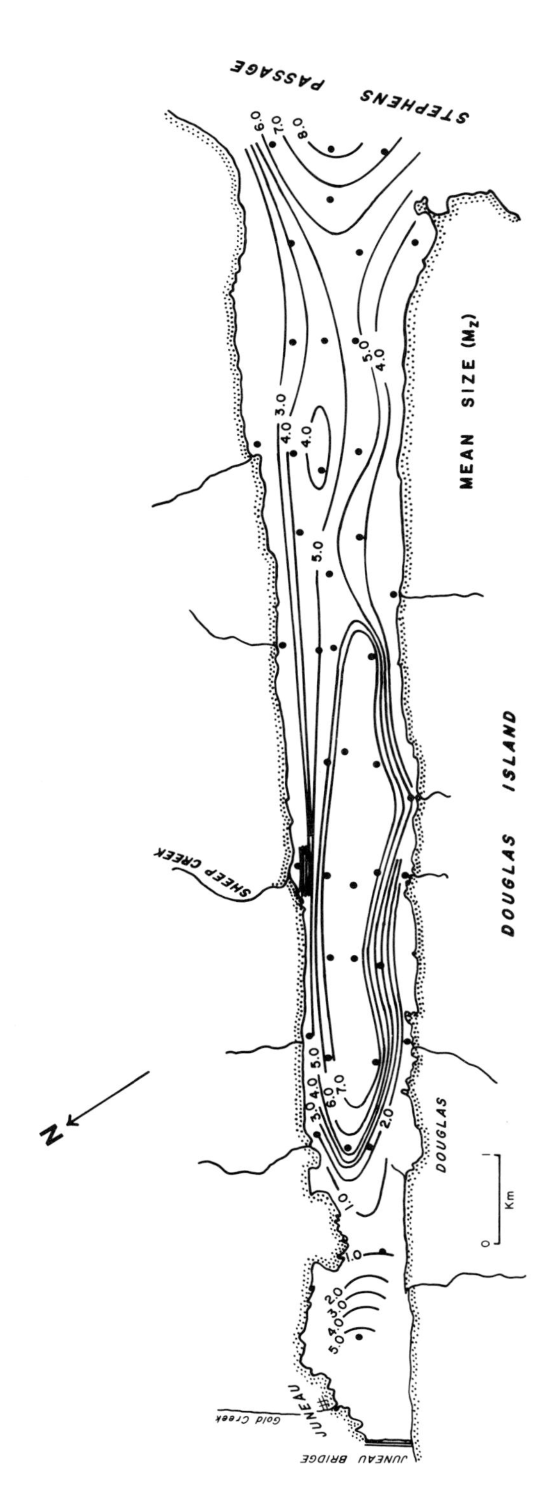

Figure 2-8. Sediment mean size distribution, Gastineau Channel.

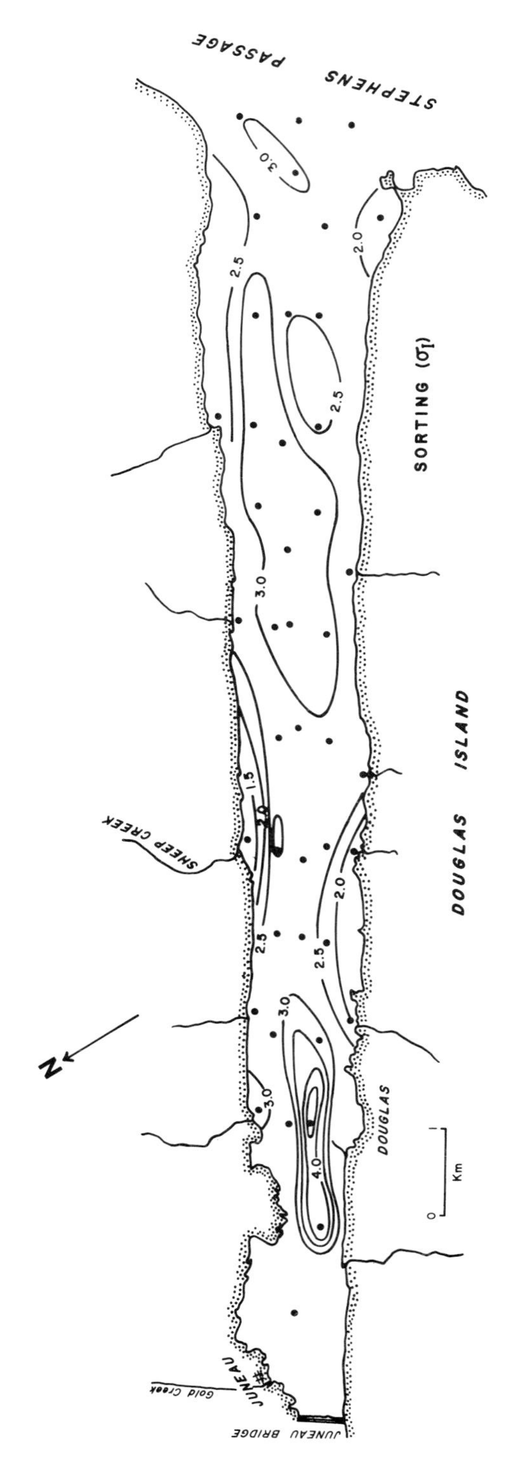

Figure 2-9. Grain size sorting in sediments, Gastineau Channel.

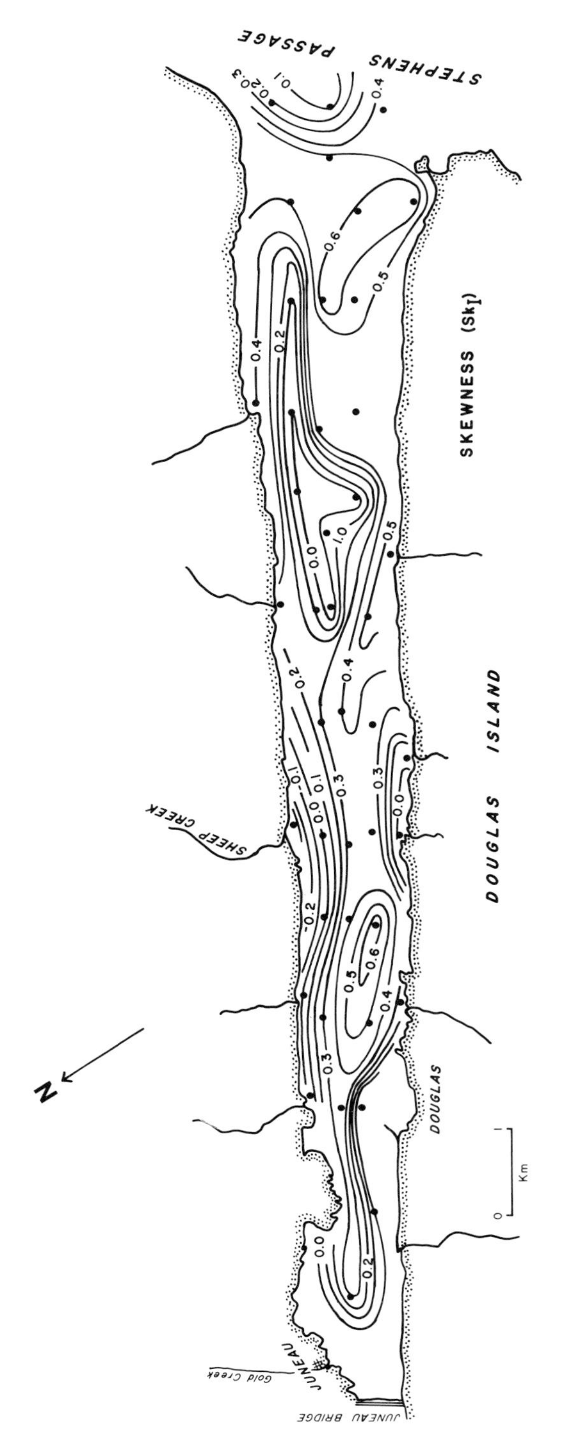

Figure 2-10. Grain size skewness in sediments, Gastineau Channel.

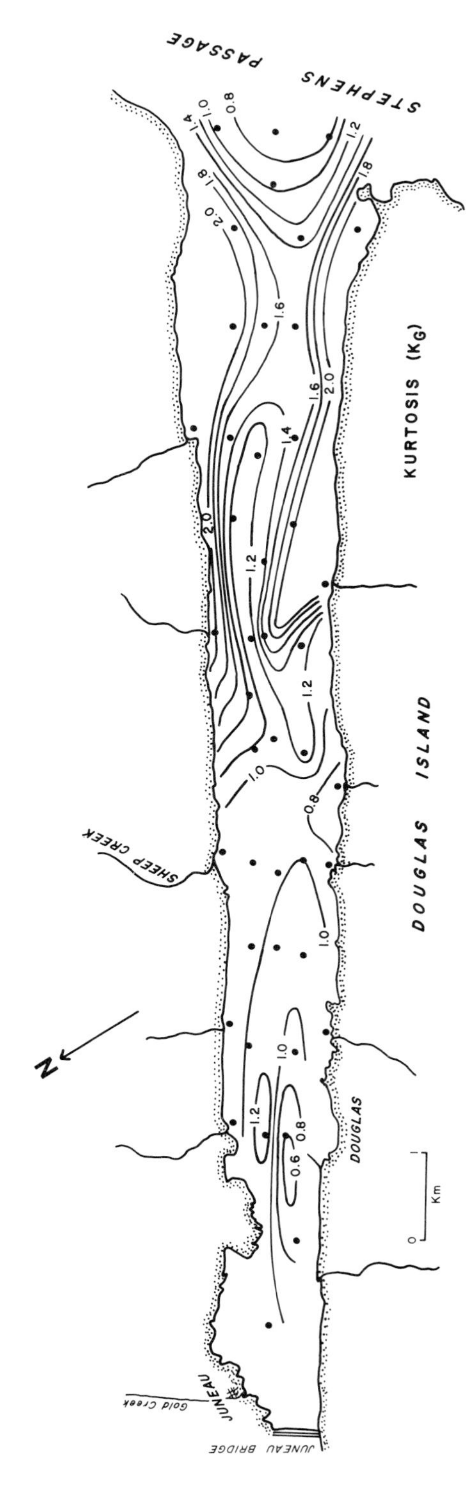

Figure 2-11. Grain size kurtosis in sediments, Gastineau Channel.

135°05′W longitude (Figs. 2-2 and 2-12) and has an area of about 70 km^2. The bathymetry is typical of a bay; depth increases gradually from the head of the bay to 170 m near the entrance in Lynn Canal. Berners, Lace and Antler rivers enter the bay at the head and built a sizable delta. A smaller stream, Cower Creek, discharges near the mouth of the bay. All these rivers contribute significant quantities of fresh water and detritus to the bay. The salinity and temperature distributions suggest that most of the freshwater input is attributable to the rivers at the head of the bay, and that it flows outward along the west shore to Point St. Mary. A subsurface flow of saline water from Lynn Canal enters the bay at depth and flows northward along the eastern shore to maintain the density slope. This deep saline water mixes with fresh water and becomes entrained by the freshwater runoff.

Sediments in the bay include reworked glacial material and detritus from the surrounding land area. Sand is deposited in the upper reaches of the bay and near river mouths, grading seaward to clayey silt deposited from suspension (Fig. 2-12). The coarse to fine size gradient from the head of the bay toward the mouth strongly suggests a major source for sediments at the head of the bay. Outside the entrance of the bay there is an abrupt increase in sand; this is perhaps due to the migration of coastal sand caused by tidal currents in Lynn Canal.

Glacier Bay. Glacier Bay is an elongated northwest-oriented indentation in the Coast Range which lies between latitudes 58°57.3′ and 58°22′N and 135°50′ and 136°56′W longitude (Figs. 2-2 and 2-13). It is bordered to the north and east by the Takinsha and Chilkat ranges and by the Brady Ice Field (650 km^2) on the west. Numerous ridges from these ranges extend toward the bay, forming between them a complex of 1- to 7-km-wide U-shaped valleys partially submerged under the sea. These U-shaped valleys generally are filled with glaciers, which head in cirques or ice fields at 1,250 to 2,000 m elevation and descent almost to tidal water. The bay has its opening at Icy Strait near Gustavus, with a sill depth at the mouth of only 69 m. Northward, halfway from the mouth, the bay bifurcates into north-south trending Muir Inlet, while the main depression extends northwest toward the United States-Canada boundary (Fig. 2-13). Overall, the bay is about 90 km long and has an average width of 11.1 km and a maximum width (about 20 km) in the middle portion. The bay covers an area of 765 km^2 and is bounded by a coastline of approximately 595 km; its area has been increasing during the past century due to recession of the glaciers.

There are numerous islands in Glacier Bay, notably Russell, Gilbert, Drake, and Willoughby Islands, as well as a group of islands known as the Beardslee Islands near the mouth (Fig. 2-13). Three major basins in Glacier Bay are formed by three bedrock sills. The first basin, with a maximum depth of 404 m, lies between the entrance sill at 69 m and a sill at 201 m protruding near Drake Island. The third sill rises to a depth of 204 m separating Tarr Inlet from the central basin.

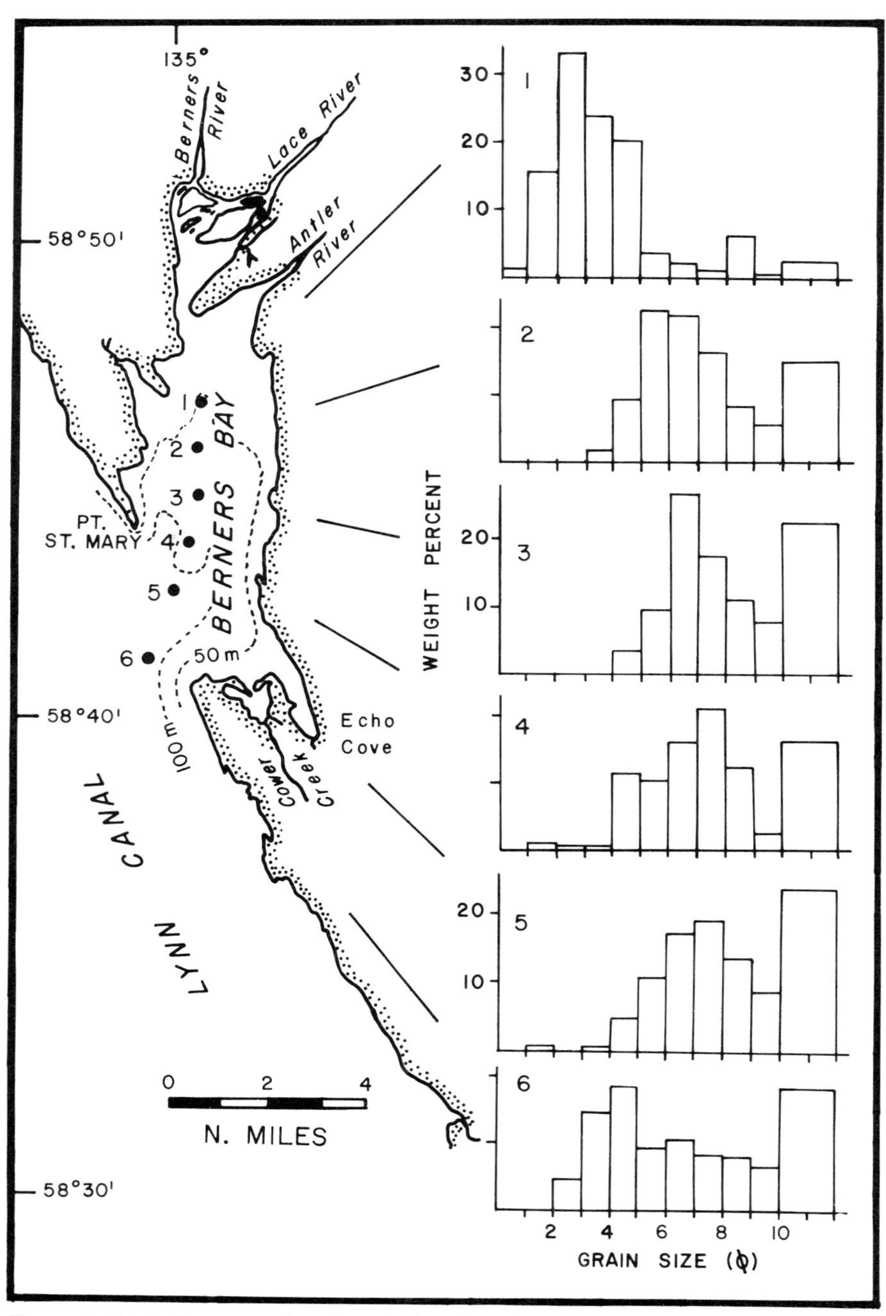

Figure 2-12. Index map showing station location and sediment distribution in Berners Bay, Lynn Canal.

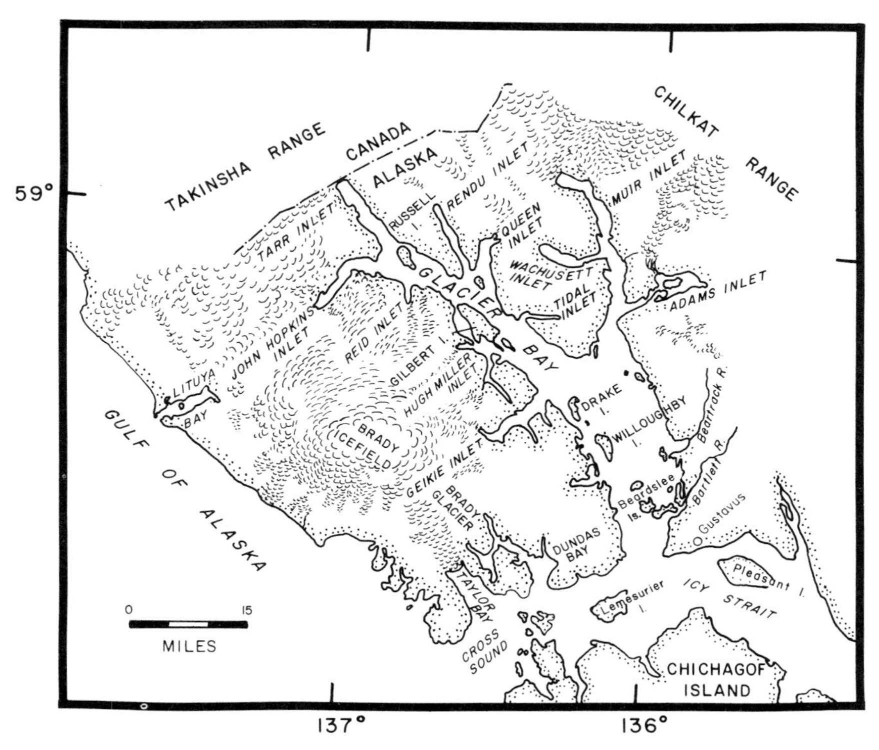

Figure 2-13. Index map of Glacier Bay, Southeast Alaska.

Glacier Bay has followed a similar general geologic evolution to and has the same structural trends as the rest of southeast Alaska. However, there are certain episodes during the evolution of the bay which are of particular importance. During the Early Paleozoic, orogenic activity prevailed throughout southeast Alaska, resulting in a marine trough in the region. From the Late Silurian to the Early Devonian the region was a relatively stable marine shelf-platform, and widespread limestone was deposited. The influx of terrestrial sediments during this period was minimal. At the end of the Devonian there was a brief episode of erosion, followed by deposition of more carbonates in the Middle Permian. There is no evidence of sediments deposited between the Late Permian and the Middle Triassic, suggesting uplift and erosion during the interval. A Late Triassic orogenic episode resulted in submergence and volcanic activity; this was relatively brief and was followed by non-deposition and erosion during the Jurassic. At the beginning of the Cretaceous intense crustal deformation and related igneous activity began, lasting until the Early Tertiary. The absence of post-Cretaceous deposits south and west of Glacier Bay indicates that Paleozoic rocks were uplifted during the orogeny and have remained largely above sea level.

The morphology of Glacier Bay suggests intense Pleistocene and pre-Pleistocene glacial scour and erosion. Evidence of early glaciation in the bay is provided by Middle Miocene tills exposed west of Fairweather Fault. Miocene-Pleistocene glaciation is well represented by the glacial-marine deposits of the Yakataga Formation, which is exposed between Icy Point and the Prince William Sound region (Miller, 1957; Rossman, 1963; Plafker, 1967). Continental ice sheets 700-1,300 m thick covered the area during the various glacial maxima. The last major glaciation in Glacier Bay occurred during the Wisconsin. Recent advances and recessions of glaciers in the bay have been studied by Goldthwait (1966) and McKenzie (1968). Most glaciers in the bay advanced from 5,000 B.C. until about 270 years ago, when a general recession began. In 1794 Vancouver observed a glacier terminus near the entrance to the Glacier Bay; by about 1860 the ice had withdrawn 33 km northward to the entrance of Muir Inlet (Lawrence, 1958). During the last century the ice in Muir Inlet has receded steadily at about 0.7 km/yr (Field, 1947) and in the west arm the ice has retreated even faster. About 1900 the major ice fronts became stabilized (Cooper, 1937). In general, the glaciers in the bay, including Muir Glacier, have retreated at a faster pace. The accelerated retreat of glaciers deposited a veneer of glaciomarine sediment of complex texture.

The bedrock exposed in Glacier Bay is mostly carbonate, plutonic granitoids and high-grade to low-grade metamorphic rocks. The Paleozoic sedimentary sequence consists of the Tidal Formation, Pyramid Peak Formation, Rendu Formation and Black Cap Limestone. Granitic masses and numerous dikes emplaced during the Cretaceous-Early Tertiary are widely distributed, and metamorphic rocks associated with intrusives are scattered throughout the region.

The climate in Glacier Bay can be classified as humid, temperate with cool summers. Temperatures recorded for eight years at Gustavus indicate that the mean monthly high of 13.3°C occurs during July and the mean monthly low of -2.1°C occurs during December. For the same period, the average annual precipitation at Barlett Cove was 1,800 mm. Maximum precipitation occurs during September, and the minimum occurs in May.

There are no major rivers draining into Glacier Bay. Most of the streams are small, have steep gradients, and drain small areas. Two modest streams, Bartlett and Beartrack rivers, discharge their loads near the mouth of the bay. Most freshwater input to the bay is precipitation, snowmelt, and ablation of glaciers. About 20 of these glaciers enter the sea as tidal glaciers. The discharge pattern in Glacier Bay is typically unimodal and is at maximum during late summer. The volume of surface runoff is significantly smaller than the tidal flux. Tidal frequencies are complex and the tidal currents are swift, particularly where masses of water are connected by narrow passages. The maximum tidal range in the bay is 7.93 m and ebb tides generally generate the strongest currents.

The rapid recession of glaciers in Glacier Bay has left a bottom cover of gravel, sand, silt and clay. These sediments have been partly reworked by the advancing sea. Tills deposited during the recession are not exposed because they have been reworked and covered by contemporary marine sediments. These contemporary sediments are light to dark greenish-gray sandy and clayey silts, with occasional silty clays (Fig. 2.14).

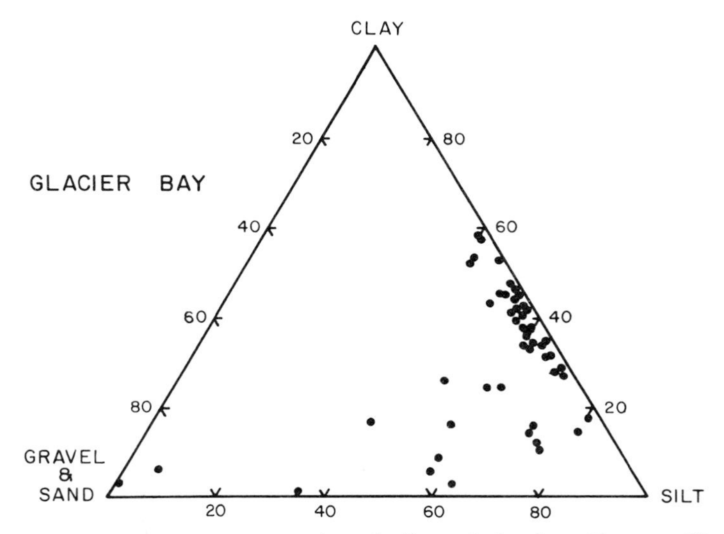

Figure 2-14. Weight percents gravel-sand, silt, and clay in sediments, Glacier Bay.

The textural characteristics of the bottom sediments in Glacier Bay vary significantly. Near tidal glaciers the sediments are mostly gravel, grading to silt farther from the head of the inlets. In Tarr Inlet at the head of Glacier Bay, for example, the sediments are predominantly sand, becoming coarser toward the glacier terminus and finer toward the mouth of the inlet. This general seaward decrease in grain size, however, is locally modified by the influx of the sediments from various sources and by basinal and sill topography. Near the sills and the sediments sources the sediments may become coarser. The normal sediment gradient is also altered by ice rafting (Ovenshine, 1970). Locally, landslides result in totally anomalous sedimentary textures. Portions of such slump masses are transported farther into the basin as turbidites. These may extend across a basin until stopped by a sill.

The interplay of all these processes and sources results in a very complex distribution of sediments and sediment texture. The coarse fraction of sediments in Glacier Bay consists primarily of feldspar (approximately 45%) and quartz with lesser amounts of hornblende, pyroxene and magnetite. Green hornblende is the most abundant of the heavy minerals, with considerably smaller amounts of diopside and actinolite, as well as trace quantities of tremolite and hypersthene.

The clay fraction (<4 μm) is composed of primary silicates with little or no apparent chemical weathering. It shows a remarkable mineralogic uniformity throughout the bay, consisting mainly of illite and chlorite (Fig. 2-15), with varying lesser amounts of vermiculite, amphiboles and feldspar. The chlorite appears to be magnesium rich (based on 001 x-ray diffraction intensities) and the illite is shown to have a trioctahedral mica component (based on 060 spacing by x-ray diffraction).

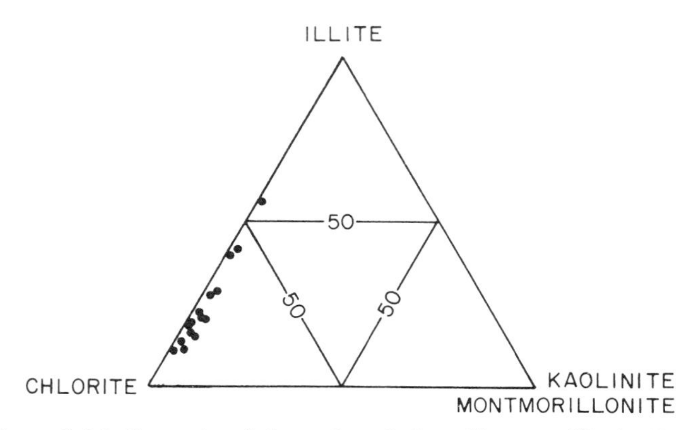

Figure 2-15. Percents of clay minerals in sediments, Glacier Bay.

In addition to the lateral textural variations observed in surficial bottom sediments, sediment texture also varies with depth in cores. Such variations were observed in a series of cores taken along the longitudinal axes of various fjords, both tidal and normal. In Muir Inlet (see Fig. 2-16 for locations), texture in the cores varied erratically with depth, as shown in Figure 2-17. Core Mui-10 is located midway between the head and the entrance of the inlet and is about 10-20 km from major sediment sources. The upper 65 cm of this core is comprised of sandy silts and clays typical of fjordal sediments. The underlying sediments, however, increase in sand with depth, and become almost 90% sand at a depth of 140 cm. This sand horizon is underlain by normal fjordal sandy clayey silt. At first glance it may appear that the sand sequence represents a deposit laid near the ice front during the retreat of the glacier. Yet this could not be the case, because the sand horizon cannot be correlated laterally in other cores. Furthermore, the underlying sediments are fine, instead of the coarse gravel which is typically deposited near a glacier ice face.

Sand lenses, such as these, occur in cores throughout the inlet and appear to be turbidite deposits, probably resulting from mass movement of water caused either by landslide or by ice calving. Masses of ice falling from the faces of glaciers generally churn up the bottom sediments, which may then be transported in suspension. Continual falling of small ice masses results in the removal of clay

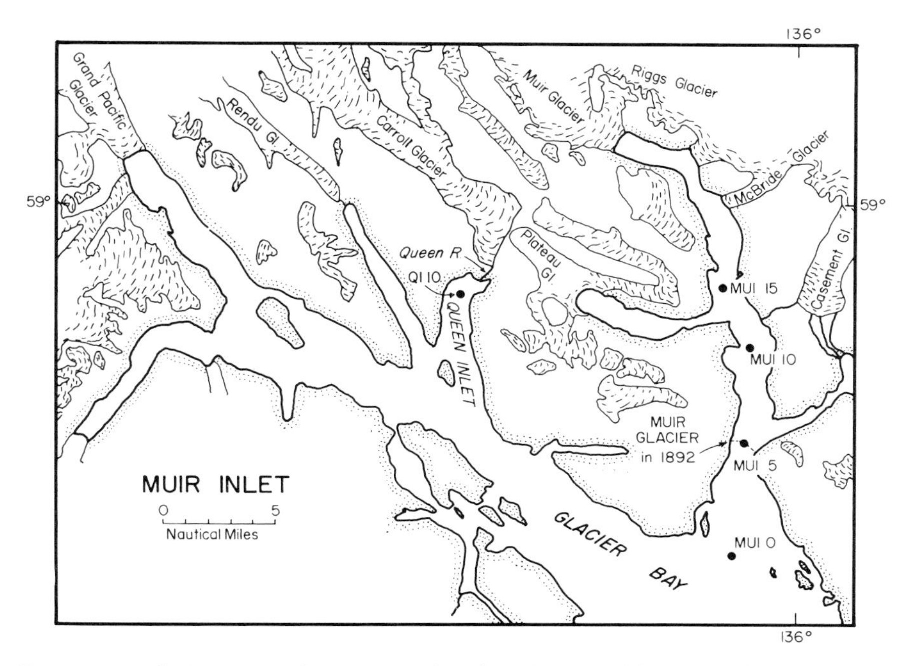

Figure 2-16. Index map showing station locations in Muir and Queen Inlets, Glacier Bay.

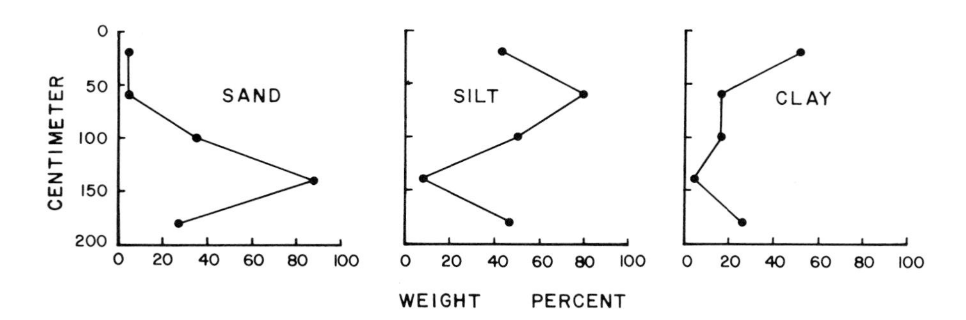

Figure 2-17. Weight percents sand, silt, and clay in sediments from Muir Inlet core Mui-10.

and silt, which are carried farther into the inlet; the calving of occasional large masses resuspends sand for transport by turbid flow. Because of this process there is an occurrence of relatively coarse sediments, mostly gravel, near the glacier face which grades into sands in adjacent areas. Grand Pacific Glacier in the north part of Tarr Inlet provides a good example of such sediment distribution. Near the terminus of the glacier, which is currently receding, the bottom consists of sandy gravel, which grades into sand in the center of Tarr Inlet.

Queen Inlet, on the other hand, is not fed by tidal glaciers and sediments are brought into the inlet by outwash streams and tidal action. Cores up to 4 m in length from all parts of this inlet consist of sandy clay-silt and show very little variation in sediment texture.

Therefore, the differences in sediment textures adjacent to tidal and nontidal glaciers suggest that falling ice can be an important factor in sediment transport. This significant aspect of fjordal sedimentation has not been discussed previously in the literature.

Biogenic Sediments

The biogenic component in southeast Alaska sediments is mainly derived from phytoplankton and benthic fauna; the latter are minor contributors. Two prolific seasonal blooms of diatoms occur each year in this region. These diatom blooms remove silica from sea water and some of this is deposited as biogenic sediments. In order to estimate the biogenic constituents, an attempt was made (Sharma, 1965-1967, unpublished) to count and weight diatoms in measured amounts of sediment. This method was soon abandoned because of the fragile nature of diatom tests and the uncertainty resulting from the dissolution of these tests soon after burial. A better estimate of the biogenic silica contribution to sediment can be obtained by studying the dissolved silica budget for the region. Such a study provides not only information about the silica cycle but also some insight on water dynamics.

Over fifty stations in southeastern Alaska were occupied and an average of 600 water samples from standard hydrographic depths were collected during each of six cruises (Fig. 2-18), at two-month intervals, 1965-1966. These cruises were scheduled to sample two annual blooms and to measure seasonal water mass properties. Salinity, temperature and dissolved silica and oxygen in the water column were measured routinely. Samples were also collected from rivers and streams draining into inlets. Silica varies with depth and season in southeast Alaskan waters as shown in Figure 2-19. Maximum seasonal variations occur in the surface layer, where photosynthesis occurs and from which silica is extracted by organisms. In the underlying layer, separated by a sharp pycnocline, soluble silica increases with depth, in general varying from 0.5 ppm at the near surface to 5.0 ppm at depth. Minimum silica concentrations were observed during summer months and bloom periods, with maximum values during winter months and between the bloom seasons. The waters in typical estuaries, with bimodal

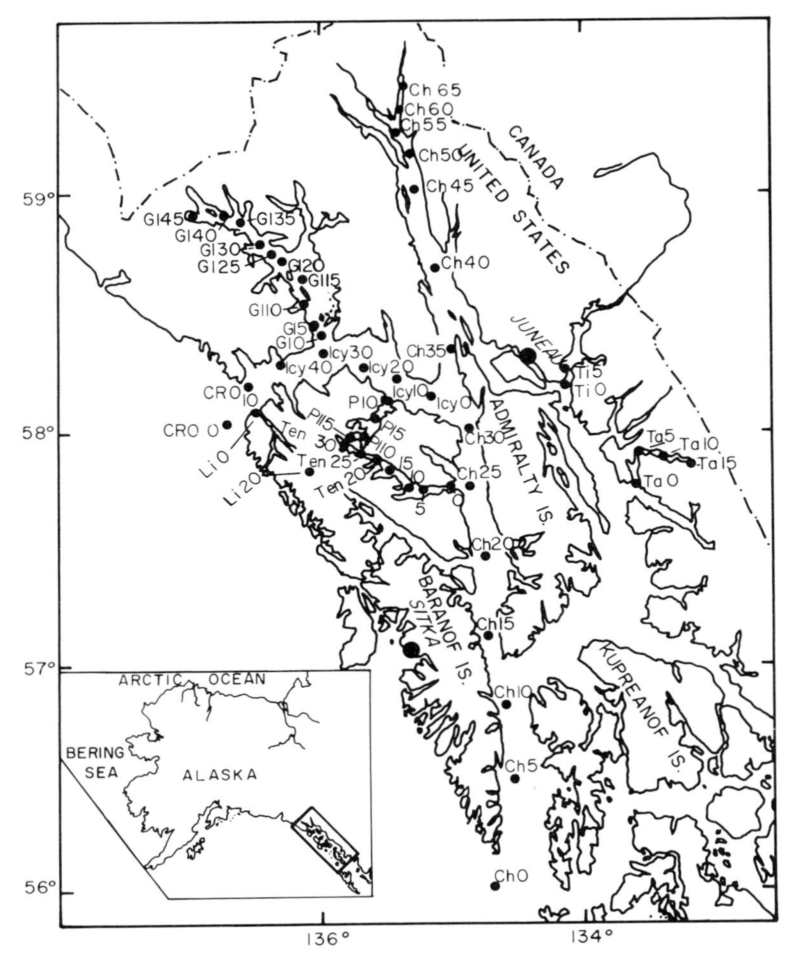

Figure 2-18. Index map showing water sample station locations, southeast Alaska.

discharge patterns, such as the Lynn Canal-Chatham Strait system and Tenakee Inlet, are stratified during most of the year, with a surface layer low in silica and a bottom layer enriched in silica. On the other hand, inlets with tidal glaciers such as Glacier Bay, have a unimodal drainage pattern and are stratified only during late summer.

Dissolved silica in surface runoff from rivers and streams is not significantly higher than that in adjacent marine water (Table 2-2). Silica-salinity diagrams for Chatham Strait and Glacier Bay also suggest that the influence of freshwater input during summer months is slight and restricted to the surface layer (Fig. 2-20).

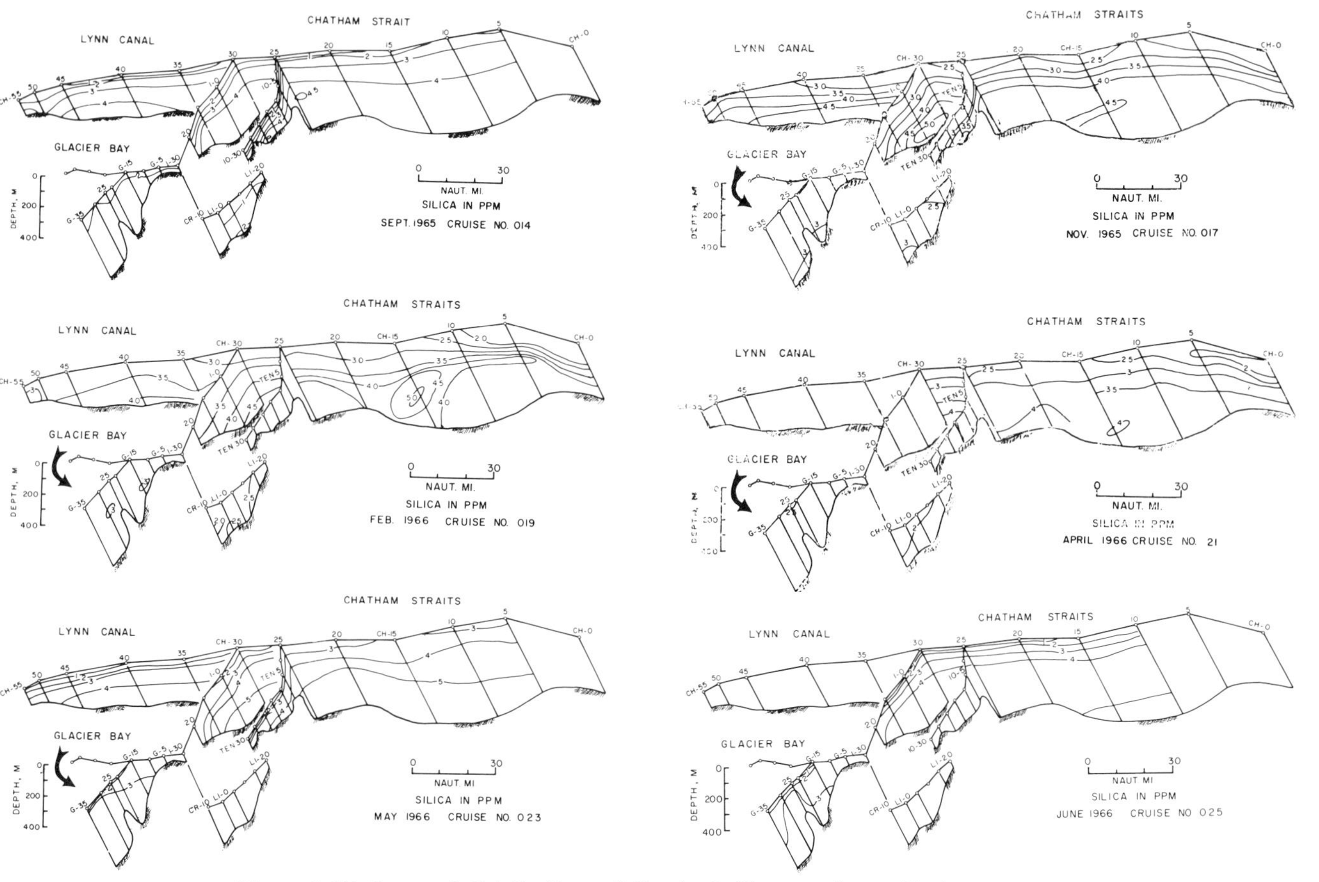

Figure 2-19. Seasonal distributions of dissolved silica, southeast Alaska.

TABLE 2-2

Silica, Salinity and Temperature of Waters from Various Environments in Southeast Alaska

Area	River			Tidal Zone			Delta			Bay		
	SiO_2 ppm	Sal ppt	Temp °C	SiO_2 ppm	Sal ppt	Temp °C	SiO_2 ppm	Sal ppt	Temp °C	SiO_2 ppm	Sal ppt	Temp °C
Tenakee Inlet, Corner Bay	4.2		2.2	4.2		2.7	3.2		6.3	3.6		6.0
Tenakee Inlet, Head	3.8	3.808	7.4	3.7	4.654	7.1	2.8	15.515	7.5	3.5	30.276	6.5
Chatham Strait, Pt. Howard	2.5	0.78	2.4	2.8	7.95	3.0	3.4	30.96	3.7			
Tanya River	6.3	6.97	3.2							3.8	28.98	2.7
Glacier Bay, Ptarmigan Creek	3.0	0.12	1.0	3.0	0.82	2.6	2.8	27.56	2.1			
Geikie Inlet	3.8	0.09		4.0	0.93							
Muir Inlet	3.1	0.14	3.2	3.1	1.60	3.2	3.2	8.90	2.5	2.5	30.23	2.3
Lisianski Inlet, Head	2.9		1.8	2.8		4.2	2.2		4.7			
Excursion Inlet 1	3.2	0.30	1.9	3.0	25.25	3.6						
Excursion Inlet 2	4.2	0.16	2.3	3.8	9.11	4.1						

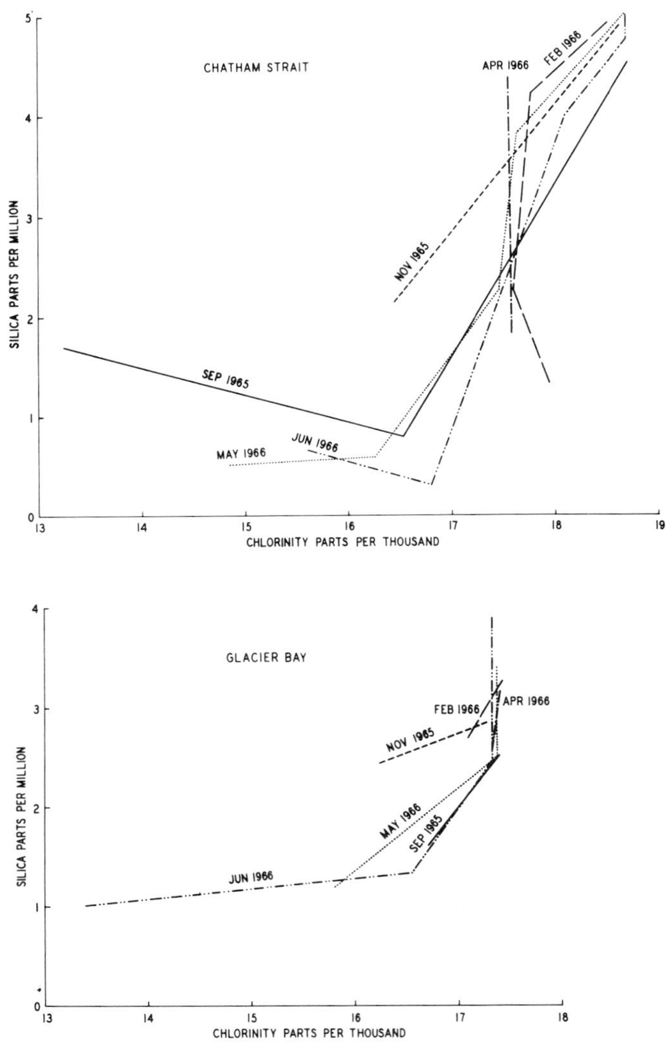

Figure 2-20. Seasonal changes in silica-salinity plots during 1965-1966 in Chatham Strait and Glacier Bay.

Dissolved silica at any point in the oceans is a result of dynamic equilibrium between rates of physical transfer (lateral and vertical mixing and turbulent diffusion) and rates of chemical, biochemical and phsico-chemical processes (solution, coagulation, sorption, desorption, etc.). In southeast Alaska, the distribution of silica is controlled mainly by biogenic processes. During diatom blooms large amounts of soluble silica in the euphotic zone are converted into siliceous tests composed of amorphous silica. Part of this is redissolved and part is deposited on the sea floor as diatomaceous sediments. As plankton with siliceous tests die and sink, gradual solution begins, resulting in an increase of silica with depth. Not all of this amorphous silica leaves the euphotic layer because part is dissolved and reextracted by phytoplankton, giving rise to a minor cycle of silica in the surface layer.

The annual net removal of silica from sea water in an area can be estimated from the seasonal fluctuations of silica in the water column. During summer months, when the seasonal pycnocline is well defined, the silica loss may be estimated by differences in concentration above and below it. In winter the concentration of silica in the surface layer reaches a maximum; this is later reduced to a minimum during summers and especially during the bloom. The difference between the maximum and minimum is therefore a measure of the yearly loss of silica due to its extraction by organisms and presumably deposited as sediment. This measure, multiplied by the volume of the surface layer in which the loss occurs, gives the total yearly loss. In southeast Alaska waters seasonal fluctuations occur in the upper 100 m. For a one square meter area with a depth of 100 m, a difference in silica concentration of 1 ppm amounts to a loss of 100 g/m^2/year. Using this reasoning, the silica loss in southeast Alaska during 1965-1966 is estimated at 120-150 g/m^2. Computations for the northern part of the Pacific Ocean yield slightly lower values, 60-120 g/m^2/year.

The sediments contributed by benthic organisms in southeast Alaska are mostly carbonates and have been described by Hoskin and Nelson (1969). Beach sediments in Pirates Cove, Three Entrance Bay, and Taiguid Island, as well as the surficial bottom sediments west of Chichagof Island and in Symond Bay at Biorka Island, consist of barnacles, echinoids, bryozoans, and mollusks.

Solitary corals occasionally have been observed in bottom grabs from southeast Alaska. Surprisingly, these corals (Fig. 2-21) can grow even in the cold waters of Glacier Bay. Large masses of coral, for example, were dredged from the shallow sill south of Berners Bay in Lynn Canal.

Thus, although terrigenous material predominates, it is apparent that biogenic sediments locally can form a significant constituent (mainly diatoms) of bottom sediments in southeast Alaska. Small numbers of diatoms and foraminifera occur in sediments throughout the region, and high concentrations of carbonate sediments occur in some beaches, shallows and bays.

GEOCHEMISTRY

Physical and chemical changes occurring during the transfer of sediments from their sources to the site of deposition have been the subjects of intensive studies. These are particularly important when transfer of terrestrial sediments to the marine environment takes place. Geochemical investigations should delineate

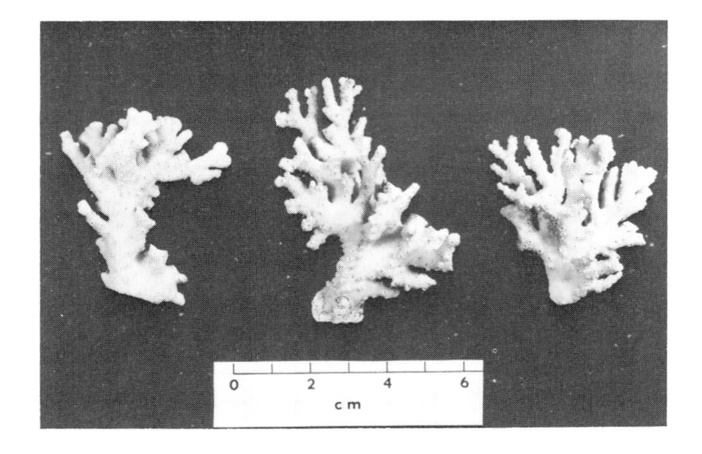

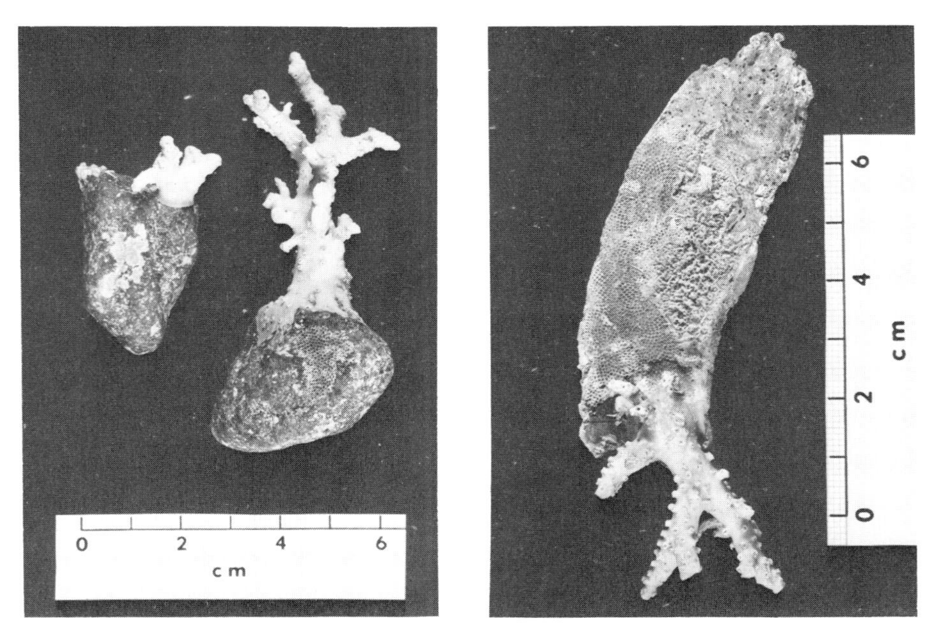

Figure 2-21. Coral specimens from southeast Alaska.

changes in elemental and mineralogic composition resulting from textural differentation as well as alterations in sediments due to change in the environment.

The young, rugged and tectonically active coast of southeast Alaska provides large amounts of sediments, mostly transported by short rivers and streams or by glaciers to the nearby sea. This region is, therefore, ideally suited to the study of geochemical alterations of terrestrial sediments as related to glacial-fluvial erosion, transport, and sediment-seawater interaction.

Sediments were collected from meltwater at the face of the Carroll Glacier in Queen Inlet (Figs. 2-13 and 2-16) in order to study the interaction between glacial sediments and seawater (Fig. 2-22). Replicate samples were thoroughly mixed with distilled and synthetic seawater and allowed to stand (Table 2-3). Similar studies of interaction were conducted between suspended sediments from Queen River and Queen Inlet (Table 2-4). It is apparent that the sediments rapidly exchange cations with seawater by taking up sodium, potassium, and magnesium, and maximum release of calcium was observed for suspended sediments from the river; suspended sediments from the inlet showed markedly less cation exchange. The amount of sodium removed from synthetic seawater by glacial sediments was identical to that removed by suspended river sediments.

Sediment-seawater interaction over an extended period can be studied by comparing the composition of the interstitial waters of the bottom sediments with the overlying seawater. Such studies in Muir Inlet (Fig. 2-16) show noticeable differences between the composition of pore and overlying waters (Fig. 2-23). In general, the total ionic concentration in pore water is less than that in the overlying water. The sodium and magnesium ion concentrations in pore water are significantly lower than in the overlying water, while potassium and calcium concentrations are higher. This maximum disparity between sodium and calcium concentrations in pore water and in overlying water occurs at the head of the inlet. Maximum magnesium ion concentration, however, is found at the mouth of the inlet and the minimum occurs at the head. These differences reflect the extent of sediment-seawater interaction. Sediments deposited near the source do not have sufficient time to absorb sodium and reach seawater-sediment equilibrium; thus they continue to remove large quantities of sodium from the interstitial fluid after deposition.

To further elucidate the nature and extent of the interaction between fluvioglacial sediments and seawater, thirteen cores were obtained from various fjords and inlets of southeast Alaska and the cationic concentrations of major ions in the pore water were determined (Sharma, 1970a, 1970b, 1971a and 1971b). The total cationic concentration in pore waters from cores is variable between the cores and within cores; in general it increases with depth. The sodium ion concentration shows the most dramatic changes with depth and tends to increase with the percentage of the silt. To some extent the sodium varies inversely with the percentage of clay, but this is expected since silt and clay contents in sediment are complementary. The concentration of major cations in the pore water is not influenced by the relative percentages of illite or chlorite or by the relative amounts of the magnesium/iron components of the chlorites. Manganese in pore water generally decreases with depth in all cores, presumably due to upward migration (Boström, 1971).

The foregoing observations indicate that fluvioglacial sediments of southeast Alaska do exchange a number of cations when transferred to marine water. It

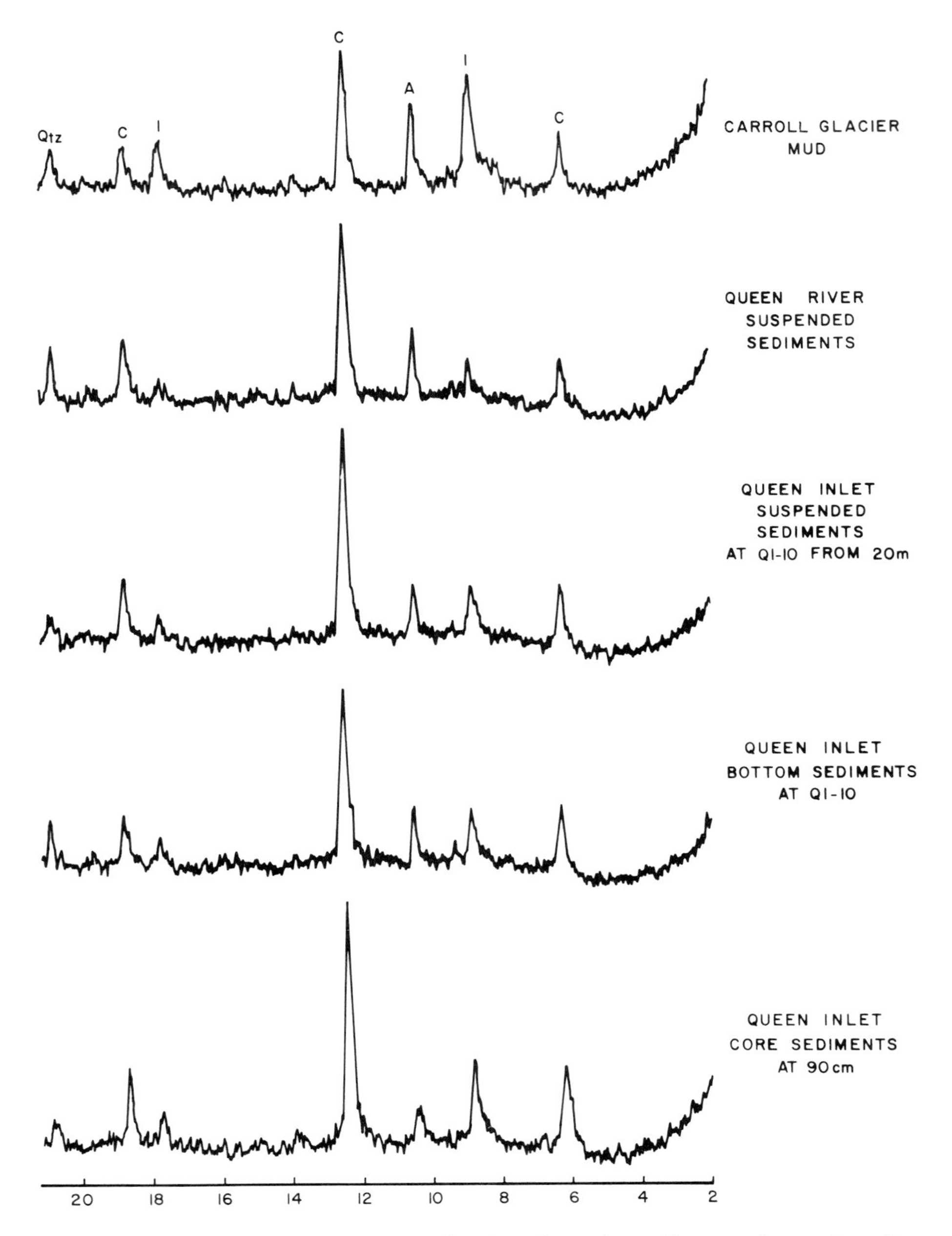

Figure 2-22. Relative distribution of mineralogy in sediments from Carroll Glacier, Queen River, and Queen Inlet. A, amphibole; C, chlorite; I, illite; Qtz, quartz.

TABLE 2-3

Changes in Composition of Synthetic Sea Water During Seven Days Contact with Sediments

Sample	Loss (in milliequivalent)			Gain (in milliequivalent)	pH
	Na+	K+	Mg++	Ca++	
Distilled water					7.3
Synthetic sea water					7.8
Distilled water & glacial sediments					7.3
Synthetic sea water & glacial sediments (well shaken)	8.59	0.89	0.0	7.48	6.95
Synthetic sea water & glacial sediments (stirred with electric mixer for five minutes)	34.78	0.51	9.46	2.99	7.1

TABLE 2-4

Changes in Composition of Synthetic Sea Water During Fourteen Months Contact with Sediments

Sample	Loss (in milliequivalent)			Gain (in milliequivalent)	pH
	Na+	K+	Mg++	Ca++	
Queen River suspended sediment	34.78	1.66	8.22	2.49	8.7
Queen Inlet, low tide suspended sediment from O m depth	17.39	1.02	4.93	0.74	9.2
Queen Inlet, low tide suspended sediment from 20 m depth	17.39	0.89	4.11	0.74	9.2

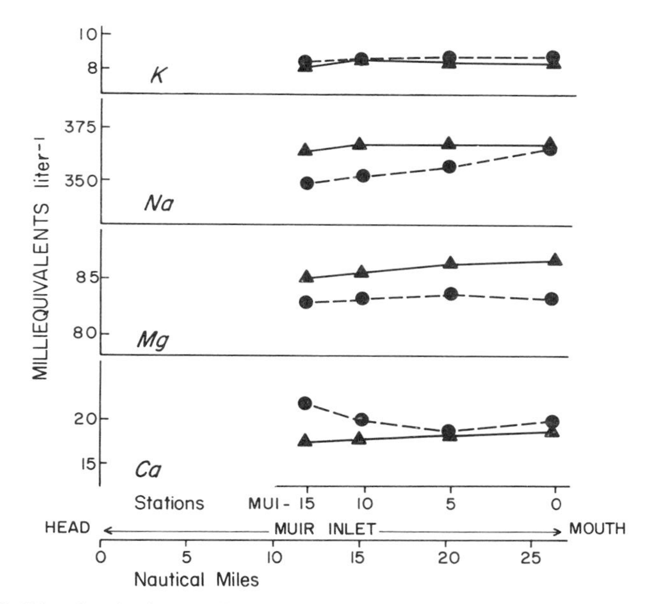

Figure 2-23. Dissolved chemical species in bottom and interstitial waters, Muir Inlet. (▲, bottom waters; ●, interstitial waters).

appears that the composition of interstitial waters is largely controlled by the exchange capacity of the sediments, which in turn is controlled by mineralogy and particle size. The gain and the loss of cations in interstitial waters by reaction with sediments also depends on the transport distance and the duration of the sediment exposure to the marine environments.

The chemical composition of sediments from a region is primarily a function of (1) the source, (2) weathering, and (3) transport. Weathering at the source generally breaks down parent rock into mineral aggregates and individual mineral grains. These, when transported, are separated by physical sorting processes according to size, shape, and density. In the basin such sorting is the major cause of differences in the bulk chemical composition of sediments. Composition is further modified by the stability of hydroxide under varying Eh-pH conditions and the precipitation of minerals, such as $Fe(OH)_3$.

A detailed discussion of the geochemistry of sediments from Lynn Canal and Chatham Strait (Fig. 2-24) will serve as an example since the nature and distribution of the sediments here are typical of southeastern Alaska. In Chatham Strait the sediments are sand, silt, and clay (Fig. 2-25). Major and minor elemental compositions for fourteen bulk sediment samples from this region (Fig. 2-24) are shown in Table 2-5 and Table 2-6, respectively. Among the major elements, Al and K decrease from the head of the Lynn Canal (sample 13) to the mouth of Chatham Strait (sample 1). Magnesium and Ti appear to decrease in a

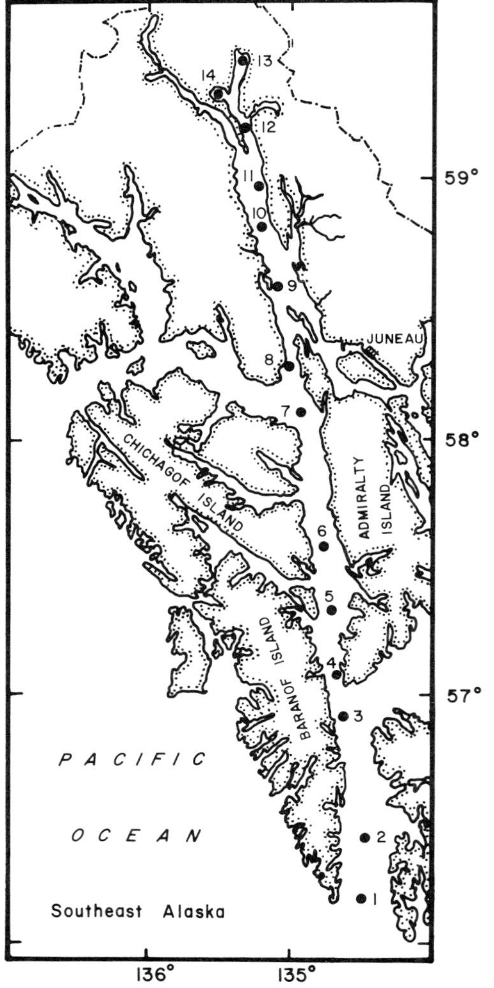

Figure 2-24. Index map showing station locations, Chatham Strait.

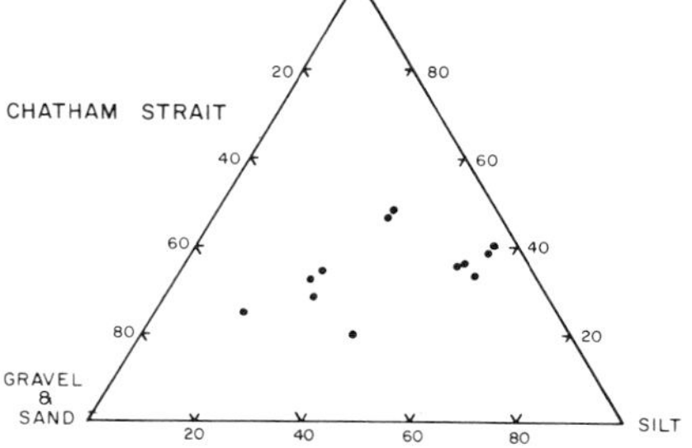

Figure 2-25. Weight percents gravel-sand, silt, and clay in sediments, Chatham Strait.

TABLE 2-5

Major Elements for Chatham Strait Bulk Sediments, in Percent

Sample Number	Latitude	Longitude	Al_2O_3	Fe_2O_3	CaO	MgO	K_2O	Na_2O	MnO	TiO_2	SiO_2
Ch-1	56°11.5′	134°30.0′	10.40	5.76	3.57	2.02	1.72	4.19	0.088	0.57	68.56
Ch-2	56°26.5′	134°28.8′	10.40	4.82	4.90	2.09	1.60	4.21	0.073	0.57	64.15
Ch-3	56°55.5′	134°37.8′	13.70	7.74	4.76	3.15	2.00	4.08	0.098	0.73	58.65
Ch-4	57°05.3′	134°41.8′	10.40	5.91	5.25	2.72	1.69	4.99	0.079	0.63	57.34
Ch-5	57°20.8′	134°42.5′	10.40	10.85	3.99	2.85	3.13	3.92	0.077	0.59	56.82
Ch-6	57°35.5′	134°44.7′	8.51	5.19	4.13	3.29	1.75	7.69	0.089	0.63	55.40
Ch-7	58°07.2′	134°56.8′	14.62	7.32	4.62	3.65	1.94	4.86	0.175	0.72	52.01
Ch-8	58°17.0′	135°00.0′	11.34	6.84	4.62	3.39	2.06	4.99	0.156	0.67	56.01
Ch-9	58°35.4′	135°05.8′	12.76	5.42	5.04	3.85	2.18	4.59	0.277	0.72	57.27
Ch-10	58°49.3′	135°11.6′	13.70	8.35	5.60	4.45	2.30	4.99	0.193	0.88	52.87
Ch-11	58°58.8′	135°13.0′	13.70	8.02	5.18	4.13	2.30	4.25	0.121	0.90	55.98
Ch-12	59°11.0′	135°18.0′	14.18	6.61	5.46	3.59	2.54	4.72	0.113	0.86	58.82
Ch-13	59°26.9′	135°20.1′	14.36	3.08	3.50	1.19	2.60	4.21	0.038	0.37	67.25
Ch-14	59°18.7′	135°31.3′	14.18	3.96	4.41	3.59	2.66	4.21	0.131	1.03	62.41

TABLE 2-6

Minor Elements for Chatham Strait Bulk Sediments, in PPM

Sample Number	Latitude	Longitude	Barium	Cobalt	Chromium	Copper	Nickel	Strontium	Zinc
Ch-1	56°11.5′	134°30.0′	310	20	430	120	15.5	220	57
Ch-2	56°26.5′	134°28.8′	400	20	410	190	18.5	234	78
Ch-3	56°55.5′	134°37.8′	460	20	510	250	29.5	247.5	105
Ch-4	57°05.3′	134°41.8′	350	16.5	220	250	29.5	220	109
Ch-5	57°20.8′	134°42.5′	240	21	480	190	20.0	261	72
Ch-6	57°35.5′	134°44.7′	370	18	480	320	20.0	278	93
Ch-7	58°07.2′	134°56.8′	580	21	510	350	34	277	110
Ch-8	58°17.0′	135°00.0′	610	27.5	550	320	35.5	275	126
Ch-9	58°35.4′	135°05.8′	640	24	550	320	37	330	109
Ch-10	58°49.3′	135°11.6′	580	29	460	390	35.5	316.5	116
Ch-11	58°58.8′	135°13.0′	420	29	440	390	34	330	109
Ch-12	59°11.0′	135°18.0′	680	26	420	290	31	440	85
Ch-13	59°26.9′	135°20.1′	680	12	110	60	-	440	41
Ch-14	59°18.7′	135°31.3′	1020	21	450	290	28	440	130

similar fashion, although the decrease is not as regular and pronounced. Of the minor elements, Sr and Ba also decrease from the head to the mouth of the Lynn Canal-Chatham Strait system. Silica gradually decreases from the head to Icy Strait (sample 7) and then increases toward the mouth of Chatham Strait.

Manganese and the related minor elements Co, Cr, and Cu increase southward in Lynn Canal (Tables 2-5 and 2-6); however, in the lower basin (Chatham Strait) the distribution of these elements is irregular. The other minor elements analyzed, Ni and Zn, do not show any systematic distribution pattern.

The Al, Si, Fe, Ti, and Mn contents of sediments near the head (north end) are similar to that of acid igneous rocks, again reflecting the lack of chemical weathering. The typical chemical composition of these ice-eroded sediments reflects their mineral composition, which is characteristically low in quartz with a high content of feldspar, micas, pyroxenes, amphiboles, and epidote. The low maturity of sediments resulting from physical abrasion and lack of chemical weathering is also reflected in the high Al_2O_3/SiO_2 values. This important ratio ("maturity value") is a useful indicator of the paths of sediment migration. The values of the ratios gradually decrease from a high of 0.23 near the head of Lynn Canal to 0.15 near the mouth of Chatham Strait (Fig. 2-26). A slight southward increase in samples 12, 11, and 10 is due to particle differentiation (enrichment of clay) with increasing water depth.

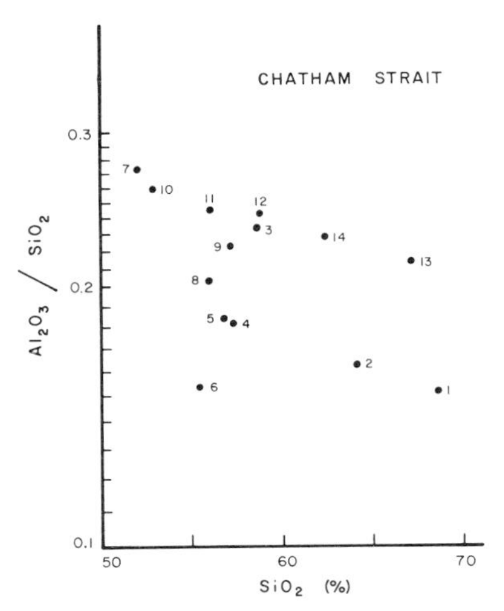

Figure 2-26. Al_2O_3/SiO_2 - percent SiO_2 of sediments, Chatham Strait.

Stations 7 and 3 lie near the mouths of Icy Strait and Frederick Sound, respectively; these channels are large and significant water and sediment movements are evident from the variations in chemical composition. Aluminum in sediments from station 6 is strikingly low, probably a result of high water energy that removes silicates. This shallow station is located on a broad sill, over which large amounts of water must pass during each tidal cycle.

Source and sediment movement in the Lynn Canal-Chatham Strait system can be deduced from various elemental distributions and interelement relationships (Fig. 2-27). The major sediment source for the system appears to lie in the drainage basins located at the head of Lynn Canal, where sediments are carried to the inlet by numerous rivers and streams. Sediments deposited near their source in Taiya Inlet and Lynn Canal (stations 14-7, Fig. 2-24) have high Al_2O_3/SiO_2 (Fig. 2-26). The decrease in silica is mostly compensated by an increase in Na, K, Fe, Ca, and Mg. Sediments in the Chatham Strait basin also show a decrease in Al, but near the mouth there is an increase in silica with decreasing Na, K, Fe, Ca, and Mg. Sediments near the mouth may be mixtures of two sources with different mineralogies (Fig. 2-28). The apparent decrease in Al, accompanied by a decrease in Ca, Mg, Na, K, and Fe, suggests a relative increase of Si. Infusion of Si in sediments without related elements from the erosion of igneous rock can only be made by biologic contribution or chemical precipitation. Near the mouth of Chatham Strait, upwelling of nutrient-rich water from a depth of approximately 300 m occurs and causes prolific diatom blooms in this region (see "Biogenic Sediment" section). Diatom tests in surface sediments were observed from this region. The addition of biogenic silica, therefore, can explain the increase in the relative amount of silica in the sediments.

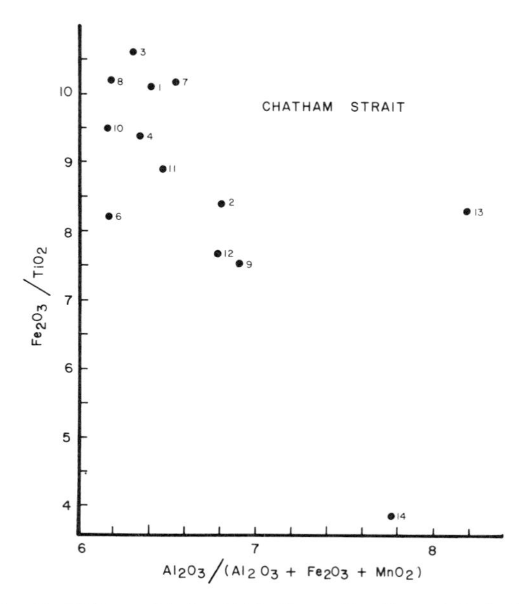

Figure 2-27. Fe_2O_3/TiO_2 - $Al_2O_3/(Al_2O_3 + Fe_2O_3 + MnO_2)$, Chatham Strait.

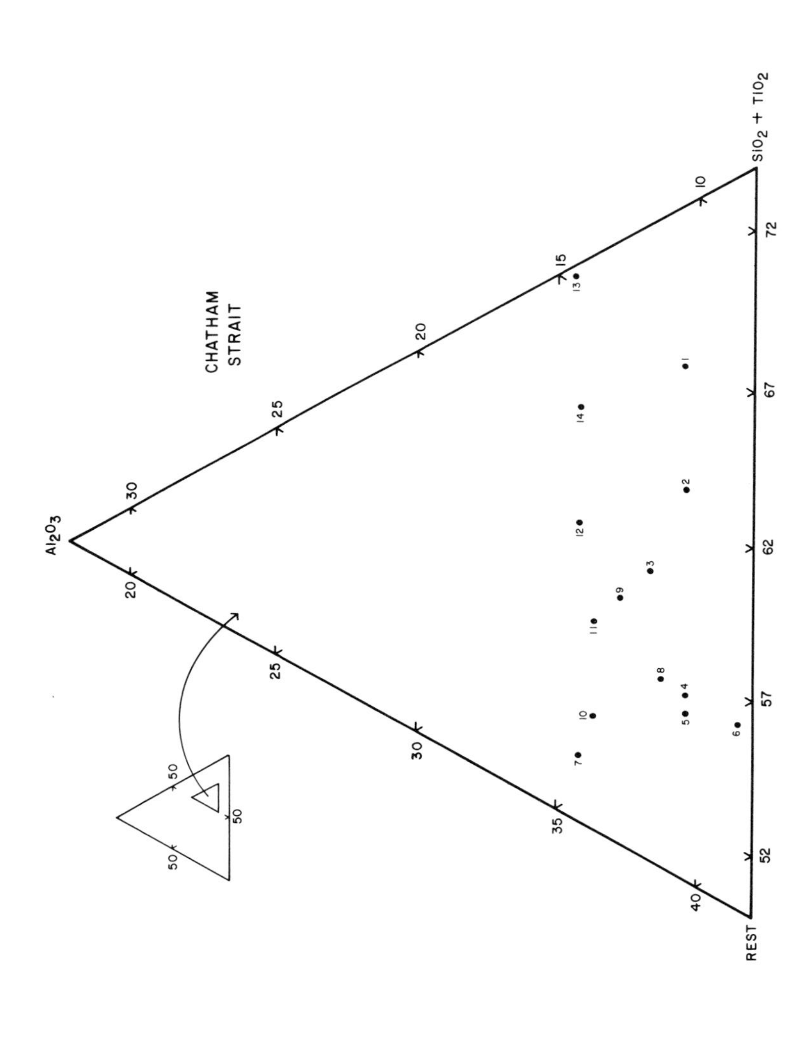

Figure 2-28. Triangular plot with percents Al_2O_3, SiO_2 + TiO_2, and Rest (Fe_2O_3, CaO, MgO, Na_2O, K_2O, MnO_2) in sediments as end members, Chatham Strait.

Changes in elemental ratios show a general decrease in K_2O/Na_2O and Al_2O_3/Na_2O with increasing (northward) latitudes, whereas Al_2O_3/TiO_2 and CaO/MgO increase (Fig. 2-29). It is known that K_2O/Na_2O and Al_2O_3/Na_2O values in sediments commonly decrease with increasing distance from the source, whereas Al_2O_3/TiO_2 and CaO/MgO values increase. These relationships and the Al_2O_3/SiO_2 distribution strongly suggest that the major source for the sediment deposited in Lynn Canal-Chatham Strait lies at the northern end of the system. Slight variations, of course, are introduced by bottom topography and hydrodynamic conditions in the basin.

Glacier-derived sediments, in comparison to the nonglacial sediments from regions of either arid or humid climates, generally contain higher K, Na, Al, and Ti and a somewhat lower content of Si. This is primarily due to lack of prolonged chemical weathering, which gives rise to low maturity coefficient in glacial sediments. Characteristically, the arkosic sands reflect a low degree of weathering of source material. Sediments from southeast Alaska show low values of maturity coefficient (SiO_2/Al_2O_3 and quartz feldspar), particularly those sediments which are deposited close to their sources (Fig. 2-29). Also their chemical composition is similar to the arkosic sands. The low maturity coefficient of these sediments is thus indicative of their glacial origin, and their chemical composition suggests that they have undergone only the initial stages of weathering (mainly physical).

7It is abundantly clear that the sediments in southeast Alaska are of glacial origin. It is also obvious that sediments have not undergone prolonged chemical weathering prior to their deposition. The persuance of the discussion in this case was to establish that the chemical parameters are indeed related to sediment weathering and transport. These parameters thus can be effectively used as the indicators of sediment weathering and transport in other regions.

SEDIMENT SOURCE AND TRANSPORT

Sediments in southeast Alaska fjords and inlets are mostly transported in suspension. Suspended matter includes terrigenous particles brought into the inlets by rivers and wind, material resuspended by tides, organic matter (living or dead), and material formed from inorganic precipitation. Distribution and composition of suspended matter vary with distance from shore, bottom topography, season, prevailing wind pattern, and biologic productivity. The terrigenous particles form the dominant part of suspended matter observed in southeast Alaska. Next in abundance is biogenous matter.

Suspended matter derived from rivers and streams draining into the inlets is carried in surface layers (plumes) of relatively low-salinity water. The Secchi depths mostly were less than 1 m near the heads of inlets (Pickard, 1967). Pickard also showed a relationship between low salinity and Secchi depth in southeast Alaskan inlets, suggesting sediment suspension and movement in

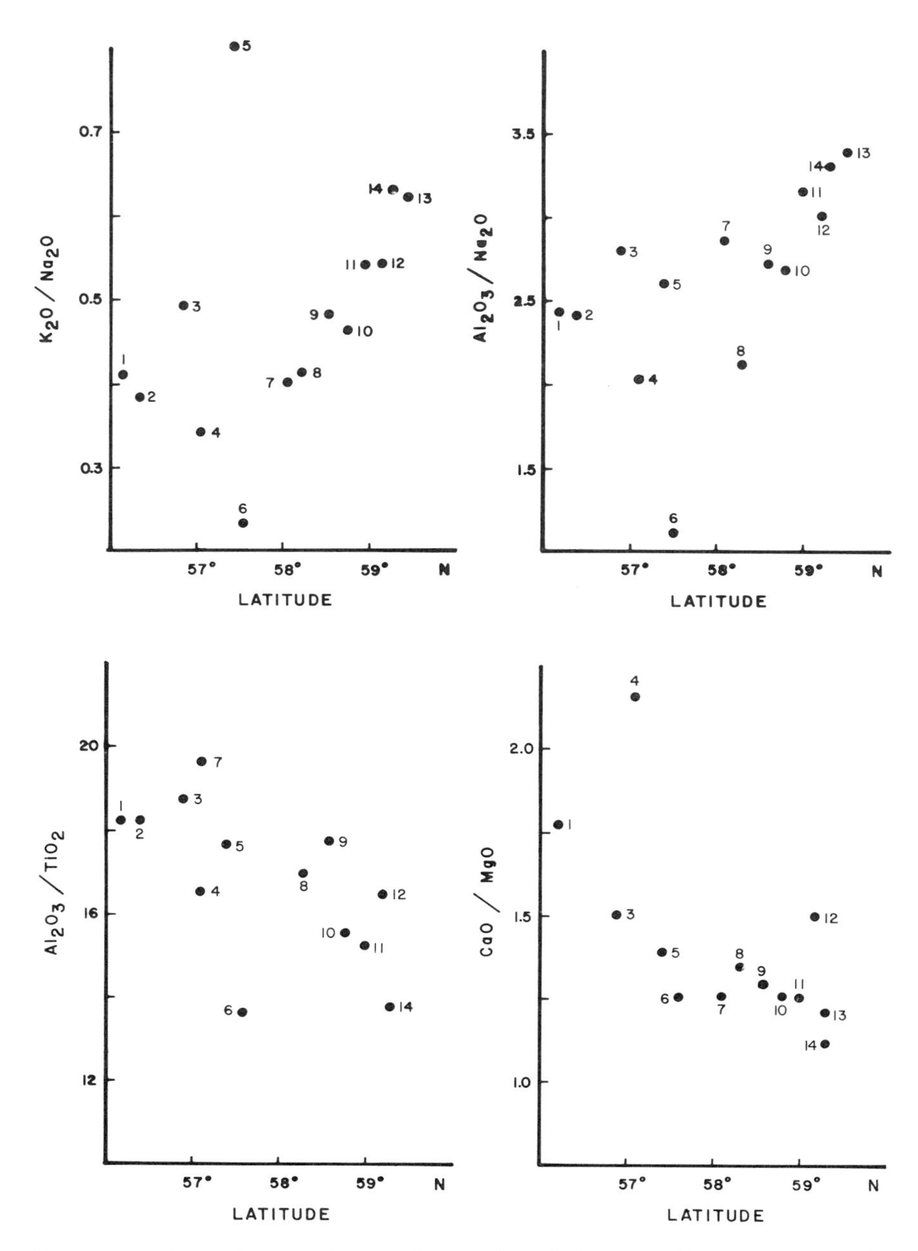

Figure 2-29. Variations in elemental ratios in relation to sediment transport distance (latitude).

freshwater plumes. This is due to the slow rate of mixing between fresh and marine waters, and a strong pycnocline which persists at depth and with distance is formed at the interface. This pycnocline retards settling of sediment particles at its lower boundary, forming a sediment-laden freshwater prism (plume) at the surface. Near the source the prism is thickest and it retains its identity as it extends seaward for several miles. Surface plumes are sometimes destroyed by wind, which generates turbulence in the surface layers so as to cause mixing and flocculation of sediments.

Normally, two or more plume prisms are stacked on top of one another near the source by oscillating tides. These layers are generally restricted to near the surface, with depths ranging from 19 to 40 m. Such plume prisms, however, do not extend very far into the inlet and do not persist beyond a few tidal cycles. Various investigators have observed discrete water layers at depth (nepheloid layers), with sediment concentrations higher than underlying and overlying layers. This suggests that some material is indeed transported within the water column. The waters of these discrete subaqueous layers show normal temperatures and salinities at depth but have higher suspended loads. These layers are oriented such that the depth near the head (source) is relatively shallow but becomes increasingly deeper seaward.

Suspended load in turbid layers varies from a few milligrams per liter to about 1000 mg/ℓ. Concentration of sediments exceeding 1000 mg/ℓ in Taku Inlet (Wright, 1971) and 115 mg/ℓ in Endicott Arm (Buckley and Loder, 1968; Wright and Sharma, 1969) have been reported. The nepheloid layer sediments generally consist of varying mixtures of silt and clay.

The spatial distribution of the surface suspended load is a function of several variables, including: river input; tidal-, geostrophic-, and wind-induced circulation; and biologic production. The amount of river-derived detritus varies significantly and is dependent primarily on precipitation. In general, the surface suspended sediment load increases in proportion to the increase in discharge of the rivers and streams draining into the inlet. Inlets with large river discharges, therefore, have a variable influx of sediments in suspension. In the Taku River and the tidal regions of the estuary, for example, the suspended load reached 15 and 10 g/ℓ, respectively, after two days of precipitation in the drainage area. Farther from the river, the suspended load varied between 0.1 and 1.5 g/ℓ. During drier periods, however, the concentration of sediments in suspension in the Taku Estuary becomes considerably less. Sediments in the upper layers of water during dry seasons are generally resuspended from the tidal flats by tidal currents. During spring and fall, when phytoplankton blooms, the biogenic matter contributes significantly to the suspension load.

Inlets which are fed by glacial meltwater generally have a steady input of sediments in suspension. The glacier ablation begins in early summer and reaches its peak during late summer and early fall. There are three possible ways glacial material can be transported into an inlet:

1. by a tidal glacier which introduces its ablation directly;

2. by glacial melt carried by a stream which flows over an outwash fan;
3. by a stream from a glacial meltwater lake formed near the glacier terminus.

Most sediments carried into an inlet by tidal glaciers generally are transported within the ice, at or near the sole of the glacier. Lesser amounts are carried by subsurface subglacial streams. The lateral streams which flow along the glacier and bedrock generally do not carry sediments in appreciable quantities. Therefore, tidal inlets do not have sediment plumes of any appreciable significance at the head.

Glacial streams, originating either near the terminus or from a meltwater lake, transport large amounts of sediments, both in suspension and as bed load. The sediments in suspension may exceed 10 g/ℓ. An average of 12.8 g/ℓ sediments in suspension in the meltwater stream draining Carroll Glacier and discharging into Queen Inlet was reported by Slatt (1970). Suspended load in the braided streams of Taku Glacier also varied between 1 and 10 g/ℓ in 1964. The sediment load fluctuates significantly with the tidal range and varies as a function of the geomorphologic configuration of the stream. Sediments transported in suspension are silt and clays. Generally the suspended sediment is finer than medium silt; occasionally it contains coarse silt and fine sand.

High-energy turbidity currents are also prevalent in the area, a result of frequent slumping. Trough-like fjord basins are characterized by shallow constrictions and steep slopes and commonly abut against mountainous shores. Short mountain streams and glaciers bring much coarse material to the basin, the majority of which is deposited as alluvial fans. This gives rise to an unstable accumulation of coarse-grained sediments along the shores. Due to soil creep and submarine slump, and often during earthquakes, an abundance of coarse material is torn from these shoreline scarps and falls into the basin. Such masses of material possess tremendous kinetic energy and slide rapidly downslope, at times moving great distances from shore. Some of the sedimentary features of these slump deposits are distinctive; and they commonly have graded bedding. Marine earthquakes affect the sediments in the basin through the reworking of bottom sediments and the removal of fine fractions.

SUMMARY

Southeast Alaska lies in an active tectonic belt. The region is distinguished by rugged mountains, which support numerous glaciers and small streams that discharge into trough-like fjordal basins. These basins are typically narrow, with steep walls that extend well below the sea level and have flat floors. These floors commonly consist of a series of troughs, separated by sills that may or may not rise above sea level. Glacial erosion of tectonically mobile regions give rise to a number of characteristic modes in the grain size distribution of sediments of southeast Alaska. The polymineralic silt and sand deposited in these intermontane troughs are typical of active tectonic belts.

Gray clayey, sandy silts or sandy and silty clays are the characteristic sediments of southeast Alaska. Striking uniformity in the mineralogy of sediments exists throughout the region. The light mineral suite consists of andesine, oligoclase and orthoclase, with minor quantities of microcline and quartz. The feldspar-quartz ratio is generally more than 3 : 2; this immaturity is due to the nature of the source materials, abetted by the relative lack of chemical weathering and short transport distance. Typically the sands are arkosic in composition. Concentrations of heavy minerals vary with texture, being greater in the finer sediments. Minerals commonly found are hornblende, diopside, hypersthene, actinolite, chlorite, and magnetite. Traces of sphene, garnet, sillimanite, apatite, tourmaline, epidote and staurolite also occur. The clay size fraction is dominated by "illite" (trioctahedral mica) and chlorite, with occasional smectite. Kaolinite has not been detected. Electron photomicrographs of clay particles and optical photomicrographs of sand and silt show well-defined angular borders, again suggesting minimal chemical weathering or abrasion during transport.

The chemical composition and mineralogy of sediments near the head of the Lynn Canal-Chatham Strait system suggest their source in acidic igneous rocks; they resemble arkoses in composition. The maturity index of sediments is low near the source and improves steadily with increasing distance of transport. Interelement relationships generally indicate major sources for sediments at the heads of inlets and bays.

In southeast Alaska the basins generally receive more of their detritus from the proximally adjacent land, much of which is or has been extensively glaciated. Thus the sediments transported into the inlets are, for the most part, primary or reworked glacial detritus. Many glaciers, either receding or advancing, form lag deposits near their termini and are drained by outwash streams, which carry sediments into marine inlets. This detritus generally consists of a mixture of sand, silt, and clay in various portions, with occasional gravel.

A major portion of the sediment is carried into the basin as suspended and as bed load. Tidal glaciers carry sediments directly to the marine environment and part of this becomes ice rafted. Silt and clay are generally carried in a distinct freshwater plume that may extend several kilometers from the source. Turbid layers with high sediment concentrations occasionally have been observed within the water column. Such subaqueous turbid layers may be an important agent of sediment transport.

REFERENCES

Boström, K. (1971). Origin of manganese-rich layers in arctic sediments. Second International Symposium on Arctic Geology, San Francisco, February 1-4, 1971 (abstract).

Buckley, D. E., and T. C. Loder. (1968). Particulate organic-inorganic geochemistry of a glacial fjord. Unpublished report to U.S. Atomic Energy Commission, Part 1. D. C. Burrell and D. W. Hood, Eds., Inst. of Mar. Sci., Univ. of Alaska, Fairbanks, p. 93.

Cooper, W. S. (1937). The problem of Glacier Bay. *Alaska Geol. Rev.* 37(1).

Field, W. O. (1947). Glacier recession in Muir Inlet, Glacier Bay. *Alaska Geog. Rev.* 37(3).

Goldthwait, R. P. (1966). Evidence from Alaska glaciers of major climatic changes. Royal Meteorological Proc. of the Intl. Symp. on World Climate from 8000 to 0 B.C., Royal Meteorological Soc. London, p. 229.

Hood, D. W., and D. D. Wallen. (1971). Oceanography of southeast Alaska, Inst. Mar. Sci. Rept. R. 71-1, University of Alaska, Fairbanks, Alaska, p. 198.

Hoskin, C. M., and D. C. Burrell. (1972). Sediment transport and accumulation in a fjord basin, Glacier Bay, Alaska. *J. of Geol.* 80(5):539-551.

Hoskin, C. M., and R. V. Nelson, Jr. (1969). Modern marine carbonate sediment, Archipelago, Alaska. *J. Sed. Petrol.* 39(2):581-590.

Hoskin, C. M., and R. V. Nelson, Jr. (1971). Size modes in biogenic carbonate sediment, southeastern Alaska. *J. Sed. Petrol.* 41(4):1026-1037.

Jordan, G. F. (1962). Redistribution of sediments in Alaskan bays and inlets. *Geog. Rev.* 52:548-558.

Kunze, G. W., L. I. Knowles, and Y. Kitano. (1968). The distribution and mineralogy of clay minerals in the Taku Estuary of southeastern Alaska. *Marine Geol.* 6:439-448.

Lawrence, D. B. (1950). Glacier fluctuation for six centuries in southeastern Alaska and its relation to solar activity. *Geog. Rev.* 40(2).

Lawrence, D. B. (1958). Glaciers and vegetation in southeastern Alaska. *Amer. Sci.* 46(2):20.

McKenzie, G. D. (1968). Glacial History of Adams Inlet, Southeastern Alaska. Ph.D. Dissertation, Ohio State Univ.

Miller, D. J. (1957). Geology of Southeastern Part of the Robinson Mountains, Yakataga District, Alaska. U.S.G.S. Oil and Gas Inv. Map OM-187.

O'Brien, N. R., and D. C. Burrell. (1970). Mineralogy and distribution of clay size sediment in Glacier Bay, Alaska. *J. Sed. Petrol.* 40(2):650-655.

Ovenshine, A. T. (1970). Observations of iceberg rafting in Glacier Bay, Alaska, and identification of ancient ice-rafted deposits. *Bull. Geol. Soc. Amer.* 81:891-894.

Ovenshine, A. T., and D. A. Brew. (1972). Separation and history of the Chatham Strait fault, southeast Alaska, North America. 24th Intl. Geog. Congress, Sect. 3, Tectonics, Montreal, pp. 245-254. J. E. Gill (Ed.), 24th Int. Geol. Cong., 601 Booth St., Ottawa, Canada. Printed by Harpell's Press Co-operative, Garden Vale, Quebec, Canada.

Pickard, G. L. (1967). Some oceanographic characteristics of the larger inlets of southeast Alaska. *J. Fish. Res. Bd. Canada* 24(7):1475-1506.

Plafker, G. (1967). Geologic Map of the Gulf of Alaska Tertiary Province, Alaska. U.S.G.S. Map I-484.

Pritchard, D. W. (1952). The Physical Structure, Circulation, and Mixing in a Coastal Plain Estuary. Chesapeake Bay Inst., The Johns Hopkins Univ., Tech. Rept. 3, Ref. 52-2.

Pritchard, D. W. (1967). Observation of circulation in coastal plain estuaries. In *Estuaries.* G. H. Lauff, Ed., Publ. 83, Amer. Assoc. Adv. Sci., Washington, D.C., pp. 37-52.

Rossman, D. L. (1963). Geology and petrology of two stocks of layered gabbro in the Fairweather Range, Alaska. *U.S.G.S. Bull.* 1121-F, pp. F1-F50.

Sharma, G. D. (1970a). Sediment-seawater interaction in glacio-marine sediments of southeast Alaska. *Geol. Soc. Amer. Bull.* 81(4):1097-1106.

Sharma, G. D. (1970b). Evolution of interstitial waters in recent Alaskan marine sediments. *J. Sed. Petrol.* 40(2):722-733.

Sharma, G. D. (1971a). Geochemical evolution of interstitial waters in arctic and subarctic marine sediments. Proceedings of IUGG, XV General Assembly, Moscow, U.S.S.R., 1971. Symposium on Air and Water Pollution, p(19)36, Abstract.

Sharma, G. D. (1971b). Sediments. Chapter 5 in *Impingement of Man on the Oceans.* D. W. Hood (Ed.), Wiley & Sons, Interscience Series, New York, 738 p.

Slatt, R. M. (1970). Sedimentological and Geochemical Aspects of Sediment and Water from Ten Alaskan Valley Glaciers. Unpublished Ph.D. Dissertation, Univ. of Alaska, Fairbanks, p. 125.

Slatt, R. M., and C. M. Hoskin. (1968). Water and sediment in the Norris Glacier outwash area, upper Taku Inlet, Southeastern Alaska. *J. Sed. Petrol.* 38(2):434-456.

Spencer, A. C. (1906). The Juneau gold belt, Alaska. *U.S.G.S. Bull.* 287:161.

Suess, E. (1892). *Das Antlitz der Erde.* I. Vienna.

U.S. Geological Survey, Preliminary Report to U.S. Army Corps of Engineers. Unpublished Report. February 1966.

Wright, F. F. (1971). Suspension transport in southern Alaska coastal waters. Third Annual Offshore Technology Conference, Houston, Texas, April 19-21, 1971.

Wright, F. F., and G. D. Sharma. (1969). Periglacial marine sedimentation in southern Alaska. In. Proc. VIII Cong. INQUA, Paris, 1969, Théme II Méthodes d'etude due Quaternaire Sous-Marin. pp. 179-185, Univ. of Paris, France.

CHAPTER 3
Gulf of Alaska Shelf

INTRODUCTION

The Gulf of Alaska forms the northeastern part of the Pacific Basin. The central part of the gulf, lying between Kayak and Kodiak islands, represents the seaward expression of a bight, where the northwest trending North American Cordillera bends to southwest at least 90° and merges into the eastern end of the Aleutian Arc and the trench. Tectonically the entire region has been active during and since the Tertiary and presently is a site for recurrent earthquakes. This tectonic activity is a result of the release of structural stress caused by the strong relative movement between the Pacific and American plates.

The continental shelf which adjoins the Pacific Basin is semicircular and of variable width. In the southeastern corner, it is about 100 km broad and widens, with some irregularity, to a width of 250 km in the southwest. In the far southwest Aleutians it becomes very narrow. The shelf is traversed by numerous sea valleys.

Three regions constitute the Gulf of Alaska Shelf. These regions do not have geologic or geomorphic significance and are mainly devised to better describe the sediments and structure of the shelf. They are as follows:

1. Northeastern Gulf of Alaska Shelf, which lies between Cross Sound in the southeast and Kayak Island in the northwest.
2. Central Gulf of Alaska Shelf, which is a broad semicircular shelf extending from Kayak Island to Resurrection Bay and includes Prince William Sound and its many islands.
3. Northwestern Gulf of Alaska Shelf: (a) Kodiak Shelf – this region surrounds Kodiak Island, including Shelikof Strait, (b) Peninsular Shelf – narrow shelf lies between Kodiak Island and Unimak Pass.

GEOLOGY

The coastal terrain along the Gulf of Alaska is dominated by an arcuate mountain range. The northeastern and northern regions of the gulf are flanked by the Chugach-St. Elias mountains and the northwestern region is bordered by the Kenai Mountains and the Aleutian Range. The Aleutian Range, a probable extension of the Alaska Range, to some distance parallels the Kenai Mountains and then continues farther southwest to form the Alaska Peninsula. An elongated depression separates the Kenai Mountains from the Aleutian Range. The southwestern part of this depression, Cook Inlet, is submerged under the sea, whereas northward it forms the Matanuska Valley.

The geologic evolution of the Alaskan Pacific margin is complex, and the early history of the geosynclinal basin is somewhat obscure. The surficial and structural geology of the region has been described by many investigators, notably Miller *et al.* (1959), Gates and Gryc (1963), Burk (1965), Moore (1969), and Plafker (1967 and 1971). The Paleozoic rocks exposed in southeast Alaska have not been observed in this region; consequently, little is known about this era. It is known that the Paleozoic era closed without major orogeny, and that the geosynclinal deposition continued into the Mesozoic era. The absence of Paleozoic strata could suggest that the shelf region is a Mesozoic accretion to the margin of the older continental crust, which lies further north. In the Early Mesozoic era two distinct geosynclines were formed: the Chugach Mountain Geosyncline and Matanuska Geosyncline. Clastic sediments comprising graywacke and sandstones were deposited during the Jurassic and Cretaceous periods. At the close of the Mesozoic era, the geosyncline was uplifted and strongly deformed.

Along the Alaskan Pacific margin, which includes some onshore regions as well as the shelf, a large elongated continental margin prevailed during the Tertiary. The northern margin of the arcuate Tertiary Basin is exposed in the Prince William Sound and Yakutat Bay regions. This basin was filled with a thick sequence of several thousand meters of clastic and volcanic sediments (Plafker, 1971). The first episode, culminating in the Oligocene, led to emplacement of granitic stocks and metamorphism of sedimentary sequences. The later episode of uplifting and faulting began in the Middle Miocene time and these processes presently continue in some regions of the basin.

Glaciation on the Gulf of Alaska Shelf and the adjacent land has not been investigated in detail. Glacial detritus interbedded with marine sediments indicates that the region was repeatedly glaciated from Middle Miocene time onward (Miller, 1953; Plafker, 1971; Plafker and Addicott, 1976). A thick glacial ice sheet continued to scour the region intermittently during the Pleistocene epoch. The morainal deposits of the Malaspina and neighboring glaciers suggest an impressive glacial advance which reached its peak between 700 and 1,400 years ago (Plafker and Miller, 1958). This advance is believed to be one of the major advances and glacial ice probably covered parts of the shelf and the slope. About 600 years ago, a warming trend initiated a minor interglacial period. The last ice advance started about 275 years ago and lasted only about 200 years. At present, with a few exceptions, most glaciers are either stagnant or receding. The coastal region, as well as the shelf, show extensive glacial erosion and deposition. The sea valleys on the shelf may have been scoured by ice and served as main channels for seaward ice flow.

HYDROLOGY

The entire coastline of the Gulf of Alaska lies in a maritime climate zone. The hydrographic data obtained from sparsely located stations over many years have been summarized by the Inter Agency Technical Committee for Alaska (IATCA) (1970). The region has a large range of temperatures, precipitation and stream flow extremes. The records indicate that annual precipitation varies from 300 to 4,500 mm, and extreme temperatures range from -36°C to 28°C.

The largest river draining into the Gulf of Alaska, the Copper River, extends well back into the coastal range hinterland. It has a drainage area of about 7,200 km^2 and an average runoff of approximately 1,500 m^3/sec. The annual discharge of the Copper River is mostly confined to six summer months from May to October. During the winter months the flow is restricted and precipitation is largely stored as snow and ice at higher altitudes. During summer peak months, the Copper River discharges up to 1,700 mg/ℓ of sediments.

The mean annual discharge in the coastal region is approximately 0.15 m^3/km^2. For some of the smaller drainage areas the peak discharge may reach to 1.2 m^3/km^2. The seasonal variations in suspended load of the stream generally depends on whether the runoff is of glacial or non-glacial origin. The sediment discharge in a glacial stream reaches its maximum during June, July, and August, and this accounts for most of the annual sediment discharge. Non-glacial streams carry sediments throughout the year, with a maximum during June. The sediment load of these streams varies between 4 and 30 mg/ℓ.

HYDROGRAPHY

The first recorded observations in this region were made in 1778 by Captain Cook. Almost a century later further oceanographic information was gathered during the H.M.S. Challenger expedition (1872-1905). This was followed by three cruises of the R/V Albatross (1888-1905), the Canadian Arctic Expedition (1913-1924), and the Carnegie Expedition (1929). At the close of World War II, the strategic importance of the North Pacific and Arctic regions led to an intensive investigation of the physiography and the water properties. Later the need for optimum harvest of renewable marine food resources from the gulf region by various nations also stimulated scientific investigations.

Acquisition of oceanographic observations from the Gulf of Alaska is impeded by its remoteness and stormy weather, particularly during fall and winter months. Between 1922 and 1952, a total of 59 severe storms and 15 very severe storms were recorded in the region (Danielsen *et al.,* 1957). These storms appear to follow a twelve-year cycle, with a cyclic peak anticipated in 1983.

The synopsis of data obtained periodically from the North Pacific and sporadically from the shelf suggests that the shelf water can be classified into three water masses. A well-defined coastal water mass with salinities of less than

32.5‰ and temperatures between 2 and 11°C is commonly found in surface layers. The second water mass, generally termed intermediate water, with salinities between 32.5 and 33.8‰ and temperatures between 5.8 and 6.5°C, underlies the surface layer. Oceanic water at depth, with salinities greater than 33.8‰ and temperatures between 2 and 6.4°C, forms the third water type.

The shelf water characteristics and related movements in the northern Gulf of Alaska have been investigated by Royer (1975), who suggests that local winds over the shelf largely determine the water properties and the subsequent water flow pattern. In the near-shore area, the North Pacific Current at the shelf break is of secondary importance to the water movement on the shelf. Locally, river runoff may dominate the near-shore circulation.

The seasonal water mass formation and water movement on the shelf is highly dependent on the meteorologic conditions. Seasonal wind patterns control the surface circulation in the gulf. A seasonal halocline is formed on the shelf during summer as a result of surface runoff. The summer high-pressure cell prevailing over the eastern Pacific produces southwesterly winds resulting in Ekman drift offshore and coastal divergence. The renewal of shelf bottom water by divergence in summer probably also replenishes inland fjords with high-density oxygenated waters. Whereas annual flushing on the northern shelf has been suggested by Royer (1975), only occasional flushing is postulated along the eastern shelf.

During winter, the large atmospheric low-pressure cell occupying the gulf generates strong easterly and southeasterly winds. These winds cause northward Ekman transport (onshore direction) resulting in coastal convergence accompanied by downwelling and flushing of deep water from the shelf to offshore. The water at depth on the shelf is replaced by low-temperature and low-salinity surface waters during winter.

REFERENCES

Burk, C. A. (1965). Geology of the Alaska Peninsula-island arc and continental margin. *Geol Soc. Amer. Mem.* 99(1):250.

Danielsen, E. F., W. V. Burt, and M. Rattray. (1957). Intensity and frequency of severe storms in the Gulf of Alaska. *Trans. Am. Geophys. Union* 38(1):44-49.

Gates, G. D., and G. Gryc. (1963). Structure and tectonic history of Alaska. *Am. Assoc. Petrol. Geol. Mem.* 2:264-277.

IATCA (Inter Agency Technical Committee for Alaska). (1970). Alaska ten year comprehensive plan for climatologic and hydrologic data, 3rd ed., August.

Miller, D. J. (1953). Late Cenozoic marine glacial sediments and marine terraces of Middleton Island, Alaska. *J. Geol.* 61(1):17-40.

Miller, D. J., T. G. Payne, and G. Gryc. (1959). Geology of possible petroleum provinces in Alaska, with an annotated bibliography by E. H. Cobb. *U.S.G.S. Bull.* 1094:1131.

Moore, G. W. (1969). New formation on Kodiak and adjacent islands, Alaska. *U.S.G.S. Bull.* 1274-A:A27-A35.

Plafker, G. (1967). Geologic map of the Gulf of Alaska. U.S. Geol. Survey Misc. Geol. Inv. Map I-484, scale 1:500,000.

Plafker, G. (1971). Possible future petroleum resources of Pacific-Margin Tertiary basin, Alaska. *Am. Assoc. Petrol. Geol. Mem.* 15:120-135.

Plafker, G., and W. O. Addicott. (1976). Glaciomarine deposits of Miocene through Holocene age in the Yakataga Formation along the Gulf of Alaska margin, Alaska. U.S. Geol. Survey, Open-file Rept. 76-84.

Plafker, G., and D. J. Miller. (1958). Glacial features and surficial deposits of the Malaspina district, Alaska. U.S. Geol. Survey Misc. Geol. Inv. Map L-271.

Royer, T. C. (1975). Seasonal variations of waters in the northern Gulf of Alaska. *Deep-Sea Res.* 22:403-416.

CHAPTER 4
Northeastern Gulf of Alaska Shelf

INTRODUCTION

The slightly arcuate northeastern Gulf of Alaska continental shelf extends from Cross Sound in the southeast to Kayak Island in the northwest (58°-60°N and 137°-144°W) (Fig. 4-1). For the most part the width of the shelf varies between 60 and 100 km. It is widest (100 km) in the southeast and narrows to 30 km northwestward.

The bordering land to the east is a rugged, glaciated, mountainous terrain. The coastline is indented with three large embayments; Lituya, Yakutat, and Icy bays. The remaining shoreline is formed by a narrow coastal plain which is partly covered by glaciers. Two of the so-called "piedmont glaciers," the Malaspina and the Bering, are among the world's largest.

The seaward edge of the shelf is bordered by a steep slope which descends steeply to the Alaska Abyssal Plain. The slope is also transected with several canyons, which are generally aligned with those on the shelf.

GEOLOGY

The structural geology of the region is dominated by the arcuate Chugach-St. Elias-Fairweather fault system (Plafker, 1967) and a probable northwest-southeast trending fault zone at the base of the slope (Fig. 4-1). The Chugach-St. Elias-Fairweather Fault separates the pre-Tertiary metasediments of the St. Elias Mountains from the relatively narrow band of Mesozoic to Recent sediments of the coastal plain and shelf. Considerable vertical strike-slip displacement has occurred along this fault plane (St. Amand, 1957). Landward of this fault lies a rugged coastal mountain range which provides large amounts of sediments to the shelf. Seaward lies a coastal plain which continues under the sea. Geomorphologically, therefore, the region can be divided into two major provinces:

1. The hinterland, a complex of intrusives and metamorphosed Mesozoic rocks, and
2. The coastal plai' belt and shelf, a predominantly sedimentary sequence of Cenozoic rocks '..ıh some rugged Tertiary outcrops.

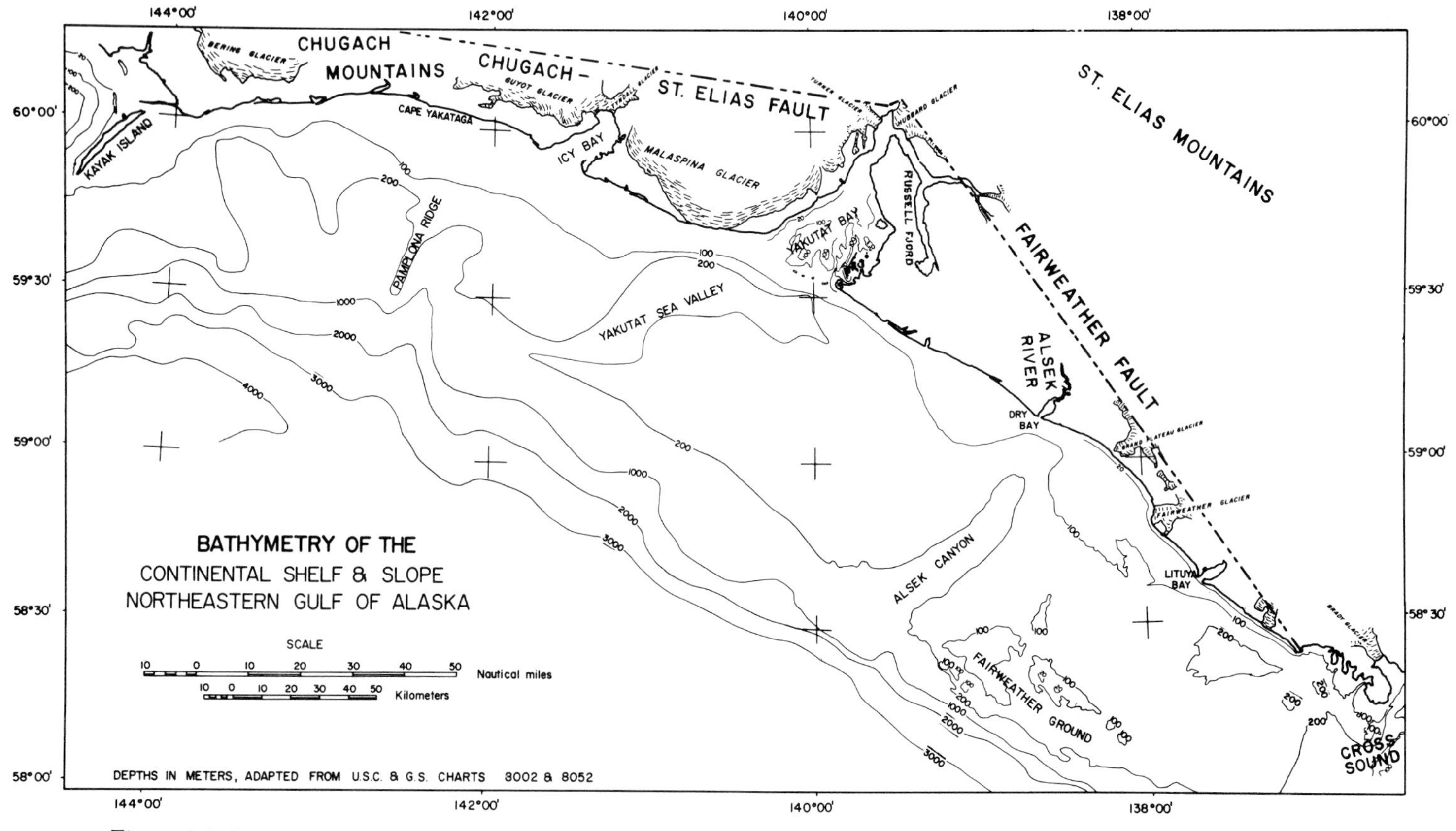

Figure 4-1. Index map showing general physiography of the northeastern Gulf of Alaska Shelf and the adjacent coastal region.

The considerably thick eugeosynclinal facies materials, deposited during the Cretaceous and through the Pleistocene, underlie the coastal plain and the shelf. Interestingly, the combined width of the coastal plain and the shelf between the Chugach-St. Elias Fault and the shelf break is surprisingly uniform, about 80 km, along the entire shelf length. The coastal plain and the inner shelf are relatively smooth and featureless in comparison with the outer shelf. The origin of some of the bathymetric features of the outer shelf has not been investigated. It appears, however, that the coastal plain and the shelf have similar structural evolution and depositional histories. The region is presently undergoing structural deformation and a recent local uplift of more than 5 m was probably caused by the Yakutat earthquake of 1899, reported by Tarr and Martin (1912).

What has taken place in recent history is well manifested in the sediments deposited along the coastal plain and the shelf. The plain, the shelf and part of the slope were covered by glacial ice during most of the Wisconsin glaciation. A major recession of ice began at about 1,100 years B.P. (Hopkins, 1972). This recession, however was interrupted by minor advances. For example, a considerable advance of the Malaspina Glacier culminated about 1000 ± 50 years B.P. (Sharp, 1958). Most glaciers at present are receding or are stagnant but a few, such as Hubbard Glacier (Fig. 4-2), are advancing. Studies of contemporary glaciers in this region indicate that 82 percent are receding, 15 percent are in equilibrium, and 3 percent are advancing (Miller, 1964). The coastal plain is mantled by Holocene marine terraces and glacial deposits, and the topography is marked with moraines and meltwater streams.

BATHYMETRY

The eastern Gulf of Alaska Shelf can be divided into inner and outer shelves. The inner shelf extends to a depth of about 100 m and is mostly featureless, with a relatively steeper slope. The outer shelf, between the 100-m isobath and the shelf margin, has a gentler slope and is traversed by various sea valleys and topographic highs (Fig. 4-1).

Two broad U-shaped bottom scours, Yakutat Sea Valley and Alsek Canyon, are incised across the middle and southeastern shelf, respectively, and a complex of three finger-like depressions separated by two topographic highs transects the northwestern shelf. The most easterly feature, the Alsek Canyon, is a prominent southwest trending submarine valley which heads at the mouth of the Alsek River. Its average width and depth are about 16 km and 75 m, respectively. Westward, the Yakutat Sea Valley trends west to southwest across the continental margin from the Malaspina Glacier and Yakutat Bay vicinity. This sea valley is about 100 km long and 10 to 16 km wide. Another broad depression, somewhat ill defined, lies seaward of the Bering Glacier.

The most prominent topographic high, the Pamplona Ridge, is a steep-sided finger-like projection on the shelf and lies west of the Yakutat Sea Valley. It is

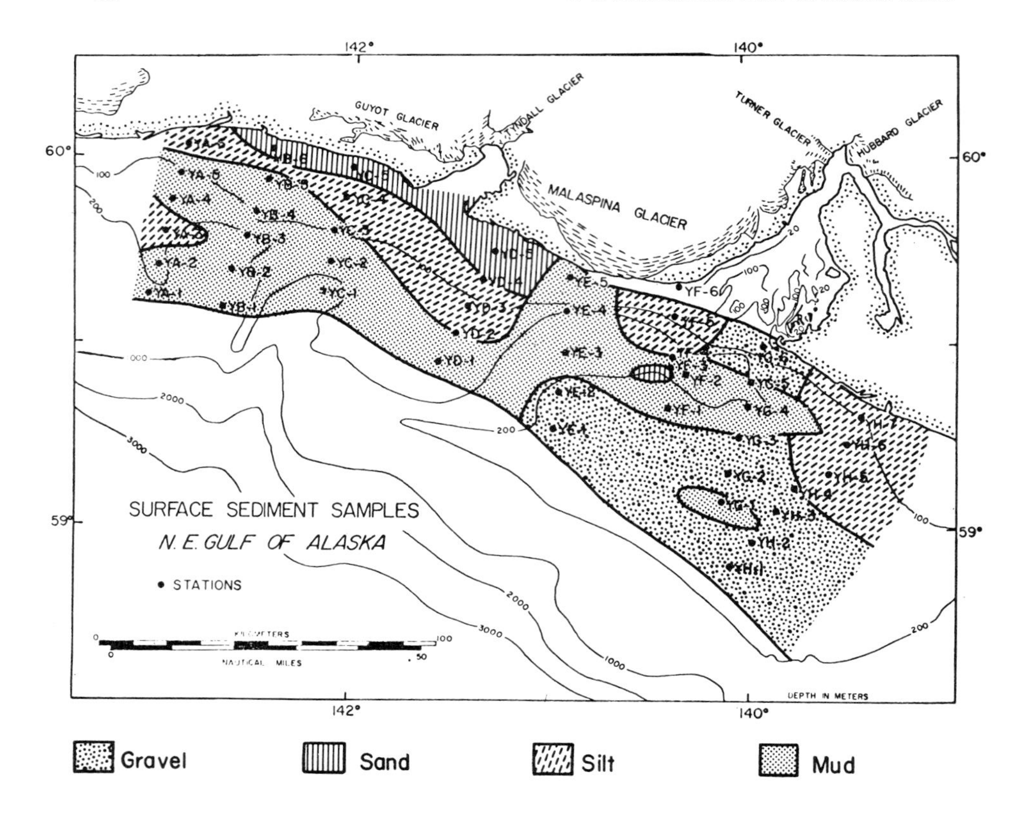

Figure 4-2. Sediment distribution of the shelf adjacent to Icy and Yakutat bays (Wright and Sharma, 1969).

approximately 65 km long and extends across the shelf and southwestward onto the continental slope. It shoals to less than 125 m.

A large shallow area, the Fairweather Ground, lies near the slope south of Alsek Canyon and 55 km southwest of Lituya Bay. It is an unusual feature on the otherwise relatively smooth eastern shelf. The 100-m isobath encompassing the bank covers about 650 km^2. In some parts the bank floor rises to 15-m water depth.

HYDROLOGY

The climate of the region is alpine-maritime, similar to that of southeastern Alaska. The general seasonal atmospheric pattern is related to Aleutian winter lows and summer highs; but the near-shore weather is significantly modified by the coastal mountains with their large ice fields. Noticeable influence on the climate is also exerted by the omnipresent North Pacific Current. Water parameters on the shelf, therefore, are continually affected by meteorologic conditions prevailing over the mountainous region along the coast and the northwest moving current along the slope. Anomalous local waters are intermittently formed due to intense mixing during storms. Because the shelf lies between the continental high and the Aleutian low, it is frequently ravaged by storms. These storms generally bring a humid air mass, which, after cooling, results in precipitation. Mean annual precipitation is over 3,300 mm in this region.

HYDROGRAPHY

Normally there are two definable water masses on the shelf. The coastal water resulting from mixing of runoff and shelf water is generally found adjacent to the coast but is sometimes carried offshore by Ekman transport. The saline shelf water is the predominant water mass and covers most of the shelf. Water circulation on the shelf is primarily controlled by the northwest moving North Pacific Current. The current velocity on the shelf off Yakutat Bay is generally between 75 and 105 cm/sec. Near-surface waters mostly flow with the prevailing winds. The average tide in the area is 3.3 m and the tides may set up a local water circulation.

SEDIMENTS

Shelf

The shelf surficial sediments consist of gravel, sand, silt, and clay (Wright and Sharma, 1969). The near-shore sediments are mostly sand and offshore generally grade into silt and clay, with the exception of an anomalous region south and southwest of Yakutat Bay which is covered by gravel (Fig. 4-2). Sediments are poorly to very poorly sorted and generally polymodal-bimodal in distribution. Fragments of corals, pelecypods, and echinoderms were occasionally observed in sediments. West of Icy Bay, the sand fraction of one sample consisted primarily of biogenous carbonate sediments.

Feldspars, quartz, pyroxene, and amphiboles make up the bulk of the sand and silt fractions, while chlorite with illite dominates the clay fraction of the sediments. The clay minerals are characteristic of ice-abraded particles, with little apparent chemical weathering. The heavy mineral contents in the sands are low (4-10%) with few exceptions.

Subsurface studies in this region have been conducted by Wright (1968). He described two subsurface structural trends, both normal and parallel to the shoreline. The trend which parallels the shoreline is characterized by folding and possible thrust faulting. Folds and possible horst and graben structures are the main features of the trend normal to the shoreline. Pamplona Ridge, Icy and Yakutat bays, and the sea valleys all are manifestations of the latter trend (Fig. 4-1). The course of Yakutat Sea Valley is complex because it follows both trends. Inshore, where it parallels the shoreline, the depression lies on a fault. Its seaward extension, however, occupies a synclinal fold filled with a sediment wedge which thickens landward.

Seismic profiles also show a pronounced anticlinal structure along the continental margin (200-m isobath) of the northeastern shelf. Subbottom profiles suggest that the region is mantled by rapidly deposited contemporary sediments. With the exception of a few exposed gravel areas (relict sediments), the shelf is covered with poorly sorted and poorly stratified sediments, which vary in thickness between 0 and 300 m. These sediments were presumably deposited after the ice retreat (sometime about 7000 years B.P.) and therefore the rate of sedimentation in this area appears to be extremely high (10-15 mm/year) (Molnia and Carlson, 1978). The high rate of sedimentation in this region has completely blanketed the paleoshorelines.

Yakutat Bay

Yakutat Bay is a broad and complex indentation in the rugged St. Elias Mountains located between 59°30′and 60°05′N latitude and 139°30′ and 140°20′W longitude (Fig. 4-3). The bathymetry of the bay has been described in detail by Wright (1972). Seaward of the bay lies the complex Yakutat Sea Valley, which cuts across the shelf. The floor of the bay is quite irregular and consists of mounds and small basins. A shallow sill extending across the bay near Blizhni Point forms a shallow elongated basin, Disenchantment Basin, near the head of the bay. The interesting features of the bay floor are three ridges: the two smaller ridges extend from shore to the middle of the bay; the third ridge, the largest and most well defined, is near the entrance. This large arcuate ridge extends across the bay from Ocean Cape to Point Manby. The ridge rises to an average of 30-m depth of water.

Sediment cover in Yakutat Bay consists of relict moraines (gravel) and contemporary sediments (sand, silt, and clay). The detailed sediment distribution in the bay is also described by Wright (1972) and is shown in Figure 4-3. The shallow areas and the ridges in the bay are covered with gravel, whereas the depressions contain sand, silt, and clay. Silt prevails in the northern part whereas clay mantles the central basin. The silt and clay deposits are separated by a narrow belt of sand and gravel. Seaward, beyond the mouth of the bay, sand is predominant.

The sediment size distribution in Yakutat Bay, in general, is related to water depth. The seaward lateral textural variation in sediments, however, is of interest; silt near the head grades into sand in the mid-bay and, subsequently, seaward the sand grades directly into clay. The bulk of the sediment is introduced by various glaciers near the head of the bay, where coarser fractions are deposited

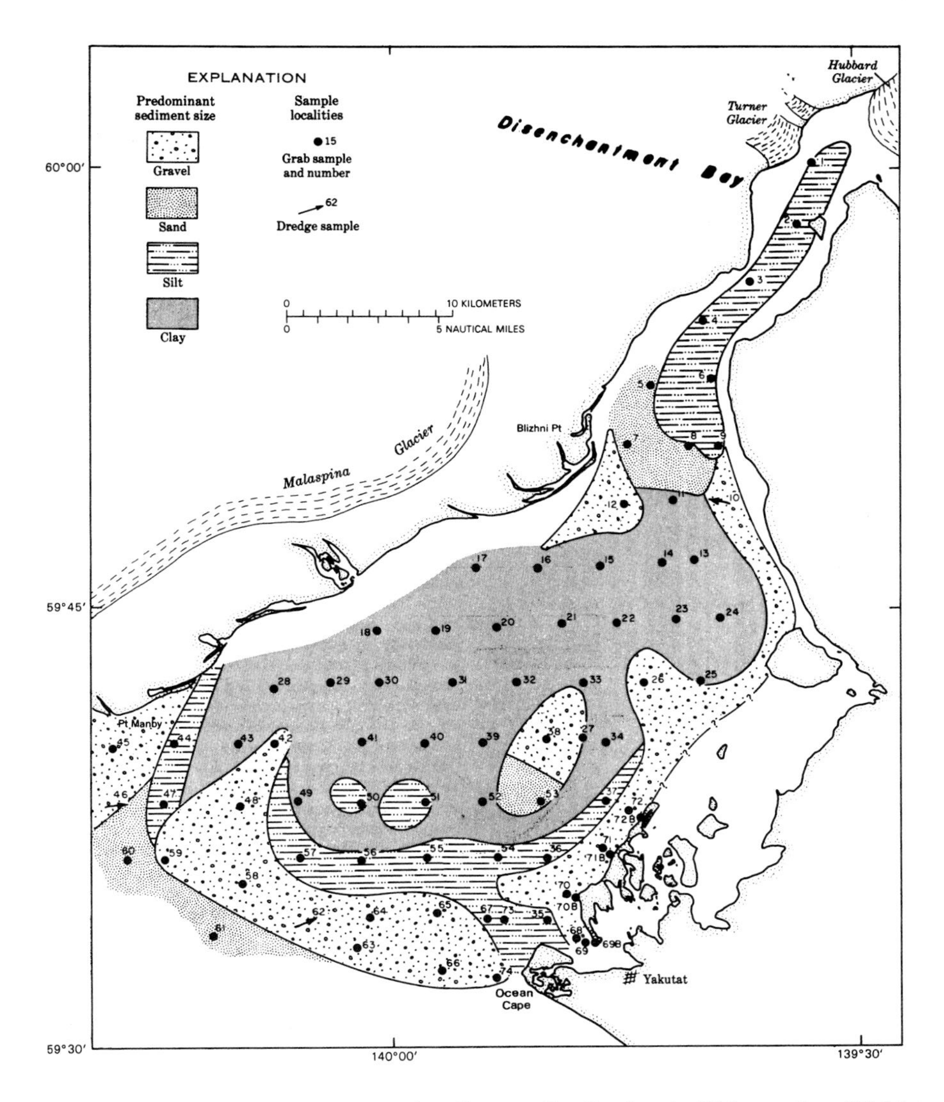

Figure 4-3. Station locations and sediment distribution in Yakutat Bay (Wright, 1972).

while fines continue to move seaward. Sand and silt settle out in Disenchantment Bay and clays are carried into the central basin. A sand deposit at the entrance of Yakutat Bay probably is due to a littoral drift on the shelf. The gravel ridges in the bay are undoubtedly submerged moraines deposited by the receding glaciers. Elutriation of fine sediments from the crusts of the ridges by tidal currents probably also contributes sediments to the basin and the shelf. The sand deposit of the mid-bay is probably a glacial outwash which has been reworked by marine processes.

The mineral composition of the sediment is characteristic of the hinterland. The heavy mineral suite consists of chlorite, amphibole, pyroxene, magnetite and ilmenite (Wright, 1972).

Subbottom seismic profiles obtained by Wright (1972) show acoustically opaque subsurface topographic highs separated by poorly stratified sediments in the depressions. The topographic highs do not show stratification and are typical of morainal deposits. In the depressions, the layered sediments reach thicknesses of 250 m as observed in Disenchantment Bay. These layered sediments must have been deposited after the last ice recession, which occurred approximately 1000 years ago. The rate of deposition in the basinal areas of the Yakutat Bay, accordingly, is about 25 cm/year.

Cores obtained recently from the Russell Fjord (Fig. 4-1), which is a part of Yakutat Bay, show cyclic sedimentation. The textural analyses of the sediments from distinct horizons suggest the presence of annual varves (Figs. 4-4 and 4-5). Assuming that each cyclic episode was deposited annually, then the extremely high rate of sedimentation in this region should completely blanket the paleo-shorelines.

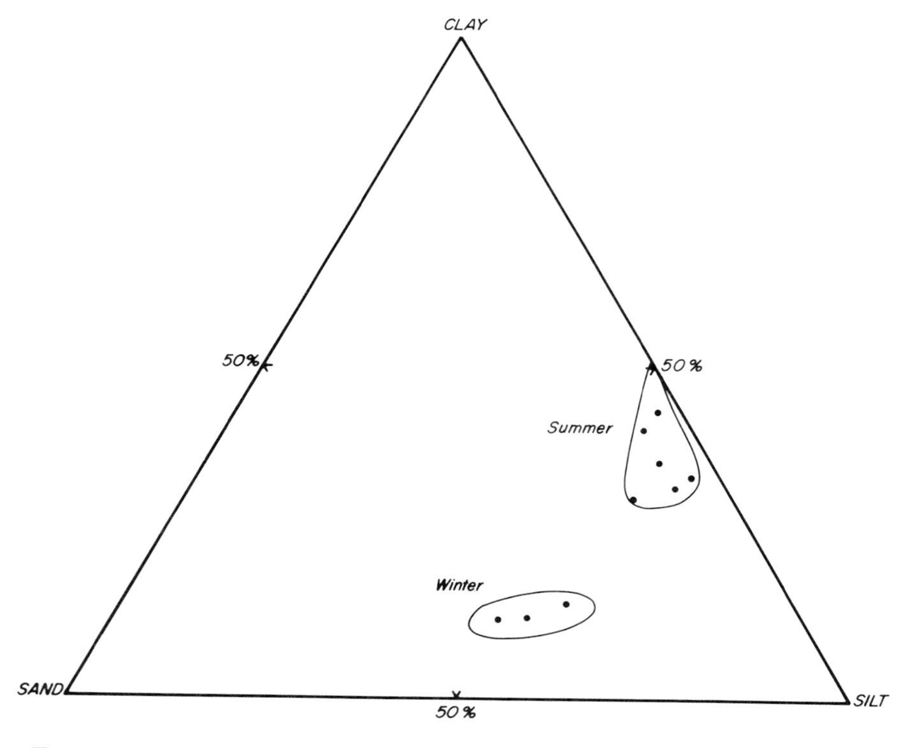

Figure 4-4. Weight percents sand, silt, and clay in core sediments from Russell Fjord, Yakutat Bay reflecting summer and winter deposition.

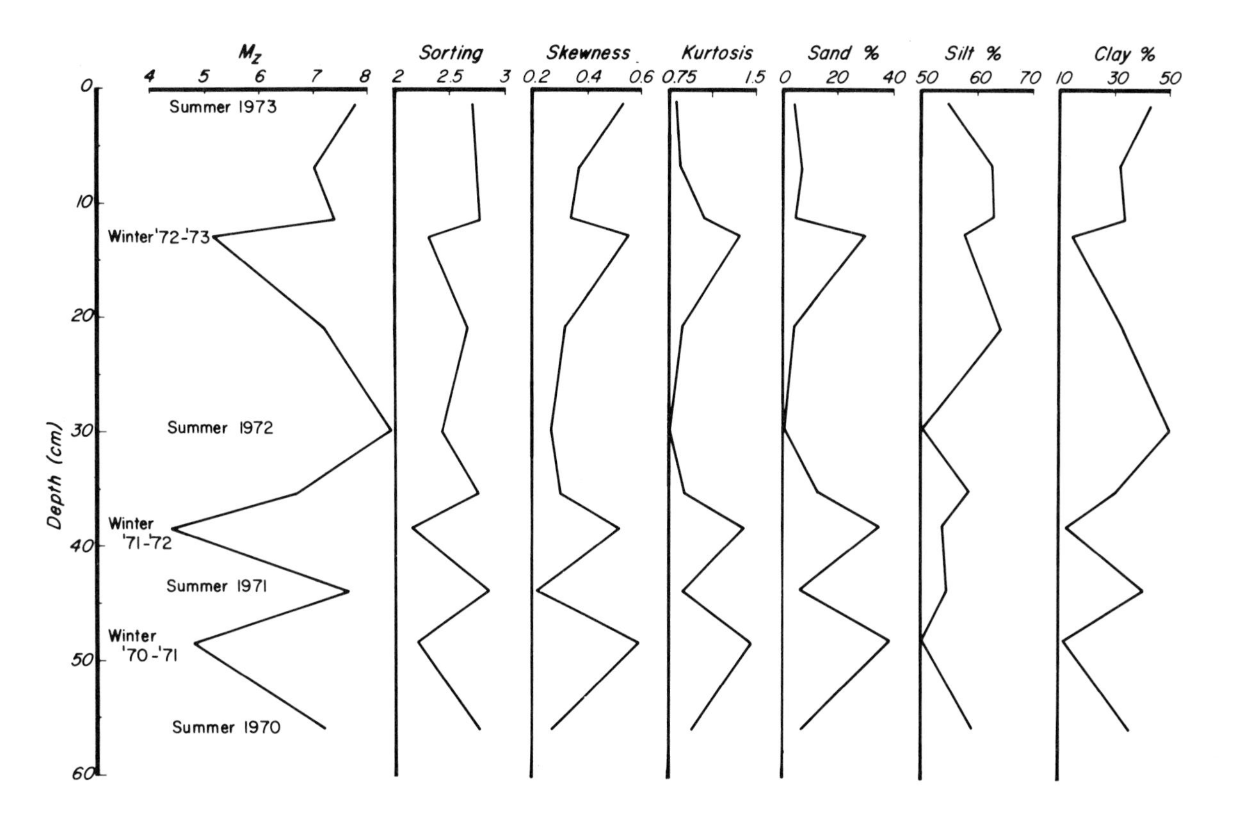

Figure 4-5. Depth profiles of sediments in core from Russell Fjord, Yakutat Bay.

GEOCHEMISTRY

No geochemical investigations have been conducted in this region.

SEDIMENT SOURCE AND TRANSPORT

The rugged hinterland, scoured by large glaciers, provides the major influx of the sediment, which is subsequently carried by numerous small rivers and streams to the shelf. The coastal morainal deposits also are actively reworked under the influence of the extreme maritime climate. Winds and waves remove sediments from the coastal plain as well.

The extent of detritus discharged to the shelf is best illustrated by Earth Resources Technology Satellite-1 (ERTS-1) imagery (Fig. 4-6). It must be pointed out that data obtained from ERTS-1 imagery are synoptic and even repeated coverage often will miss short-lived phenomena. From the imagery, it is apparent that tidal glaciers and meltwater streams draining glacier termini provide large influxes of sediment along almost the entire coast. Density slicing of satellite images has been a very useful tool for obtaining relative sediment distribution in surface waters over large areas (Sharma *et al.,* 1974). ERTS-1 imagery, using the density-slicing technique, provided bands of gray density intervals such that each band represented specific ranges of suspended load. Sediment distribution obtained through density slicing of ERTS-1 imagery (Fig. 4-6) is shown in Figure 4-7. The sediment plume, for example, in Yakutat Bay is extremely dense near the head and extends seaward onto the shelf. The progressive seaward decrease in reflectance of surface water suggests that most sediments settle near the head of Disenchantment Bay. Rapid settling of sediment in the upper part of the bay is further supported by the textural distribution and thickness of the sediments described earlier.

The outflow from Yakutat Bay, as displayed by a distinct and large plume, carries sediment to the shelf. After leaving the bay, the sediments proceed westward along the coast, and subsequently merge with the plume originating from Guyot Glacier (Fig. 4-8). It is noteworthy that offshore waters south of Yakutat Bay contain relatively little sediments in suspension: indeed, as mentioned earlier, the sea floor in this region is covered by relict gravel. Because of the absence of significant sediment sources to the southeast, combined with the westward transport of Yakutat Bay sediment, this region lacks input of contemporary sediments, so relicts are left exposed. The westward movement of sediments observed from satellite imagery conforms well with the long-term currents and also with distribution of relict and contemporary sediments.

Offshore transport of sediments in suspension is, to some extent, evident from the ERTS-1 imagery. Various clockwise and counterclockwise gyres, as well as tongues of sediment-laden waters in offshore regions, are often noticed in the ERTS images. The offshore movement of these surface waters is a result of

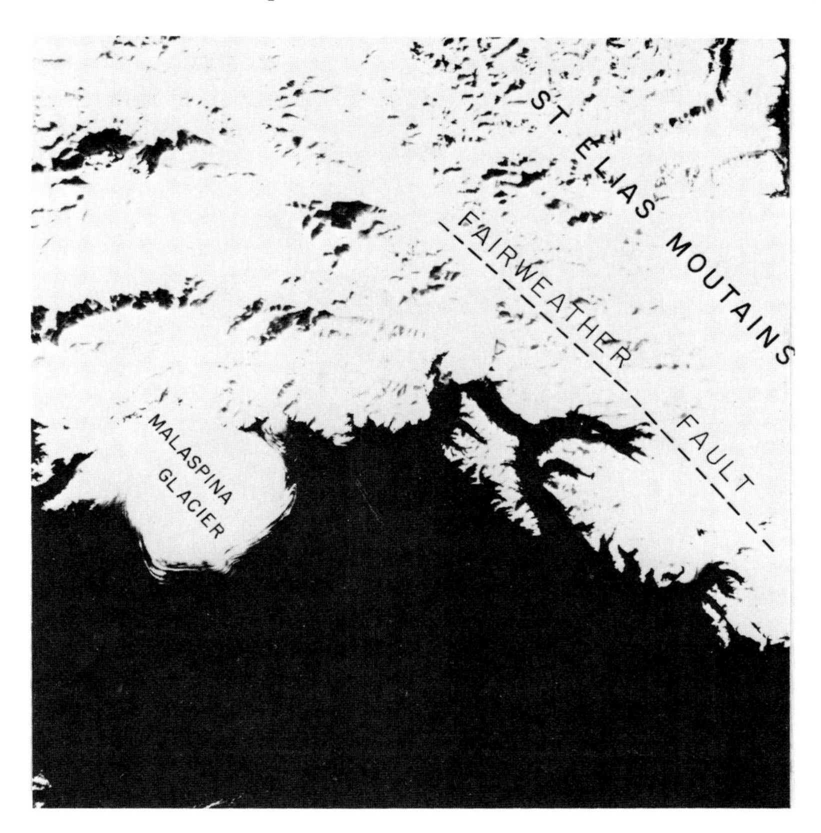

Figure 4-6. ERTS-1 image showing sediment plumes in Yakutat Bay and adjacent shelf during July and August peak runoff. Dark regions near-shore depict waters with heavy suspended load.

Ekman transport. The floors of sea valleys generally contain finer sediments than those deposited on the adjacent shelf. The relative thickness of these sediments (contemporary) is also greater in the sea valleys. It therefore appears that large masses of sediments are channeled offshore along the near-bottom regions of various canyons and sea valleys. These elongated depressions also served as major pathways for ice and sediment movements in the past.

In summary, the influx of sediments from the various glaciers and streams along the coast is generally carried westward under the influence of the North Pacific Currrent, but Ekman transport contributes to the small offshore transport of surface suspended sediments. With the exception of a nominal lateral diffusion, the relatively unmixed sediment-laden runoff is mostly retained close to the coast. The lack of fine sediment in some offshore regions, such as Fairweather Ground, is primarily due to shunting of detritus along the coast and

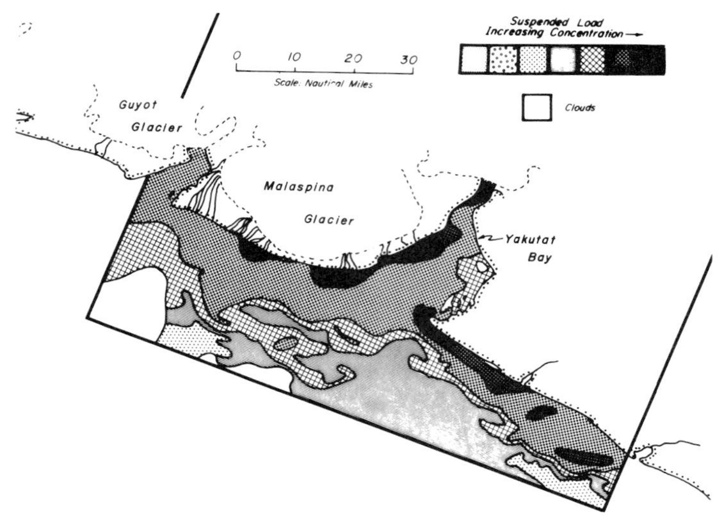

Figure 4-7. Isodensity distribution in satellite imagery showing relative suspended loads in near-surface water of Yakutat Bay and adjacent shelf, on 28 October 1972.

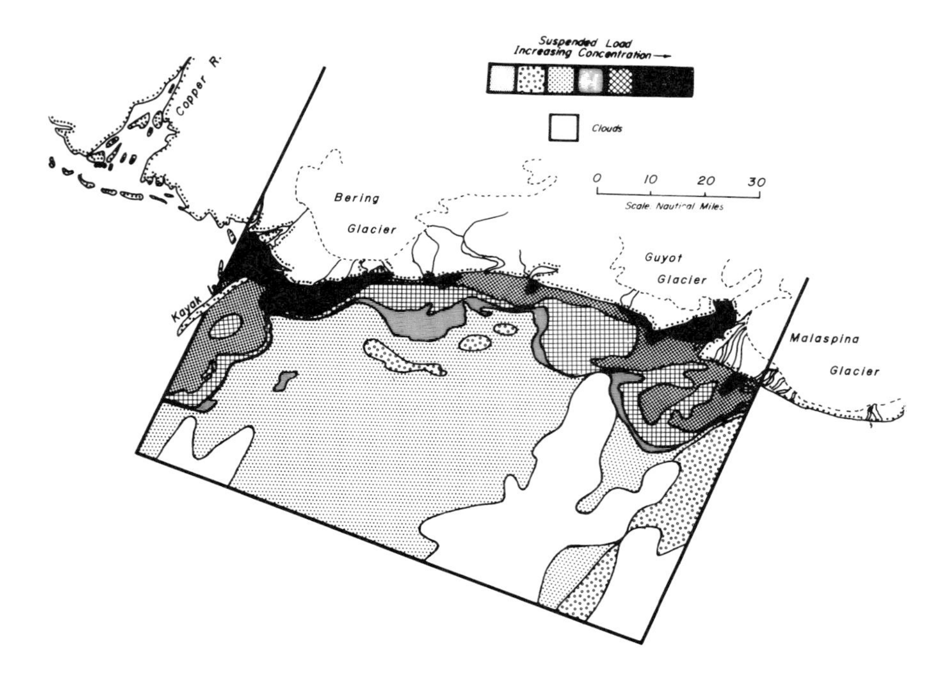

Figure 4-8. Isodensity distribution in satellite imagery showing relative suspended loads in near-surface water of the shelf between Kayak Island and Guyot Glacier, on 22 September 1972.

winnowing of fines from the shallow region by the long-period swells and the bottom currents of the North Pacific Gyre. Offshore movement of sediments through Alsek Canyon, combined with a lack of major sediment source west of Alsek Canyon, severely limits the contemporary sediment influx in the region between Alsek Canyon and Yakutat Sea Valley. In spite of offshore shunting of sediments through Yakutat Sea Valley, the fine contemporary sediment cover on the shelf west of this sea valley is due to a greatly increased influx of sediments from the bay itself, as well as from the large glacier to the west.

Near-shore deposition of clayey silt which grades offshore into sand on the eastern shelf is unusual and contrary to normal seaward sediment size grading on a shelf. This unusual sediment distribution is caused by the combined effects of the movement of coastal water and the inordinate sediment input along the coast. As the water moves westward, there is a successive addition of sediments along the coast. The coast-parallel cumulative sediment increase in the narrow coastal belt becomes excessive for the water to transport it either along the coast or offshore. This results in deposition of sand as well as silt and clay. The occasional Ekman transport of surface water is probably an effective mechanism of transporting fine sediments offshore to the outer shelf and beyond. Anomalously coarse deposits (gravel) on the shelf may be remnants of moraines. These relict glacial sediments have been reworked by the advancing sea and the winnowing action of the storm waves and North Pacific Current.

REFERENCES

Hopkins, D. M. (1972). The paleogeography and climatic history of Beringia during late Cenozoic time. Interod 12, Centre d'Etudes, Arctiques, Sorbonne, Paris, France, pp. 121-150.

Miller, M. M. (1964). Inventory of terminal position changes in Alaskan coastal glaciers since the 1750's, *Proc. Am. Phil. Soc.* 108:257-273.

Molnia, B. F., and P. R. Carlson. (1978). Surface sedimentary units of northern Gulf of Alaska continental shelf. *Amer. Assoc. Petrol. Geol.* 62:633-643.

Plafker, G. (1967). Geologic map of the Gulf of Alaska. U.S. Geol. Survey Misc. Geol. Inv. Map I-484, Scale 1:500,000.

Sharma, G. D., F. F. Wright, J. J. Burns, and D. C. Burbank. (1974). Sea-surface circulation, sediment transport, and marine mammal distribution, Alaska Continental Shelf. Final Report, Contract NAS5-21833, Geophysical Inst., Univ. of Alaska, Fairbanks, p. 77.

Sharp, R. P. (1958). Malaspina Glacier, Alaska. *Geol. Soc. Am. Bull.* 69:617-646.

St. Amand, P. (1957). Geological and geophysical synthesis of the tectonics of portions of British Columbia, the Yukon Territory, and Alaska. *Geol. Soc. Amer. Bull.* 68:1343-1370.

Tarr, R. S., and L. Martin. (1912). The Earthquake at Yakutat Bay, Alaska in September, 1899, with a preface by G. K. Gilbert. U.S. Geol. Survey Prof. Paper 69, Washington, D.C., p. 135.

Wright, F. F. (1968). Sedimentation and Heavy Mineral Distribution, Northeastern Gulf of Alaska Continental Shelf. Inst. of Mar. Sci., Univ. of Alaska, Fairbanks, Report, p. 23.

Wright, F. F. (1972). Marine Geology of Yakutat Bay, Alaska. U.S. Geol Survey Prof. Paper 800-B, pp. B9-B15.

Wright, F. F., and G. D. Sharma. (1969). Periglacial marine sedimentation in southern Alaska. Proc. VIII Congress, INQUA, Univ. of Paris, France. Thème II, Méthodes D'Etude du Quaternaire Sous-Marin, pp. 179-185.

CHAPTER 5
Central Gulf of Alaska Shelf

INTRODUCTION

The arcuate central Gulf of Alaska Shelf extends from Kayak Island in the southeast to Resurrection Bay in the southwest (Fig. 5-1). The shelf is only 30 km wide in the southeast and broadens gradually to over 100 km in the southwest. An interesting feature of the shelf is the presence of a small island,

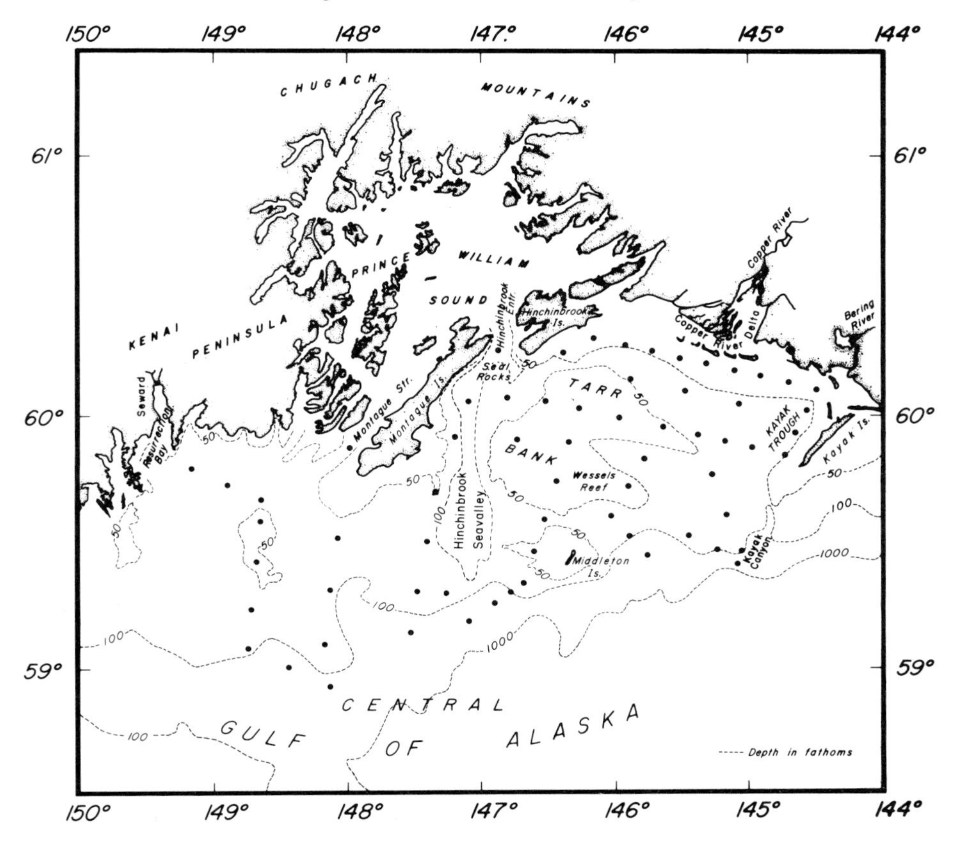

Figure 5-1. Physiographic features of the central Gulf of Alaska Shelf.

Middleton Island, located near the shelf slope. In the northeast, the shelf receives the discharge of the Copper River, one of the large Alaskan rivers, and the Bering River. These rivers have formed an unusual delta of Holocene sediments. The delta not only progrades seaward but also extends parallel to the shore over a distance of 100 km and is separated from the open shelf by a chain of barrier islands.

This portion of the shelf differs tectonically and hydrodynamically from the adjacent eastern and western shelves. Tectonically it is an area where the northwest trending eastern shelf and the northeast trending western shelf merge. This merging of structural trends has resulted in three distinct features which form fundamental regional boundaries on the shelf. Landward the shelf is bounded by an arcuate broad fault zone, which separates the Kenai-Kodiak and St. Elias blocks. Seaward, a large basin, Prince William Sound and the shelf along the Gulf of Alaska, are separated by a fault zone. In particular, the sound is a unique large basin with a depth of over 400 m. Finally, there is a broad complex anticlinal arch near the shelf break.

It is due to these shelf features that unusual hydrodynamic conditions prevail in this region. A large tidal flux flowing over the shelf and into the sound sets up a water circulation which is unique to this region. Because of this unusual water movement and the resulting sediment transport, the central Gulf of Alaska and Prince William Sound are described separately.

GEOLOGY

The character of Paleozoic and Early Mesozoic rocks from the central Gulf of Alaska Shelf is not known. The entire shelf and the region adjacent to the gulf during the Paleozoic was a geosyncline filled with sedimentary deposits. No evidence for any significant pre-Jurassic orogeny has been observed. It appears that this region is a mid-Mesozoic crustal accretion to an older continental margin (Triassic) which lies farther inland. During the mid-Jurassic, a major orogeny occurred, and as a result the present tectonic configuration began to evolve and has continued to evolve since then. The most important features of this orogeny were folding, faulting, and emplacement of numerous plutonic masses. This was followed by two relatively minor orogenies during Late Cretaceous and Early Tertiary (Oligocene), succeeded by a major orogeny which began during Late Tertiary (Pliocene) and is presently active in this region (Plafker, 1969 and 1971).

The Late Tertiary orogeny had a profound effect on the evolution of the central shelf. During this orogenic episode a major fault of northeast-southwest orientation developed along the trend occupied by Hinchinbrook, Montague, and Kodiak islands. Due to continual offset along this fault the seaward shelf—the central Gulf of Alaska—had a significantly different geologic history than the landward shelf—Prince William Sound.

The seaward basin has a broad structural arch near the shelf break, which began rising during the Miocene or Pliocene (von Huene and Shor, 1969). The rising of this arch and concurrent sinking of the Tertiary Basin is associated with the sliding of the oceanic plate under the continental mass along the adjacent Aleutian Trench. This sliding caused uplift near the continental margin and tilting of the entire continental shelf to the northwest. The tilting and subsidence of the shelf has resulted in entrapment of sediments on the shelf. Subsurface sediments and structure off Kodiak have been described by Shor (1965) and von Huene and Shor (1969). They found that the shelf is filled with 3 to 4 km thick Tertiary sediments off Kodiak. Between Middleton Island and Hinchinbrook Entrance the sediments are at least 1 km thick. Due to the extreme thickness of sediments deposited, the lower boundary and configuration of the basin remain obscure.

The thick Paleozoic and Mesozoic sedimentary and volcanic rocks were metamorphosed and highly deformed by the Middle Jurassic orogeny and therefore cannot be easily differentiated. Overlying Tertiary rocks, however, can be divided into three lithologic units, each representing three major depositional environments which prevailed on the shelf (Plafker, 1971). The lower Tertiary unit consists of continental pillow lava, tuff, and tuffaceous sandstone and siltstone. The middle Tertiary unit is a marine sequence with mudstone, siltstone, and occasional sandstone beds. The thick bedded upper Tertiary unit (over 5,000 m) is comprised of characteristic shallow water deposits, consisting of mudstone, muddy sandstone and glacial detritus. Parts of the Tertiary sequence are exposed around Prince William Sound and on Kodiak Island.

The Pleistocene epoch throughout the region is generally associated with glaciation. Although Miocene glacial sediments deposited 10 million years ago have been observed on the central shelf and on land, these are not widespread. During the Pleistocene epoch, extensive ice field and piedmont glaciers repeatedly covered the shelf. The extent of glaciers and ice sheets in this region before and during the early part of the Wisconsin are poorly understood. It appears that at the height of Late Wisconsin glaciation (16,000 - 20,000 years ago) ice covered much of the continental shelf and part of the slope (Hopkins, 1972). The warming trend began about 13,000 to 14,000 years B.P. and this trend initiated the recession of ice from the shelf with attendant rise in sea level from its minimum of -125 m to about -35 m. During this last interglacial period the ice retreated rapidly inland to higher altitudes. This glacial retreat was largely completed by 11,000 or 12,000 years ago. Karlstrom (1964) reported seven advances in the Kenai Peninsula and Cook Inlet during the "Alaska Glaciation."

BATHYMETRY

The relatively broad shelf has a complex bathymetry. In the southern and eastern extremities there are two notable islands; Middleton and Kayak islands. The central shelf has a large circular shallow (<50 m) area, Tarr Bank, which lies between Middleton and Hinchinbrook islands (Fig. 5-1). The bank occupies an area of about 5,200 sq km, with its minimum depth <100 m. A smaller elongated shallow bank is located in the southwestern portion of the region. Submerged rocks, Wessels Reef and Seal Rocks, are located on Tarr Bank and south of Hinchinbrook Entrance, respectively (Fig. 5-1).

A moat-like depression with a depth of more than 100 m lies between Tarr Bank and the mainland, and extends southwest along Hinchinbrook and Montague islands and south along Kayak Island.

The shelf edge lies along the 200-m isobath. Seawards the continental margin drops rapidly, forming a very steep slope which is traversed by two short submarine canyons, Hinchinbrook Canyon and Kayak Canyon. These canyons are restricted to the upper reaches of the slope and do not extend to the base.

HYDROLOGY AND HYDROGRAPHY

The properties and circulation of waters on the central Gulf of Alaska Shelf are influenced by:

1. the Copper River discharge,
2. the permanent counterclockwise gyre of the Alaska Current and
3. the seasonal wind direction.

Surface water input into the region is mostly brought by the Copper River. In addition, the shelf also receives input from numerous smaller streams draining the coastal area, in particular the Bering River in the northeast. The fresh water discharge has been described in Chapter 3, Hydrology.

The westerly flowing Alaska Current, with velocities of about 25 cm/sec near the continental margin, is a dominant feature of water circulation and water characteristics in this region. On the shelf proper, the intensity of this current is reduced to about 8 to 10 cm/sec.

Inshore and offshore movement of water on the shelf is predominantly controlled by meteorologic conditions prevailing over the shelf and the Northern Pacific Ocean. Seasonally, during winters the "Aleutian Low" which occupies the gulf produces strong easterly and southeasterly winds on the shelf, causing Ekman drift toward the shore with concomitant coastal downwelling. Similarly, a summer high-pressure cell results in Ekman drift toward the offshore, with coastal divergency.

The deflection of westerly moving water by Kayak Island forms a vortex on the leeward side. The Ekman drift generally moves the vortex offshore during summer. Large eddies with diameters exceeding 20 km are well defined by suspended sediment and can easily be seen on ERTS-1 imagery.

Salinity and temperature measurements of surface waters from this region during 24-28 February 1973 (Figs. 5-2 and 5-3) show two distinct water masses; the high-salinity water near the continental margin and the low-salinity coastal water near the mouth of the Copper River and in the vicinity of Middleton

Island. Similar water masses are indicated by isotherm distributions. The salinity temperature distribution also indicates onshore movement of saline water through the channel between Kayak and Middleton islands.

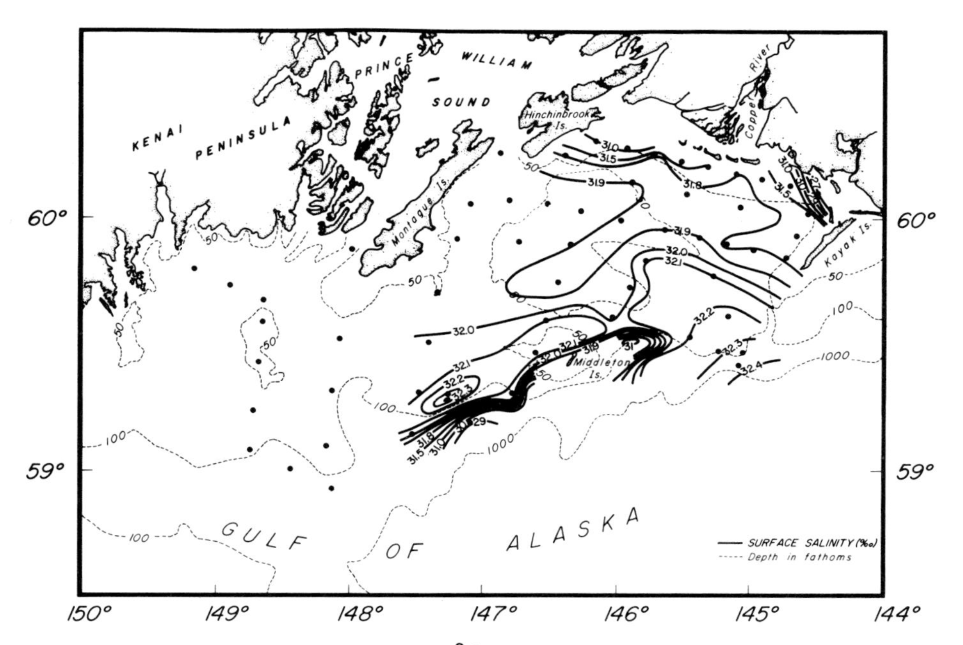

Figure 5-2. Surface water isohalines ($^{\circ}/_{oo}$) during 24-28 February 1973, central Gulf of Alaska.

Along the shore, the coastal water is formed by mixing of surface runoff and the shelf saline water. The mixed water originating near the mouths of the Bering and Copper rivers flows in a northwesterly direction close to the shore. The westerly extension of the coastal mixed water formed near the mouth of Copper River is a consequence of advection by the easterly near-shore remnant of the Alaska Current.

SEDIMENTS

Sediment distribution on the central Gulf of Alaska Shelf has an extreme textural variation. Most of the shelf is mantled with silt and clay (Figs. 5-4, 5-5, and 5-6). Clay predominates in sediments from the channels and the Kayak

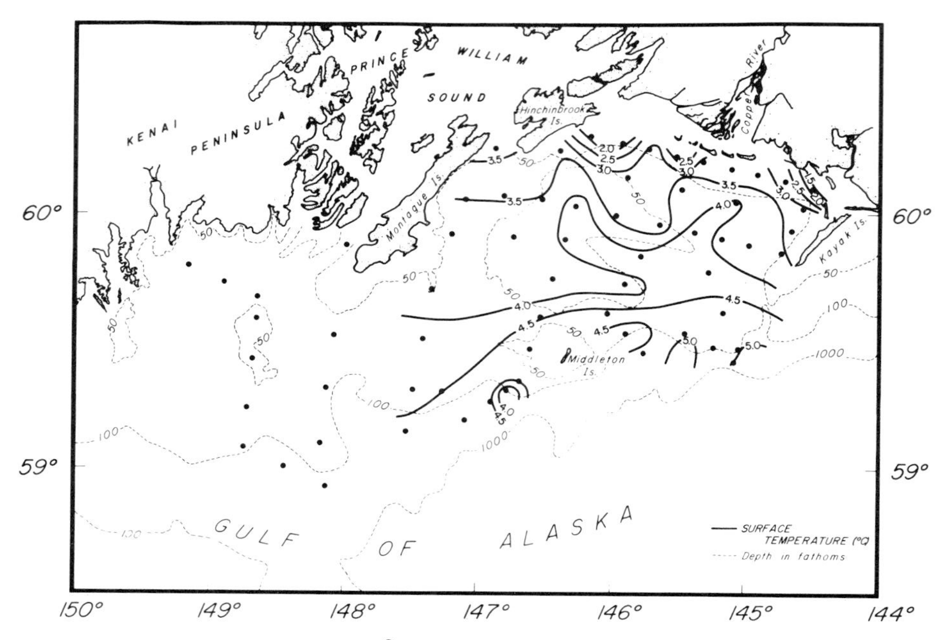

Figure 5-3. Surface isotherms (°C) during 24-28 February 1973, central Gulf of Alaska.

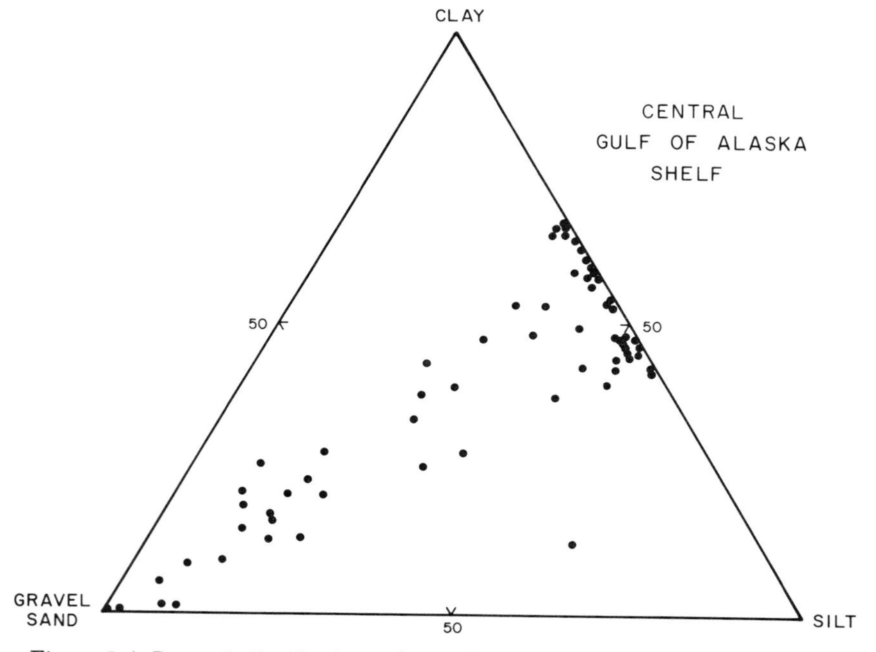

Figure 5-4. Percent distribution of gravel-sand, silt, and clay in sediments from central Gulf of Alaska, showing a wide textural variation.

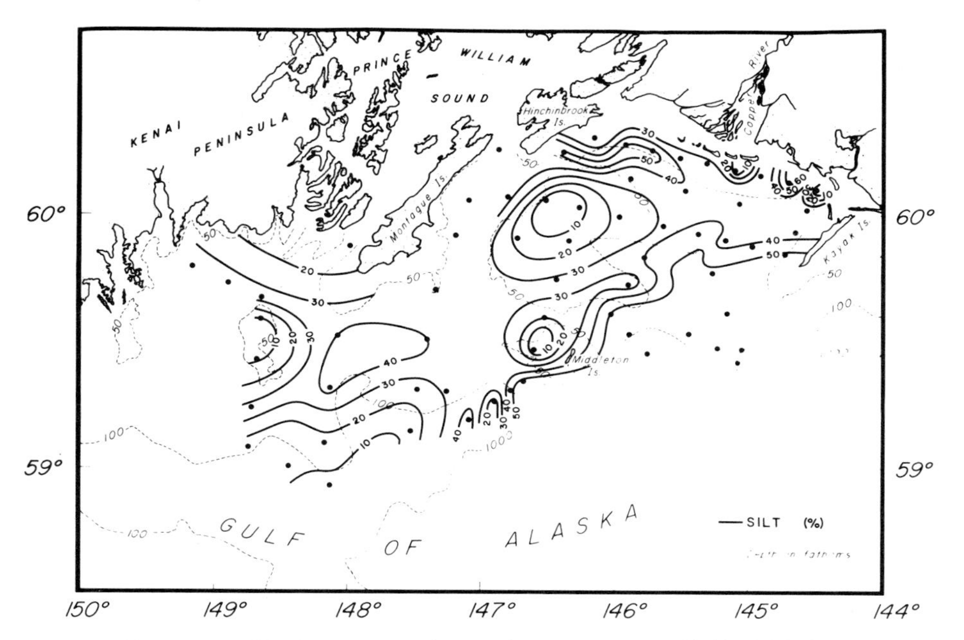

Figure 5-5. Weight percent silt in sediments, central Gulf of Alaska.

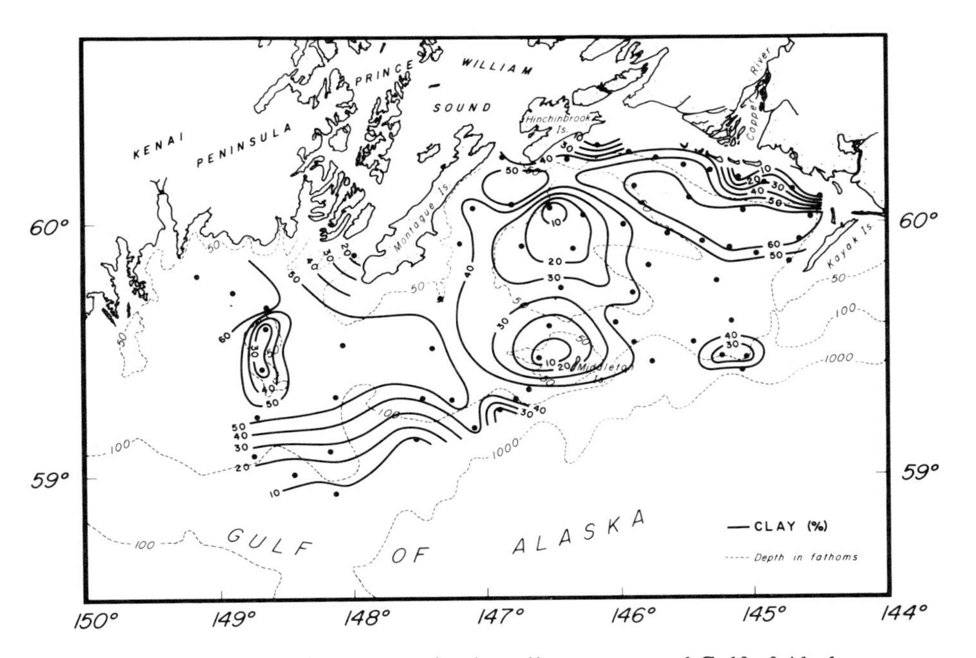

Figure 5-6. Weight percent clay in sediments, central Gulf of Alaska.

Trough. Near-shore sediments are silt and sand. Sand is mostly deposited along the northeastern shore (Copper River delta), in Montague Strait, west of Middleton Island, and along the slope southwest of Middleton Island (Fig. 5-7). Tarr Bank is covered with coarse sediments with calcareous shell and coral fragments. Large patches of coarse sediments, mostly gravel, also occur near the edge of the shelf and on the shallow bank located in the west (Fig. 5-8). The coarse sands of Tarr Bank contain as much as 50% gravel. The gravel fraction mostly consists of biogenous carbonate skeletal fragments. The gravel and coarse sand from the continental margin and from the bank in the southwest, however, contain pebbles, cobbles, and boulders characteristic of glacial till.

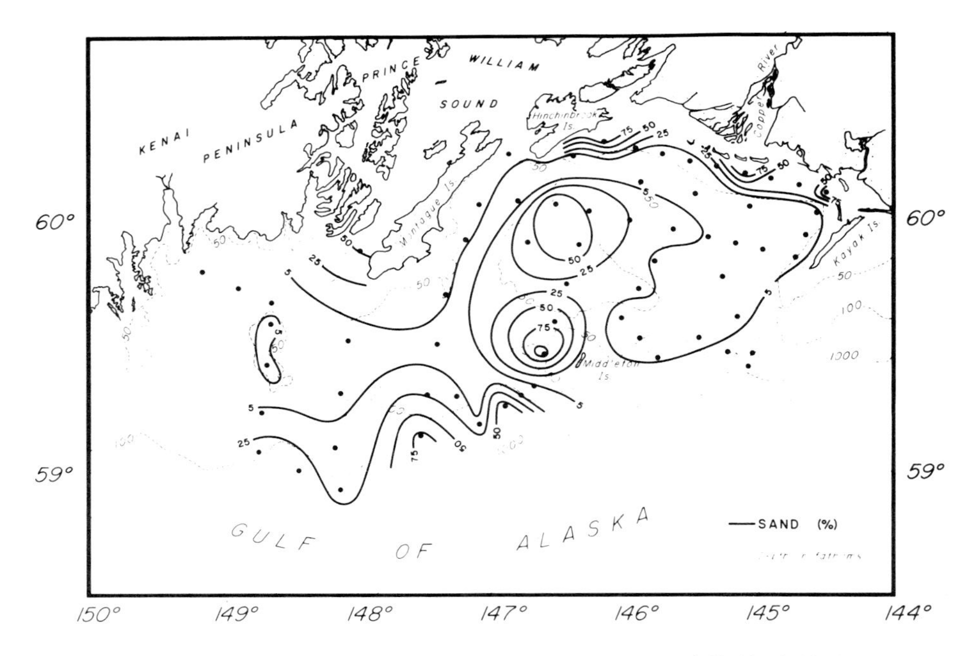

Figure 5-7. Weight percent sand in sediments, central Gulf of Alaska.

Sorting in the sediments is a function of sediment mean size: Sorting displays a sinuoidal relation with mean size. The sand from the Copper River delta is well sorted (Figs. 5-9 and 5-10), but the silty clay and gravel are extremely poorly sorted. The sorting in gravelly sediments is characteristic of glacial till. The mean size and sorting in sediments deposited on the shelf do not reflect any relation to the water depth (Fig. 5-9). Extremely coarse sediments are found on the shallow bank as well as along the continental margin where depths exceed 200 m. The sediments are negatively skewed to extremely positively skewed, and platykurtic to leptokurtic (Figs. 5-11 and 5-12).

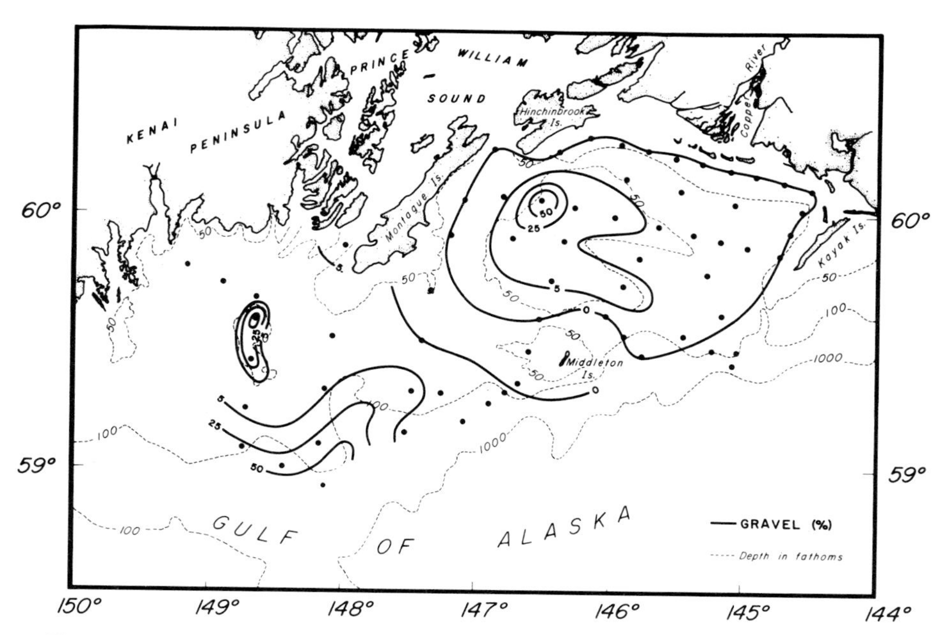

Figure 5-8. Weight percent distribution of gravel in sediments, central Gulf of Alaska.

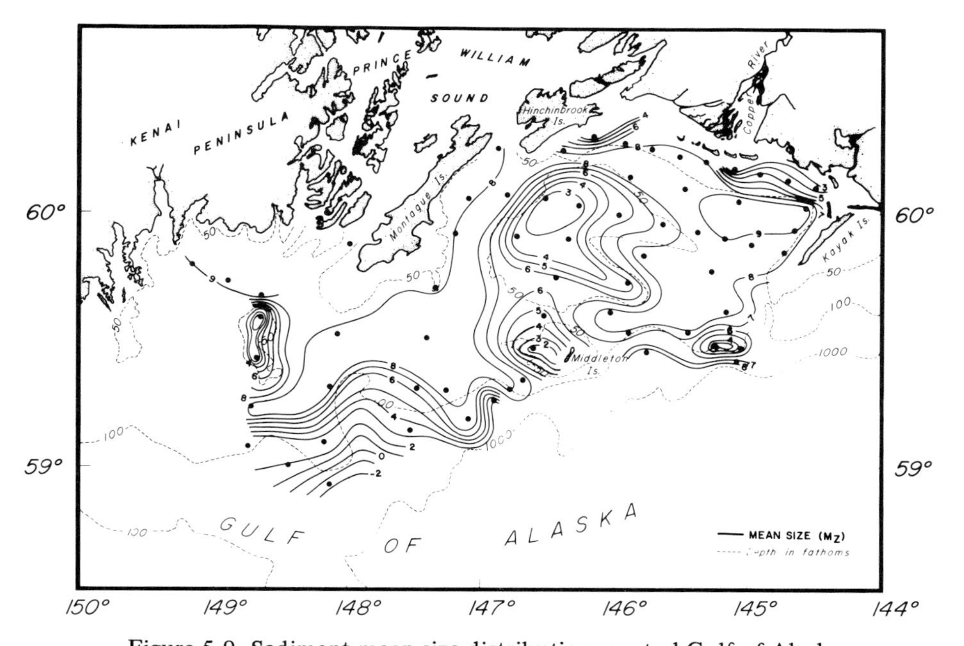

Figure 5-9. Sediment mean size distribution, central Gulf of Alaska.

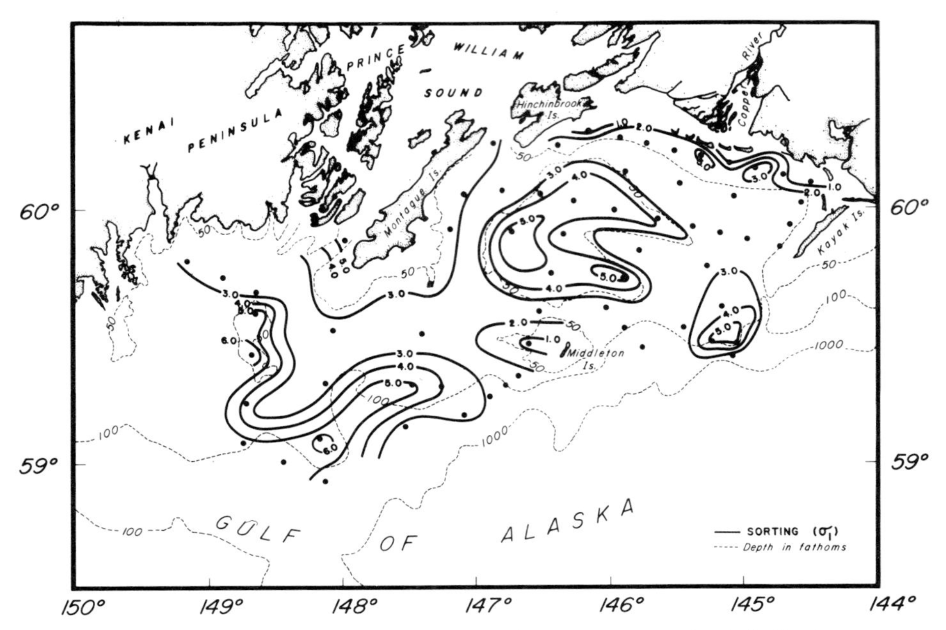

Figure 5-10. Grain size sorting in sediments, central Gulf of Alaska.

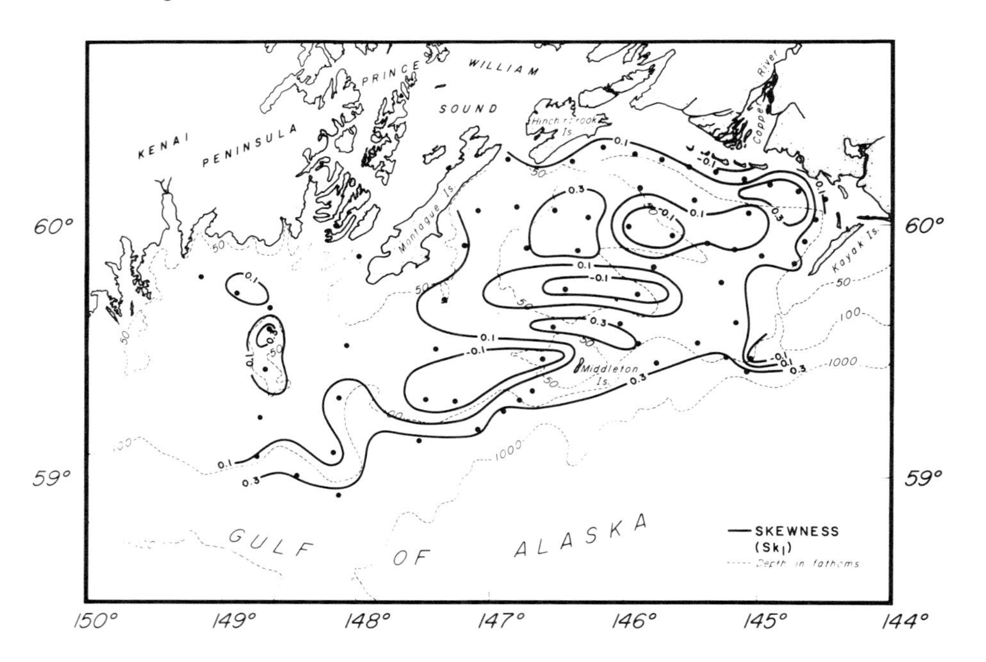

Figure 5-11. Grain size skewness in sediments, central Gulf of Alaska.

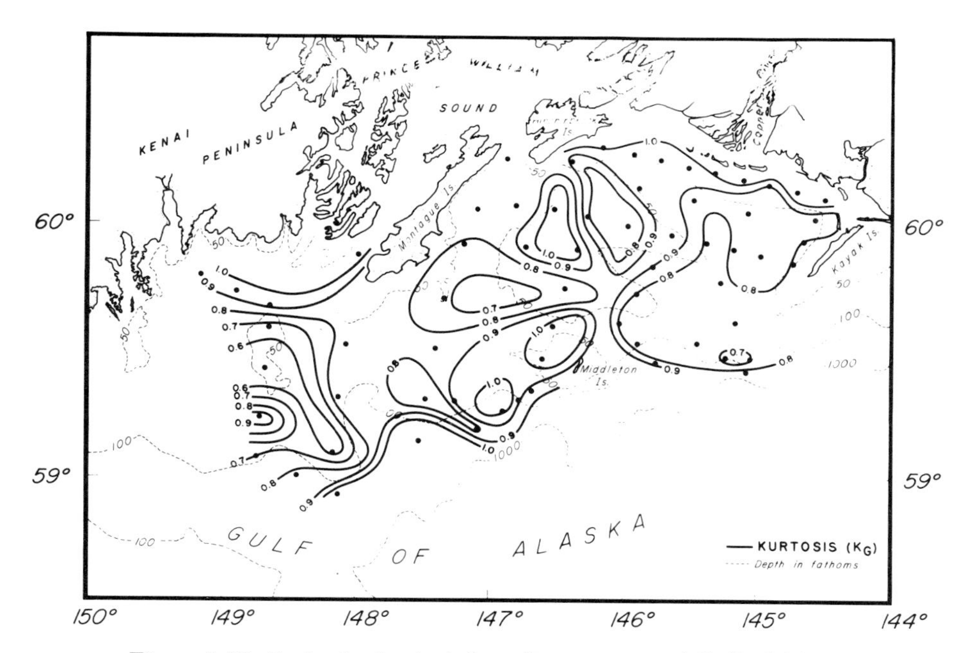

Figure 5-12. Grain size kurtosis in sediments, central Gulf of Alaska.

In general, the shelf sediments display two unusual characteristics:

1. the presence of gravel in offshore deposits, and
2. the large contribution of biogenous sediments on the shallow banks.

The textural distribution of sediments suggests that the coarser fractions of the contemporary sediments discharged from the Copper River and other coastal drainages are deposited along the northeastern coast as well as being partly carried into Prince William Sound. Fine contemporary sediments are transported offshore; portions are deposited in the channels surrounding Tarr Bank and part is carried as plume to the west and southwest along the coast. A significant amount of fine sediment is also carried into Prince William Sound through the various passages.

Large shallow Tarr Bank is a structural feature and is covered with coarse sediments of biogenic derivation. The small bank to the southwest appears to be an end moraine. The glacial origin for these sediments is suggested by coarse mean size, presence of gravel, extremely poor sorting, and an extremely small sand fraction, which is typical of till from Alaskan glaciers. Sediment cover here also includes biogenic component. Glacial sediments are also deposited along the continental slope. At the close of the Wisconsin glaciation, the entire shelf was probably covered with glacial sediments. In some areas these remain exposed, while in others these relicts have been overlain by contemporary sediments.

Organic carbon content in the sediments varies within a very narrow range, averaging about 0.5% by weight. Maximum organic carbon content (3.0%) was observed in the near-shore deposits east of Copper River (Fig. 5-13). These organic-rich sediments appear to have their origin in the Bering River. The distribution of organic carbon is not solely related to the clay content of the sediments (Fig. 5-14).

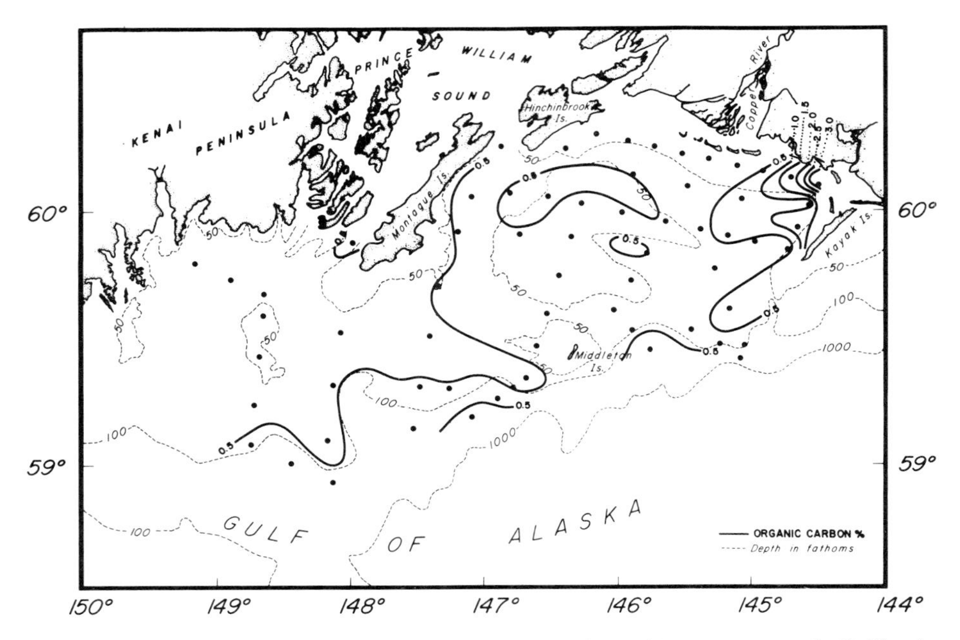

Figure 5-13. Weight percents organic carbon in sediments, central Gulf of Alaska.

GEOCHEMISTRY

The percentage of aluminum in sediments from the central Gulf of Alaska varies over a wide range (<5.0% and >9.0%). The distribution of aluminum, as usual, is related to sediment texture (Fig. 5.15). High aluminum content occurs in silty clays deposited in the trough and channels surrounding Tarr Bank. The contemporary sands along the coast and near Middleton Island generally contain 7% or less aluminum. These sands contain relatively more quartz than feldspar grains. This enrichment of quartzitic sand presumably is primarily due to the susceptibility of feldspar to abrasion during transport. During transport and wave action feldspar grains part along the cleavages and become smaller in size; then hydraulic action removes these smaller (lighter) grains to form a lag deposit rich in resistent quartz grains. Glacial sediments also contain approximately 7% aluminum, which is comparable to the amount found to be the average content of acid igneous rocks. The average distribution of aluminum in glacial deposits

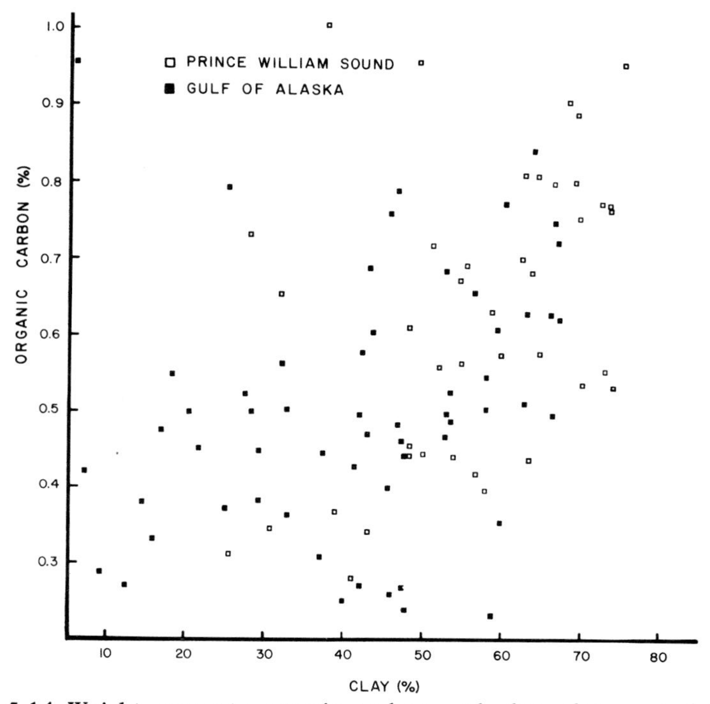

Figure 5-14. Weight percents organic carbon and clay plot, central Gulf of Alaska.

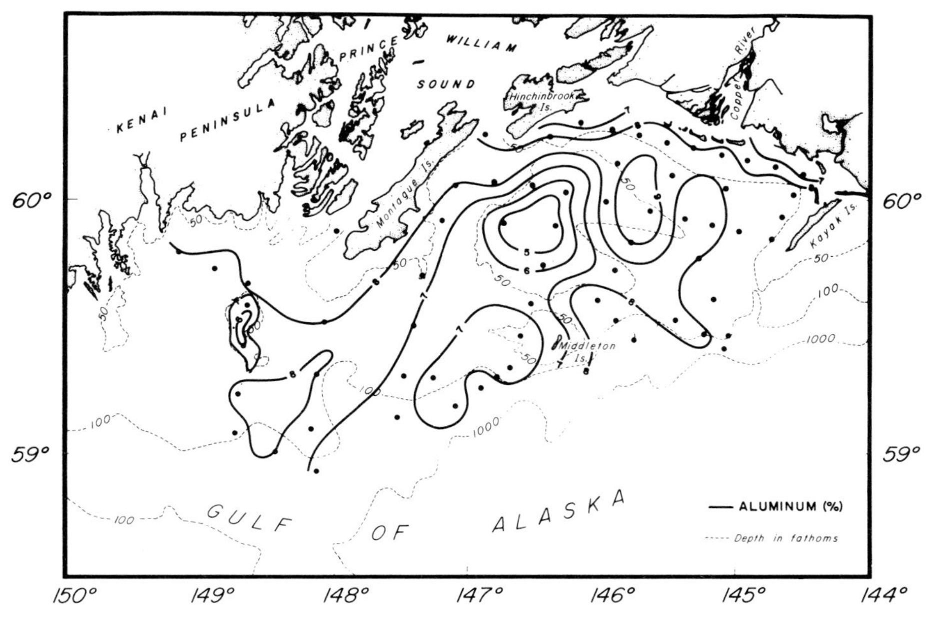

Figure 5-15. Percent aluminum in sediments, central Gulf of Alaska.

reflects the dominance of mechanical weathering, as well as the absence of marine transport with its accompanying mineralogic differentiation. The lower aluminum content in sediments deposited on Tarr Bank is a result of the appreciable biogenous calcareous fraction, with commensurately greater amounts of calcium (Fig. 5-16).

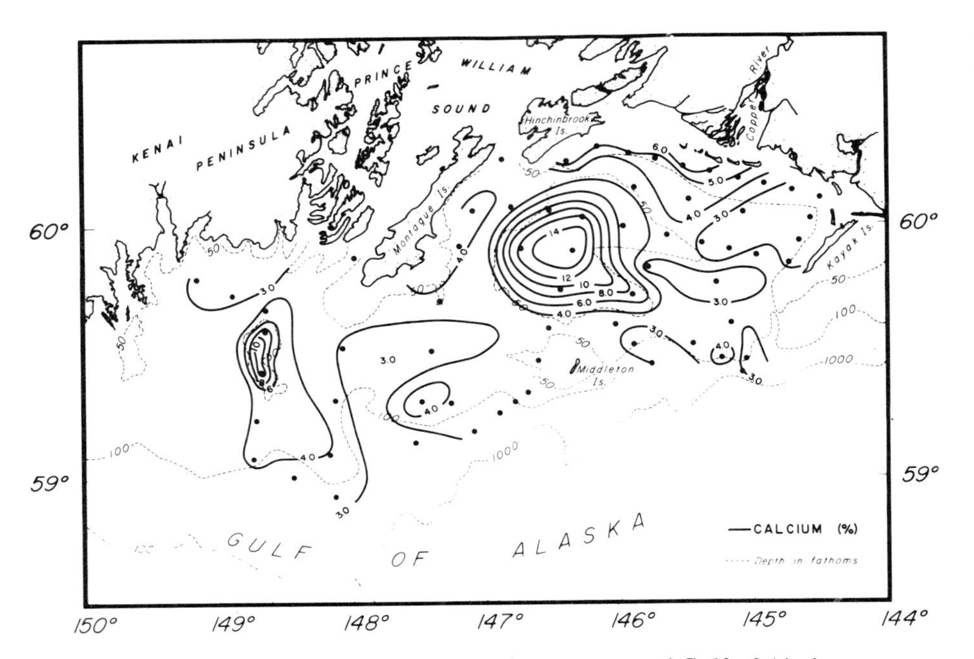

Figure 5-16. Percent calcium in sediments, central Gulf of Alaska.

The distribution of iron appears to be analogous to the sediment size distribution (Fig. 5-17). Coarse sediments, relict glacial as well as contemporary, contain between 4 and 5% iron, whereas finer sediments deposited in the channels contain more than 5% iron. Biogenous coarse sediments contain the least iron.

The concentrations of other major elements–magnesium, potassium, sodium and titanium–in these sediments also seem to be inversely related to the sediment mean size. In contemporary sediments, magnesium varies from 1.0% in sand to 3.0% in silty clays. The relict glacial deposits contain between 1.5 and 2.0% magnesium (Fig. 5-18). Invariably, the magnesium in these sediments varies directly with the clay content, suggesting that the magnesium is mostly associated with clay minerals, most likely chlorites and trioctahedral micaceous phases.

The distributions of titanium, potassium, and sodium in these sediments are similar to that of aluminum and show steady increases with decrease in sediment size (Figs. 5-19-5-21). The increase of these elements is directly related to increased percentages of silt and clay. During transport, the monomineralic quartz

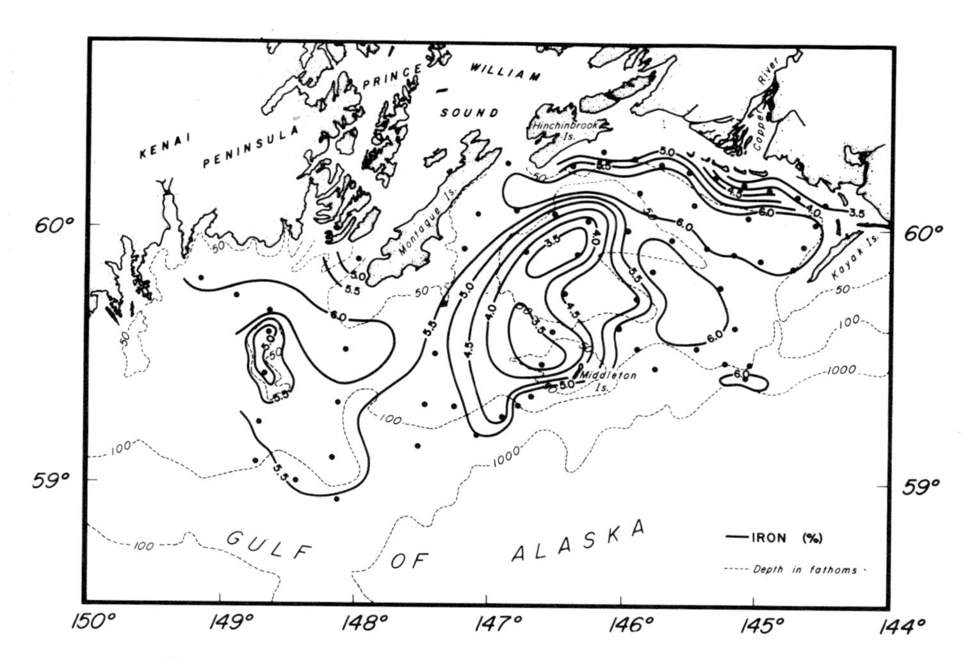

Figure 5-17. Percent iron in sediments, central Gulf of Alaska.

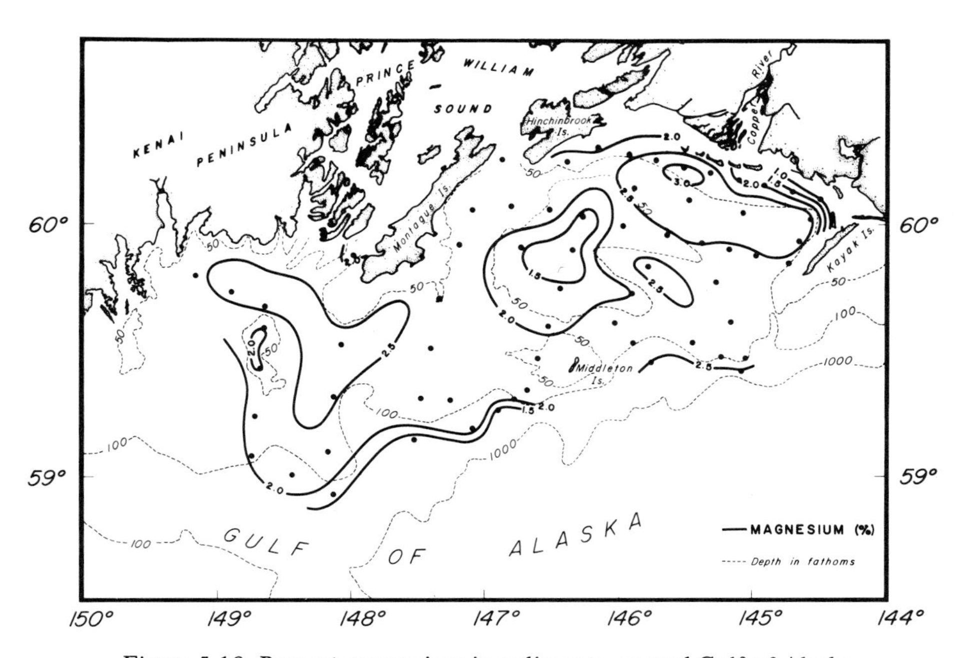

Figure 5-18. Percent magnesium in sediments, central Gulf of Alaska.

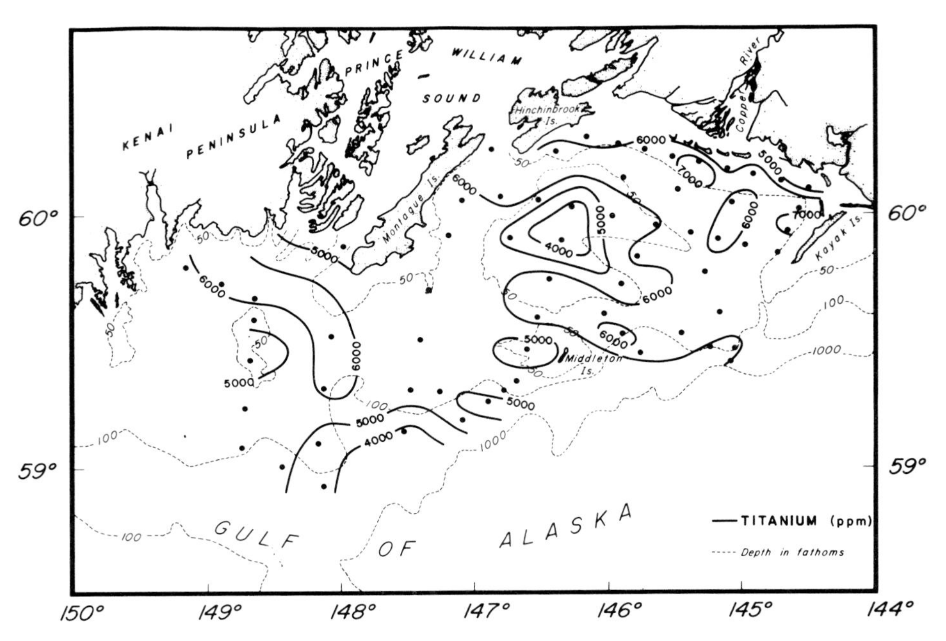

Figure 5-19. Titanium (ppm) in sediments, central Gulf of Alaska.

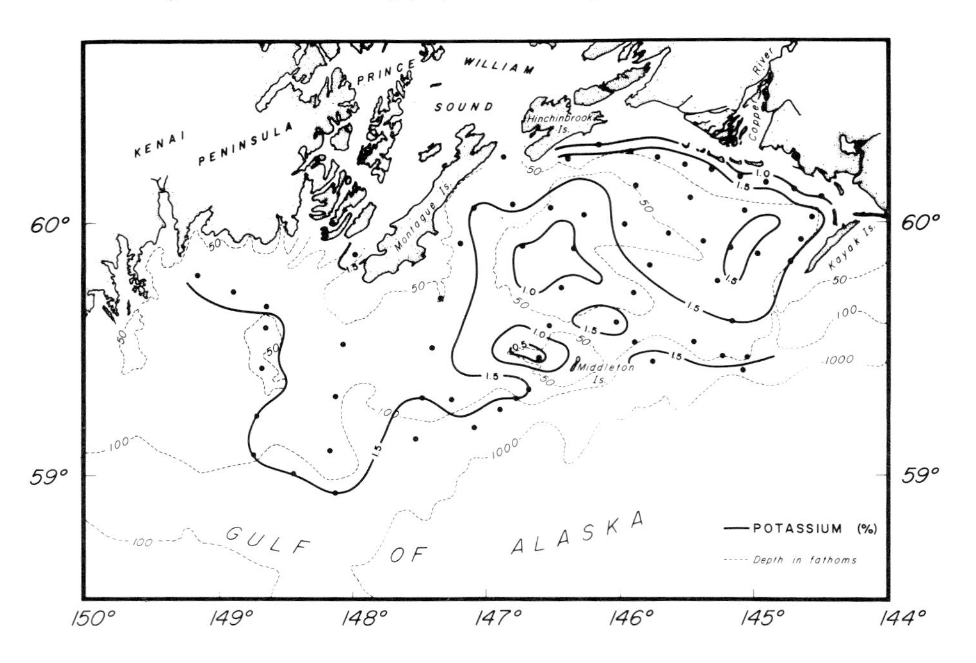

Figure 5-20. Percent potassium in sediments, central Gulf of Alaska.

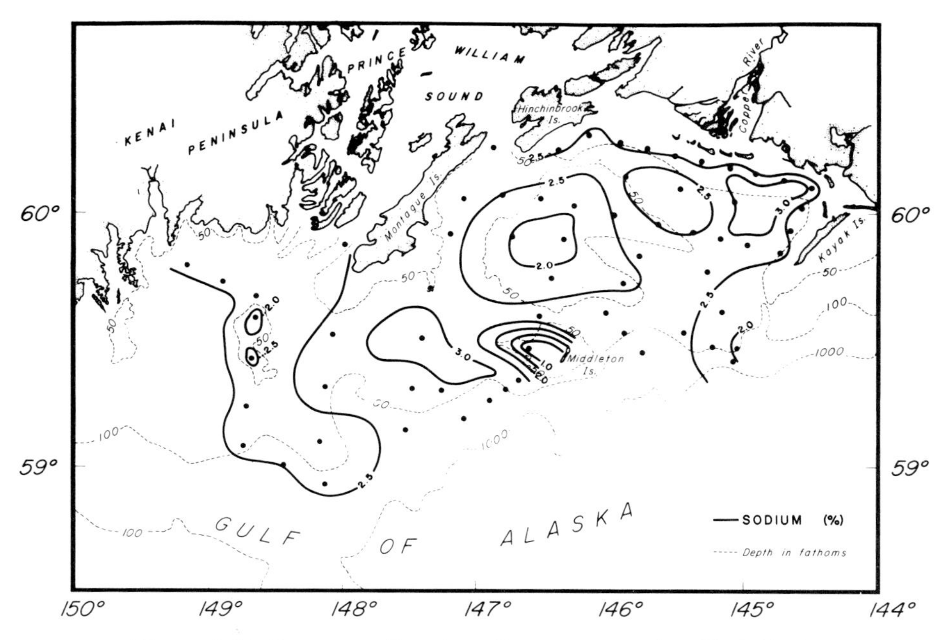

Figure 5-21. Percent sodium in sediments, central Gulf of Alaska.

particles are more resistant to abrasion than arkosic and oligomict minerals. Therefore, sediments deposited closest to the source tend to be coarser and more quartz rich, since the abrasion and transport effects cause segregation of feldspars and micas in the silt and clay fractions of the sediments. The increasing content of these latter minerals contributes to higher major elemental concentrations in finer sediments, with the exception of silicon. The glacial deposits, of course, have not been subjected to appreciable hydraulic transport, and thus do not undergo such mineralogic and chemical differentiation. Glacial deposits therefore contain potassium, sodium, and titanium in amounts similar to the average contents in the source rocks.

Calcium content in most sediments varies between 3 and 4%. The sediments from the shallow banks, however, contain considerably more calcium (Fig. 5-16), due to increased calcareous skeletal fragments deposited on the bank. Calcium displays an affinity with strontium; the distributions of both these elements are similar (Fig. 5-22). Strontium varies between 250 and 850 ppm, the coarse and glacial sediments containing about 400 to 500 ppm, while the silty clays contain only 250 to 300 ppm strontium. The biogenous sediments, as with calcium, show a dramatic increase in strontium. On the other hand, the distribution of barium in sediments is antipathetic to that of calcium and strontium (Fig. 5-23); near-shore coarse and glacial sediments contain relatively low barium, in comparison with offshore fine marine sediment.

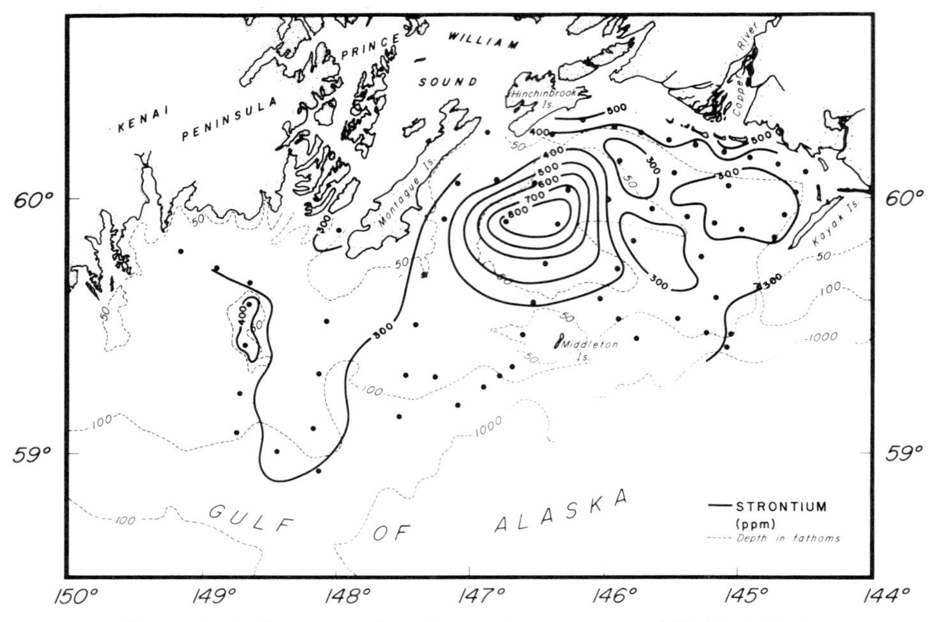

Figure 5-22. Strontium (ppm) in sediments, central Gulf of Alaska.

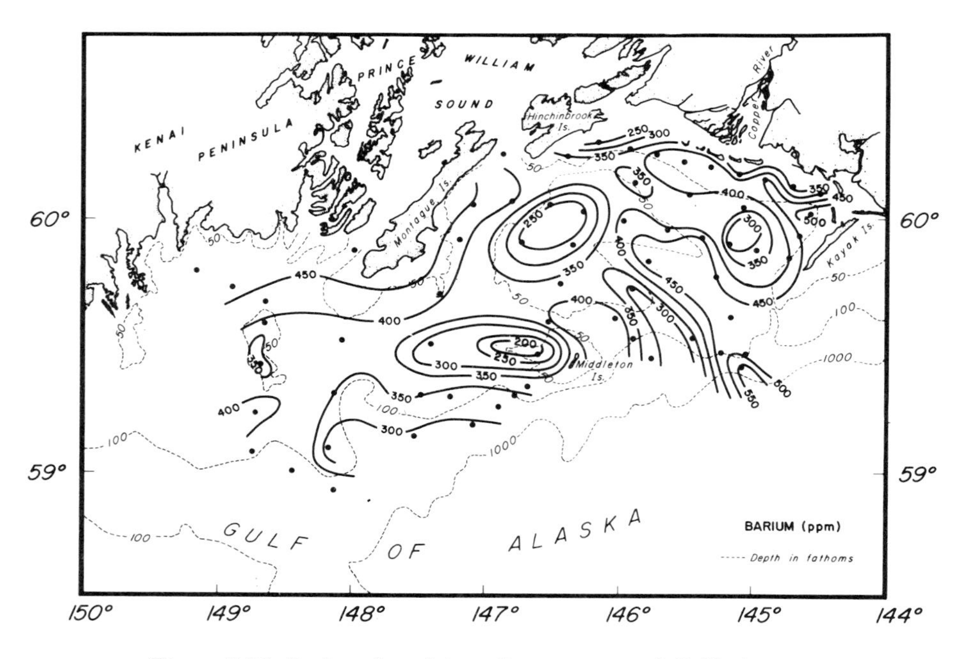

Figure 5-23. Barium (ppm) in sediments, central Gulf of Alaska.

The distribution of cobalt, chromium and nickel is directly related to the clay distribution in these sediments (Figs. 5-24, 5-25 and 5-26); an increase in clay content generally parallels an increased content of these elements. The migration and deposition of these elements appear to be associated with clay-sized sediments, which suggests that these elements may be adsorbed on the clay particles or, perhaps more likely, are incorporated in the basic framework of the clay mineral-mica structures.

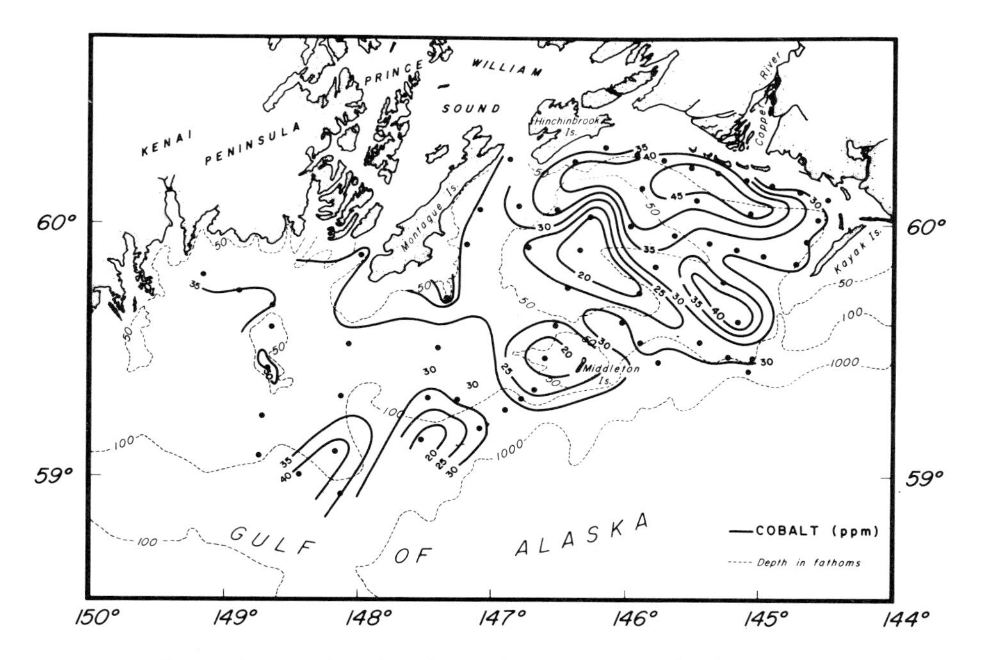

Figure 5-24. Cobalt (ppm) in sediments, central Gulf of Alaska.

Copper mostly varies between 30 and 40 ppm (Fig. 5-27); however, coarse sands contain less (20 ppm) and silty-clayey sediments contain more copper (60 ppm). Manganese, with a somewhat similar distribution pattern, varies between 800 ppm and 1,400 ppm (Fig. 5-28). Both elements are highly mobile, and therefore their distributions are controlled by sediment movement as well as by water dynamics. Thus it is not surprising to observe some similarity between the distributions of these elements and the amounts of clay in the sediments. Manganese content in fine sediments is also affected to some extent by the upward migration of hydrated manganese in the interstitial waters.

Zinc content varies over a wider range (from <50 to >200 ppm) and its distribution is complex (Fig. 5-29). With the exception of two small regions of high zinc concentrations, in general the gravelly and sandy deposits contain between 50 and 100 ppm, while silty and clayey deposits contain more than 100 ppm zinc.

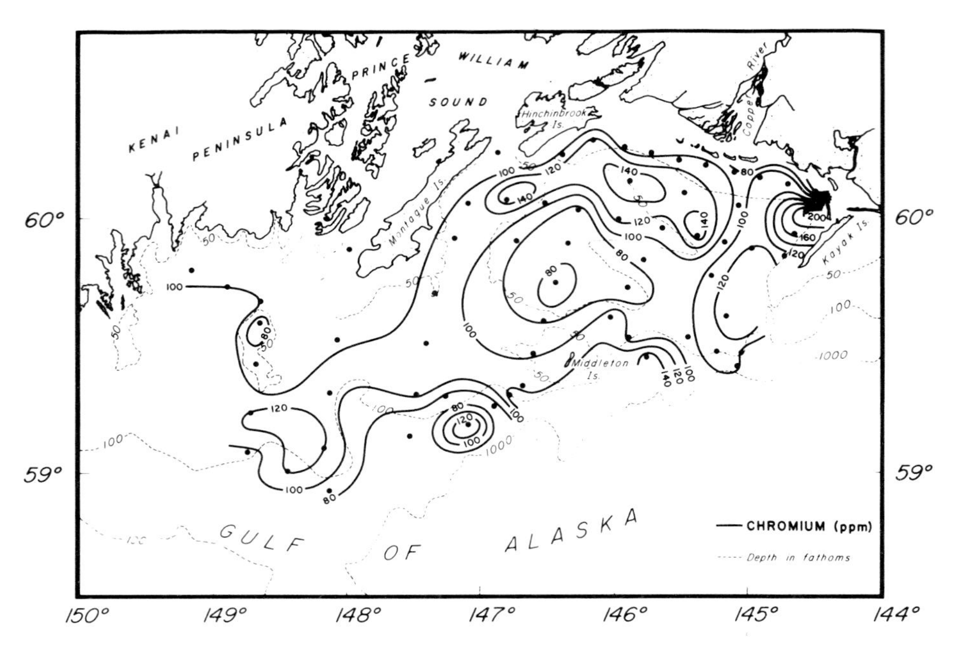

Figure 5-25. Chromium (ppm) in sediments, central Gulf of Alaska.

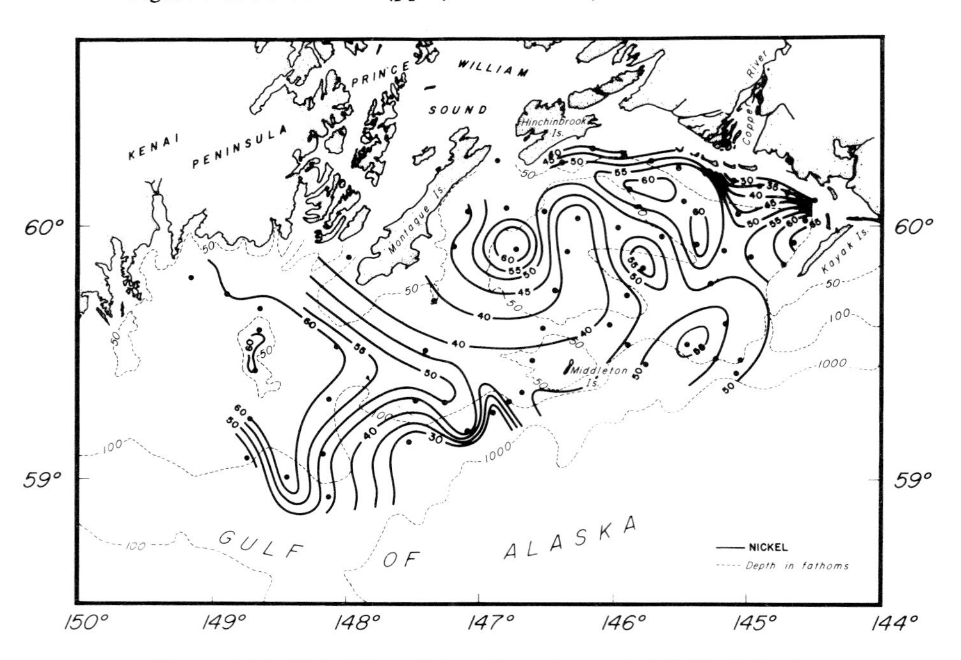

Figure 5-26. Nickel (ppm) in sediments, central Gulf of Alaska.

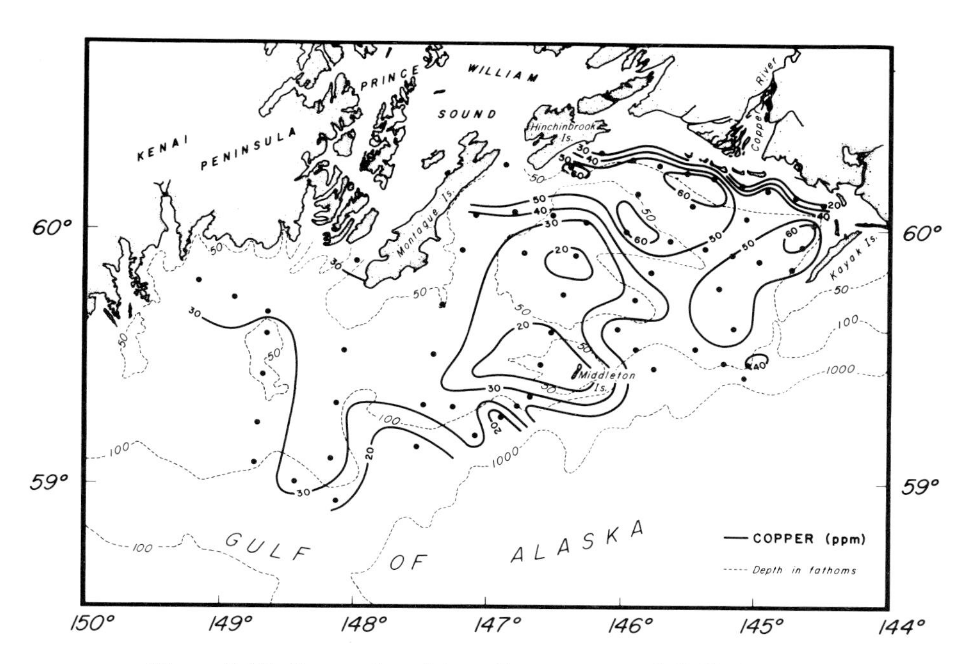

Figure 5-27. Copper (ppm) in sediments, central Gulf of Alaska.

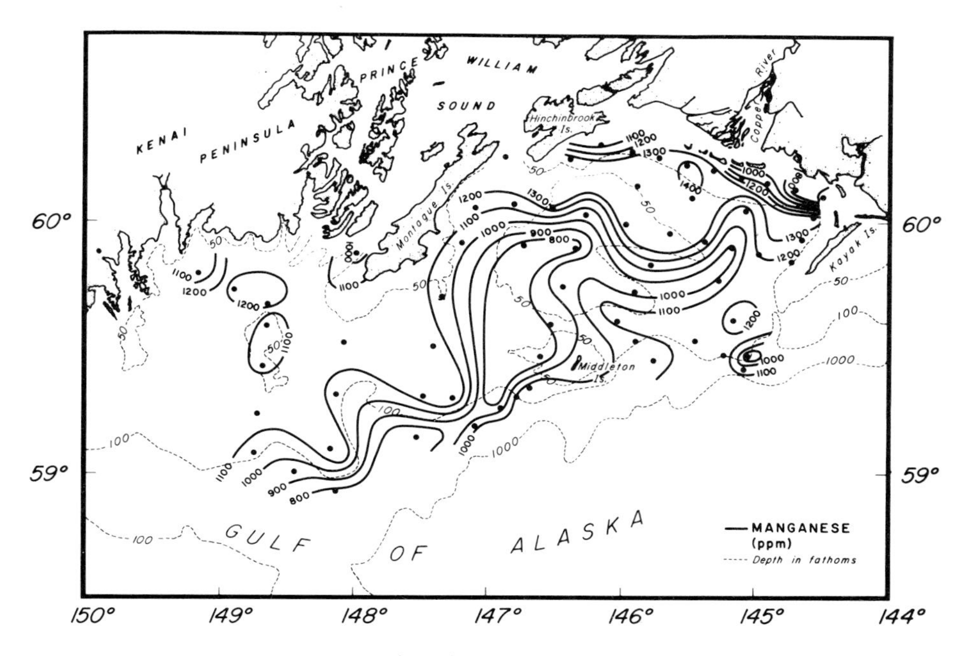

Figure 5-28. Manganese (ppm) in sediments, central Gulf of Alaska.

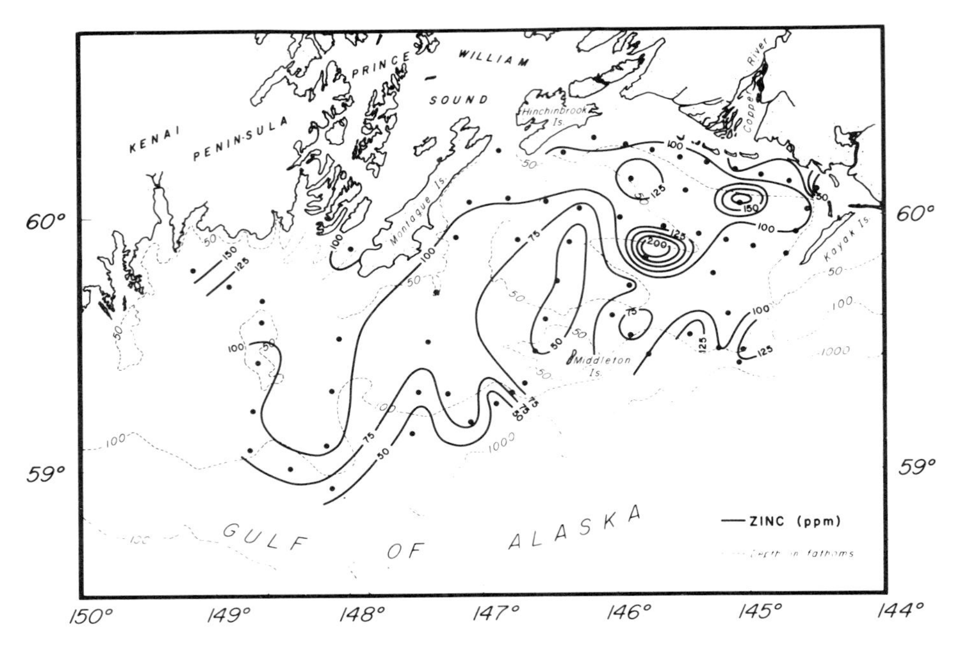

Figure 5-29. Zinc (ppm) in sediments, central Gulf of Alaska.

SEDIMENT SOURCE AND TRANSPORT

The major source of sediments in this region is the Copper River. Smaller rivers, in particular the Bering River, and local coastal drainage also provide sediments to the adjacent shelf. The Copper River discharge alone reaches 1,500 m^3/sec during summer peak months, and it carries up to 1,700 mg/ℓ sediment in suspension. A considerable amount of sediments is also carried seaward as bed load. The large sediment input of the river is evident from the impressive delta which has been built on the shelf, together with an almost continuous string of barrier islands deposited offshore. Important sources of sediments in the region are well defined by the surface suspended sediment load distribution on the shelf during 24-28 February 1973 (Fig. 5-30). Relatively higher sediment concentrations in suspension occur near the mouth of the Copper River and along the entire northern shore. Noticeable concentrations also occur in offshore waters southwest of Kayak Island. The source for these sediments lies east of Kayak Island, with westward transport by the Alaska Current. The suspended load distribution in the surface waters just described (Fig. 5-30) represents a period of low river discharge and minimal glacial melt; nevertheless, these concentrations are relatively higher than usually found in shelf water. Even during periods of low discharge the importance of the Copper River and the drainage east of

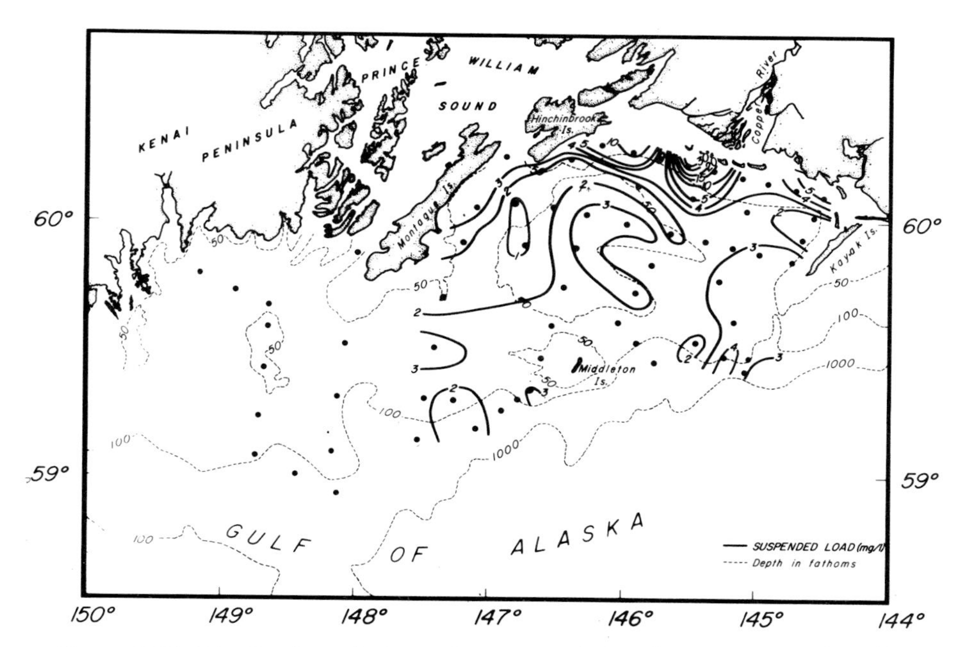

Figure 5-30. Distribution of the surface suspended sediment during 24-28 February 1973, central Gulf of Alaska.

Kayak Island as sediment contributors to the region is quite apparent from the suspended load, salinity, and temperature distributions resulting from mixing with water discharged from these sources. The sediments from both sources are mostly carried westward along the coast, and only a portion of the fine sediments is carried offshore.

During summer months the synoptic relative sediment distribution in near surface waters of this region is well expressed on ERTS-1 imagery. A series of images obtained on successive days provides a mosaic which shows the major sediment sources and pathways of sediment in suspension (Fig. 5-31). Obviously significant amounts of sediment in suspension are introduced into the region from the east. Several glaciers (Fig. 4-5, Chapter 4), particularly Malaspina and Bering glaciers, melt streams, and rivers draining the coast east of Kayak Island discharge their load in the coastal water as dense plumes. As these plumes extend offshore, the Alaska Current deflects them to the west. The westward moving sediments, particularly those originating from Bering Glacier, are deflected to the north and south near Kayak Island. Sediments passing through the narrow passage north of Kayak Island join the larger plume emanating from the Copper River and continue westward along the shore. The sediments moving southwest along Kayak Island form a large gyre which encircles the southern half of the island.

Figure 5-31. ERTS-1 imagery showing sediment plumes in central Gulf of Alaska and Prince William Sound.

Large eddies in offshore waters with relatively high suspended load are observed southwest of Kayak Island. These large clockwise rotating gyres may have been formed from a combined action of the westward moving Alaska Current, tidal movement, and coriolis force and been carried offshore by Ekman drift. These offshore moving gyres thus may be an important mode for offshore transport of fine sediments in this region.

The large sediment input of the Copper River, with its westward drift toward Hinchinbrook Entrance, is best seen in density-sliced ERTS-1 imageries (Figs 5-32, 5-33 and 5-34). The extension of the river plume farther into Prince William Sound through Hinchinbrook Entrance and northern passages is well illustrated in an image obtained on 24 September 1973 and shown in Figure 5-32 which is drawn from the density-sliced ERTS image. Beyond Hinchinbrook Entrance some of the Copper River sediments are carried farther southwest along the eastern coast of Montague Island (Fig. 5-35). Near the southwestern tip of Montague Island the remnant of the plume is deflected northward into Montague Strait and is carried into Prince William Sound. By this stage, however, the plume has lost most of its sediment and has become quite diffused.

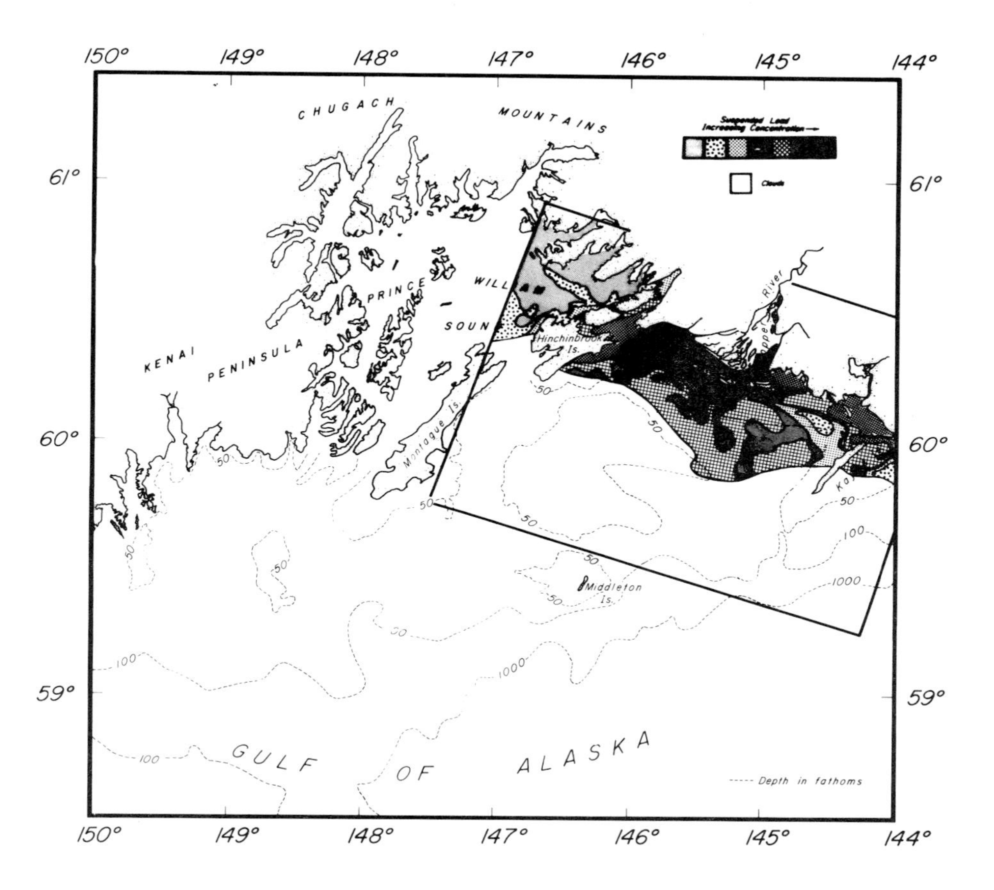

Figure 5-32. Isodensity distribution of reflectance in satellite imagery showing relative suspended loads in near-surface water of central Gulf of Alaska on 24 September 1973.

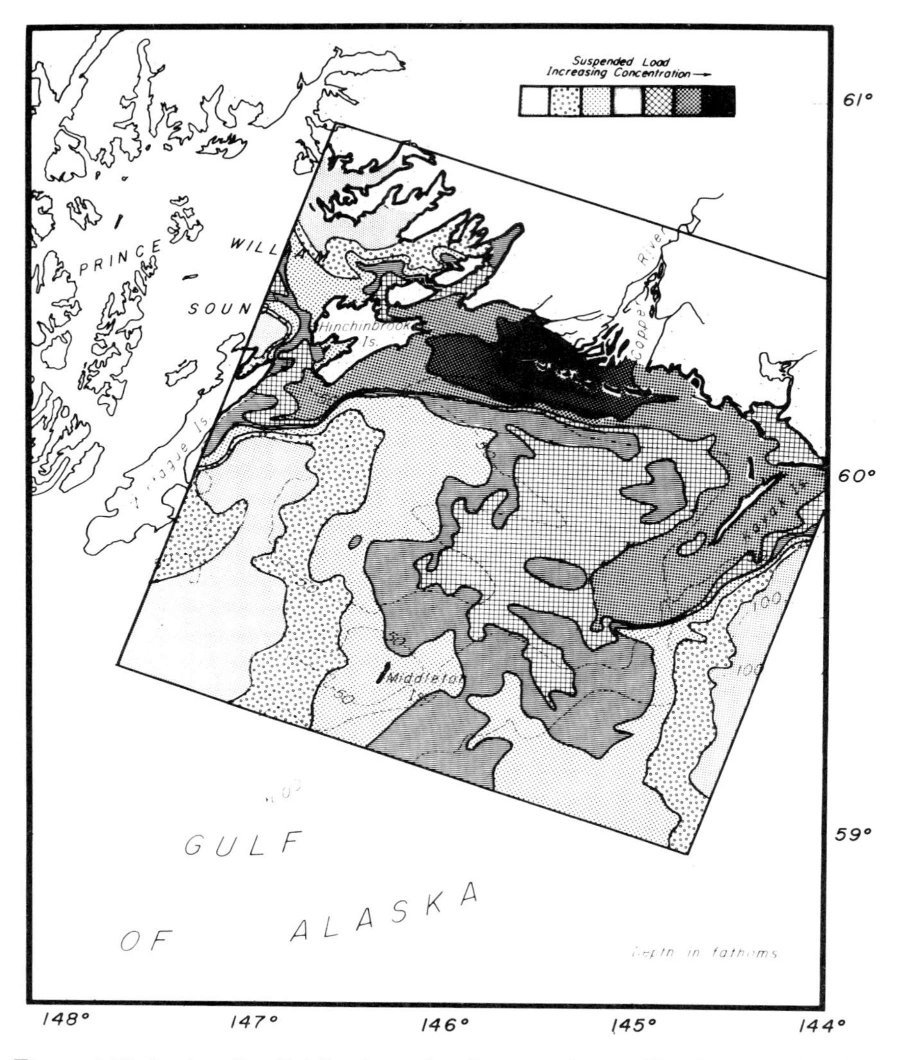

Figure 5-33. Isodensity distribution of reflectance in satellite imagery showing relative suspended loads in near-surface water of central Gulf of Alaska on 12 October 1972.

Measurements of various water parameters, including suspended load distribution during winter and the plumes in the central Gulf of Alaska, suggest that most of the sediments originating in the Copper River and east of Kayak Island are deposited in the near-shore region. The sediments brought by the Copper River form an east to west oriented delta, part of which extends into Prince William Sound. Only small amounts of fine sediments are carried offshore.

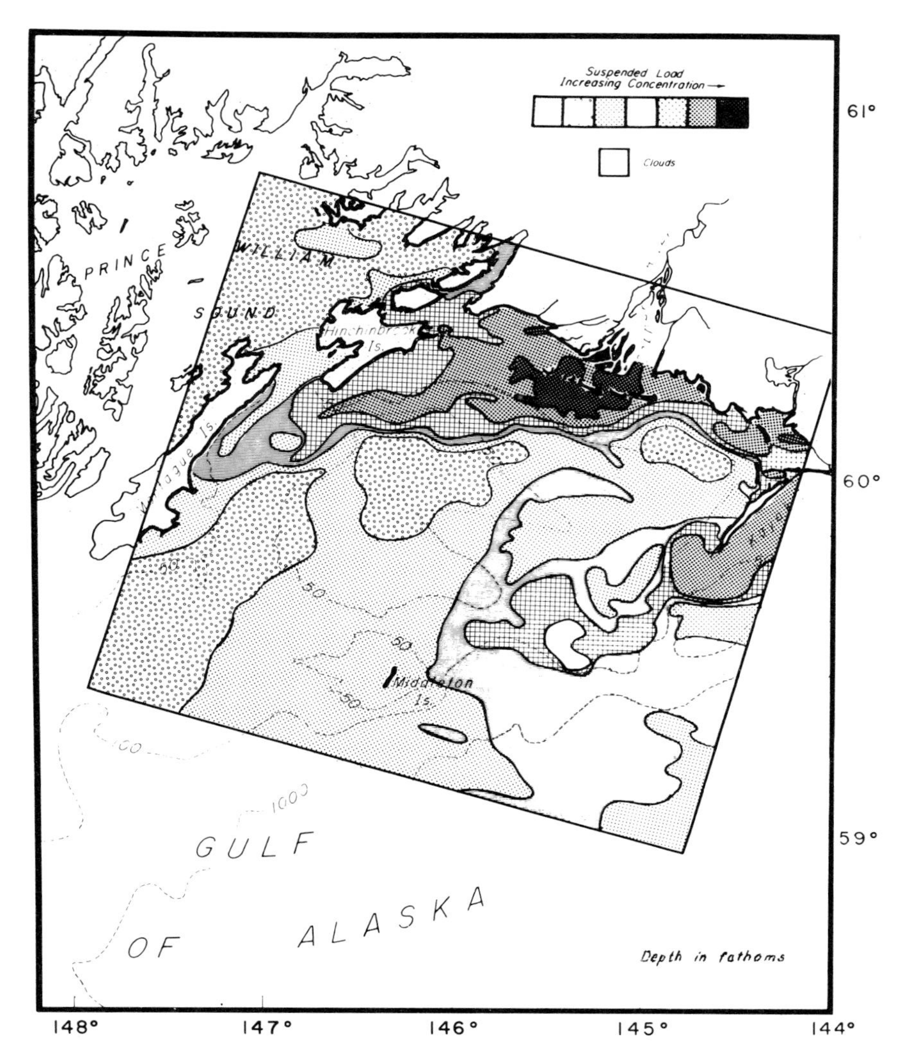

Figure 5-34. Isodensity distribution of reflectance in satellite imagery showing relative suspended loads in near-surface water of central Gulf of Alaska on 14 August 1973.

Comparisons of minerals chlorite and illite peak intensities in x-ray diffractograms of sediment particles of $<4\ \mu m$ and $<2\ \mu m$ indicated an increase of chlorite particles in the smaller size class. It appears that the chlorite particles, on an average, are finer than illite grains in this region. Because of their smaller size the chlorite particles remain in suspension longer and are carried farther from their source than illite particles. The relative distribution of these minerals

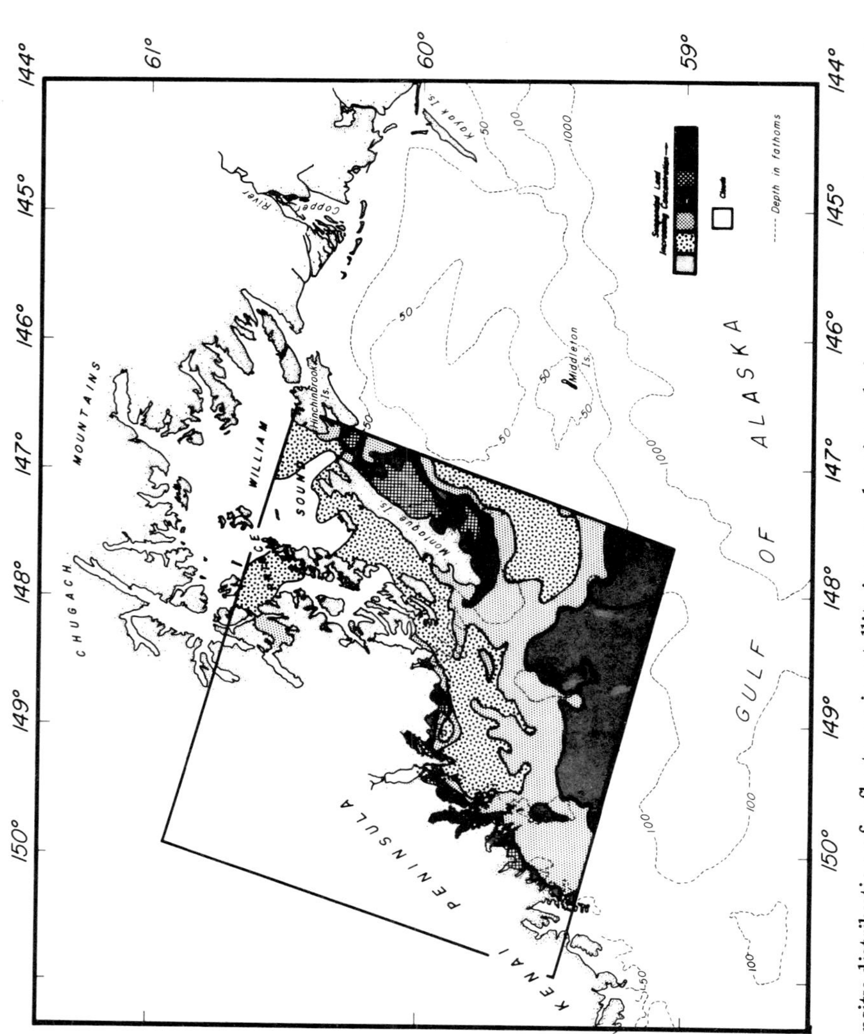

Figure 5-35. Isodensity distribution of reflectance in satellite imagery showing relative suspended loads in near-surface water of central Gulf of Alaska on 16 August 1973.

in clay size particles, therefore, can be an effective indication of sediment pathways in some regions. The chlorite-illite distribution in the Gulf of Alaska (Fig. 5-36) supports the general sediment movement suggested by the sediment texture and ERTS imagery.

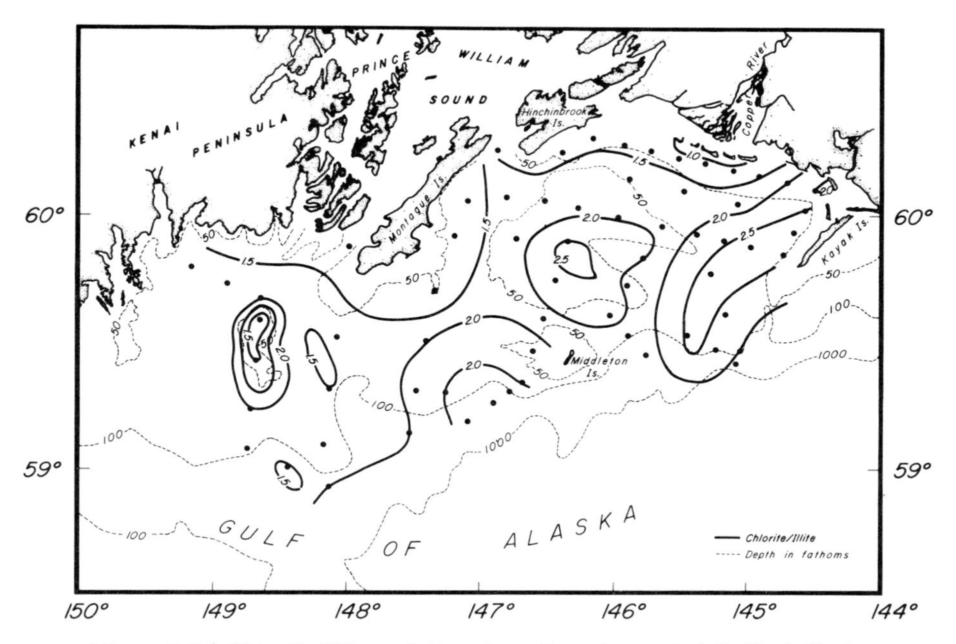

Figure 5-36. Chlorite-illite variations in sediments, central Gulf of Alaska.

The sediment transport and distribution on the shelf is complex. In particular, the sediment textural parameters correspond neither to the bathymetry nor to the distance from source. Despite the large detritus input, offshore glacially deposited coarse relict material remains exposed. In some areas, however, the relicts are covered with contemporary sand, silt, and clay. These contemporary sediments are deposited in near-shore regions and in the channel surrounding Tarr Bank. This depression serves as a sediment trap and also as a major artery for offshore sediment transport.

The complex distribution of bottom sediments in the central Gulf of Alaska Shelf can be better understood when it is discussed in conjunction with the suspended load distribution. It should be noted that both the ship measurements and density-slicing analyses of the ERTS-1 imagery indicate a similar distribution of sediments in the surface waters. Sediment plumes generally flow in the direction of water movement, and their definition and extension are mostly controlled by the hydrography of the region. The orientation of a large plume south of Kayak Island (Fig. 5-34) suggests onshore movement of water. West of Kayak

Island, the waters continue to move northward, as indicated by surface temperature and salinity distributions. Farther north, the Copper River plume is diverted westward partly because of westerly movement of Alaska Current and partly because of the Copper River discharge. The northward and westward movement of plumes is mostly channeled through the circular moat-like depression which surrounds Tarr Bank.

The counterclockwise movement of water on the shelf best explains the bottom sediment distribution on the central Gulf of Alaska Shelf as well as in Prince William Sound. The northerly moving water west of Kayak Island carries sediments shoreward. These sediments are deposited in the trough and the channel. The water then turns west and thus confines the Copper River discharge along the shore and the channel and moves it toward the Prince William Sound. Although most fine sediments are carried westward as a surface plume, some silt and clays diffuse offshore and are deposited in the channel north of Tarr Bank. As the water moves westward, part of the sediments are carried into the sound and the rest move along the eastern shore of Montague Island and are finally deposited along the channel southwest of Tarr Bank.

Although most sediments of coastal origin are shunted to the west, some fine fraction is carried offshore by Ekman transport and partly deposited on the Tarr Bank. The flood tide, it appears, resuspends these sediments and carries them into Prince William Sound where they settle out rapidly. The westward shunting of Copper River sediments, tidal flushing, and resuspension of sediments therefore prohibit deposition of contemporary sediments on the Tarr Bank and other topographic highs. The predominant movement of sediment along the shore and into Prince William Sound has also left some glacial relict of the outer shelf and slope exposed.

For the most part, the shelf is covered by contemporary marine sediments; however, some portions of the shelf are dominated by the glacially derived relict sediments. Differences in mode of transport for the contemporary and relict glacial sediments has somewhat restricted the utility of elemental ratios in delineating the sources and the migratory routes of sediments. Because glacially deposited sediments do not undergo textural and chemical differentiation, their composition is generally similar to that of the source rock. The interpretation of elemental ratio in sediments from this region therefore should be carried out with some caution.

A triangular compositional diagram obtained by plotting as end members percentages of Al_2O_3, SiO_2 + TiO_2, and Rest (total percentages of other major element oxides) shows some interesting distribution of Gulf of Alaska sediments (Fig. 5-37). The average content of SiO_2 + TiO_2 in igneous rocks is about 60% (Clark and Washington, 1924; Goldschmidt, 1954). Upon weathering and during marine transport this source rock follows two lines of sedimentary differentiation, which lead to end members as quartz sand (lag deposit of resistant mineral) and clay (mechanical and chemical breakdown of minerals). It is suggested that those sediments with combined SiO_2 and TiO_2 percentages of about 60

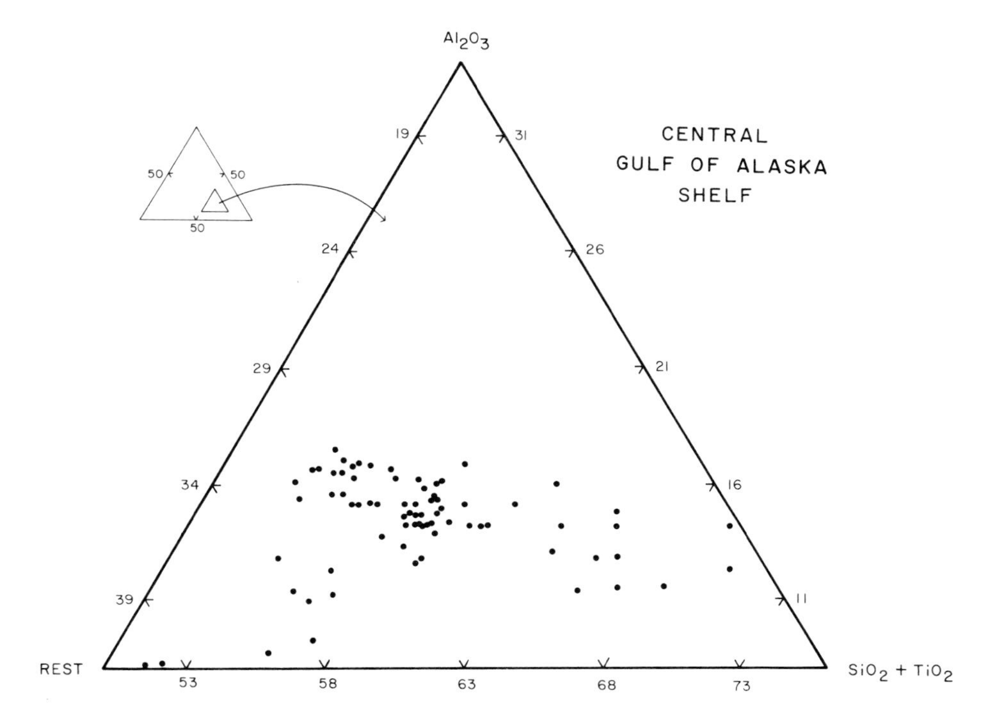

Figure 5-37. Triangular plot of percents Al_2O_3, $SiO_2 + TiO_2$, and Rest as end members, central Gulf of Alaska.

represent material which has undergone the least mineralogic-elemental differentiation. Additionally, it might further be suggested that those sediments showing significant departures from this value of $SiO_2 + TiO_2$ content might be interpreted as materials which have been subjected to an appreciable degree of elemental differentiation. Examples of the former include materials from the northern coastal area, in the vicinity of Copper River, as well as the glacial sediments near the shelf break, thus supporting the foregoing contention. Differentiation of elements by marine processes leads to maturity of sediments represented by the Al_2O_3 and SiO_2 relationship (Fig. 5-38). Hydraulic sorting and lateral transport are the most effective processes contributing toward the development of the end members quartz sand and clay. A few samples that contain low amounts of Al_2O_3 as well as SiO_2 represent the calcareous deposits of the Tarr Bank.

Further definition of sources perhaps can be sought by interpretation from the distributions of other elemental ratios. The Al_2O_3/TiO_2 variations are probably the most useful because they not only reflect sediment source and migratory path but also seem suggestive of hydrodynamic conditions prevailing in the region. These variations indicate a distinct source for sediment to the east of

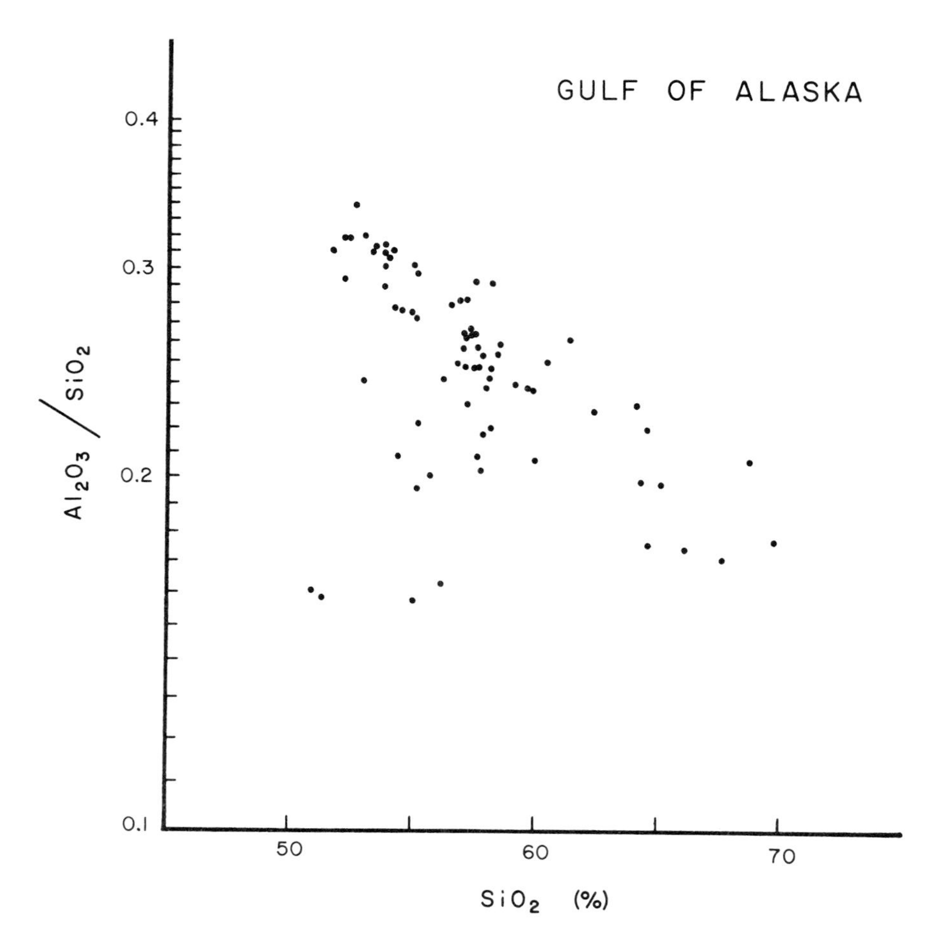

Figure 5-38. Al_2O_3/SiO_2 and SiO_2 plot for sediments, central Gulf of Alaska.

Kayak Island, and sediments from this source are deposited southwest and west of the island (Fig. 5-39). The Copper River is an apparent source for sediments deposited as an elongated delta along the northern shore. The relatively low Al_2O_3/TiO_2 values in the deltaic sediments are due to the high content of TiO_2 in heavy sands deposited in this area. An increase in this ratio reflects westward drift of Copper River sediments. The Copper River also serves as a source for some sediments deposited on the northeastern Tarr Bank. The southern and northwestern parts of the bank, however, appear to contribute sediments to the surrounding area. Outer shelf sediments display a range of Al_2O_3/TiO_2 values, the glacial relicts showing low values and contemporary sediments having high values (Fig. 5-39).

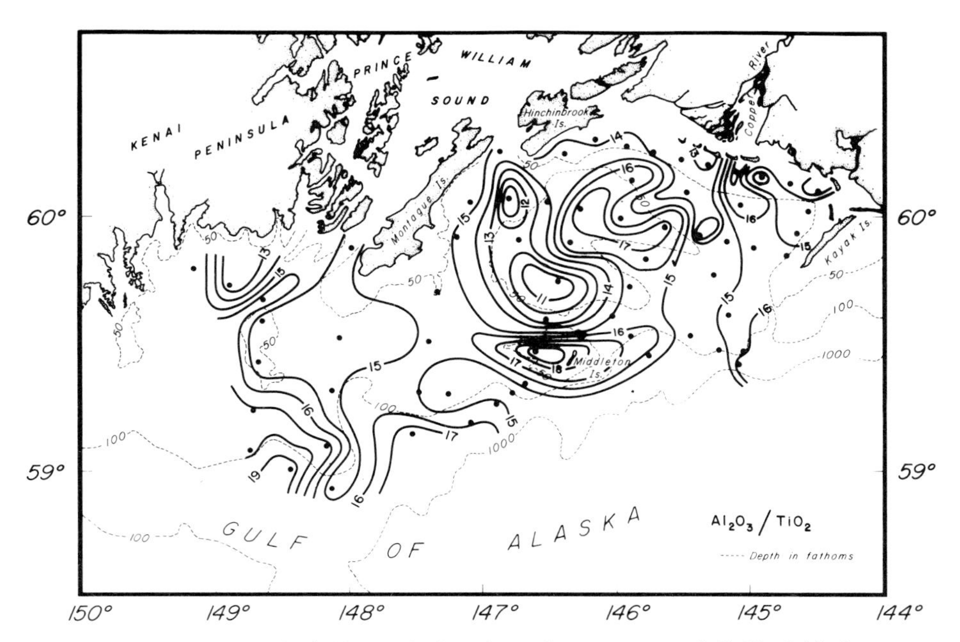

Figure 5-39. Al_2O_3/TiO_2 variations in sediments, central Gulf of Alaska.

The Al_2O_3/Na_2O distribution also indicates the same sediment sources and sediment movement on the shelf (Fig. 5-40) similar to that described earlier. Sediments from the shallow bank in the southwest and the adjacent area have high values, which are typical of source sediments. The bank is covered with glacial relict material and biogenous calcareous skeletal fragments, however, the adjacent area is mantled with silty clay, which is a contemporary deposit. The distribution of isopleths on the bank suggests that the fine sediments from the topographic high may have been winnowed and transported to the adjacent offbank area. Most of the sediments deposited on the shelf surrounding the bank, however, apparently are from the Copper River.

Delineation of sources is further confirmed by the K_2O/Na_2O distribution. The variations display, even more clearly, the dispersal pattern for sediments discharged by the Copper River (Fig. 5-41). Westward transport of sediments around Kayak Island (north and south) is self-evident. High values, characteristic of source sediments, do occur in the southwest region and somewhat complicate the regional interpretation.

The CaO/MgO variation shows quite high values on the shallow banks. The increased ratio in these areas is primarily due to large fragments of calcareous skeletal material, which constitute a significant proportion of sediment (Figs. 5-16 and 5-42). The progressive offshore decrease in CaO/MgO values near the Copper River mouth refers to the seaward transport of sediments. It is interesting to note that coarse contemporary near-shore deposits vary between 1.5

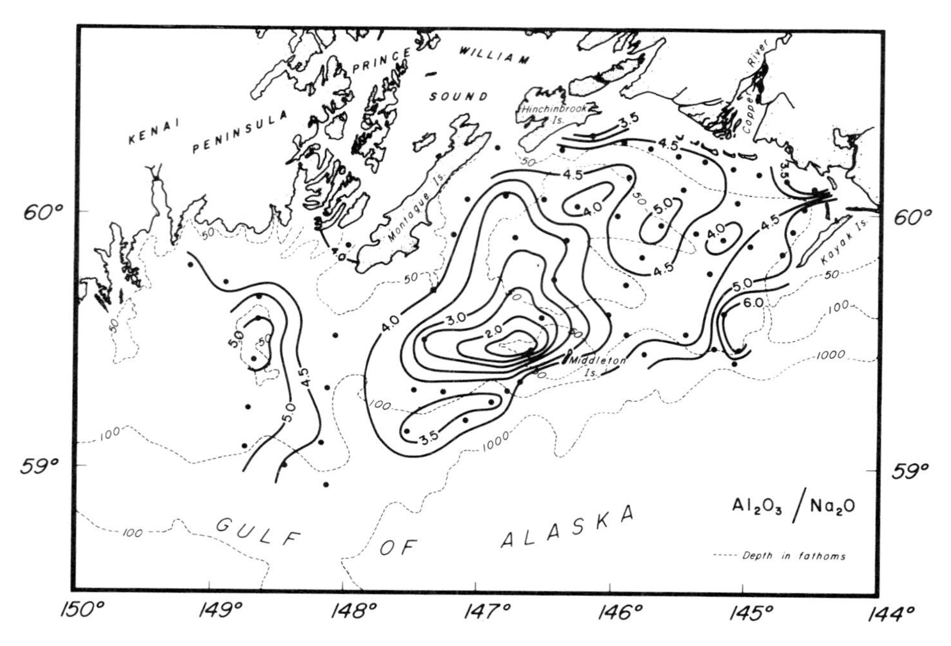

Figure 5-40. Al_2O_3/Na_2O variations in sediments, central Gulf of Alaska.

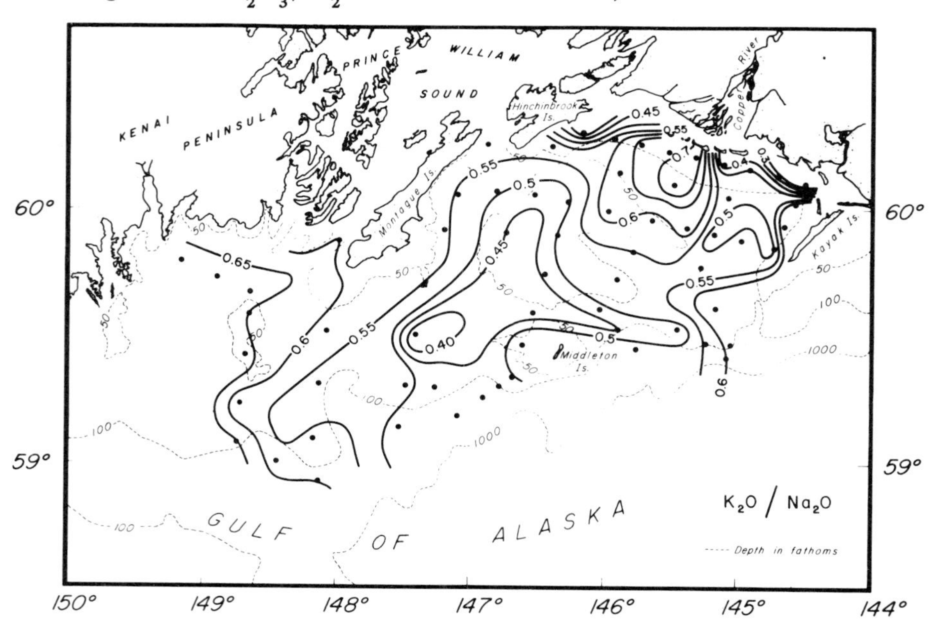

Figure 5-41. K_2O/Na_2O variations in sediments, central Gulf of Alaska.

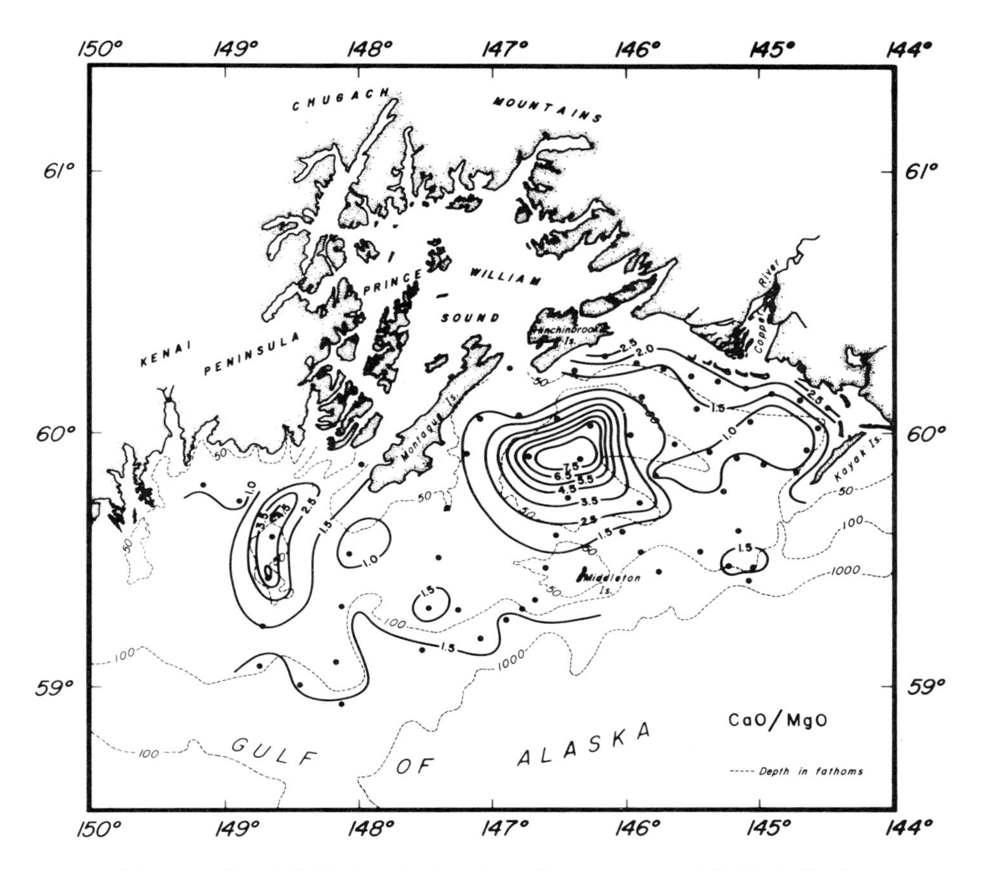

Figure 5-42. CaO/MgO variations in sediments, central Gulf of Alaska.

and 2.5 and the fine silty clay has a ratio close to 1.0. Coarse glacial relict deposits near the shelf margin have CaO/MgO ratios close to 1.5, which closely approximates the average crustal value. These sediments, therefore, apparently have not been weathered appreciably during their transport.

Based on elemental ratio distributions in sediments, suspended load distributions, and hydrographic parameters, a generalized scheme for the movement and sources of sediments in the central Gulf of Alaska Shelf and Prince William Sound regions is proposed in Figure 5-43. The arrows simply show the movement of sediments, and probably the movement of overlying water as well. The size and length of the arrow, however, are not intended to reflect the quantity of the sediment source input.

Suspended load distribution, together with the texture and geochemistry of the bottom sediments from the central shelf, suggests that the detritus presently brought by the glaciers, streams, and rivers is mostly deposited close to these

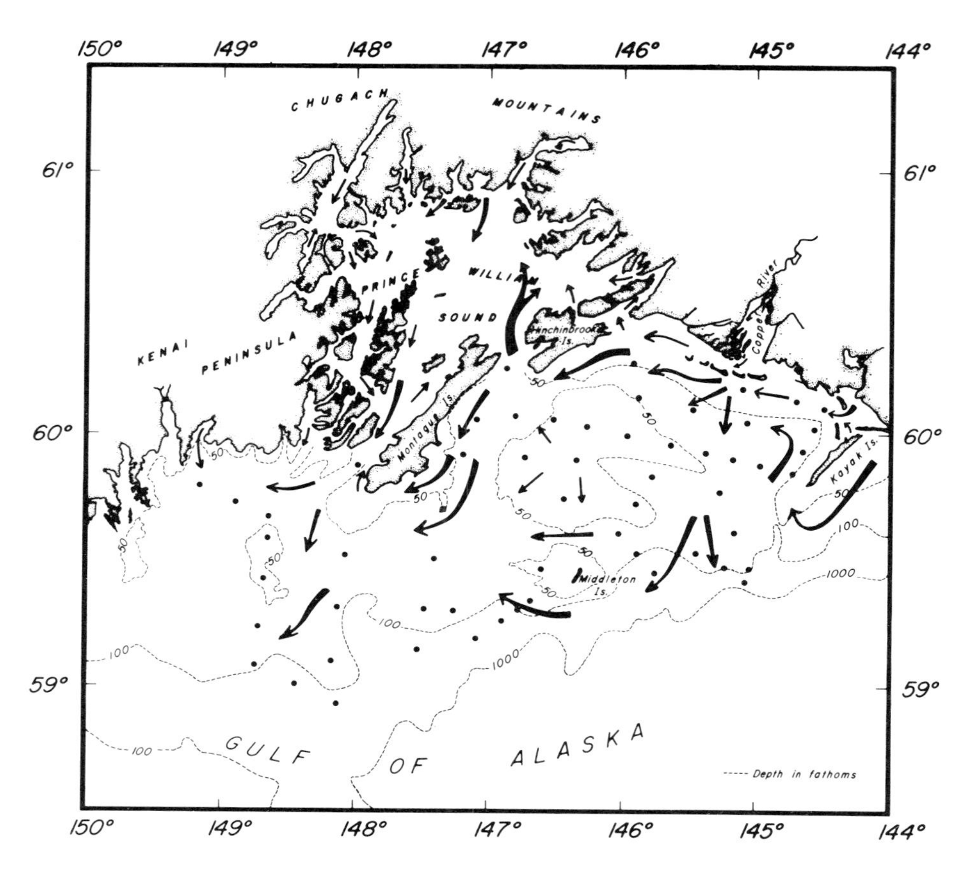

Figure 5-43. Sediment migratory pathways in central Gulf of Alaska and Prince William Sound.

sources and it seems that only a small portion of the fine fraction is transported offshore. During the height of earlier glaciation, however, the sediments were more actively eroded and were carried and deposited beyond the shelf break. Some areas near the outer shelf are covered with Pleistocene glacial sediments and are presently non-depositional, while in other areas the relicts are overlain by contemporary sediments. The reason for the lack of contemporary sediment cover in the former areas may be the onshore movement of waters in these regions inhibiting deposition. The absence of fine contemporary detritus on the shallow banks also may be attributed to winnowing by tidal currents and storm waves.

REFERENCES

Clark, F. W., and H. S. Washington. (1924). The composition of the earth's crust. U.S. Geol Survey Prof. Paper 127, 117 p.

Goldschmidt, V. M. (1954). *Geochemistry*. Oxford University Press, Fair Lawn, N.J.

Hopkins, D. M. (1972). The paleography and climatic history of Beringia during late Cenzoic time. Interod 12, Centre d'Etudes Arctiques, Sorbonne, Paris, France, pp. 121-150.

Karlstrom, T. N. V. (1964). Quaternary geology of the Kenai Lowland and glacial history of the Cook Inlet region, Alaska. U.S. Geol Survey Prof. Paper 443, p. 69.

Plafker, G. (1969). The Alaska earthquake, March 27, 1964, regional effects-tectonics. U.S. Geol. Survey Prof. Paper 543-I, pp. 1-74.

Plafker, G. (1971). Possible future petroleum resources of Pacific-Margin Tertiary basin, Alaska, *Am. Assoc. Petrol. Geol. Mem.* 15:120-135.

Shor, G. G., Jr. (1965). Structure of Aleutian Ridge, Aleutian Trench and Bering Sea (Abstract). *Trans. Amer. Geophy. Union* 46 (March):106.

von Huene, R., and G. G. Shor, Jr. (1969). The structure and tectonic history of the Eastern Aleutian Trench. *Geol. Soc. Am. Bull.* 80:1889-1902.

CHAPTER 6
Prince William Sound

INTRODUCTION

Prince William Sound, a semicircular basin, lies between 59°55′ and 60°58′ N and 145°30′ and 148°45′W. It is bordered in the north by the Chugach Mountains and in the east by the Kenai Mountains (Fig. 6-1). The principal large basin covers an area of 3595 km^2. Numerous fjords, bays and inlets extend northward from the main basin. A list of the principal water bodies is given in Table 6-1. The water characteristics in these semi-isolated elongated fjords, inlets, and bays are partly controlled by the water in Prince William Sound. Such inlets and fjords are also areas where intense mixing of sediment-laden runoff and sea water occurs, and these mixed waters ultimately flow into Prince William Sound.

Almost all fjords and islands in the sound are elongated in a northeast-southwest orientation and follow the dominant regional structural trend. There are numerous islands scattered throughout the sound; three of the largest, Montague, Hinchinbrook and Hawkins islands, separate the sound from the Gulf of Alaska. This string of three islands is aligned northeast-southwest and forms the southern margin of the sound. The water exchange between the sound and the Gulf of Alaska is restricted to shallow, narrow passages located between the islands. Two of these passages, Hinchinbrook Entrance in the south and Montague Strait in the southwest, are the major arteries of water movements. Both passages have restrictive bottoms with sills (140 m and 267 m) which prevent exchange of water at depth.

GEOLOGY

Reconnaissance studies of the surficial geology in the Prince William Sound region have been conducted by Grant and Higgins (1910) and Moffit (1954). Large-scale damage caused by tectonic movement during the Great Alaskan Earthquake in 1964 necessitated a thorough geologic investigation to minimize future damage. Extensive studies of surface and subsurface structural geology and rock deformation in the region were conducted by Case *et al.* (1966) and Plafker (1969).

The oldest exposed bedrocks covering the northern and western periphery of the sound are of Mesozoic age and are classified as the Valdez Group. They consist of metamorphosed marine graywacke and dark gray to black slate with

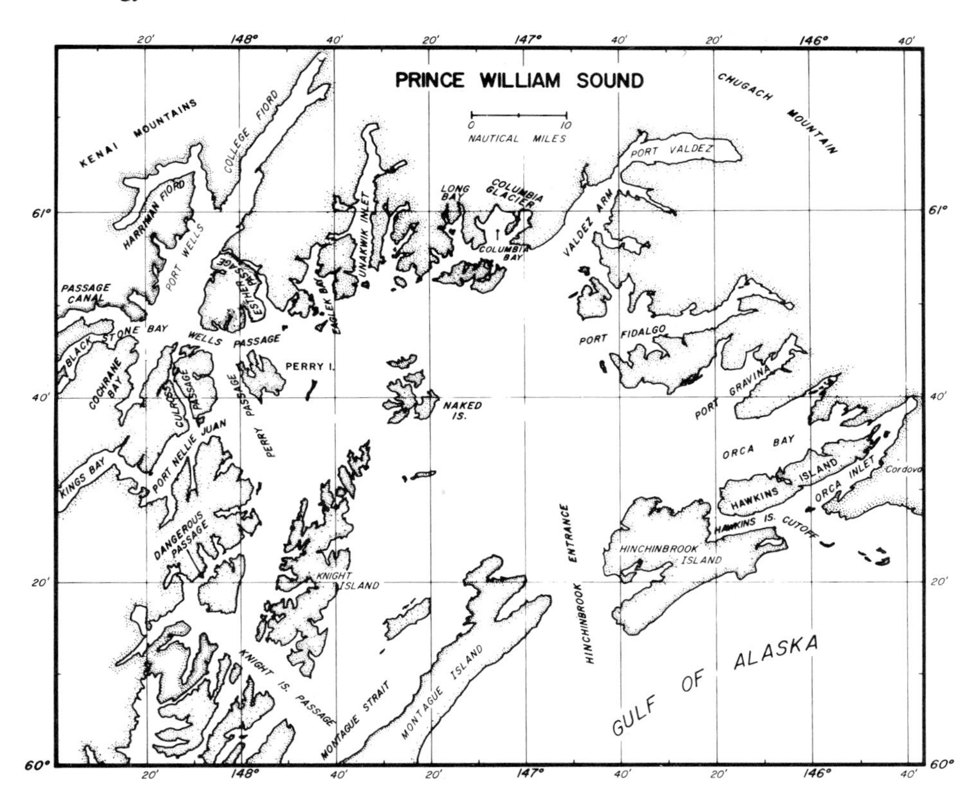

Figure 6-1. Index map showing major water bodies of Prince William Sound.

argillite and conglomerate. The Valdez Group is widely exposed in this region and covers over 75% of the shoreline. The overlying early Tertiary rocks, the lower Orca Group, consist of greenish-black basaltic flows (greenstone) with massive marine sediment (conglomerate, mudstone and sandstone); they are only sparsely exposed on a few islands, notably Knight Island. In the southwest, the upper Orca Group of folded marine sediments forms a major island chain (Montague, Hinchinbrook and Hawkins islands), and also extends northeastward into the hinterland. Intrusions of Eocene granite and quartz diorite stocks are also exposed in a few localities.

During the Pleistocene glaciation the entire area was covered with an ice sheet as much as 1,000 m thick. Ice scoured the valleys and formed elongated basins which generally conform to the underlying bedrock structure.

The exposed rocks in Prince William Sound region are intensely folded, fractured, and faulted. The geologic structure of the region is very complex. Prince William Sound is bordered by three major structural trends which bound a triangular area. At the northeast apex of the triangle, the northwest-southeast oriented Chugach Mountains and the southwest trending Kenai Mountains merge.

TABLE 6-1

Names, Locations and Some Characteristics of the Major Physiographic Features in Prince William Sound
(J. B. Matthews, unpublished data)

Name	Location Latitude Longitude		Length (km)	Mean Width (km)	Surface Area (km^2)	Maximum Depth (m)	Sill Depth Mouth → Head (m)			Number of Basins
Blackstone Bay	$60^\circ 47.5'$ $148^\circ 23.7'$	$60^\circ 39.0'$ $148^\circ 42.5'$	23.1	2.8	41.26	356	305	164		2
Cochrane Bay	$60^\circ 46.0'$ $148^\circ 16.0'$	$60^\circ 37.0'$ $148^\circ 26.8'$	17.6	3.5	49.74	285	285	181		2
College Fiord	$61^\circ 16.5'$ $147^\circ 41.0'$	$60^\circ 58.3'$ $148^\circ 4.2'$	37.4	3.7	139.67	245	111	53	93	3
Columbia Bay	$61^\circ 0.8'$ $146^\circ 58.2'$	$60^\circ 56.5'$ $147^\circ 8.7'$	8.1	6.1	37.42	292	292			1
Culross Passage	$60^\circ 46.0'$ $148^\circ 9.2'$	$60^\circ 36.6'$ $148^\circ 15.5'$	18.0	1.3	25.82	151	60	9	109	2
Dangerous Passage	$60^\circ 25.0'$ $147^\circ 58.0'$	$60^\circ 12.0'$ $148^\circ 12.3'$	28.0	3.3	64.12	491	491	36		2
Eaglek Bay	$60^\circ 56.7'$ $147^\circ 38.6'$	$60^\circ 48.9'$ $147^\circ 49.7'$	13.3	3.5	57.58	182	120	85		2
Ester Passage	$60^\circ 56.1'$ $147^\circ 51.1'$	$60^\circ 48.7'$ $148^\circ 4.2'$	18.5	1.3	28.37	168	107	20		1

TABLE 6-1 (Continued)

Name	Location Latitude Longitude		Length (km)	Mean Width (km)	Surface Area (km^2)	Maximum Depth (m)	Sill Depth Mouth → Head (m)		Number of Basins
Harriman Fiord	61° 5.3′ 148° 9.9′	60°58.0′ 148°27.0′	20.4	2.4	44.65	155	29	18	2
Hawkins Cutoff	60°30.1′ 146° 6.0′	60°24.1′ 146°23.7′	18.1	3.7	55.31	21	1	7	1
Hinchinbrook Entrance	60°28.3′ 146°37.5′	60°12.4′ 146° 6.0′	22.2	16.1	556.86	396	267	292	1
Kings Bay	60°35.8′ 148°19.7′	60°27.0′ 148°42.0′	25.2	3.5	55.88	466	378	171	2
Long Bay	61° 1.0′ 147°12.0′	60°55.2′ 147°18.3′	9.6	2.2	22.43	118	29		1
Montague Strait	60°34.4′ 147° 6.8′	59°47.3′ 148°15.0′	83.3	16.7	1630.67	360	149	140	1
Orca Inlet	60°37.9′ 145°40.1′	60°23.7′ 146° 7.6′	33.0	7.0	143.47	29	12	1	2
Passage Canal	60°50.3′ 148°23.2′	60°46.5′ 148°43.0′	18.1	2.6	35.12	349	314		1
Perry Passage	60°44.6′ 147°55.9′	60°33.7′ 148° 6.4′	14.4	7.0	196.39	616	616		1

TABLE 6-1 (Continued)

Name	Location Latitude Longitude		Length (km)	Mean Width (km)	Surface Area (km²)	Maximum Depth (m)	Sill Depth Mouth → Head (m)			Number of Basins
Port Fidalgo	60°53.2′ 146° 3.5′	60°43.9′ 146°47.3′	41.7	4.3	141.94	212	151			1
Port Gravina	60°47.9′ 146° 2.5′	60°37.4′ 146°29.5′	24.8	6.1	125.92	195	124			1
Port Nellie	60°38.9′ 148° 5.0′	60°28.0′ 148°21.7′	22.6	4.3	107.08	784	409			1
Port Valdez	61° 8.3′ 146°14.2′	61° 4.6′ 146°40.5′	23.5	4.6	90.69	252	146			1
Port Wells	61° 0.5′ 147°59.0′	60°47.9′ 148°18.6′	22.8	8.7	167.18	416	416			1
Prince William Sound	60°57.2′ 146°35.8′	60°22.8′ 148° 2.2′	63.0	64.8	3594.47	870	292			1
Unakwik Inlet	61° 9.3′ 147°30.4′	60°51.3′ 147°38.2′	32.4	3.3	95.02	385	385	5	177	3
Valdez Arm	61° 3.5′ 146°39.2′	60°50.8′ 146°58.5′	22.2	6.5	190.07	382	341	148		1
Wells Passage	60°50.1′ 147°54.2′	60°43.5′ 148°27.0′	29.6	6.1	109.84	424	336	367		2

The base of the triangle lies along the Montague, Hinchinbrook, and Hawkins islands trend. Some of the most active faults trending northeast-southwest run parallel to these islands, and frequent tectonic movements occurs along all three structural trends.

The surface and subsurface structure of the region is described in detail by Plafker (1969). Gravity anomaly contours drawn by Case *et al.* (1966) show a gradual decrease from +40 mgal on the shelf to -70 mgal in College and Harriman fjords. The contours roughly parallel the dominant arcuate structural trend. Locally, a broad gravity low covering Port Gravina and a high over Knight Island were observed. The high is attributed to dense greenstone in this area. The Port Gravina low may be the result of an adjacent granitic stock, or might be caused by the accumulation of a thick sequence of sedimentary deposits.

A series of seismic reflection profiles along the main channels of Prince William Sound was obtained and interpreted by von Huene *et al.* (1967). They inferred three subsurface acoustic units. The lower unit corresponds to the metamorphosed sediments of the Orca and Valdez groups. The upper surface of this unit is highly reflective and has an uneven topography. The unconformably overlying intermediate unit is not well defined and presumably consists of glacial sediments deposited during or since the most recent glaciation. The nearly acoustically transparent upper unit represents recent marine sediments. Abnormally thick sequences of the upper unit are deposited in Orca Bay and to the north of Hinchinbrook Entrance.

BATHYMETRY

Prince William Sound is a continuous part of the Gulf of Alaska Shelf. However, it is a basin with an unusual depth of up to 800 m (Fig. 6-2). The major bathymetric features of the sound are:

1. a large, shallow and elongated basin lying between Hinchinbrook Entrance and Port Valdez, and
2. a series of small deep basins located west of Naked Island

The basins have relatively steep slopes that may be fault scarps or more likely may be the result of glacier scour, which forms typical U-shaped valleys. The banks surrounding these basins are traversed by numerous channels and bays.

HYDROLOGY

The water characteristics of Prince William Sound have not been studied in detail. The first comprehensive study of water characteristics in Port Valdez, which was undertaken during 1971-1972, indicated a complex mixing between the inlet and the sound waters throughout the year (Sharma and Burbank, 1973). Similar mixing undoubtedly occurs between the waters from other fjords and the sound.

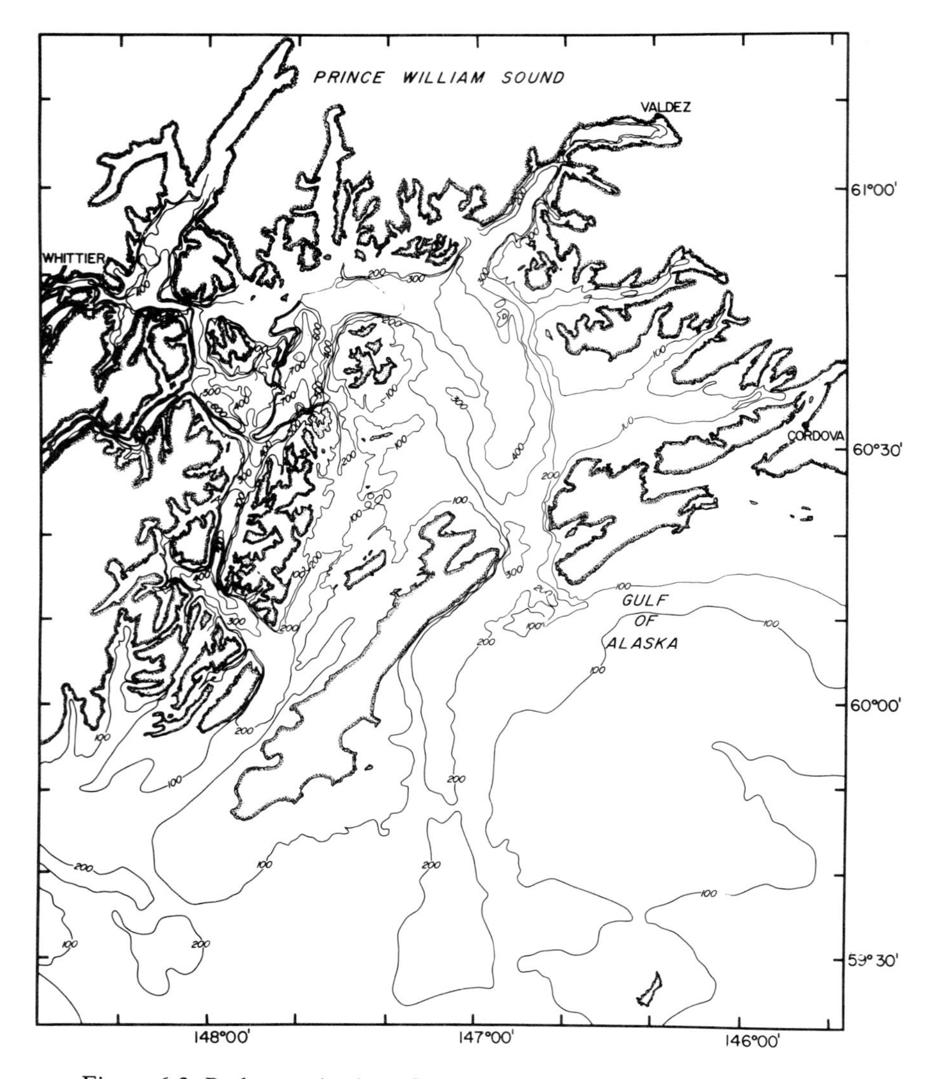

Figure 6-2. Bathymetric chart, Prince William Sound. (Depth in meters)

HYDROGRAPHY

Salinity and temperature measurements (Figs. 6-3 to 6-6) of surface waters during March 1972 indicate lower salinity and higher temperature water masses near the heads of the inlets and in the two major passages, Hinchinbrook Entrance and Montague Strait. High-salinity waters are found in regions east and west of Naked Island. During February 1973, the waters observed near the head of the inlets and in the passages were less saline but also cooler than the waters in the central basins of Prince William Sound. The salinity-temperature distribution distinctly shows that the sources for mixed waters lie predominantly near the heads of the inlets and at the passages. These incoming, less saline surface waters subsequently mix with the more saline waters of central Prince William Sound.

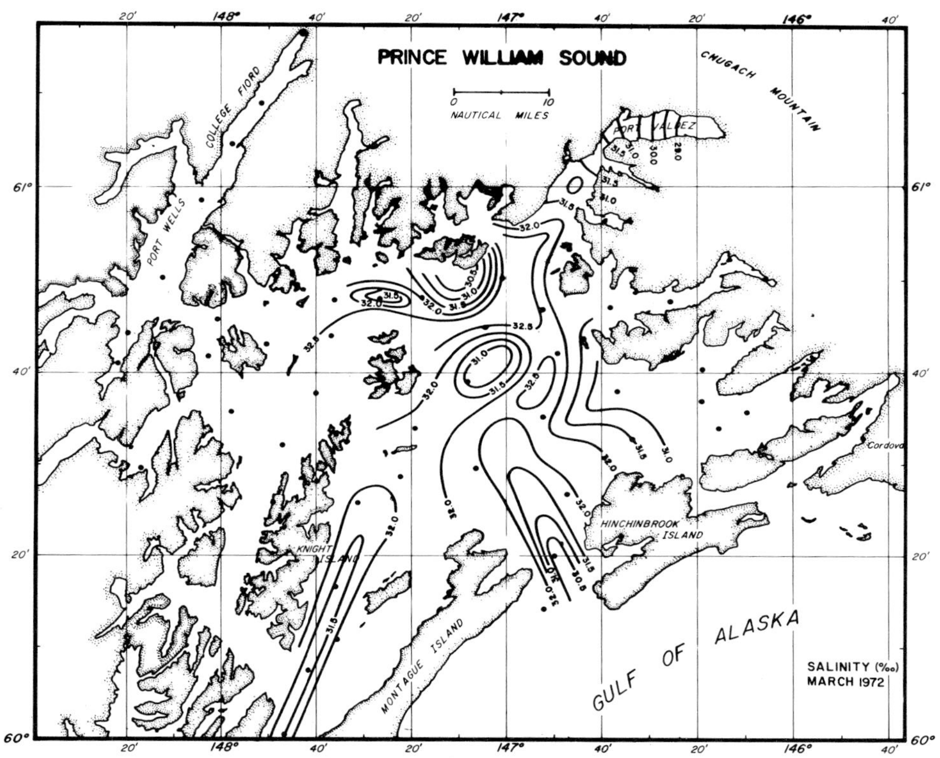

Figure 6-3. Surface water isohalines ($^{\circ}/_{\circ\circ}$) during March, 1972, Prince William Sound.

The progressive increase in surface salinity toward the center of the sound suggests that the low-salinity waters along the rim of the sound move inward and mix with basinal waters. The incoming surface water should be compensated by outgoing water at depth. A two-layered system, however, would remove salt from the basin and ultimately result in low-salinity water, contrary to the high salinity actually observed. It appears that perhaps a three-layered system prevails in Prince William Sound. The water must enter the sound at sill depths as well as at the surface while at intermediate depth water flows out of the basin. Little is known of the source for the deep waters in Prince William Sound.

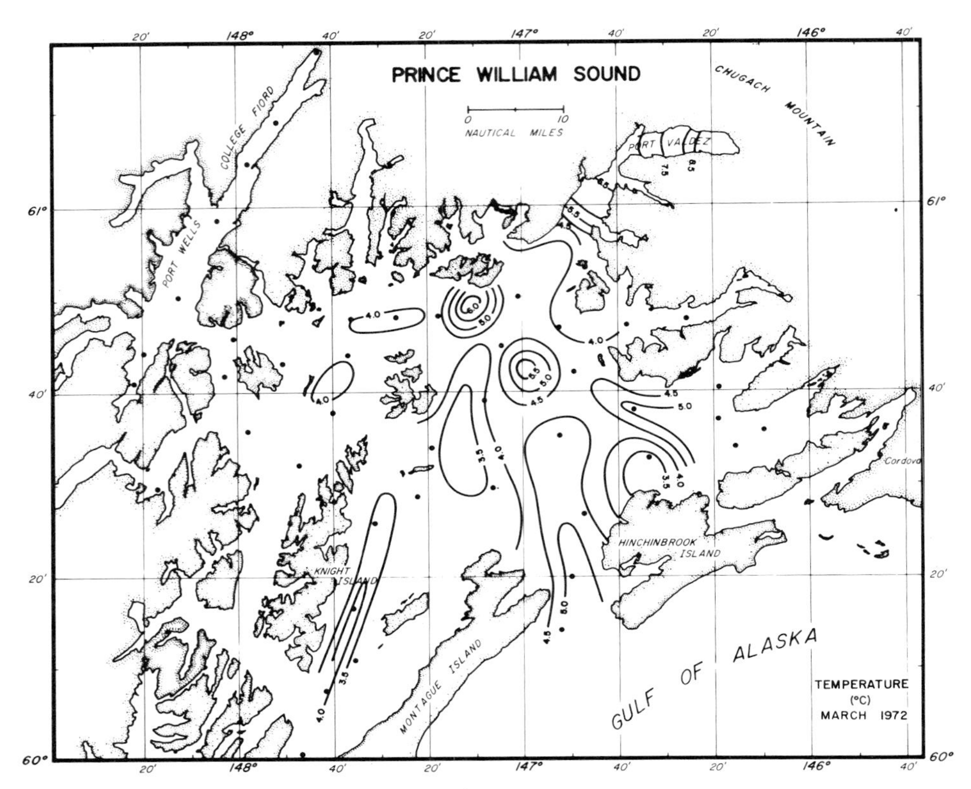

Figure 6-4. Surface water isotherms (°C) during March, 1972, Prince William Sound.

SEDIMENTS

Sediments in Prince William Sound are mostly coarse to medium clay (Fig. 6-7). Sandy silt occurs in Hinchinbrook Entrance and near the heads of the fjords and inlets. The shallow silts of the fjords are generally covered with gravelly sand (Fig. 6-8). Coarser sediments principally occur in shallows, near the head of the fjords, and in areas of extensive water movements, for instance Hinchinbrook Entrance. Finer sediments blanket the deeper portions of the fjords and the basins of the central Prince William Sound. The clays are poorly to very poorly sorted, and, in general, the sorting deteriorates with increasing grain size; silt in Hinchinbrook Entrance is very poorly sorted, while sands from the sills in College Fjord are extremely poorly sorted (Fig. 6-9). Sediments from the tidal glacier, Columbia Glacier, which are deposited in Columbia Bay and the adjacent regions in Prince William Sound, are also very poorly sorted. Sediment size distribution varies from negatively skewed to positively skewed (Fig. 6-10); however, the

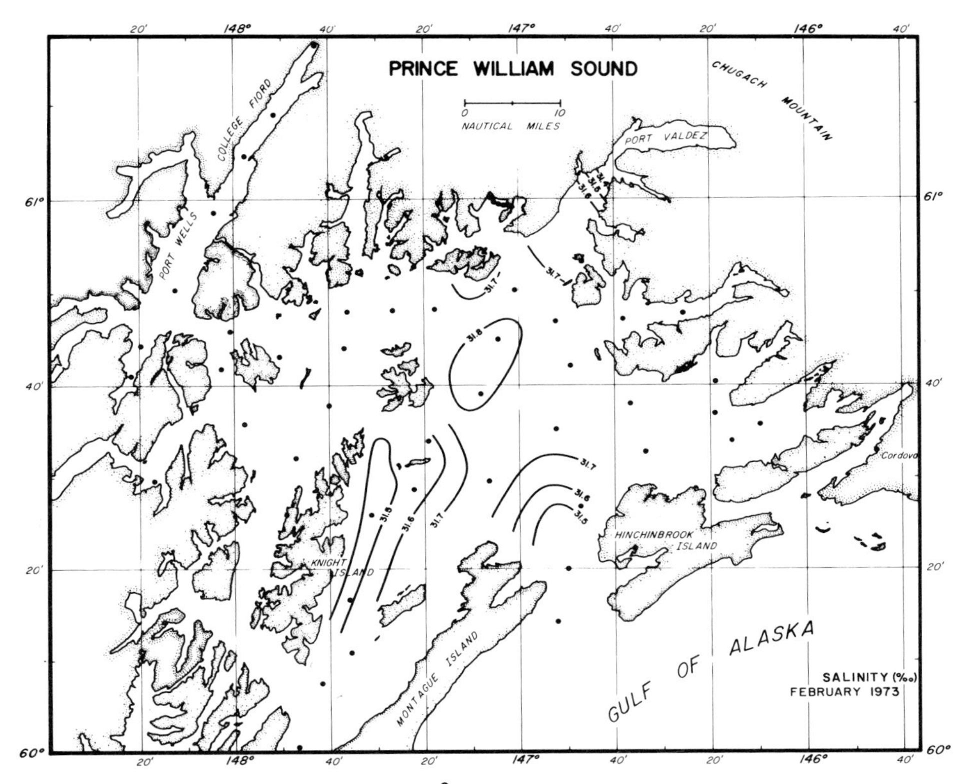

Figure 6-5. Surface water isohalines (°/oo) during February, 1973, Prince William Sound.

deep basinal sediments are symmetrical in distribution. Sediments are mostly platykurtic to mesokurtic, with some regions covered by leptokurtic sediments (Fig. 6-11).

The distribution of percent sand, silt, and clay is of considerable interest because the sand fraction in sediments of Prince William Sound is notably low. The clay fraction is most dominant in sediments, with subordinate silt (Figs. 6-12 and 6-13). Small amounts of sand are found only in Hinchinbrook Entrance and near the mouth of Columbia Bay (Fig. 6-14). The silt-sand mode is predominant in Hinchinbrook Entrance and at the mouth of Columbia Bay, whereas the clay-silt mode covers most of the floor at Prince William Sound.

Organic carbon in sediments from the sound varies from 0.28 to 1.32 weight percent (Fig. 6-15) and is directly related to the amount of clay in the sediments.

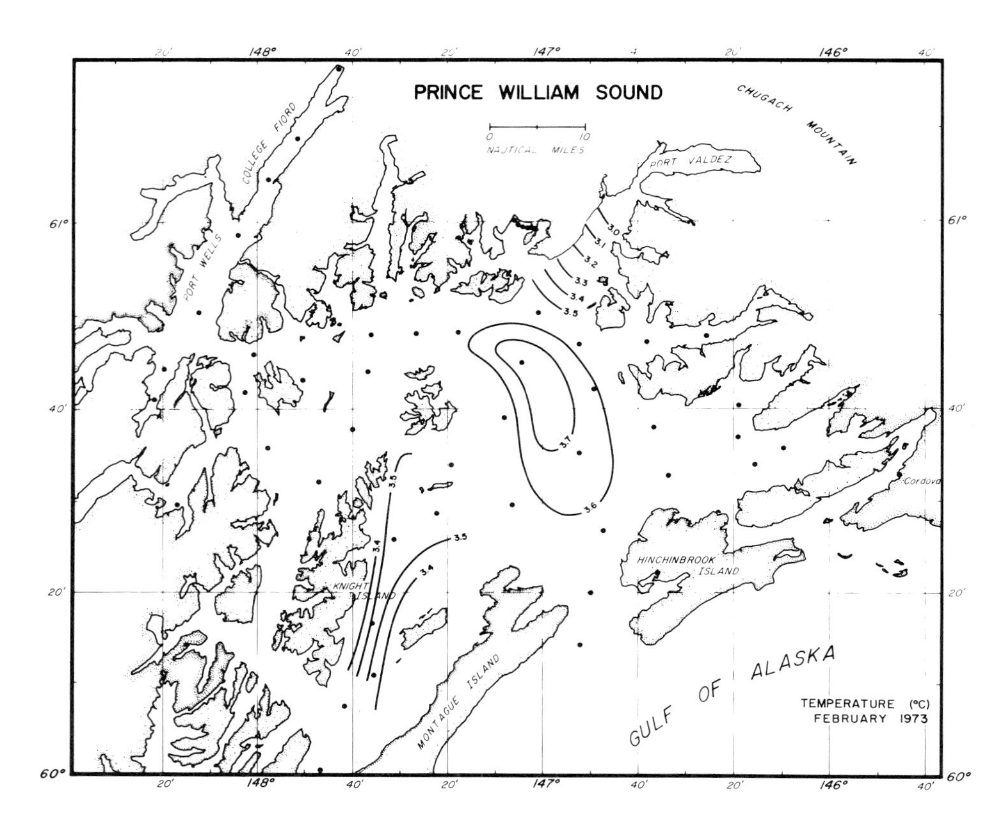

Figure 6-6. Surface water isotherms (°C) during February, 1973, Prince William Sound.

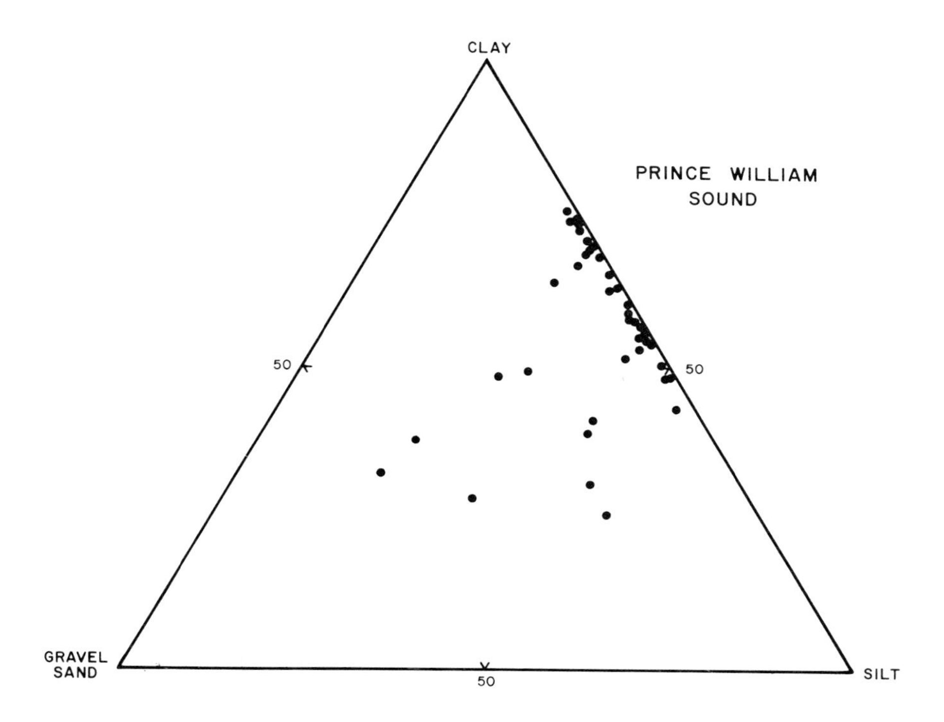

Figure 6-7. Percent gravel-sand, silt and clay distribution in sediments, Prince William Sound.

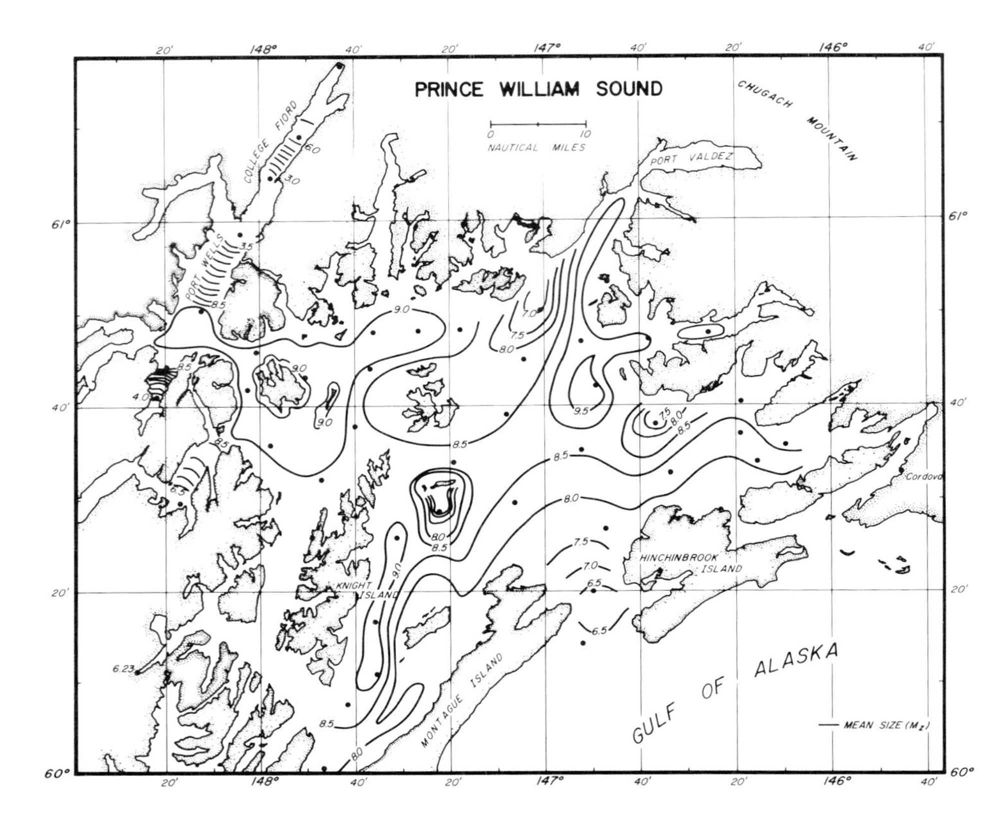

Figure 6-8. Sediment mean size (M_Z) distribution, Prince William Sound.

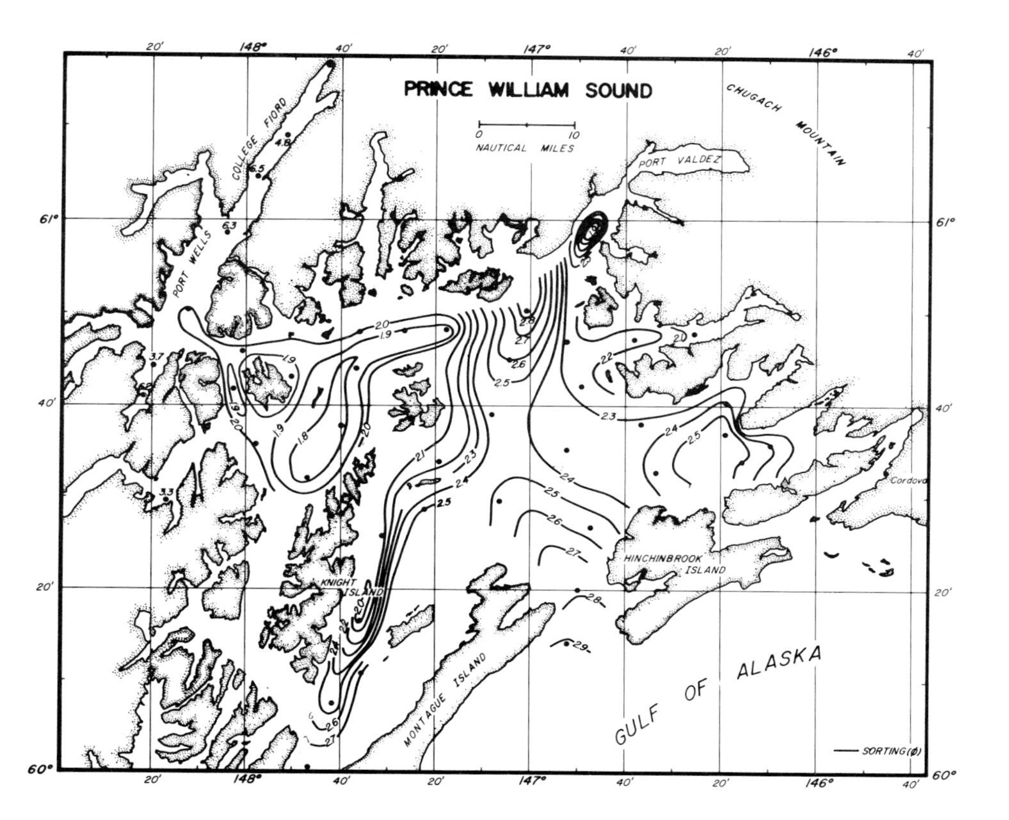

Figure 6-9. Grain size sorting in sediments, Prince William Sound.

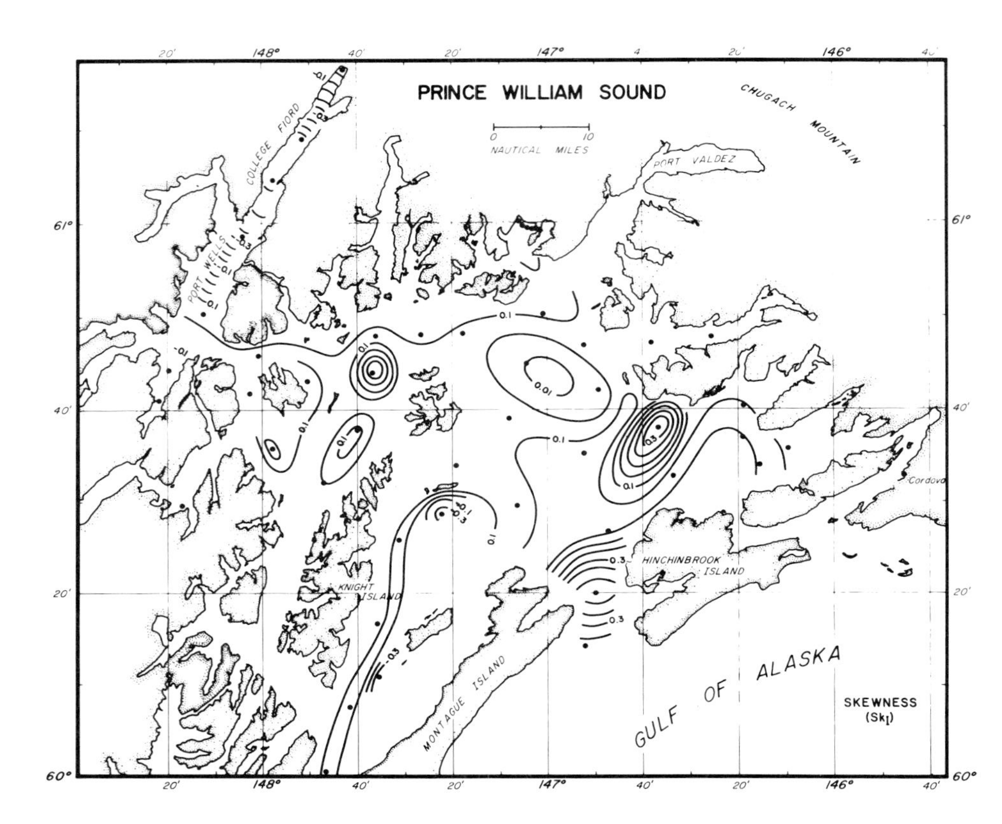

Figure 6-10. Grain size skewness in sediments, Prince William Sound.

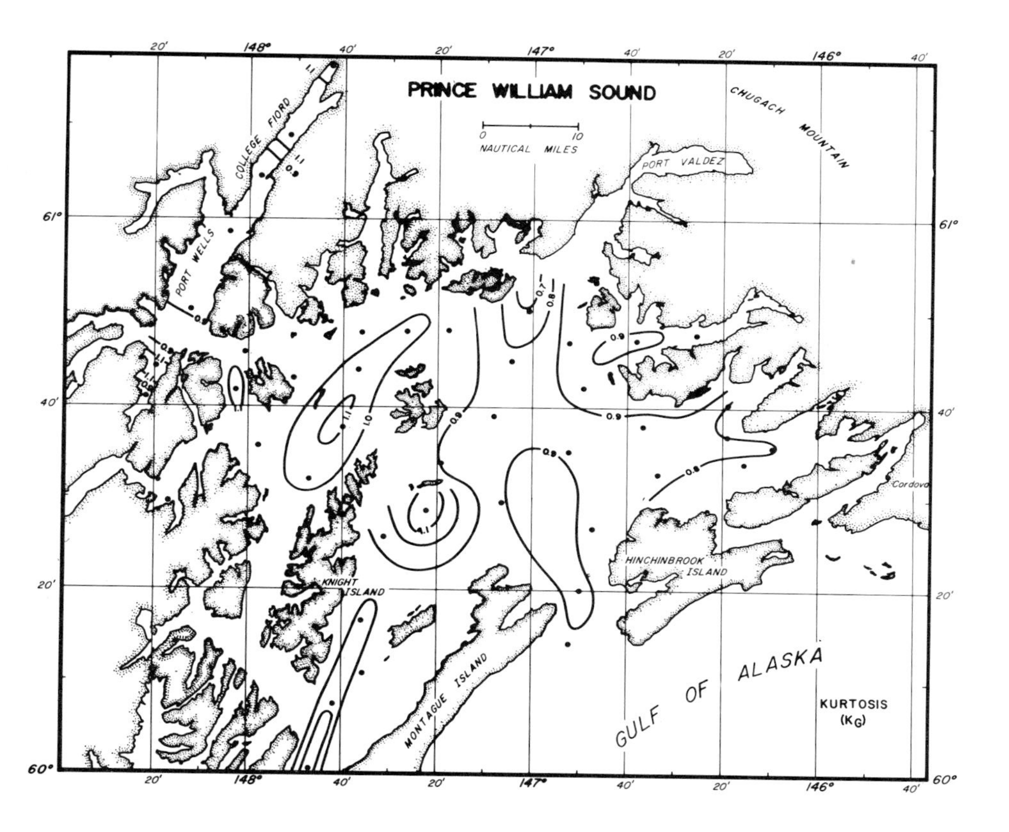

Figure 6-11. Grain size kurtosis in sediments, Prince William Sound.

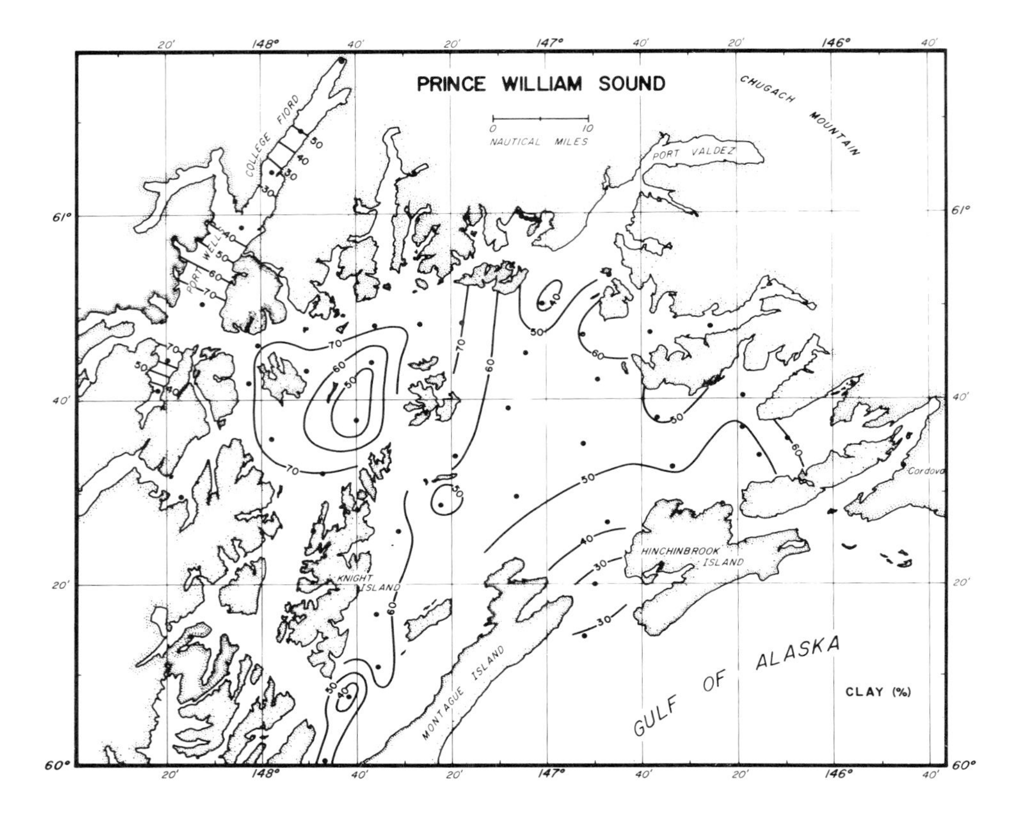

Figure 6-12. Weight percent clay in sediments, Prince William Sound.

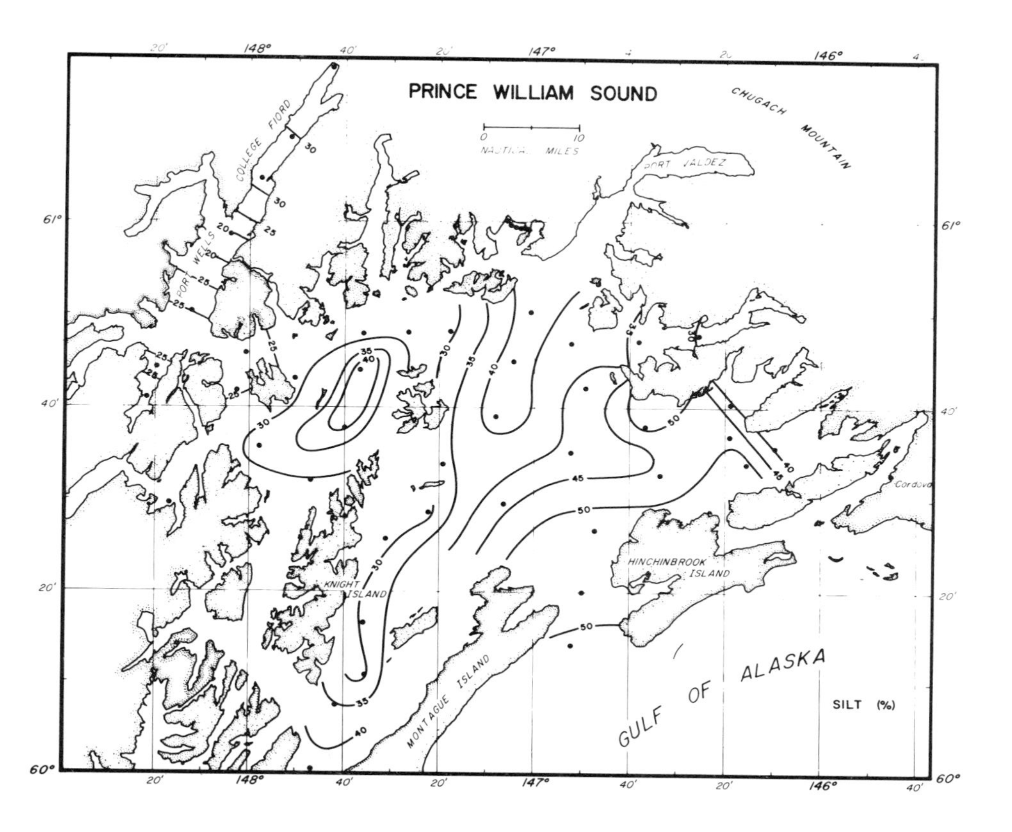

Figure 6-13. Weight percent silt in sediments, Prince William Sound.

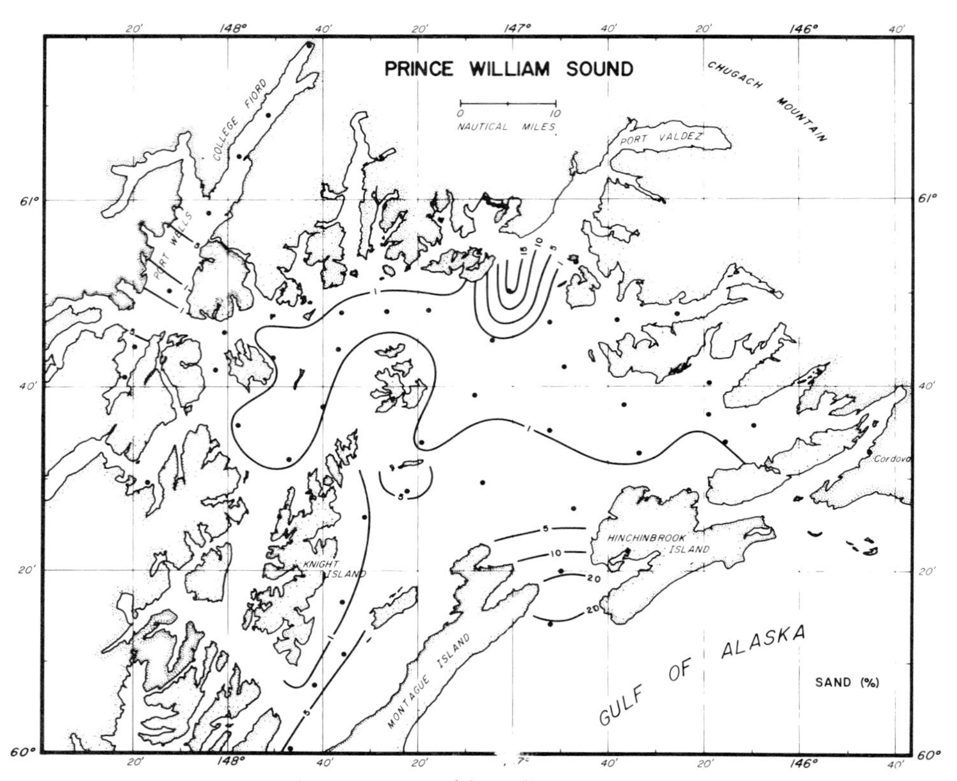

Figure 6-14. Weight percent sand in sediments, Prince William Sound.

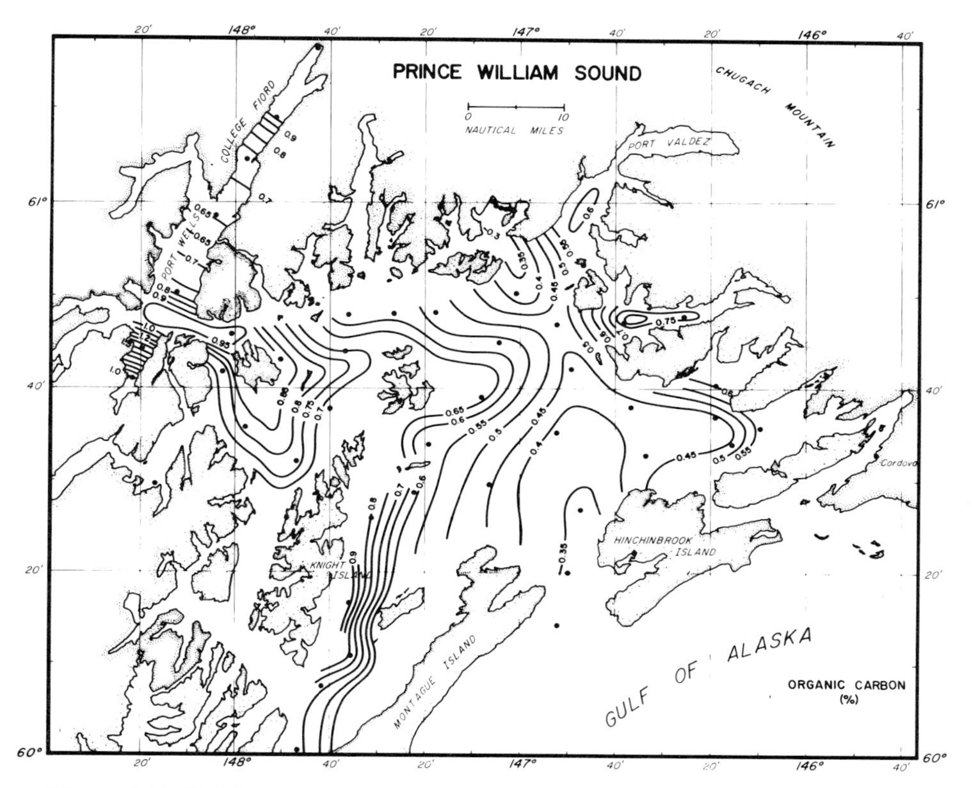

Figure 6-15. Weight percents organic carbon in sediments, Prince William Sound.

GEOCHEMISTRY

The distribution of major elements in sediments from Prince William Sound is shown in Figures 6-16 to 6-23. Aluminum in sediments is generally related to sediment mean size, so that the aluminum content increases with a decrease in sediment mean size. The maximum concentration of aluminum (9.0%) is observed in the clay deposits of the elongated deep basin north of Perry Island (Fig. 6-16), whereas coarse sediments from the Hinchinbrook Entrance contain the minimum content of aluminum. Sediments from the central basin east of Naked Island are partly ice rafted from Columbia Glacier and are mostly clay, but they contain somewhat less aluminum than the other clays. The concentration of aluminum in the finer sediments is a result of chemical and mineralogic differentiation during marine transport. Differentiation of this nature does not take place during ice rafting, and therefore such sediments are not notably enriched in aluminum.

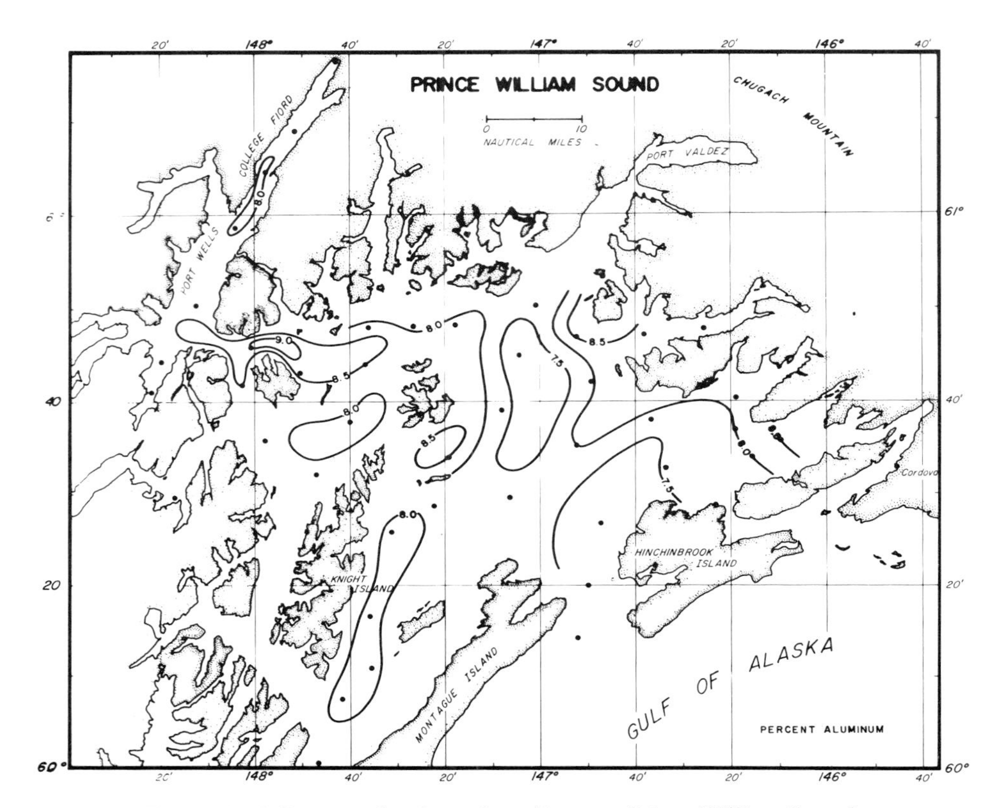

Figure 6-16. Percent aluminum in sediments, Prince William Sound.

Iron in sediments covaries with aluminum. Low iron content (4.75%) is observed near the head of College Fjord, in Columbia Bay (near Columbia Glacier), and in the vicinity of the two passages (Fig. 6-17). Coarser sediments, as well as glacially eroded sediments, in Columbia and Cochrane bays and College Fjord contain lesser amounts of iron (4.5%) than do the finer sediments of the central basins (6.0%). The maximum concentration of iron in sediments is observed in the deep channel north of Naked Island.

Calcium in sediments varies from 1.25 to 4.25% and is in antipathy with the distribution of the other major elements. Coarser sediments near the heads of the bays and the passages contain significantly higher calcium than those sediments deposited in the deeper parts of the sound (Fig. 6-18). Magnesium, on the other hand, covaries with major elements and in antipathy with calcium (Fig. 6-19).

The distribution of potassium is closely related to the amount of clay in the sediments. It ranges from a minimum of 0.95% in Hinchinbrook Entrance to a maximum of 2.5% near the mouth of Port Wells (Fig. 6-20). Sodium distribution,

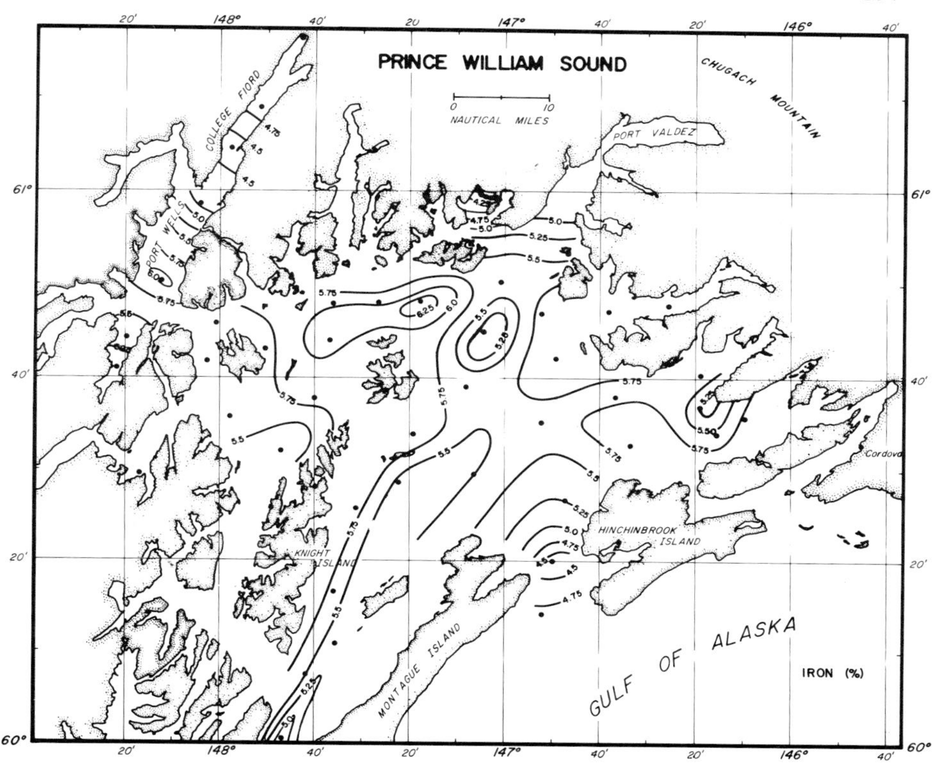

Figure 6-17. Percent iron in sediments, Prince William Sound.

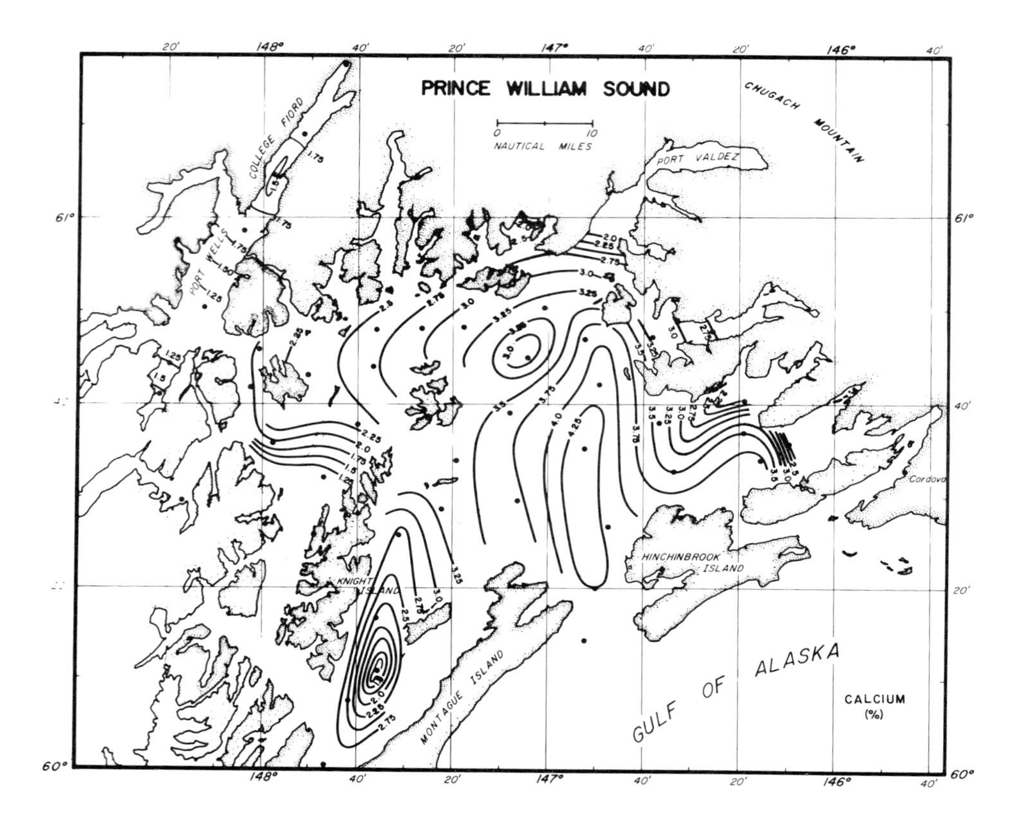

Figure 6-18. Percent calcium in sediments, Prince William Sound.

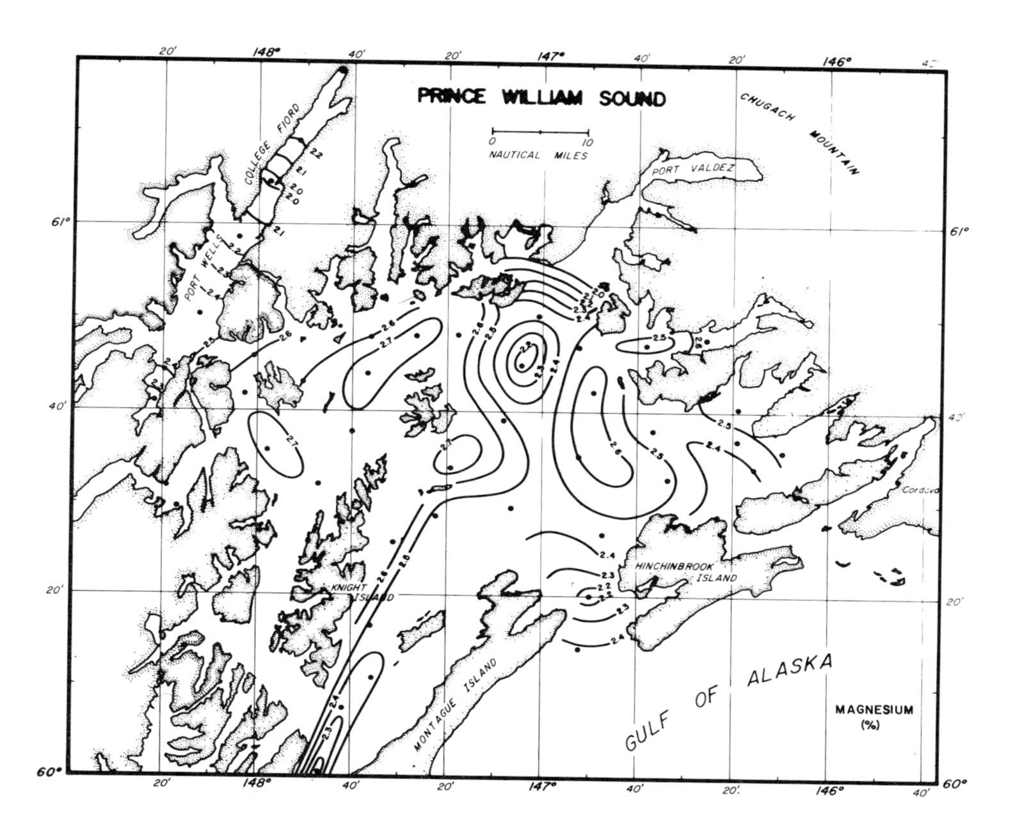

Figure 6-19. Percent magnesium in sediments, Prince William Sound.

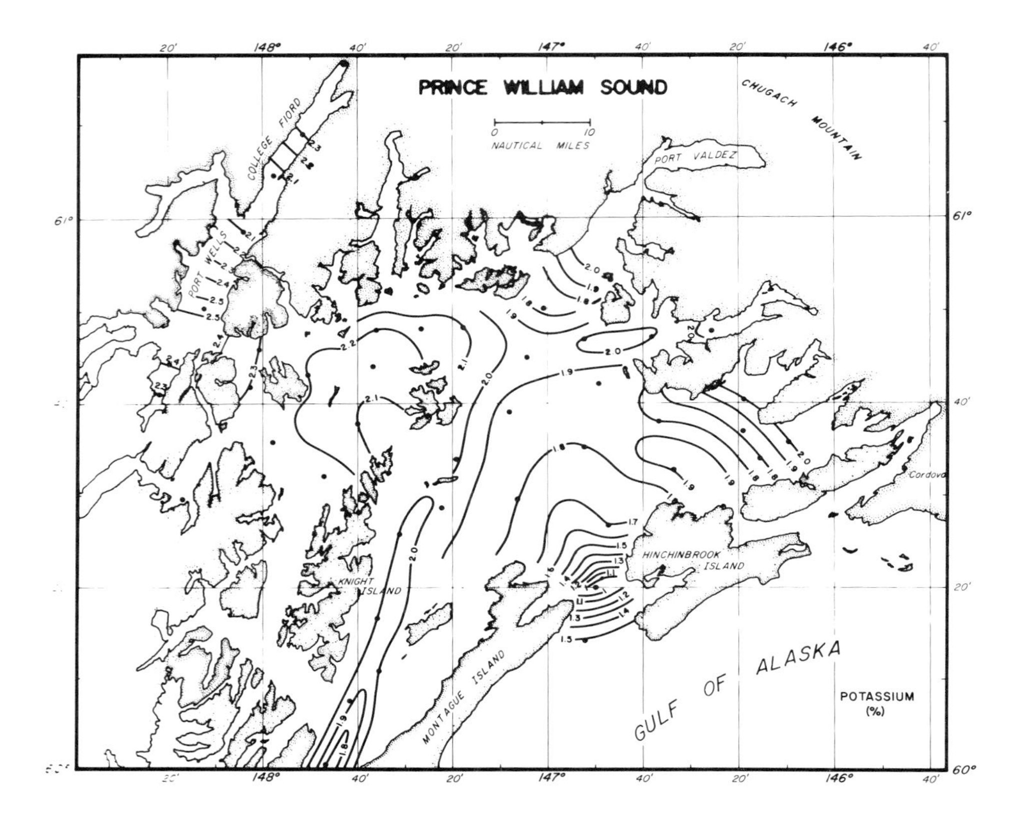

Figure 6-20. Percent potassium in sediments, Prince William Sound.

however, is more complex (Fig. 6-21). In some areas it tends to increase with the mean size, whereas in others the coarser sediments contain more sodium.

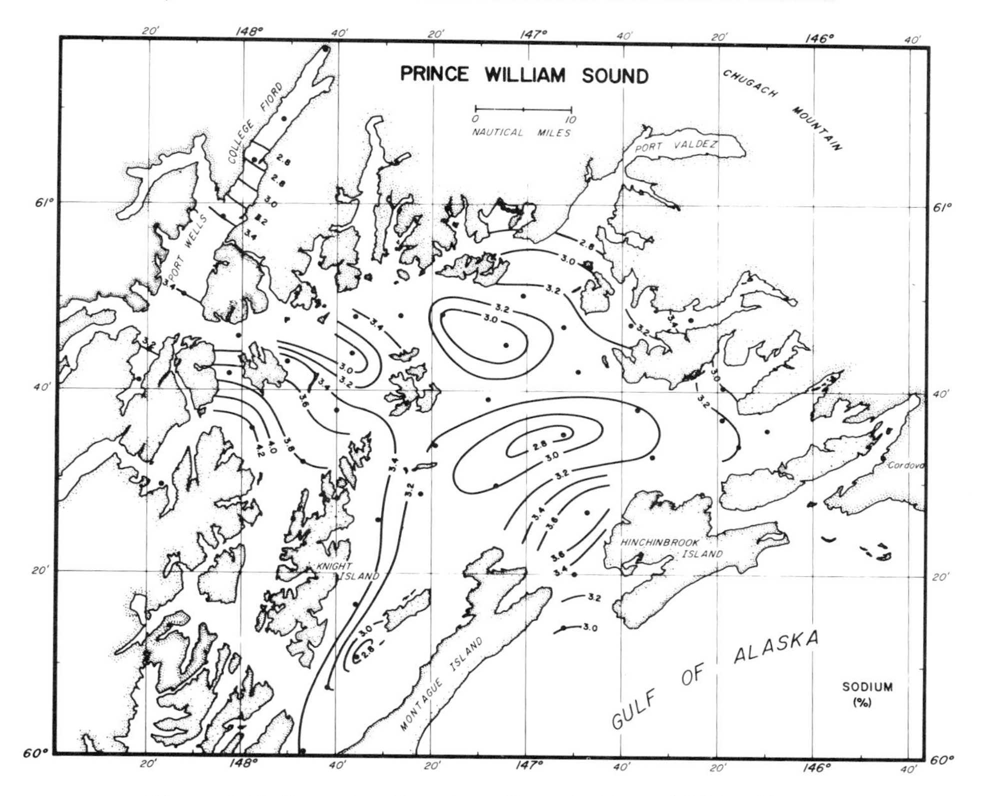

Figure 6-21. Percent sodium in sediments, Prince William Sound.

Manganese, with the exception of a high concentration in the region north of Naked Island, in general also is directly related to the clay distribution in the sound (Fig. 6-22). The relatively high values in this area, somewhat "anomalous," may be due to local mineralization in bedrock, which may have contributed unusual amounts of manganese to the sediments. Titanium varies within a narrow range and its distribution tends to be related to the sediment mean size (Fig. 6-23).

Migration and deposition of most minor elements, such as copper, cobalt, nickel, and zinc, are strongly related to clay transport and deposition. All these elements show an affinity with clay: The concentration of these elements increases with increasing percent clay in the sediments (Figs. 6-24 to 6-27). Complexity marks the distribution of chromium (Fig. 6-28). Although high concentrations of chromium in the deep basin northwest of Naked Island do coincide with the high percent clay in sediments, this direct covariance of clay and

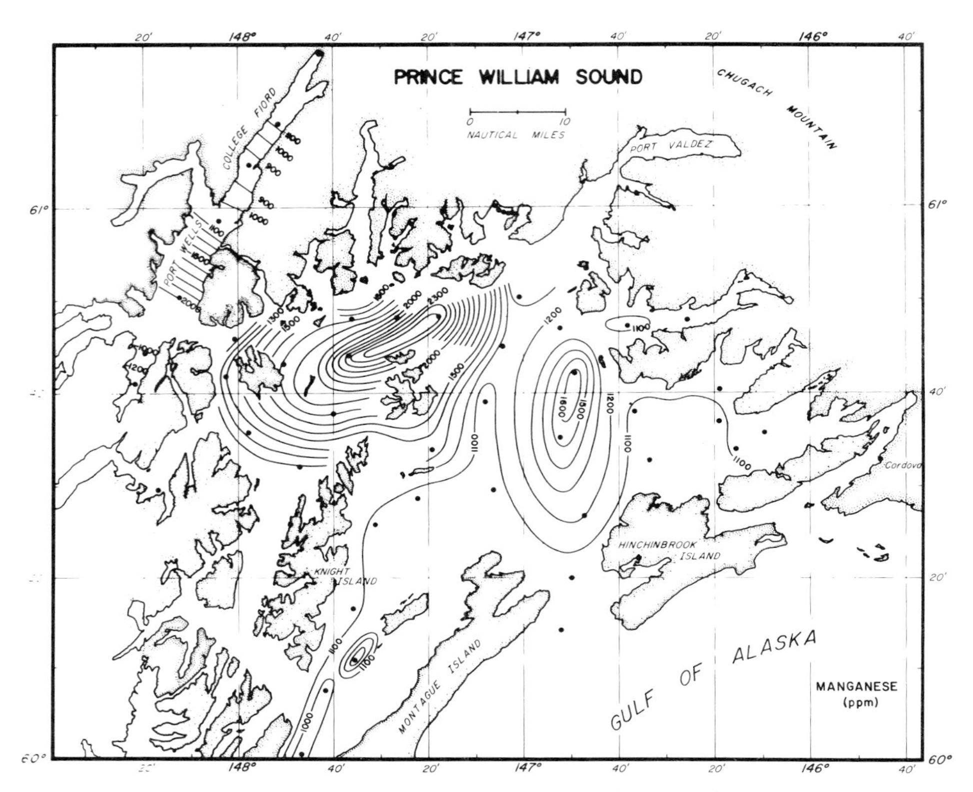

Figure 6-22. Percent manganese in sediments, Prince William Sound.

chromium contents is not observed in sediments from north and northeast of Hinchinbrook Entrance.

The distribution of barium in the sound is complex, particularly in relation to the sediment mean size (Fig. 6-29). Coarse glacial sediment deposits near the face of Columbia Glacier and near the head of College Fjord contain high amounts of barium. A high anomaly (900 ppm) is also observed near the mouth of Port Wells. The bottom sediments in this area are coarse clay and thus the observed high concentration of barium here is not compatible with the barium-sediment size relationship generally observed elsewhere throughout the sound.

The concentration of strontium in sediments varies from 225 to 400 ppm (Fig. 6-30). The concentrations are low near the heads of the inlets and are high in the vicinity of Hinchinbrook Entrance. Strontium covaries with calcium in sediments.

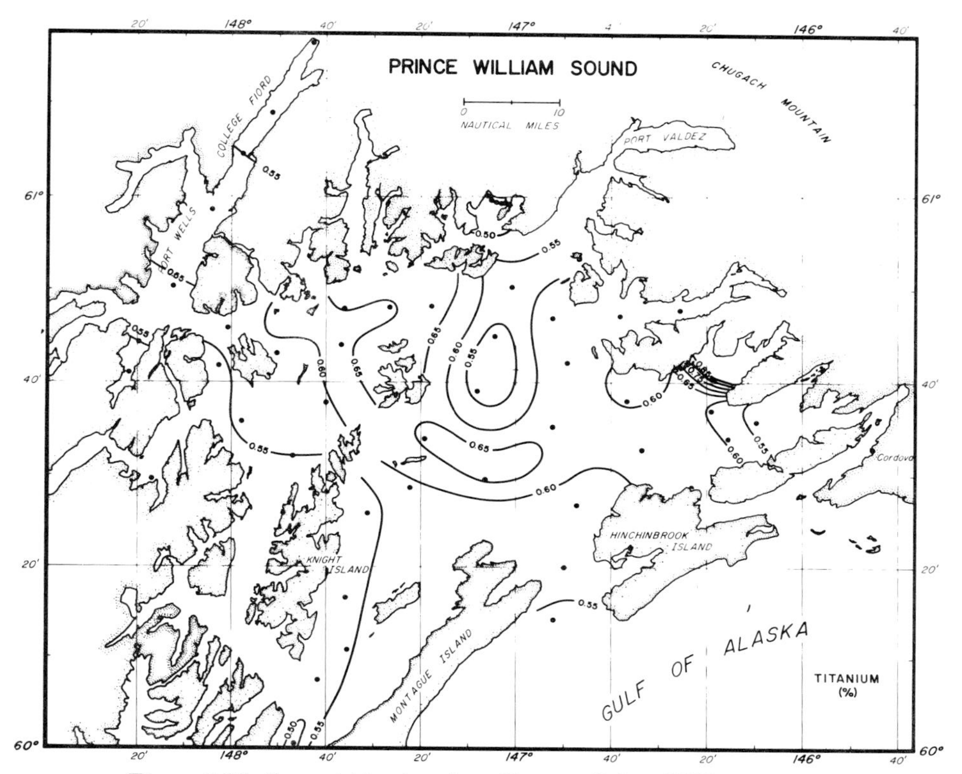

Figure 6-23. Percent titanium in sediments, Prince William Sound.

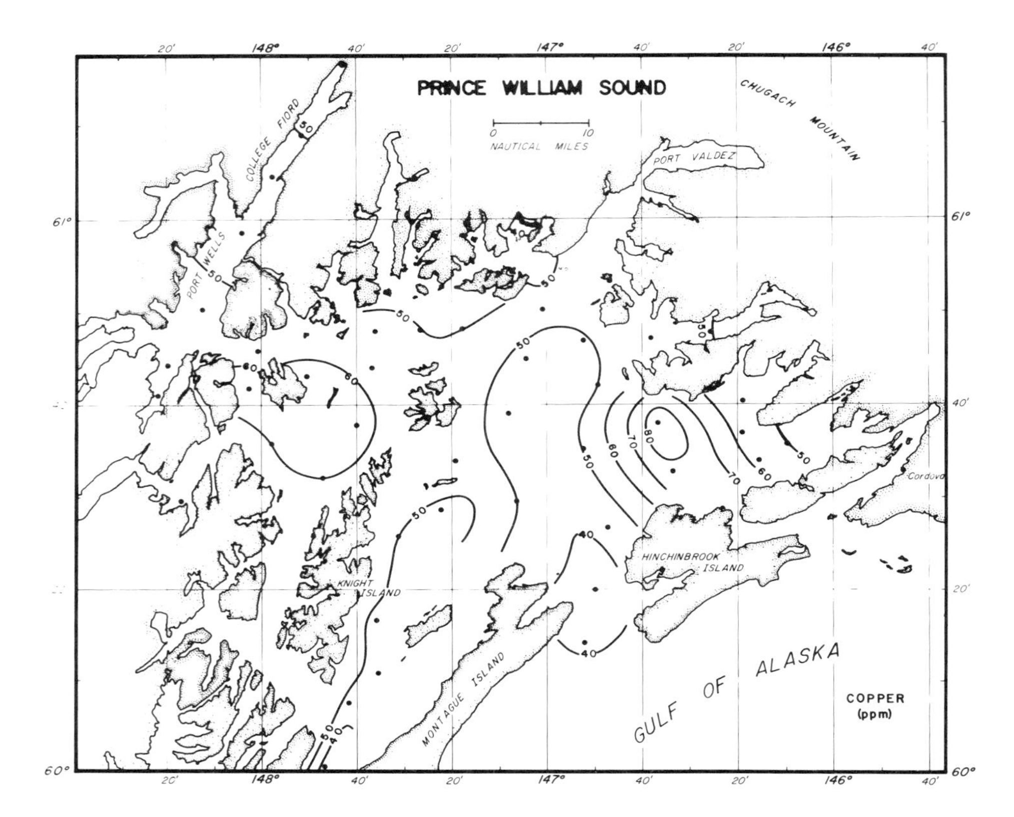

Figure 6-24. Copper (ppm) in sediments, Prince William Sound.

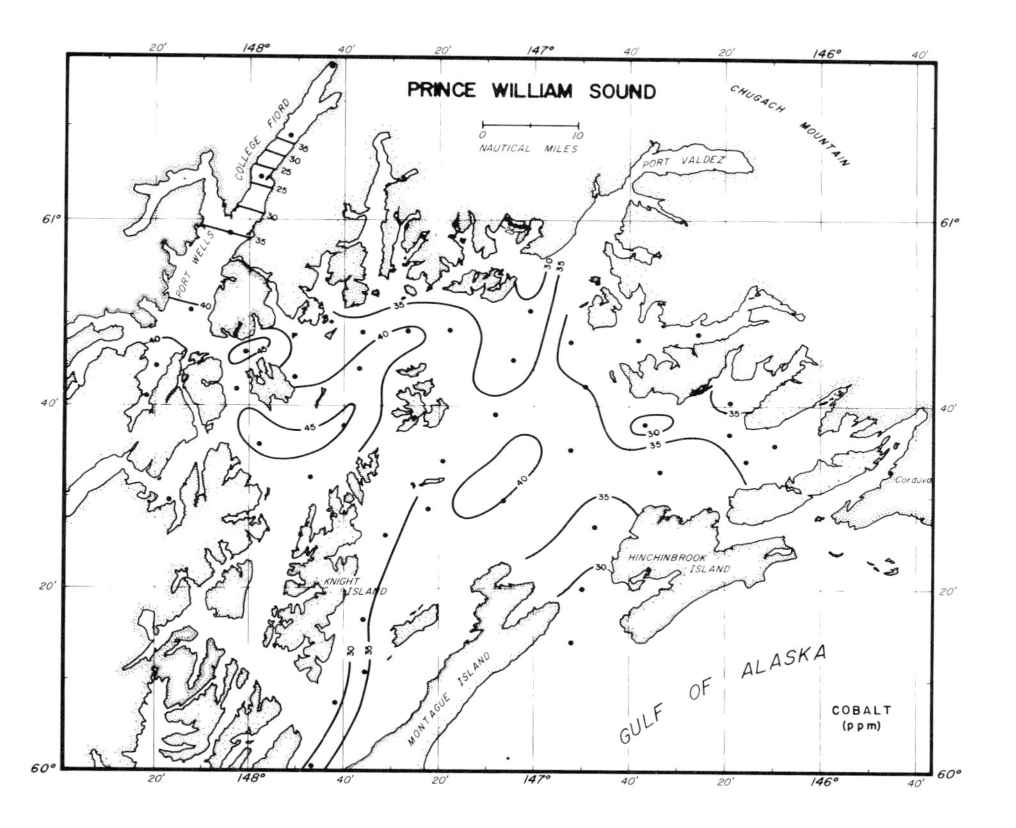

Figure 6-25. Cobalt (ppm) in sediments, Prince William Sound.

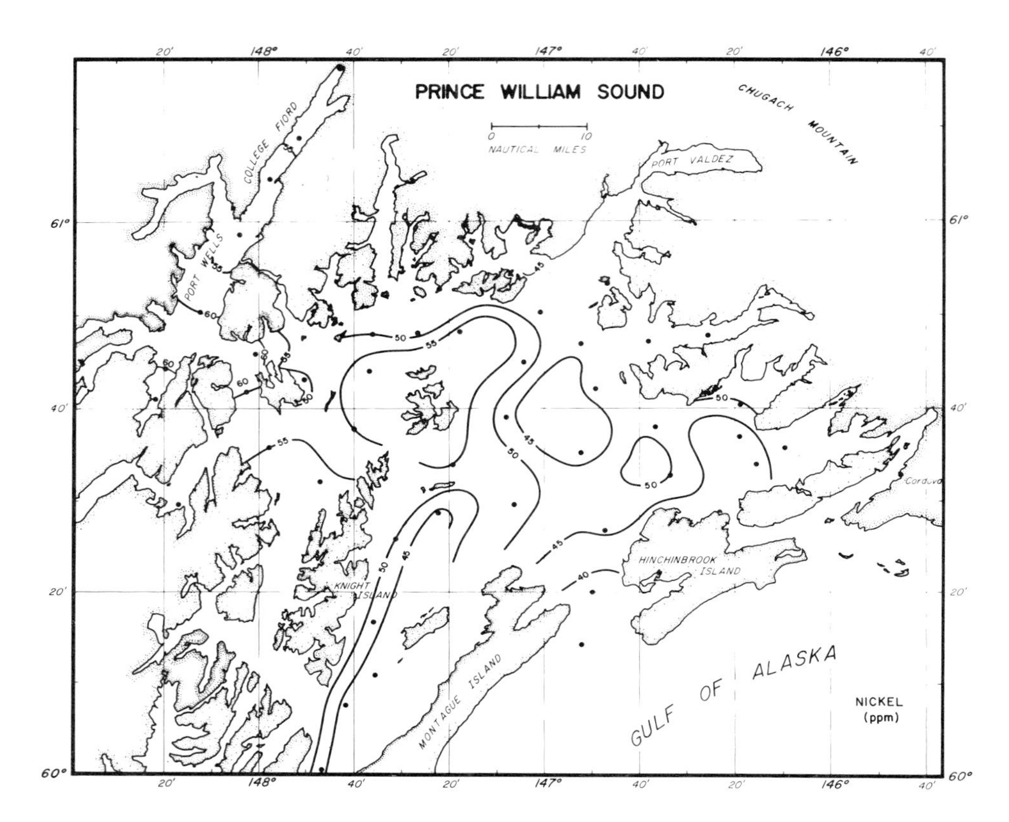

Figure 6-26. Nickel (ppm) in sediments, Prince William Sound.

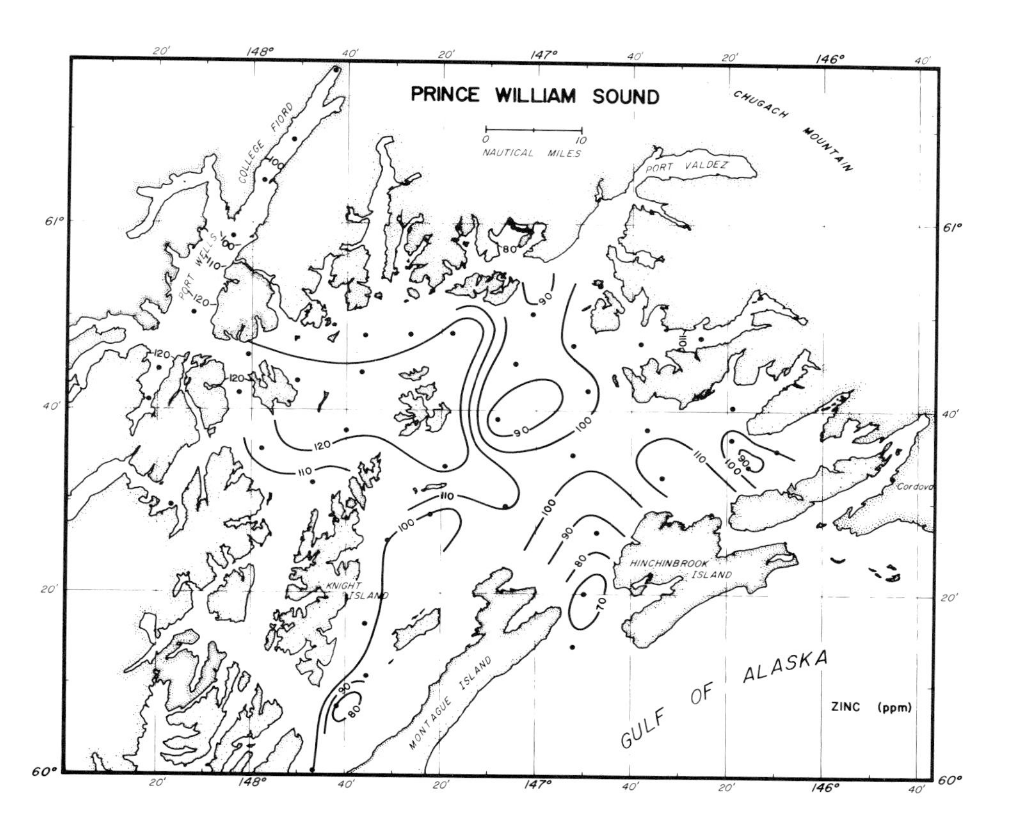

Figure 6-27. Zinc (ppm) in sediments, Prince William Sound.

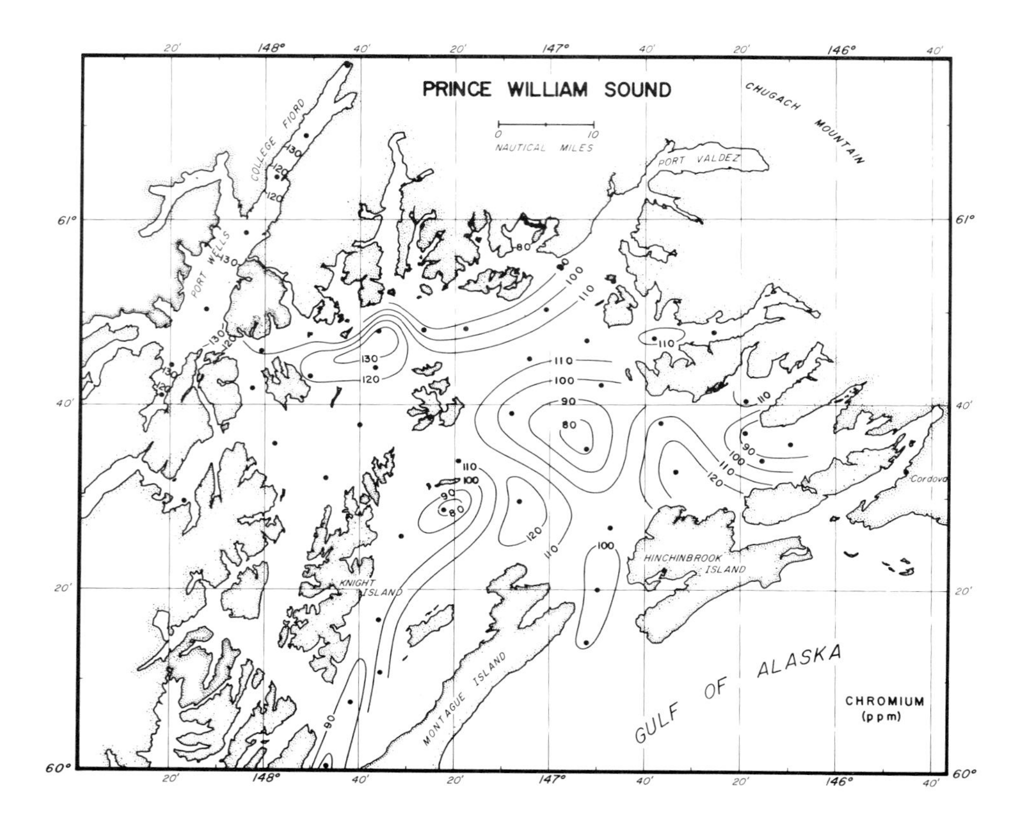

Figure 6-28. Chromium (ppm) in sediments, Prince William Sound.

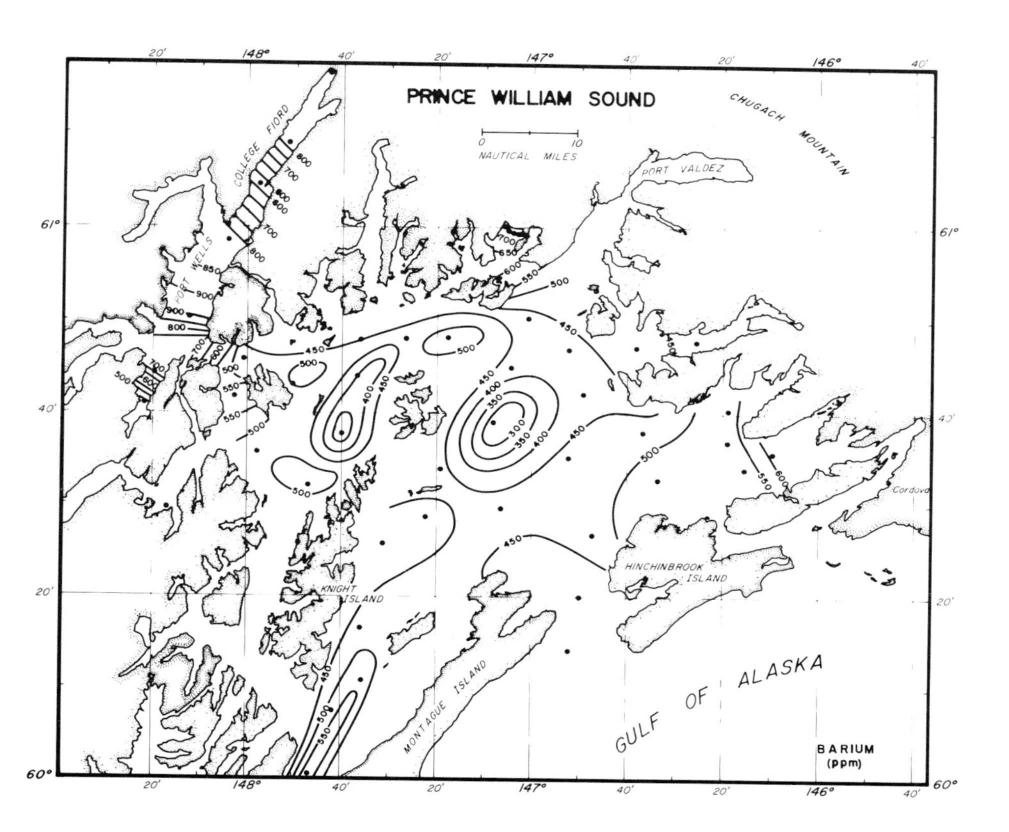

Figure 6-29. Barium (ppm) in sediments, Prince William Sound.

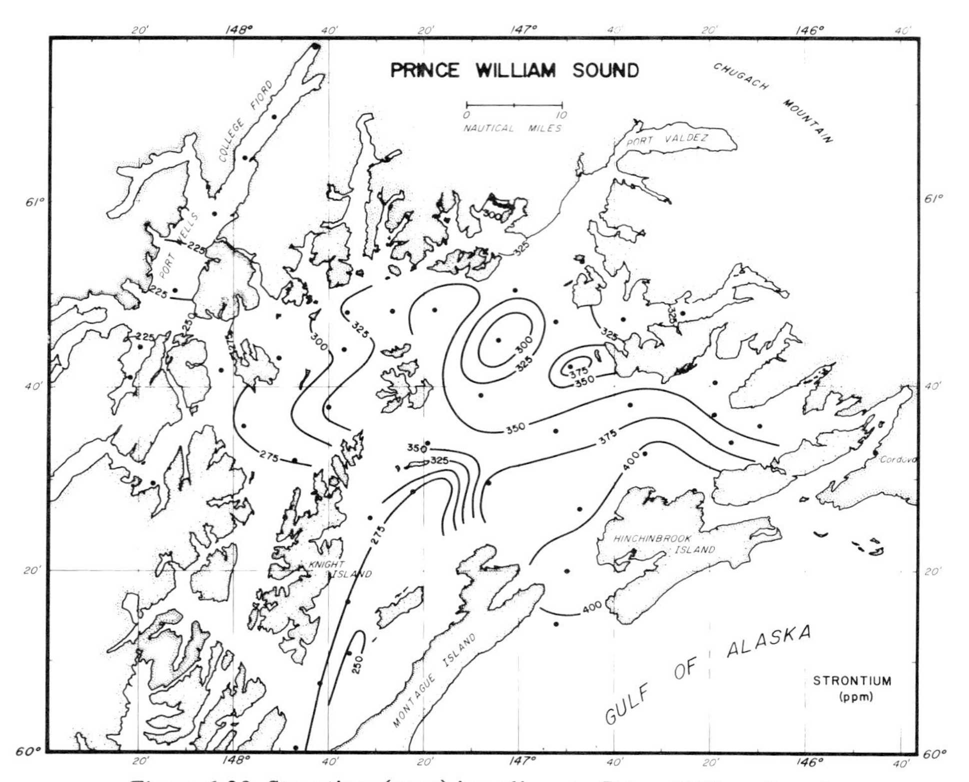

Figure 6-30. Strontium (ppm) in sediments, Prince William Sound.

SEDIMENT SOURCE AND TRANSPORT

The suspended load distributions measured in Prince William Sound during March 1972 and February 1973 (Figs. 6-31 and 6-32) were seasonally low and represented only the winter discharge. During summer, the high runoff develops more pronounced plumes and carries more sediment. The data presented here also does not cover the entire region; in particular, the evaluation of the peripheral fjords and inlets is incomplete. A detailed description of the distribution and movement of sediments in one of the fjords, Port Valdez, is presented in Chapter 7, however.

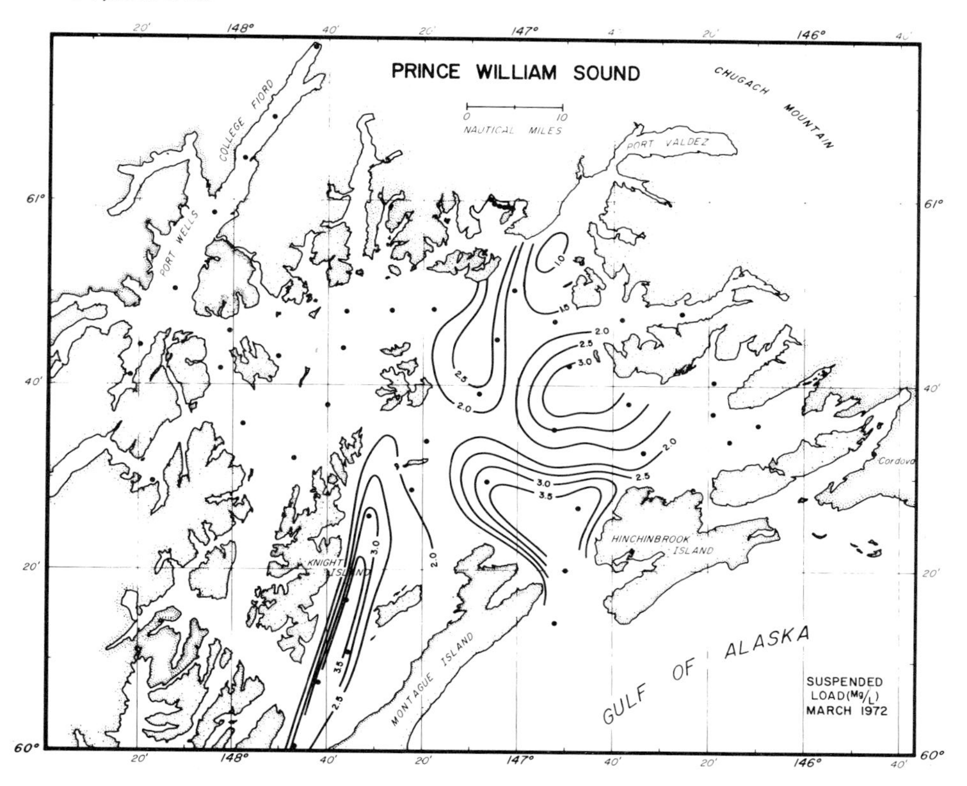

Figure 6-31. Distribution of the surface suspended sediment load during March, 1972, Prince William Sound.

During winter, the amount of detritus flowing out of the fjords and inlets and into the central basin of Prince William Sound is relatively small, compared to the significant sediment input into the sound from other areas (Figs. 6-31 and 6-32). Waters with high suspended loads enter the sound through Hinchinbrook Entrance as well as Montague Strait. The concentrations of sediment in suspension are higher outside the passages and decrease as the waters flow into the

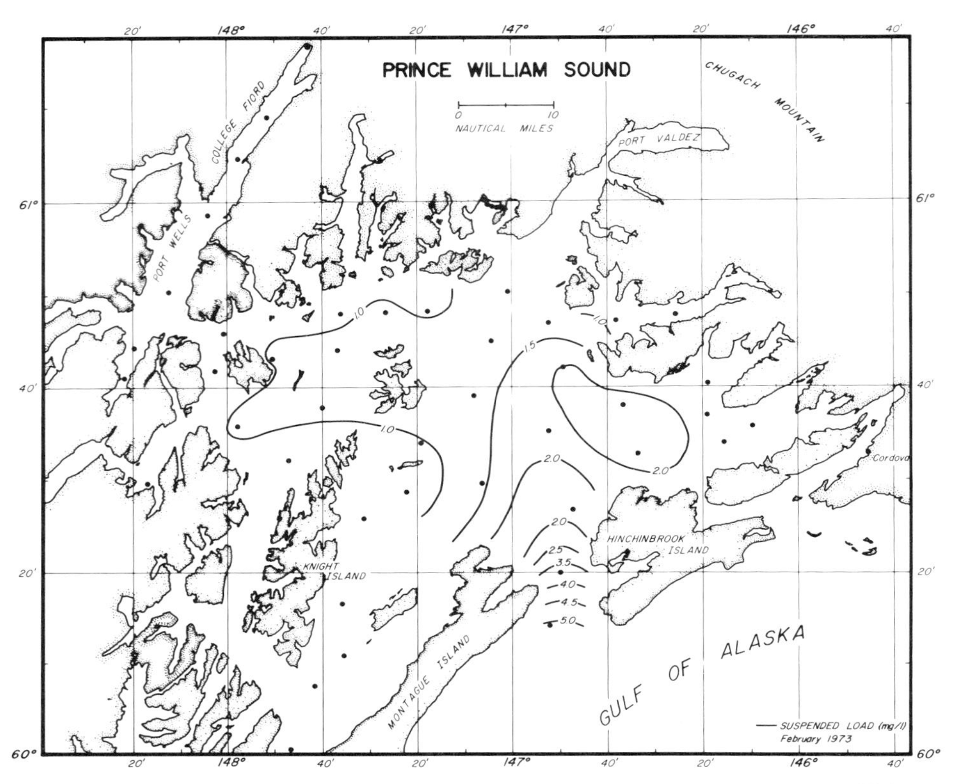

Figure 6-32. Distribution of the surface suspended sediment load during February, 1973, Prince William Sound.

sound. The shelf sediments also appear to enter from shallow passages northeast of Hinchinbrook Island. Extensive movement of water into the sound through these passages is quite evident from the salinity and temperature distributions described earlier (Figs. 6-3 to 6-6). It is interesting to note that the shelf water entering the sound has a lower salinity than the sound water. The inflowing low-salinity coastal water is formed by mixing of the Copper River discharge and the Gulf of Alaska shelf water. The Copper River's discharge not only lowers the salinity but significantly increases the suspended sediment concentration of the incoming waters. The movement of the Copper River discharge and the shelf water has been discussed in Chapter 5.

An anomalous north-south oriented water mass, with a high suspended load, originating at the terminus of Columbia Glacier, flows out of Columbia Bay and extends southward into the sound. Besides being an important source for sediments, this glacier also influences the water mass of the region. Columbia Glacier is a tidal glacier which has its terminus in Columbia Bay. It is one of the largest glaciers draining into the sound, over 65 km long, covering an area of 1,130 km^2.

The ice terminus in Columbia Bay is approximately 3.5 km wide and 60 m high. Although no accurate measurements of the ice flow have been made, estimates based on comparative movement of the surface moraines from aerial photography suggest that the glacier ice moves down the slope at a rate varying from 1 to 3.5 m/day. Rapidly moving ice causes frequent calving of large ice blocks at the glacier face. These huge icebergs float out of the bay and move southward into the sound, and most carry significant amounts of sediment. Even during the winter months, icebergs carrying sediments were frequently noticed northeast of Naked Island.

Along the sides of Columbia Glacier are several lakes, some of which drain annually. Continuous water discharge occurs on the eastern and western edges of the glacier terminus during most of the year. It is also believed that large amounts of sediments are discharged with the water flowing near the sole of the glacier.

In addition to the onboard measurements of suspended load made during winter, the relative sediment distribution in the surface waters of Prince William Sound during summer and fall was studied using ERTS-1 imagery (Fig. 6-33). Numerous satellite images were subjected to density slicing to determine the movement of sediments in suspension. The general distribution as shown by the various imageries throughout the summer is well illustrated in Figure 6-33. The image covers the entire Prince William Sound area as well as adjacent portions of the Gulf of Alaska. Because this particular image was obtained during flood tide on 2 September 1973, the movement of relatively turbid waters through the various passages into Prince William Sound is evident. Water carrying sediments enters the sound through Hinchinbrook Entrance and continues to move northwestward as far as west of Naked Island. An enlarged density slicing of the Hinchinbrook Entrance area from the same image reveals the sediment distribution in detail (Fig. 6-34). The density gradation suggests that the sediments carried through the passages are dispersed throughout the sound.

Another closeup density-sliced drawing of the northernmost region of Prince William Sound (Fig. 6-35) reveals the presence of discrete sediment plumes near the heads of various fjords and inlets. Most detritus is brought into the fjords and inlets by runoff. As this sediment-laden runoff enters the sea it forms a discrete sediment-fresh water plume. The plume retains fine sediments in suspension and carries them seaward. The inlets and fjords generally retain most of the coarse sediments and some fine sediments as well. Small quantities of very fine sediments in suspension, which cannot easily be detected on ERTS-1 imagery, nevertheless are carried into Prince William Sound. By way of example, relatively higher sediment concentration occurs at the head of College Fjord and progressively decreases toward its mouth in Port Wells; Cochran Bay also shows similar sediment distribution in surface waters. A distinct and well-defined plume originating at the terminus of Columbia Glacier extends seaward into Prince William Sound. The plume carries a significant load in suspension directly into the sound, as described earlier.

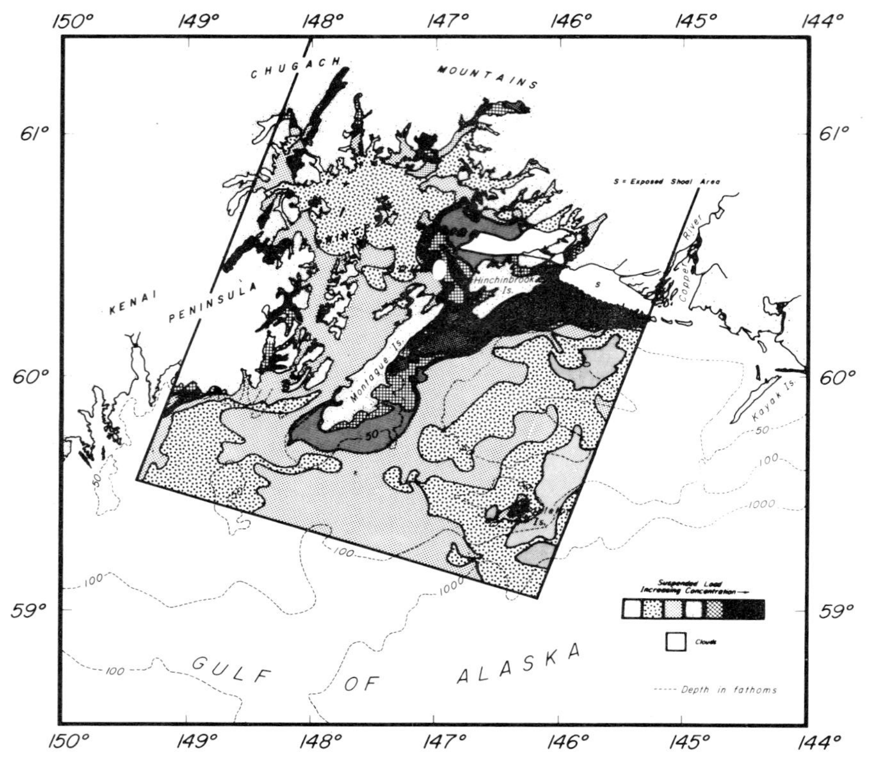

Figure 6-33. Isodensity distribution of reflectance in satellite imagery showing relative suspended loads in near-surface water of the Prince William Sound on 2 September, 1973.

Significant influx of coarse as well as fine sediments also occurs in the sound through the passages to the southwest. The fine sediments introduced in suspension and deposited in Prince William Sound therefore originate near the heads of the glaciers and fjords.

The bottom sediment grain size distribution in the sound is closely related to bathymetry: The near-shore sediments are generally coarser than those deposited in deeper water. Sediments resulting from coastal erosion are coarse and are generally in the range of sand and gravel. These sediments are deposited close to their source and are continually abraded by tides and waves, which comminute them into fine sands which are distributed in the near-shore area and are carried along the shore by longshore and tidal currents. Most shore areas in Prince William Sound are rocky, with gravel, interrupted by occasional small sandy tidal flats.

The sand fraction in bottom sediments is observed only in the regions north of Hinchinbrook Entrance and south of Columbia Bay. Hinchinbrook Entrance is one of the main arteries through which large amounts of water pass during each

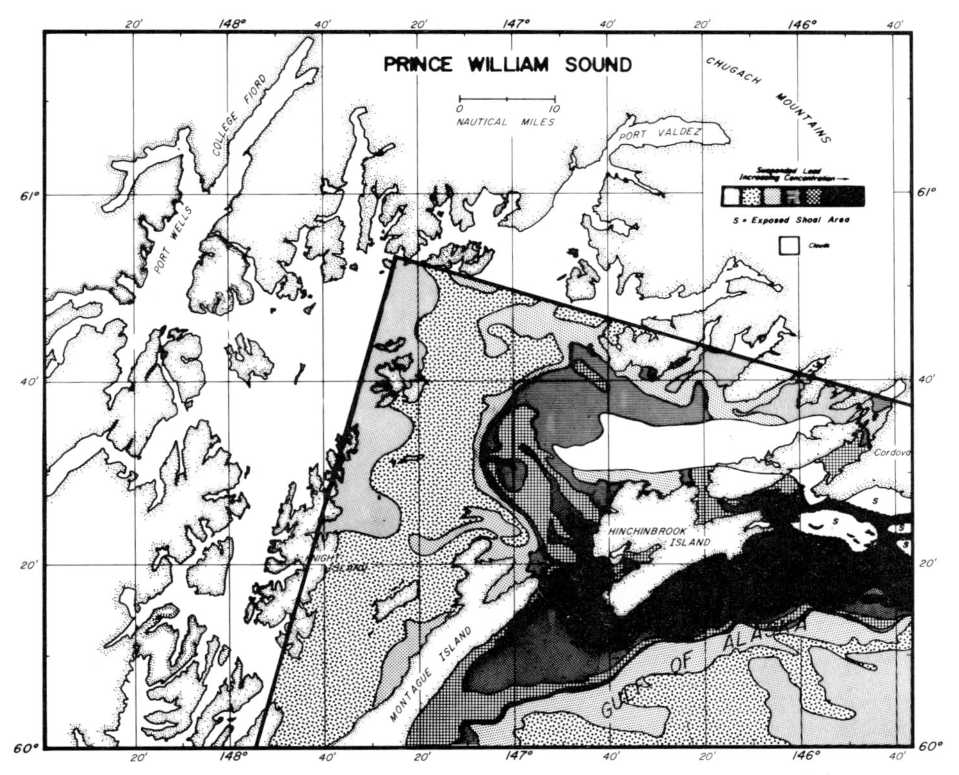

Figure 6-34. Isodensity distribution of reflectance in satellite imagery showing relative suspended load in near-surface water of Hinchinbrook Entrance on 2 September, 1973.

tidal cycle. This narrow and shallow channel has a restricted flow and causes damming of water with increasing tidal flow. The water flow through the channel is maximum during rising and falling tides and generates significant turbulence and swift currents. The water current (on the order of 50 cm/s in the channel) is strong enough to carry particles as large as pebbles. As the tidal water moves north of the entrance it gradually loses its energy and competency to transport sediments. The coarser particles settle out near the entrance whereas finer ones are carried farther into the basin. The northward progressive loss of energy in the water is further reflected in bottom sediment textural distribution (Figs. 6-8 and 6-14) and the suspended load distribution (Figs. 6-31 and 6-32).

The tongue of sandy sediments extending southward from Columbia Bay represents a combination of ice-rafted and suspension-derived materials. Normally the icebergs calving at the terminus of Columbia Glacier are not large and so generally dissipate and discharge their load before reaching the central portion of the

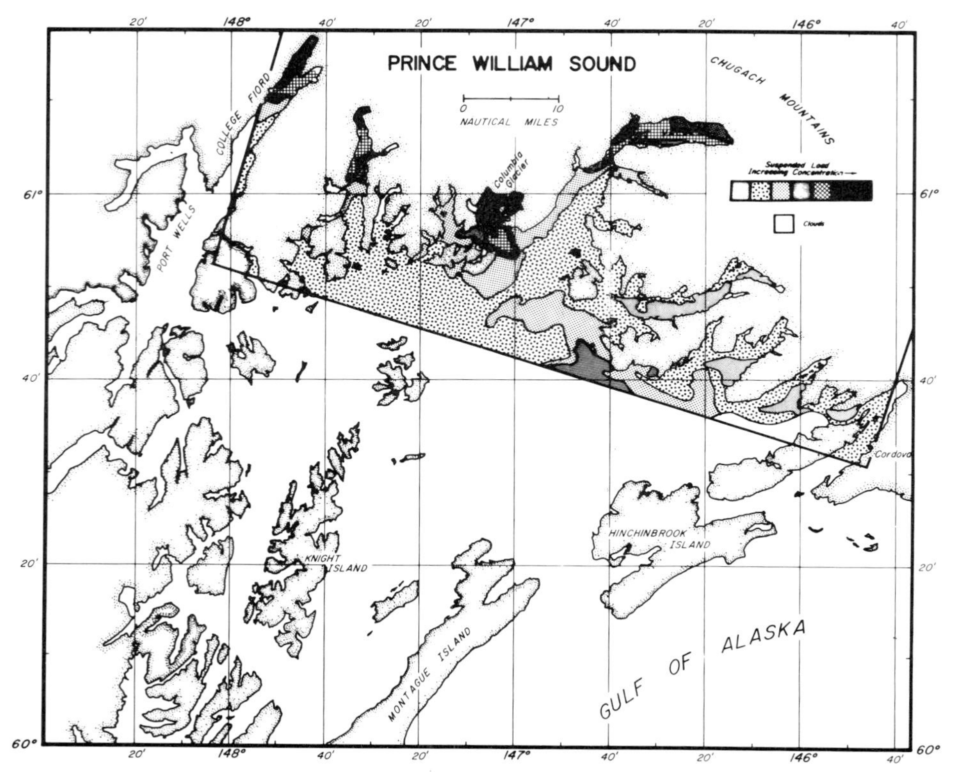

Figure 6-35. Isodensity distribution of reflectance in satellite imagery showing relative suspended load in near-surface water of northern Prince William Sound on 2 September, 1973.

sound. Exceptionally large icebergs, however, carry sediments as far south as Naked Island. The drifting and melting of icebergs in this region also affects the water characteristics, and the effect is clearly illustrated by salinity-temperature distributions.

Sediments deposited in central Prince William Sound are mostly silt and clay, and largely carried in suspension. These sediments include the very fine sediments (clay-size) which bypass the fjords and inlets and are carried into the sound in suspension, as well as silt and sand, which are carried by tidal currents and icebergs. Coastal sediments are coarse and are deposited near their source.

The Prince William Sound region lies on one of the more active tectonic belts of the earth. Numerous mild, intermediate, and severe earthquakes occur in this region. Depending upon intensity and location, these earth tremors may trigger mass sediment transport. Slumping of sediments along steeply inclined shore bedrocks may result in local turbidities. Evidence for subsurface slumps northeast of Hinchinbrook Island has been observed on seismic reflections (von Huene *et al.*, 1967).

Sediment movement in Prince William Sound is primarily associated with the coastal water masses and their movement. Streams carrying detritus from the north and east flow into various fjords and inlets, and the brackish water carries sediments offshore as surface plumes. The distribution of plumes in semi-isolated basins is controlled by the basin configuration and tidal flux. The tidal oscillation in the passages and over the crest of sills results in resuspension of bottom sediments and removal of fine sediments. The sound also receives the Gulf of Alaska Shelf sediments which are roiled by frequent storms and carried into the sound through Hinchinbrook Entrance and Montague Strait. The tide also diverts surface plumes originating from the Copper River into Prince William Sound.

The central deep basin east and west of Naked Island reflects a relatively quiescent environment. This basin is located far from major detritus sources and river discharge, and therefore receives only fine sediments. Sedimentation in the basin is primarily the result of differential particle size settling of sediment as suggested by observations of sediment texture and chlorite-illite distribution (Fig. 6-36). The chlorite particles are finer than the illite grains in this region, and therefore are carried farther in suspension. In Columbia Bay, the chlorite-illite ratio in sediments increases southwards; a similar seaward increase is seen near the head of the College Fjord. On and near the sills, because of increased turbulence, the ratio decreases. In the central sound, the low ratio is the result of deposition of undifferentiated ice-rafted sediment; the ratio increases in the surrounding areas. The extremely low values in the passages are due to high turbulence, which retains fines in suspension.

The grading of sediment mean size in various fjords, inlets, and passages suggests that the sources for sediments in Prince William Sound lie both in the rugged hinterland and in the Gulf of Alaska. The hinterland adjacent to the sound is divided geomorphologically into many drainage areas, and sediment derived from each of these areas is brought by surface flows into the fjords, to slowly settle out as it is carried farther seaward. The dominance of the sediment source at the heads of the fjords is well illustrated by the distribution of the various "maturity factors" of the sediments shown in Figures 6-37 to 6-42. The low-maturity detritus with minimal weathering is observed near the head of College Fjord, near the terminus of Columbia Glacier, and in the sediments from the southern passages. These sediments contain over 58% silica and between 13.5 and 16% Al_2O_3. As the sediments move farther from source the silica content decreases, whereas alumina increases as seen in the Al_2O_3/SiO_2-percent SiO_2 diagram (Fig. 6-37). A similar pattern is observed in the triangular diagram with end members as silica and titania, alumina, and sum of other cations (Rest). The sediment samples closer to the source area fall in the lower right-hand corner of the diagram, and as the distance from the source increases, there is a gradual loss of silica and a concurrent increase of alumina and other cations (Fig. 6-38).

Sources and migratory path of marine sediments can best be evaluated from the distribution of the elemental ratios of such sediments. For example, the sediments deposited near the source have an Al_2O_3/TiO_2 ratio similar to that of the

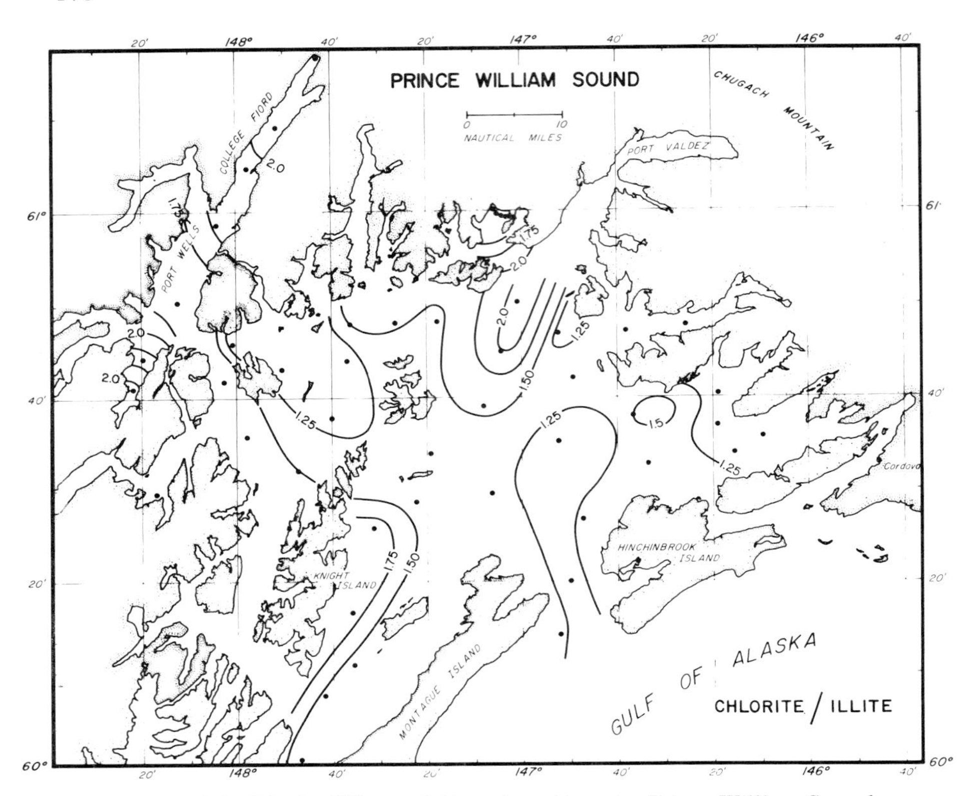

Figure 6-36. Chlorite/illite variations in sediments, Prince William Sound.

provenance (Fig. 6-39). As these sediments become subject to marine transport processes, a separation of titanium from aluminum occurs: The Al_2O_3/TiO_2 ratio increases with increased transport from the source. The distribution, however, is also affected by hydrodynamic effects, which control the distribution and accumulation of heavy minerals in sediments. In particular, the accumulation of heavy minerals in littoral sand will result in a parallel significant concentration of titanium.

The Al_2O_3/TiO_2 ratio distribution in Prince William Sound indicates various dominant sources. The main source for sediments in the sound appears to be through Hinchinbrook Entrance and the other passages to the northeast. The sandy sediment in the vicinity of the entrance has a low value because of the hydraulic action of the tidal currents. The tidal flushing removes lighter minerals from this areas and tends to enrich the lag deposit with heavy titani-ferrous minerals. Northward, however, the aluminum-titanium ratio increases in response to normal marine transport processes.

The distribution of alumina-titania ratios also indicates sources of Prince William Sound sediments in Port Gravina, Long Bay, Unakwik Inlet, College

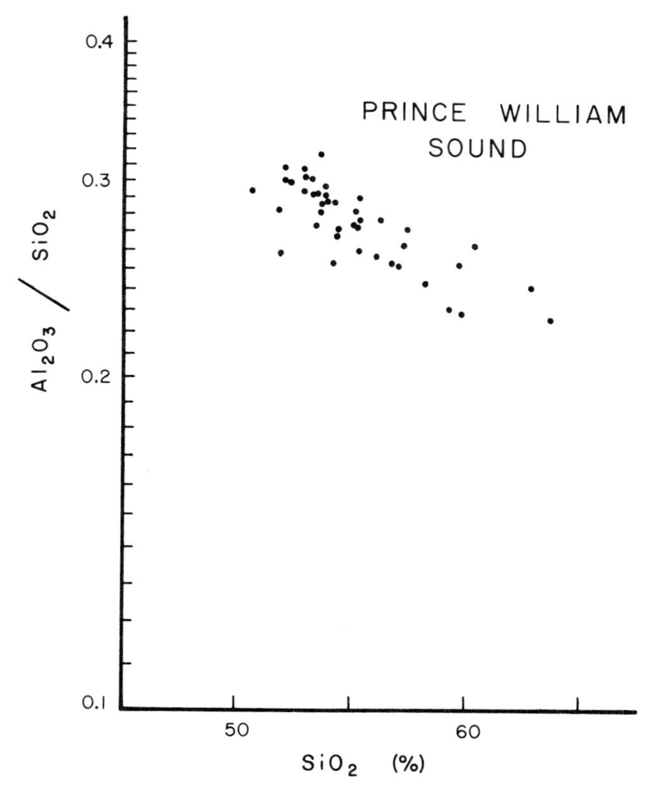

Figure 6-37. Al_2O_3/SiO_2 versus percent SiO_2 plot, Prince William Sound.

Fjord, Port Wells, and other fjords. The Al_2O_3/TiO_2 values increase as sediments from the source, mostly near the head of the inlets, migrate seaward into the basin. The distribution in the sound also indicates that the sediment migration generally follows the water circulation. During flood tide, the water enters the sound through Hinchinbrook Entrance and other passages in the northeast. The incoming water carries part of the Copper River plume into the sound. This water flows in a north and northeastward direction and mixes with outflowing surface waters from Port Valdez, Columbia Bay, and other inlets. During flood tide the sediments originating in the inlets and fjords west of Long Bay are mostly diverted to the basin west of Naked Island, and as the tide falls they continue south toward Montague Strait. Thus a counterclockwise gyre is formed in the center of the sound.

The decreasing variation of Al_2O_3/TiO_2 values in Columbia Bay, in spite of a predominant source at the head, is due to significant ice rafting of sediments in this area. Ice-rafted sediments are not subjected to Al and Ti separation, which is more typical of normal marine transport.

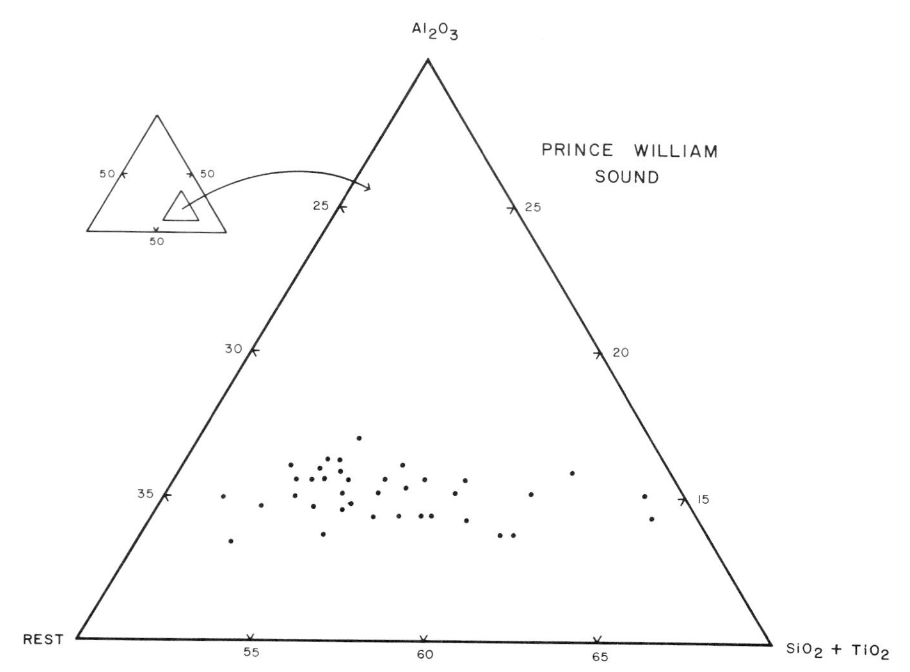

Figure 6-38. Triangular plot of percents Al_2O_3, $SiO_2 + TiO_2$, and Rest as end members, Prince William Sound.

Sediment source and sediment migration in the sound are further delineated by other elemental ratio distributions. CaO/MgO ratios for sediments near the northern periphery and the passages are high and gradually decrease toward the central basin (Fig. 6-40). CaO/MgO, K_2O/Na_2O, and Al_2O_3/Na_2O ratios decrease with increased distance of sediment transport from the source (Figs. 6-40 and 6-42). The distribution patterns of K_2O/Na_2O and Al_2O_3/Na_2O further confirm sediment sources indicated by the observed Al_2O_3/TiO_2 distribution. Large-scale movement of sediment into the Prince William Sound through Hinchinbrook Entrance since the retreat of glaciers has been inferred by von Huene *et al.* (1967). These sediments appear to have filled Orca Bay and have been spilled outside the bay to form shoals north of Hinchinbrook Island.

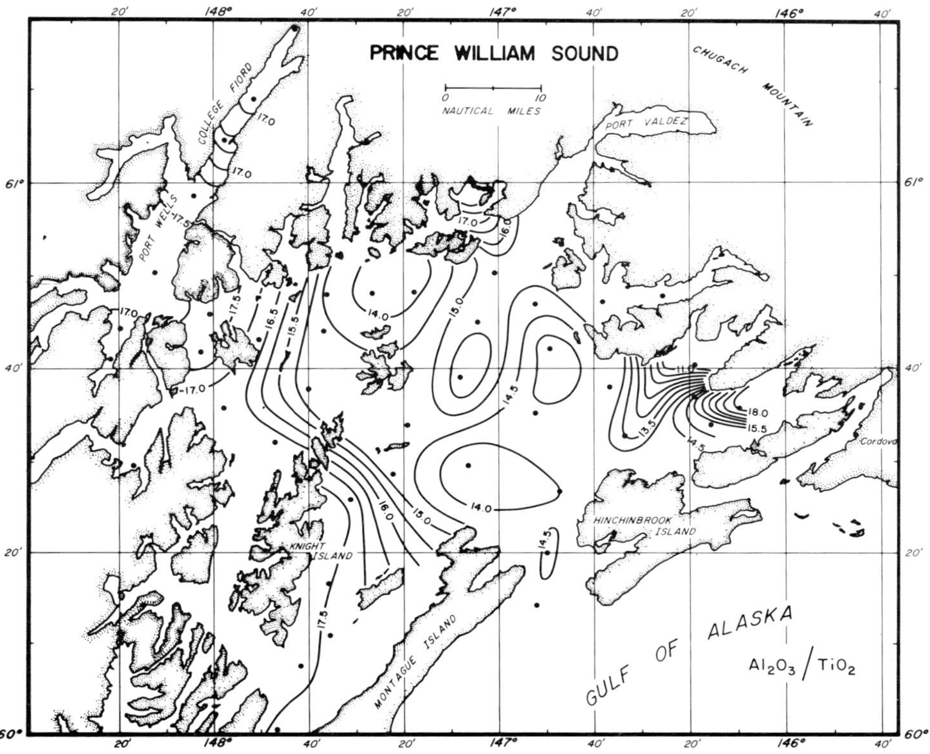

Figure 6-39. Al_2O_3/TiO_2 variations in sediments, Prince William Sound.

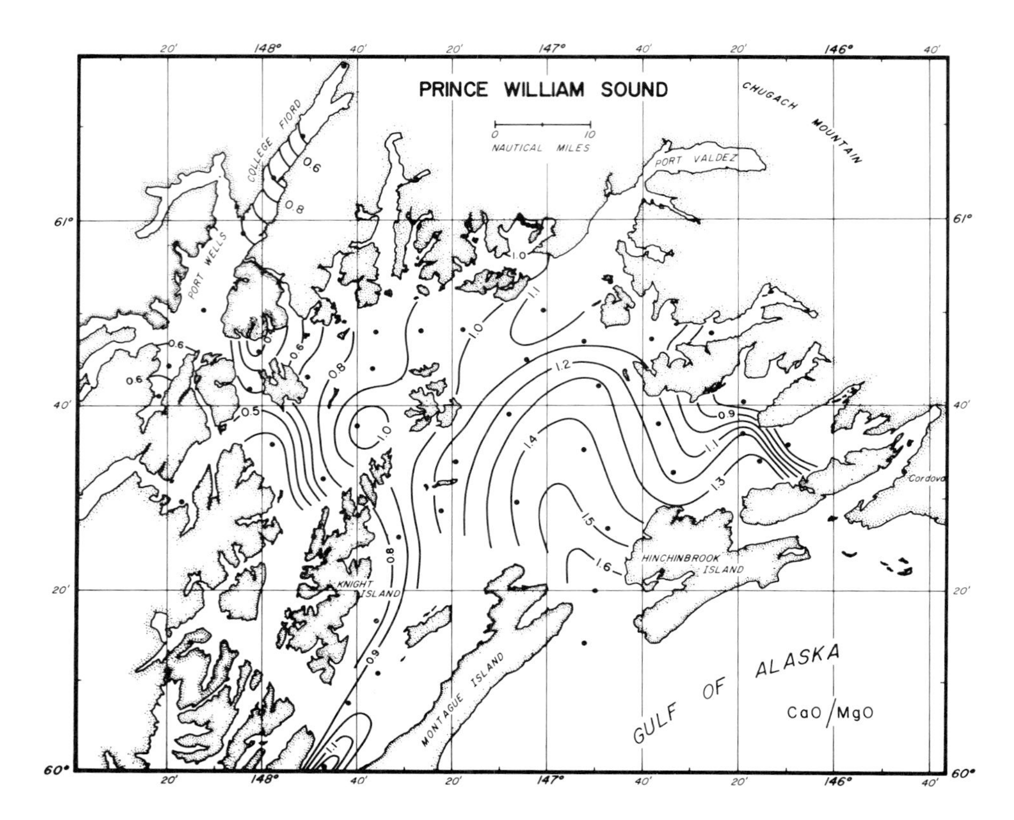

Figure 6-40. CaO/MgO variations in sediments, Prince William Sound.

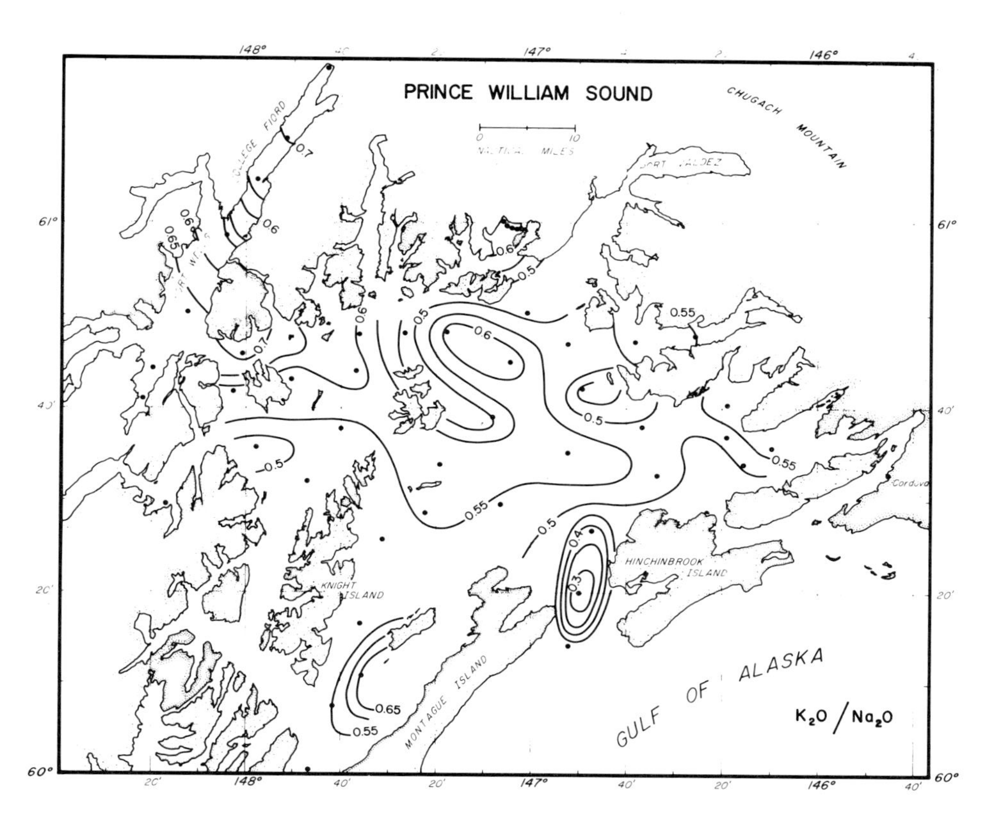

Figure 6-41. K_2O/Na_2O variations in sediments, Prince William Sound.

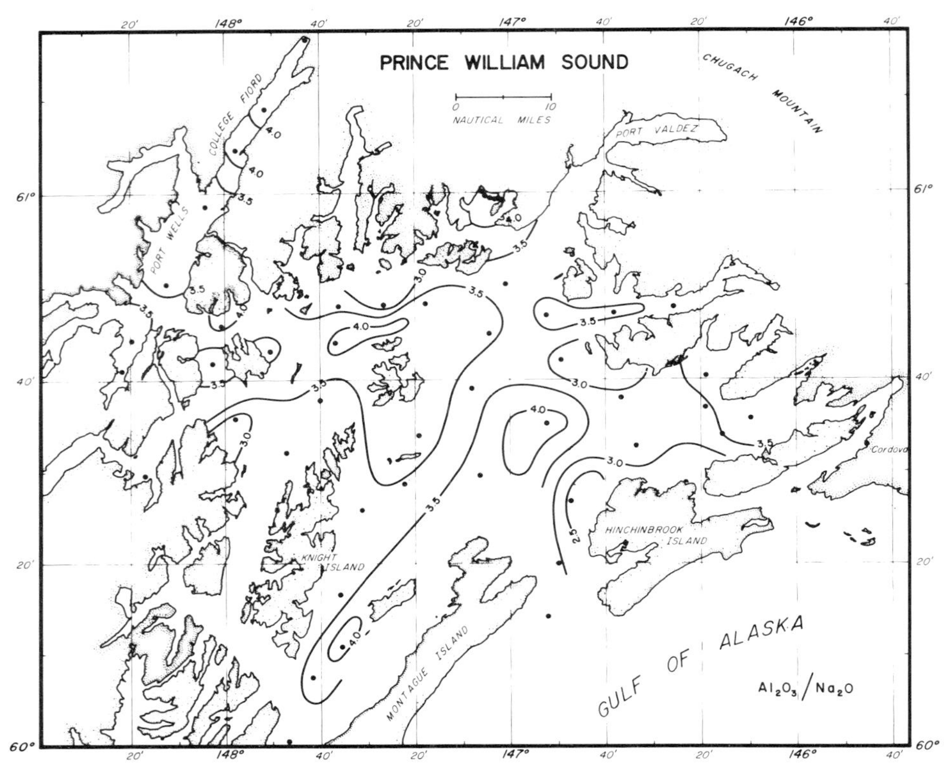

Figure 6-42. Al_2O_3/Na_2O variations in sediments, Prince William Sound.

REFERENCES

Case, J. E., D. F. Barnes, G. Plafker, and S. L. Robbins. (1966). The Alaska Earthquake, March 27, 1964, Regional Effects–Survey of Epicenter Region. U.S. Geol. Survey Prof. Paper 543-c, pp. 1-12.

Grant, U. S., and D. F. Higgins. (1910). Reconnaissance of the geology and mineral resource of Prince William Sound, Alaska. *U.S.G.S. Bull.* 443:89.

Moffit, F. H. (1954). Geology of the Prince William Sound region, Alaska. *U.S.G.S. Bull.* 989-E:225-310.

Plafker, G. (1969). The Alaska Earthquake, March 27, 1964, Regional Effects–Tectonics. U.S. Geol. Survey Prof. Paper 543-1, pp. 1-74.

Sharma, G. D., and D. C. Burbank. (1973). Geological oceanography. In *Environmental Studies of Port Valdez,* D. W. Hood, W. E. Shiels, and E. J. Kelley (Eds.), Inst. of Mar. Sci. Occ. Publ. No. 3 Univ. Alaska, Fairbanks, pp. 15-99.

von Huene, R., G. G. Shor, Jr., and E. Reimnitz. (1967). Geological interpretation of seismic profiles in Prince William Sound, Alaska. *Geol. Soc. Amer. Bull.* 78:259-268.

CHAPTER 7
Port Valdez

INTRODUCTION

Port Valdez is a relatively deep, narrow, glaciated reentrant in the northeastern part of Prince William Sound (Fig. 7-1). This east-west trending typical U-shaped fjord is approximately 21 km long and 4.5 km wide, its shores are bordered by steep walls which extend beneath the water. Seaward the depth increases rapidly to form a high-angle slope, but farther offshore the fjord is a flat-bottomed trough with a maximum depth of about 240 m. At its mouth, the Valdez Narrows, a nearly right-angle structural constriction (about 1.5 km wide) with two sills, connects the fjord with Port Valdez Arm and Prince William Sound.

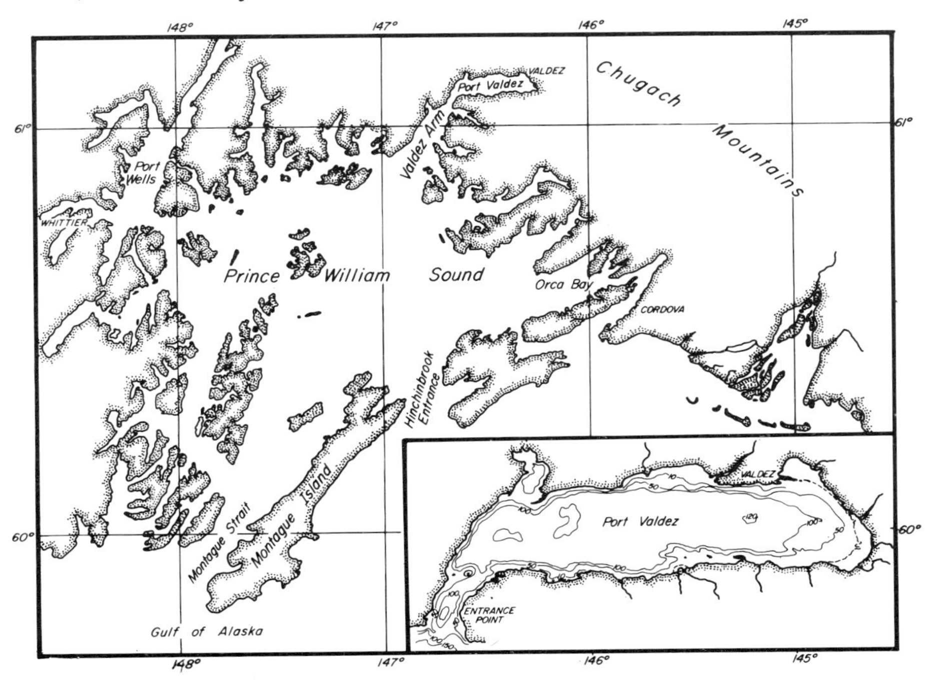

Figure 7-1. Index map showing Prince William Sound and Port Valdez. Depth in fathoms.

The Lowe and Robe Rivers and the Valdez Glacier Stream discharge their sediment loads into Port Valdez near its head and have deposited an alluvial fan. The outwash delta and the fan consist of poorly consolidated silt, sand, and gravel. A relatively large tidal range (5.5 m) has formed tidal flats adjacent to large sediment sources.

GEOLOGY

The surficial geology of the Valdez area has been described by Moffit (1954). The bedrock consists of bluish-gray phyllitic graywacke argillite, bluish-gray or black slate, and schist of the Valdez Group. These are metamorphosed muds and feldspathic sands deposited during the Late Mesozoic era. The phyllitic graywacke with quartz, orthoclase and plagioclase feldspar, muscovite and chlorite is the dominant unit and is widely exposed in this area.

Closely folded and metamorphosed beds display a remarkably uniform composition and structure. The strike of bedding in Port Valdez is east-west with an average dip of about 55° to the north, and trough appears to lie on a monocline. Rock exposures throughout the region exhibit a complex network of fractures; the most ubiquitous are oriented generally perpendicular to the strike. Extensively fractured rocks commonly are traversed by a complicated network of quartz veins.

The early geologic evolution of the region has been described in earlier chapters. The significant feature in the area is the absence of Tertiary beds in the vicinity of Port Valdez, probably because the region was uplifted during Late Cretaceous and Early Tertiary orogeny and may have remained an area of Tertiary nondeposition. Alternately the Early Tertiary sediments may have been eroded during and after the Miocene Uplift. During the Pleistocene, glaciers descended from the higher altitudes of the Chugach Mountains into the Valdez area and thence west and southwestward into Prince William Sound. The thickness of ice flow fluctuated in Port Valdez, and evidence of ice scour has been found as high as 975 m above the present sea level (Tarr and Martin, 1914). Shrinking glaciers filled Port Valdez with morainal material that blanketed the bedrock topography. Three distinct rock units–lower, middle and upper–have been detected from the seismic records (von Huene *et al.,* 1967). The highly reflective lower unit represents the slate and graywacke basement rock, which can be traced to its shore exposures. The thick unit of unconsolidated glacial drift and glaciomarine sediments represents the middle unit. The thin, evenly bedded upper unit represents sediments deposited since glaciation and marine transgression stages. The total thickness of unconsolidated sediments is about 400 m.

BATHYMETRY

The bathymetry of Port Valdez is typical of a flat-bottomed, U-shaped fjord: It is bordered by steep, rocky shores as well as by deltaic deposits with tidal flats in the east (Fig. 7-2). An east-west trending submarine valley of modest depth traverses the delta at the head. It originates in the vicinity of Old Valdez town and extends westward about 3 to 4 km to the base of the delta and terminates near two submarine fans (Fan I and Fan II). The submarine valley and various submarine fans have been named and described in detail by Sharma and Burbank (1973). With the exception of a few topographic features, Port Valdez is remarkably flat, varying between 230 and 250 m in depth. Vertical as well as horizontal constriction, characterizes the entrance of the inlet. The narrow constriction has two sills, the outer sill being the shallower (110 to 128 m).

HYDROLOGY

The mean temperature in Port Valdez ranges between 7 and 13°C during summer months, with prevailing winds from the southwest, and between 4 and 10°C during winter months, with winds from the northwest. The mean annual precipitation recorded in Valdez is 1,580 mm. The driest month is June, with an average precipitation of about 5 cm, whereas the maximum precipitation occurs during September and is in excess of 230 mm. Part of the annual precipitation falls as snow in the adjacent high terrain of the Chugach Mountains. The mean annual snowfall is about 6,000 mm. The snow-covered mountains feed glaciers which descend to within 8-16 km of Valdez.

Major sources for fresh water in Port Valdez include the Lowe River and Valdez Glacier Stream at the head of the fjord and Mineral Creek and Shoup Glacier Stream to the north. The drainage pattern of Lowe River and Mineral Creek is bimodal, with peak flows of 150 to 160 m^3/sec, and 100 m^3/sec, respectively, occurring in July and September. Valdez Glacier and Shoup Glacier streams have unimodal discharges, with their peak flows occurring during September. The peak flow of Shoup Glacier Stream is about 75 m^3/sec. Many smaller streams enter the fjords along the north and south shores, but their discharge is not significant in comparison to those described above.

HYDROGRAPHY

Preliminary studies of water characteristics and flushing in Port Valdez have been conducted by Muench and Nebert (1973) and Sharma and Burbank (1973). Water characteristics in the fjord are controlled primarily by freshwater input, tides, surface temperature, and water exchange from Prince William Sound. During the summer months, when runoff is high, the waters are stratified. The freshwater input and subsequent mixing with saline water near the head in the upper

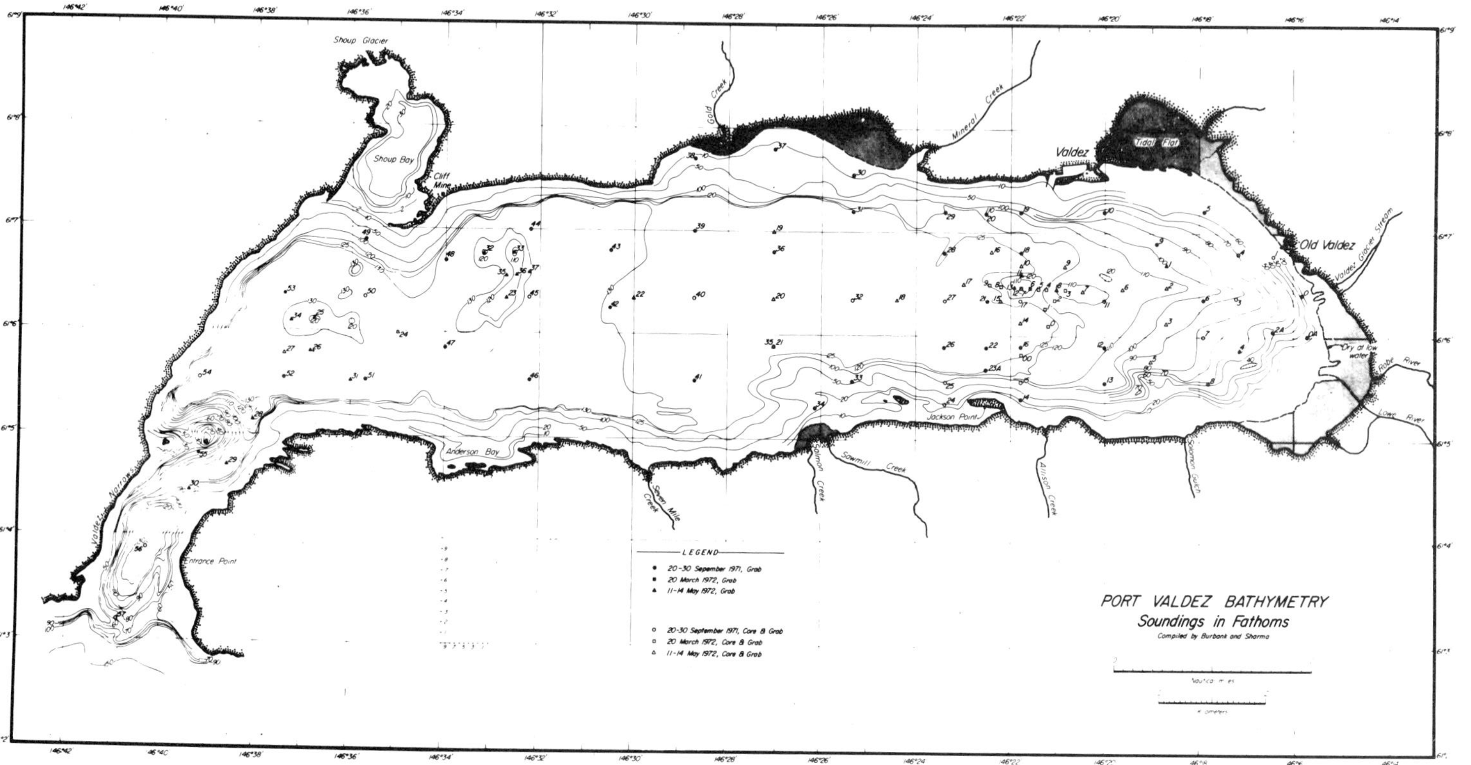

Figure 7-2. Detailed bathymetric chart (fathoms), Port Valdez.

15 m forms a prism of low-salinity (0.5 to 8°/₀₀) and warm temperature (5° to 10°C) water, which sets up an estuarine circulation. The upper water column (surface to sill depth) in the fjord oscillates laterally with tides and therefore continually mixes with outside water. The maximum tidal currents measured is about 20 cm/sec, while the non-tidal current varies between 2 and 3 cm/sec.

During winter, the lower temperature reduces surface runoff to a minimum and cools the surface waters. This lack of freshwater input combined with cooling causes vertical mixing.

Suspended load and temperature measurement from various depths between 0 and 100 m obtained from 10 evenly spaced stations (Fig. 7-3) during a single tidal cycle provide some insight to the net movement of sediments in suspension. Starting near the mouth of the port, samples were collected at each station during ebb tide, and sampling was repeated from east to west during the subsequent flood tide. The distributions of suspended load and temperature with depth along the transects A-B and C-D are shown in Figures 7-4 to 7-7. Water movement in Port Valdez, as interpreted from suspended sediment distribution, temperature, and salinity during 1972, is controlled by the freshwater input, tides, winds, and the coriolis effect. These factors set up a relatively stable counterclockwise circulation cell in the surface layer. Below the surface layer, the isotherm configurations during ebb and flood tides show a significant lateral as well as vertical movement between 10 and 100 m. Surface outward flow entrains

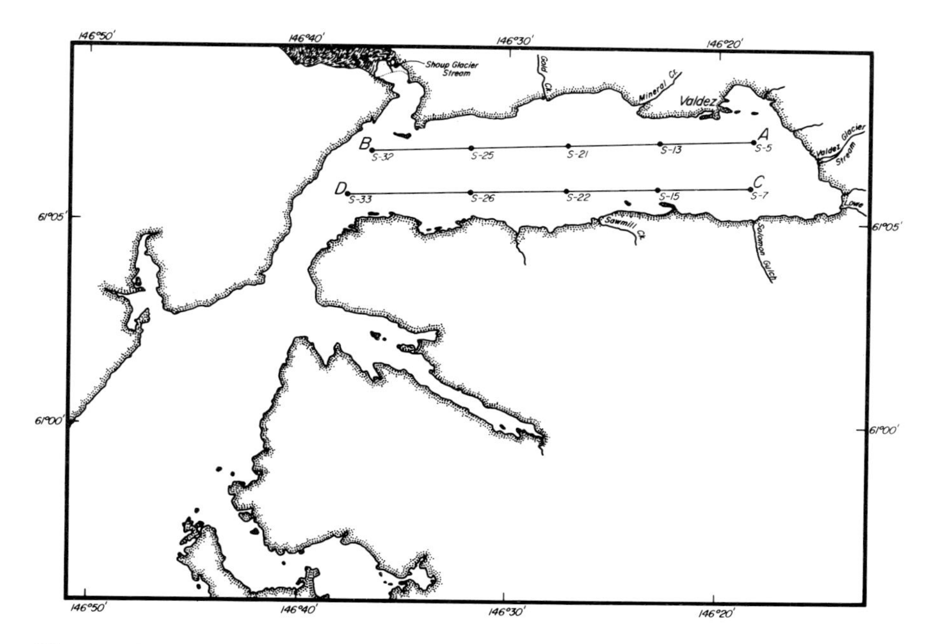

Figure 7-3. Locations of northern and southern east-west transects occupied during tidal cycle on 31 July, 1972, Port Valdez.

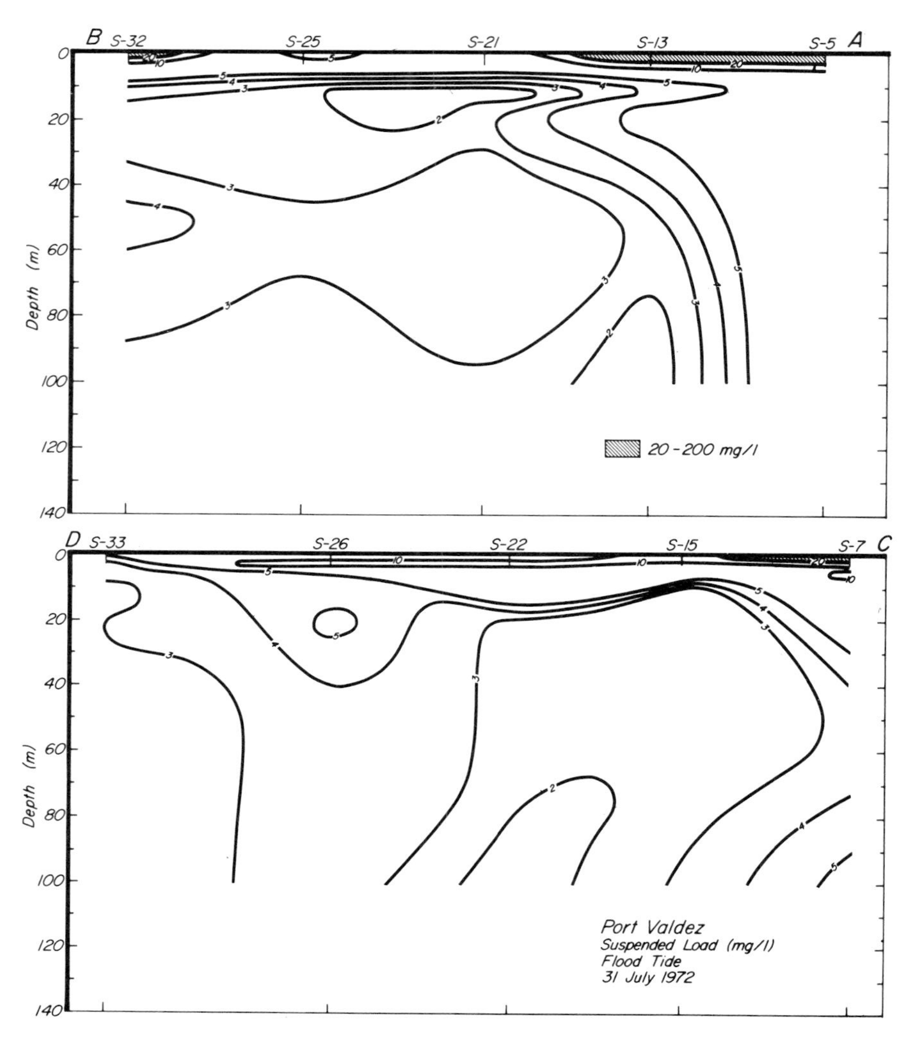

Figure 7-4. Northern and southern transects of the suspended sediment load distribution during flood tide, Port Valdez.

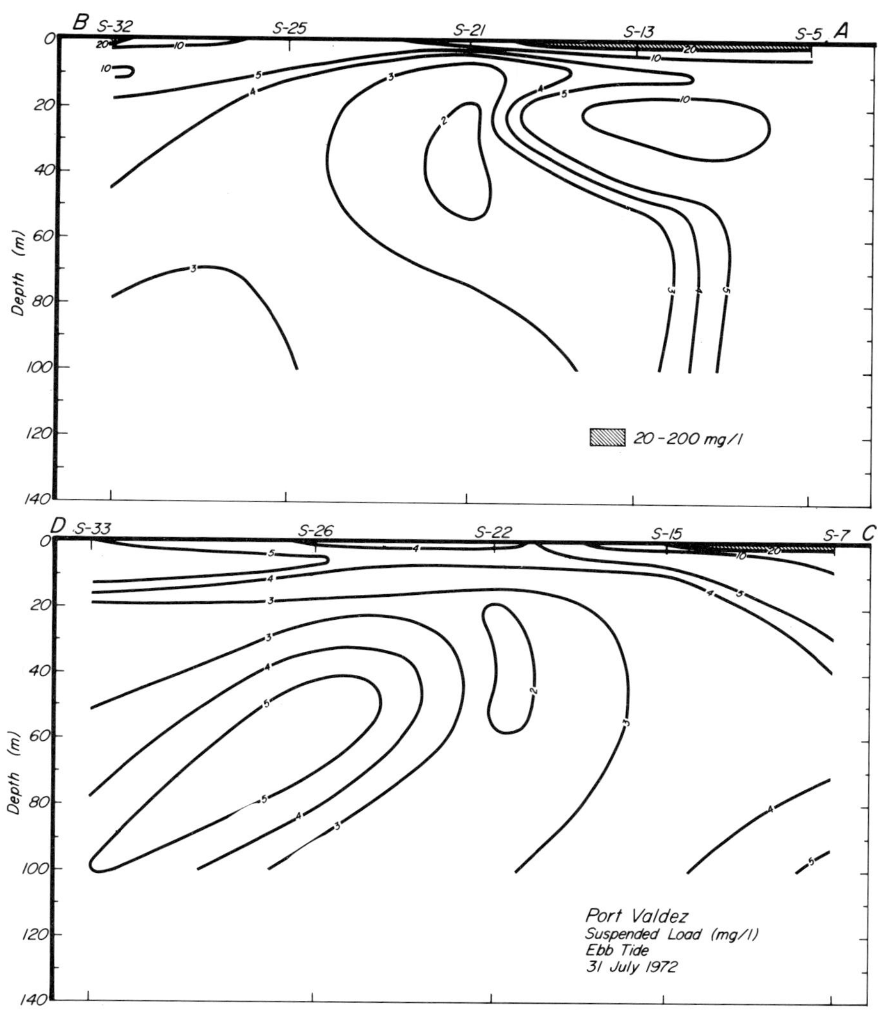

Figure 7-5. Northern and southern transects of the suspended sediment load distribution during ebb tide, Port Valdez.

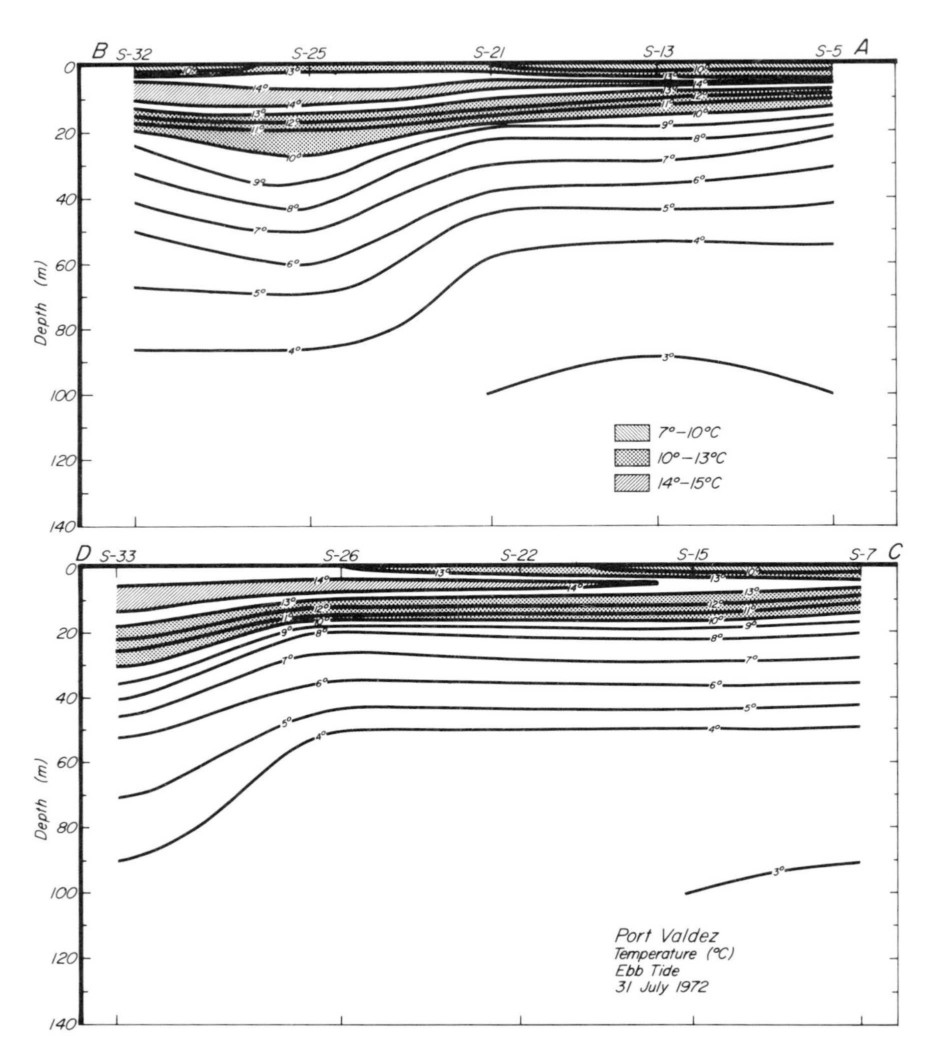

Figure 7-6. Northern and southern transects of isotherms during flood tide, Port Valdez.

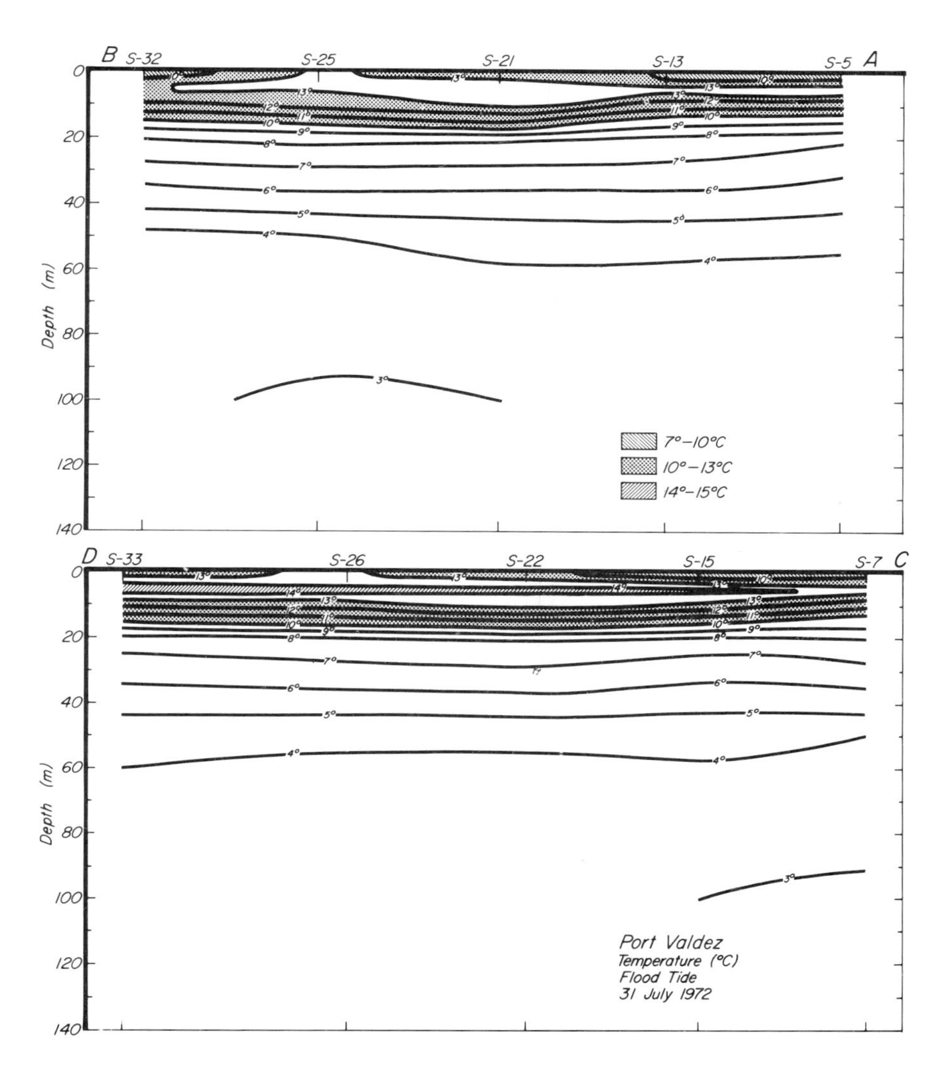

Figure 7-7. Northern and southern transects of isotherms during ebb tide, Port Valdez.

water into the fjord at the sill depth during the ebb tide. Eastward and upward movement of incoming water in Port Valdez is substantiated by measurements of temperature, as well as of suspended load, a nonconservative property.

SEDIMENTS

The surficial bottom sediments in Port Valdez are silt and clay (Fig. 7-8). Inshore sediments near the mouth of the Lowe River, Valdez Glacier Stream and Sawmill Creek, as well as the sediments in the Narrows, are gravelly sandy silt. An elongated narrow area extending west-northwest from the head of the fjord to Mineral and Gold creeks is covered by fine to very fine silt. The rest of the smooth floor of Port Valdez is covered by coarse clay. The clayey sediments are poorly sorted, and sorting becomes very poor with increasing sediment mean size. Most sediments are positively skewed and mesokurtic to leptokurtic.

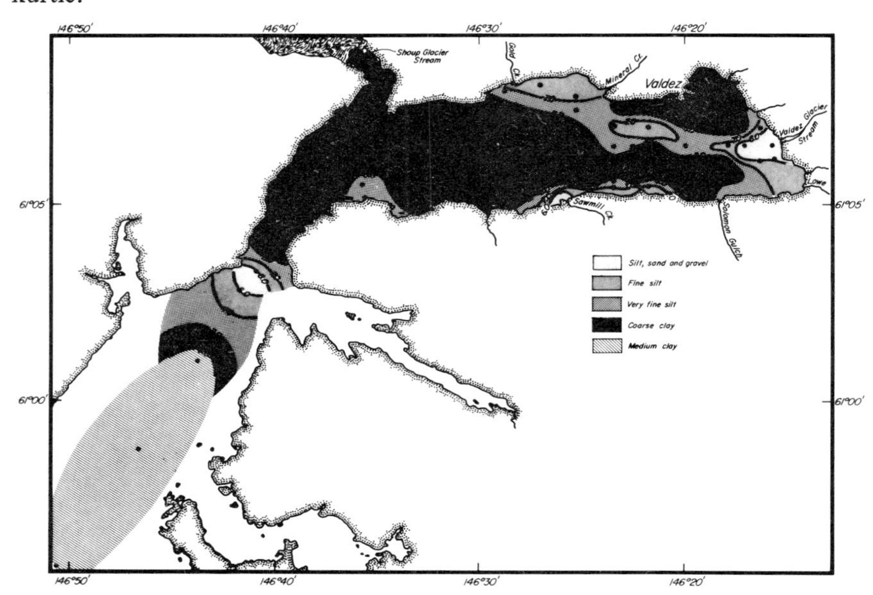

Figure 7-8. Sediment size distribution, Port Valdez.

The pH of the upper 5 cm of the surficial sediments varies from 6.6 to 8.0 throughout the port, with no discernible relation to sediment size distribution. The Eh ranges from +18 to +399 mV. Coarse sediments from shallow depths have higher Eh potential (+150 to +350 mV) than the finer sediments (+50 to +150 mV). The sediment Eh near the water-sediment interface (1 to 2 cm) was consistently higher than that measured at greater depths in sediments. Below

2 cm, the Eh values in the sediments decrease rapidly and stabilize at about 5 cm below the water-sediment interface. The Eh and pH of the overlying waters are similar to those observed in the adjacent sediments.

Eh measurements suggest that highly oxidizing conditions prevail at the depositional interface, and that these conditions change with increasing depth in sediments. The oxidation-reduction boundary in Port Valdez lies beneath the sediment surface. Near-bottom oxidizing conditions are further reflected by a thin brownish-yellow sediment layer which mantles the western part of the port. Contrary to the usual reducing environment observed in tpyical fjords of Norway and other regions, even the deepest fjord along the periphery of Prince William Sound has a oxidizing environment. The brownish-yellow coloration primarily results from the oxidized state of iron in sediments. The pH values indicate that surficial sediments are neutral to weakly alkaline and are typical of a normal coastal marine environment.

Organic carbon in sediments varies from 0.1 to 0.6% by weight. It increases with an increase in the clay size fraction (<4 μm) of sediments. Organic carbon in sediments, however, is inversely related to the Eh potential. The latter relationship suggests that the loss of organic carbon in sediments may be attributed to available oxygen and subsequent bacterial oxidation. The dissolved oxygen measurements in the bottom waters of Port Valdez suggest that the waters throughout the fjord are well oxygenated. Therefore, it appears that the organic carbon, generally associated with the clay size fraction, is mostly recycled.

Sediment samples from source areas lying adjacent to Port Valdez have a mineralogic uniformity throughout the entire region. The mineral composition of the coarser sediment fraction is mainly feldspar and quartz. x-ray diffraction analysis indicates that illite (mica) and chlorite predominate in the clay fraction, with minor amounts of quartz and plagioclase feldspar (Fig. 7-9). The illite was indicated by a strong 10 Å and less intense 5 Å peak. The chlorite (confirmed by heat treatment and slow x-ray scanning in the 25°2ϕ region to determine the presence of kaolinite) gives a strong x-ray diffraction peak at 14 Å (001), which intensified upon heat treatment from 500° to 600°C (Carroll, 1970). The thermal reaction suggests the presence of Mg-chlorite. The unaltered source rocks and the glacial deposits show similar mineral composition, suggesting minimal mineralogic alteration during transportation and deposition by the glaciers and glacial streams. The absence of chemical weathering is best indicated by the sharpness of x-ray peaks of the clay components and the well-defined grain boundaries observed in an electron microscope study. The clay size fraction of sediment delivered to Port Valdez has equal parts chlorite-illite or slightly higher in mica (average 14 Å chlorite/10 Å illite peak height ratios being 1.0 to 0.7).

There is a variation in the proportion of clay minerals in Port Valdez and adjacent land sources. As the distance from the major sources increases the chlorite-illite ratio changes from 0.4 to 0.7, *i.e.*, the relative illite content of bottom sediment is higher near the source (Fig. 7-10). Parham's study (1966) of lateral variations in clay mineral assemblages indicates that generally illite is found in

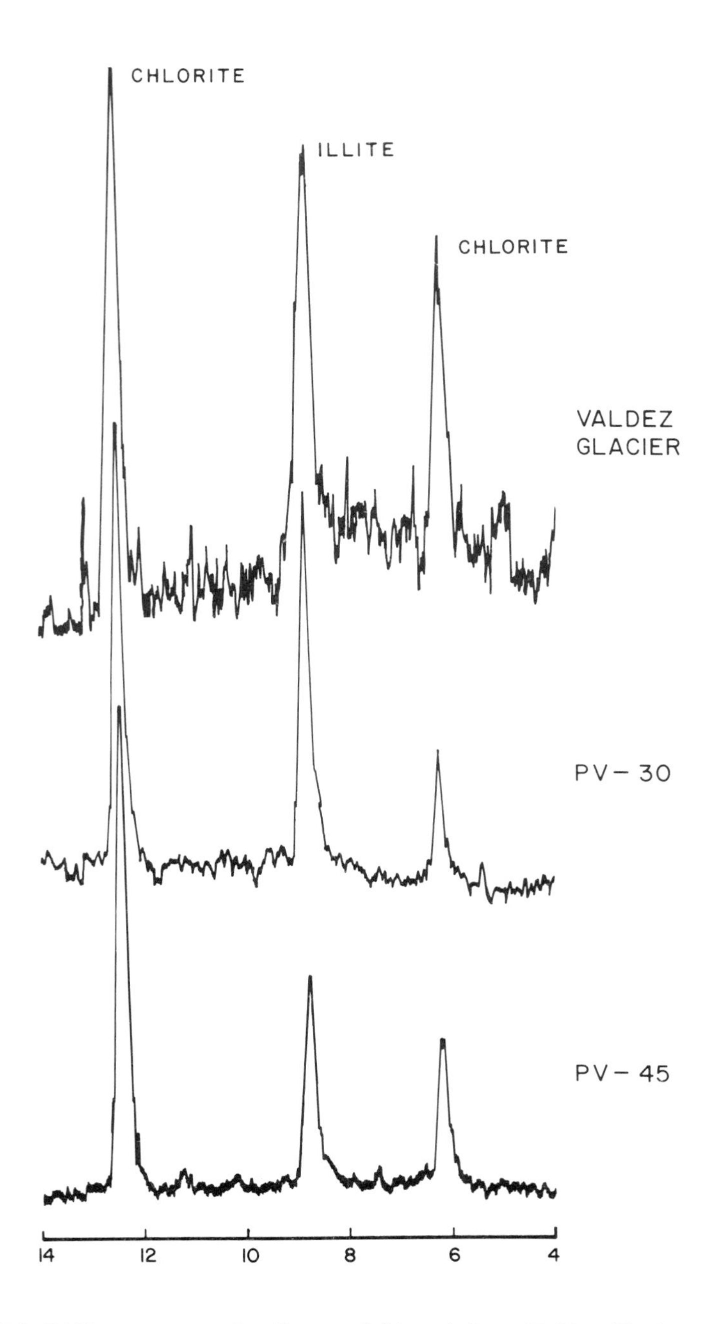

Figure 7-9. Diffractograms of sediments (<4 μm) from Valdez Glacier and Port Valdez.

greater abundance near shore, with the chlorite-illite ratio increasing basinward. Apparently in Port Valdez the clay-sized illite of relatively larger particle size settles out in the proximity of the detritus input, whereas the fine-grained chlorite remains in suspension longer, allowing farther transport.

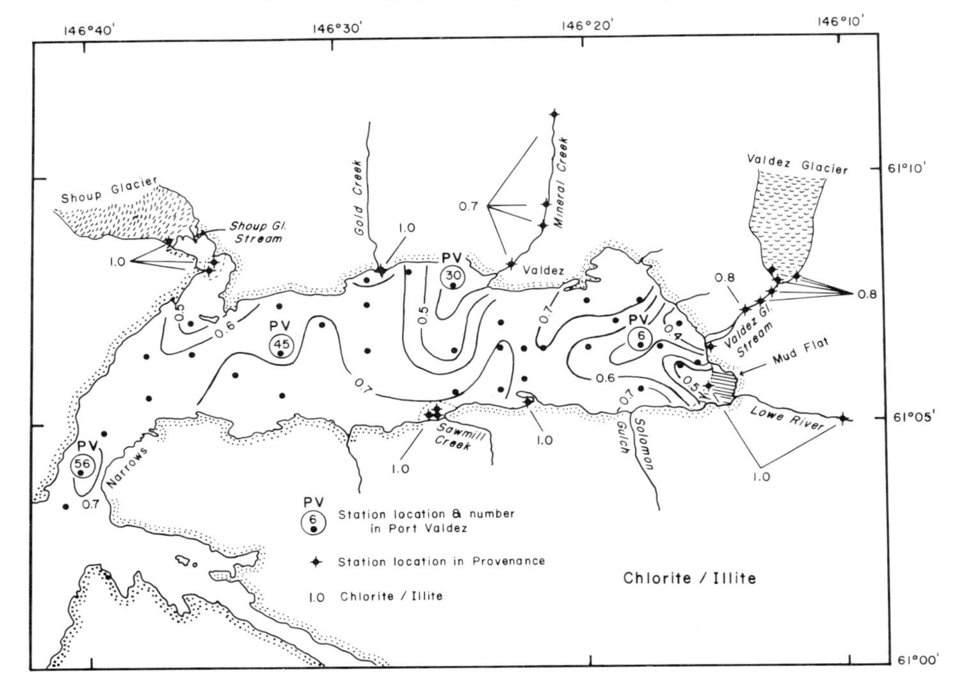

Figure 7-10. Chlorite-illite peak height ratio variation in sediments, Port Valdez.

Numerous cores taken from Port Valdez, varying from 1 to 3 m in length, contain sediments ranging in particle size from gravel to clay. A few cores taken from the top of topographic highs (submarine fans) contain a distinct subbottom horizon of gravel at shallow depth. Other distinctive characteristics of sediments from the cores were as follows:

1. Sediments are interbedded with sand and gravel.
2. Sands are moderately sorted, whereas silts and clays are very poorly sorted. This is in contrast to observed decreasing sorting with increasing grain size in surficial sediments.
3. The sediment sequence becomes finer upward.
4. Sediments often contain clay balls (1 to >5 cm in diameter).
5. The sediments in cores near the head of the fjord are positively skewed, while near the entrance sill they are negatively skewed. Sediments in cores from Valdez Arm (outside Port Valdez) have nearly symmetrical distributions.

The vertical sediment distribution in cores suggests that Port Valdez is blanketed by a thin layer of silty and clayey sediments which overlie coarse graded sediments. Thicknesses of the overlying blanket varies throughout the fjord: In some areas it is less than 15 cm (Fig. 7-11).

The cores collected in 1972 from the crests of both fans located near the base of Lowe River and Valdez Glacier Stream deltas (Fig. 7-11) display a marked unconformity. A thin layer of clayey silt overlies graded gravel. This clayey silt consists of eight cyclic sedimentary layers. The bottom section (summer sedimentation) in each of these layers contains somewhat coarser grained sediments with a high content of detrital wood, grading into a thin fine-grained layer (winter sedimentation). Each layer, therefore, represents one episode of annual deposition (*i.e.,* a varve). The lowest of these annual layers in one core was 3 cm thick, whereas the overlying layers (seven) varied between 1 and 4 cm in thickness. The unconformity and the sediment textures suggest that the underlying graded coarse sediments (gravel) were deposited as a result of a submarine slide caused by the 1964 earthquake, and that the overlying (silt) cyclic layers were deposited during each of the following 8 years.

A number of cyclic sedimentations underlain by a distinct unconformity suggests that the sediments below the unconformity have been deposited as a result of the sediment mass movement attributed to the 1964 submarine slide. Approximately 75×10^6 m^3 of gravel, sand, and silt slumped near the head of the inlet and was transported further offshore (Coulter and Migliaccio, 1966).

The sediment mass torn off from the alluvial fan near the head of the fjord was carried down the slope along the Valdez Channel. The downward and westward moving sediments encountered the topographic high, Fan I, near the toe of the channel. Fan I perhaps was deposited during earlier slumping, possibly as an aftermath of the 1911 earthquake. As the more recent sediment mass passed over Fan I, the downward and lateral flow was deflected upward, causing significant loss in energy and increased turbulence. The appearance of muddy waters on the surface in the port, as observed by some, was caused by this upward movement. The loss of kinetic energy resulted in the deposition of gravel, Fan II, while sand, silt and clay sediments proceeded toward the entrance of the fjord. Along its path, the mass movement, accompanied by turbulence, reworked the bottom sediments and carried them further offshore. The westward moving sediment mass was finally blocked by the protruding sill near the entrance of the port. A thick sequence of graded silt was observed in a core east of the sill, but no graded sediments were deposited beyond the sill in Port Valdez Arm.

Severe earthquakes in Port Valdez are generally followed by burial and breaking of submarine cables. It appears that major shocks, directly or through associated tsunamis, invariably trigger landslide and/or submarine slumping. Such slumping results in mass sediment movement, burying and breaking the submarine cables. Immediately after the Great Alaskan Earthquake of 1964 there was a major submarine slide in Port Valdez. Muddy surface waters appeared after the collapse of the dock area near old Valdez. Similar slides following the

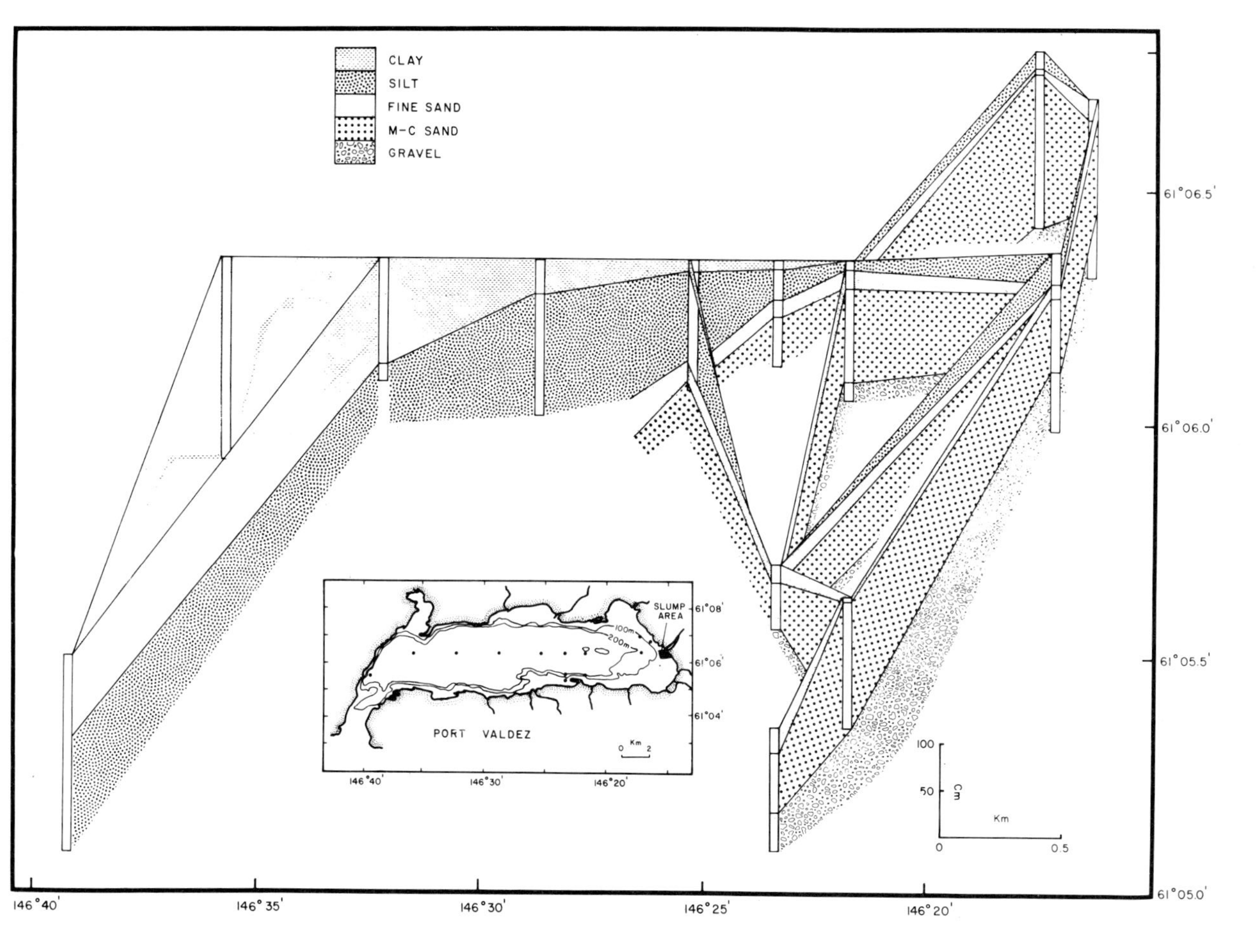

Figure 7-11. Block diagram showing lateral and vertical distribution of sediments in cores, Port Valdez.

earthquake were also inferred by Plafker and Mayo (1965) in the vicinity of the Cliff Mines near the entrance to the fjord (Fig. 7-2). Submarine cable breaks following earthquakes during the late 1800's and early 1900's have been reported from the area. A few isolated but conspicuous topographic highs on an otherwise relatively smooth bottom may be submarine slump blocks formed as a result of subaqueous slides.

GEOCHEMISTRY

The distribution of major and minor elements in surficial bottom sediments obtained from over sixty samples is given in Figures 7-12 to 7-25. With the exception of silicon, calcium and strontium, all major and minor element concentrations in these sediments increase seaward. Aluminum varies between 7 and 9% in the eastern part and averages about 10% in the western part of the fjord (Fig. 7-12). Iron is low near the head (3.5%), ranges between 3.5 and 5.5% in the near-shore area, and increases westward to over 6% (Fig. 7-13). Similar distributions are shown by magnesium, sodium, and potassium, which vary (Figs. 7-14 to 7-16) from 1.8 to 2.4%, 2.1 to 2.5% and 1.8 to 2.6%, respectively. The distributions of calcium and strontium in sediments show decreasing concentrations with increasing distance from shore (Figs. 7-17 and 7-18).

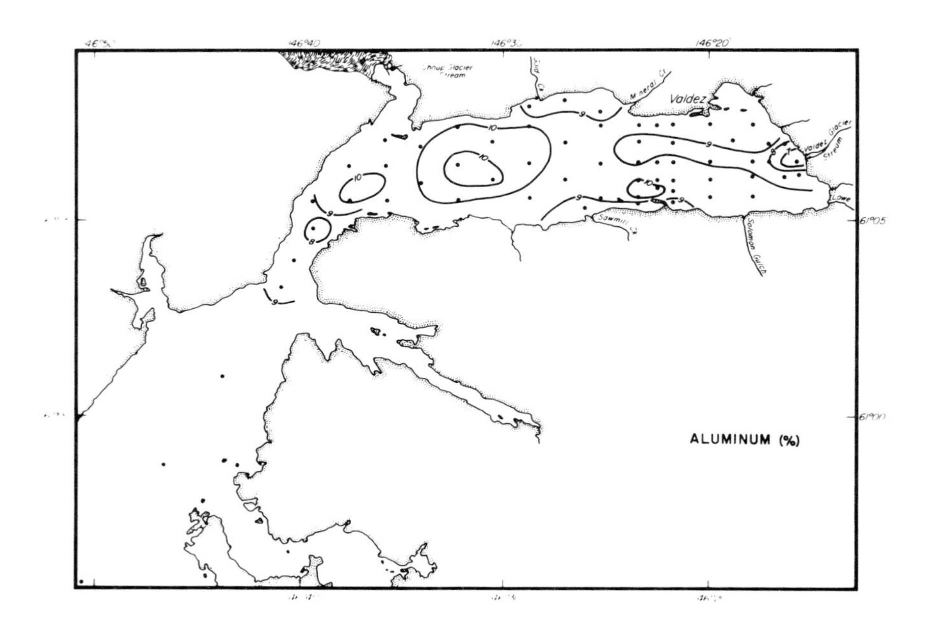

Figure 7-12. Percent aluminum in sediments, Port Valdez.

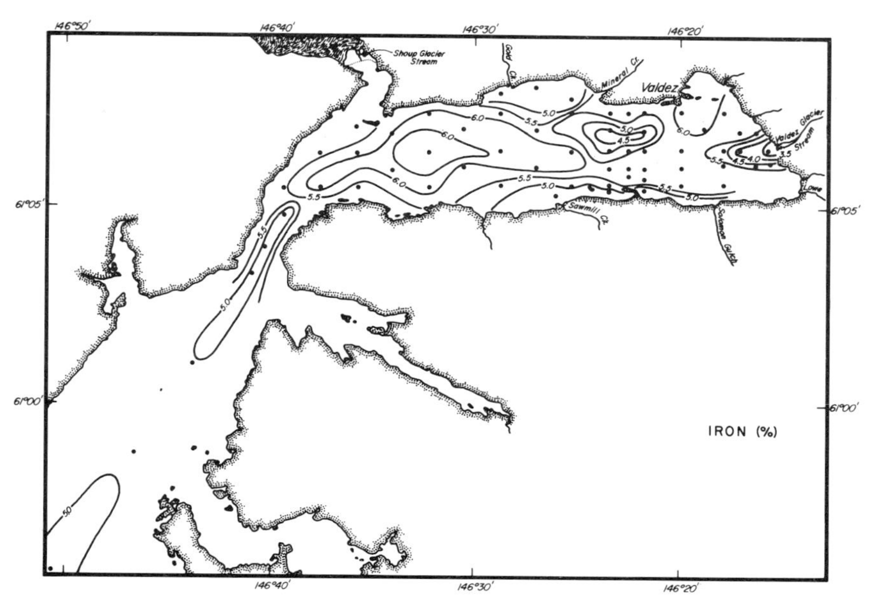

Figure 7-13. Percent iron in sediments, Port Valdez.

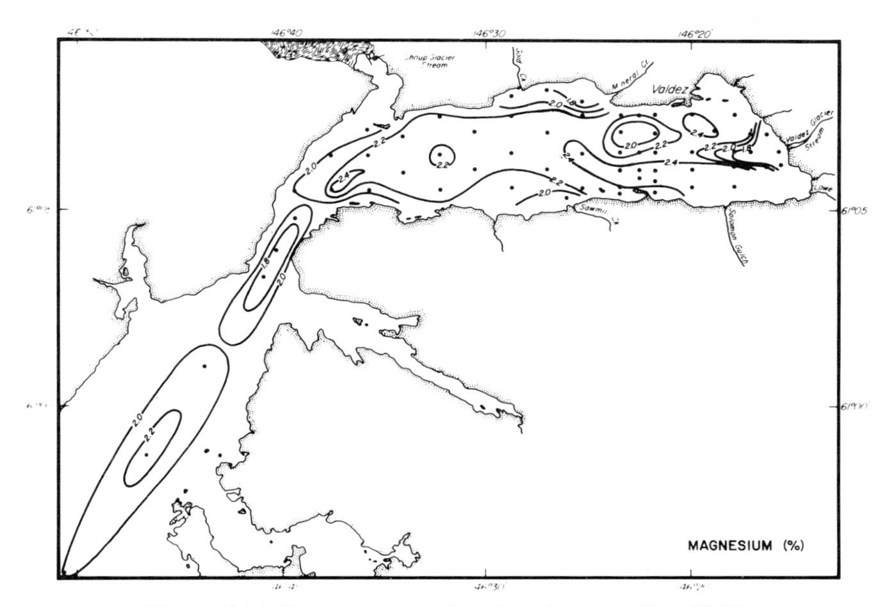

Figure 7-14. Percent magnesium in sediments, Port Valdez.

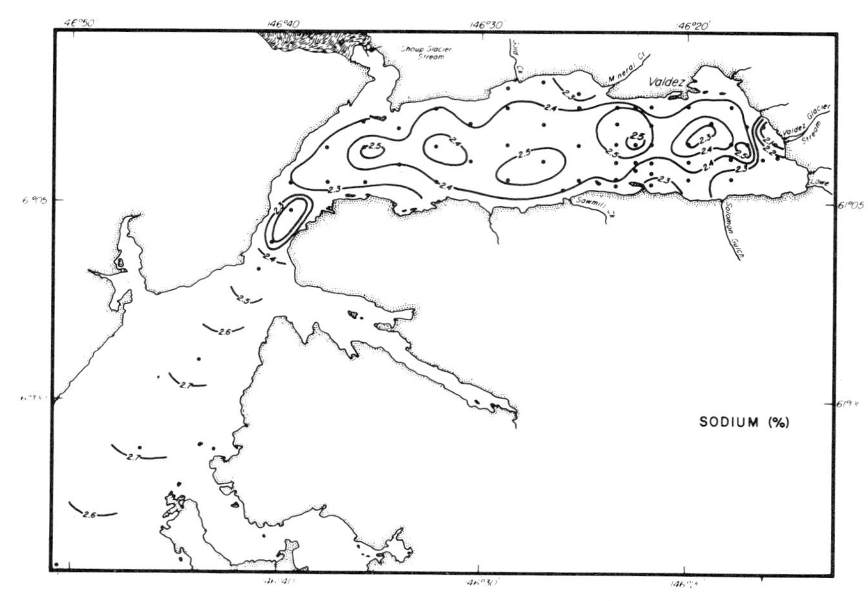

Figure 7-15. Percent sodium in sediments, Port Valdez.

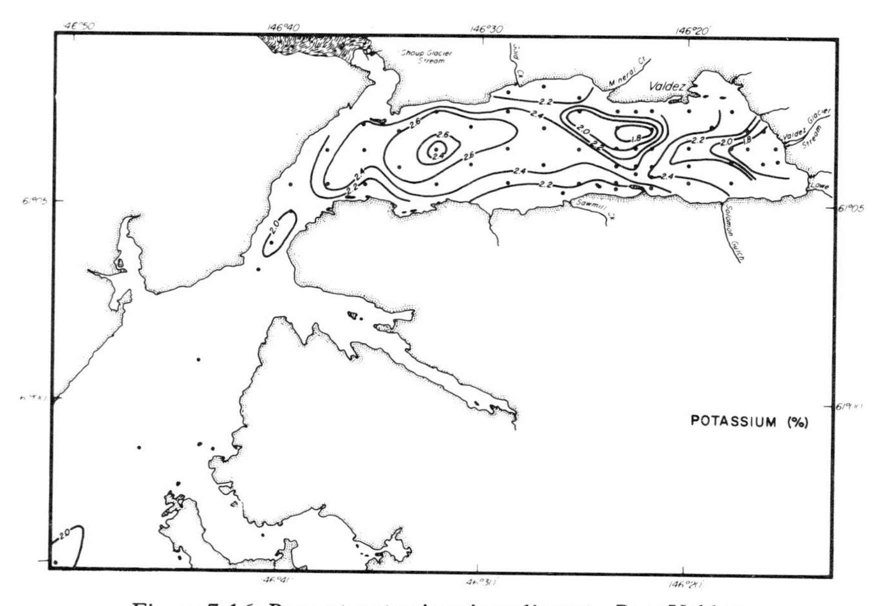

Figure 7-16. Percent potassium in sediments, Port Valdez.

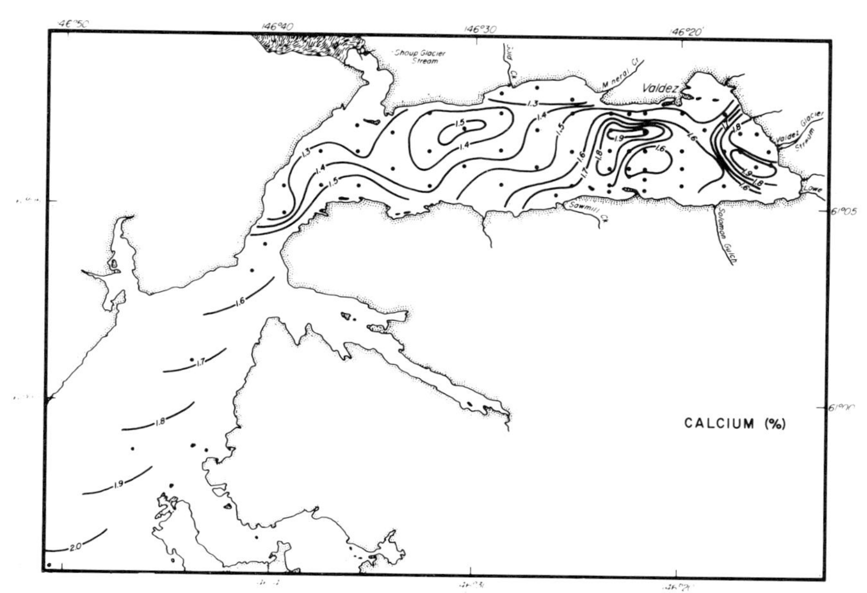

Figure 7-17. Percent calcium in sediments, Port Valdez.

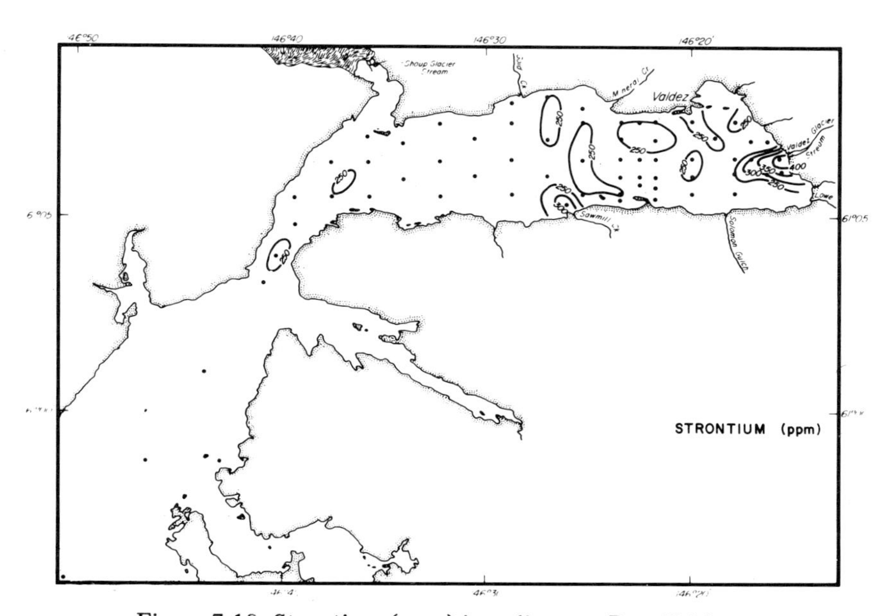

Figure 7-18. Strontium (ppm) in sediments, Port Valdez.

The distribution trends of manganese, barium, chromium, copper, cobalt, nickel, and zinc (Figs. 7-19 to 7-25) are similar to those of the major elements. The near-shore coarser sediments contain lesser amounts of these elements than the fine-grained sediments deposited further offshore in Port Valdez. Except for calcium and strontium, the distributions of major and minor elements in sediments from Port Valdez are mostly similar to sediment mean size distributions. The coarse sediments deposited in the near-shore areas and the eastern part of the fjord contain lower amounts of minor elements, whereas the finer sediments deposited seaward and in the western part contain relatively higher amounts. An increase in major element concentration is a normal consequence of marine weathering and transport processes, while the higher concentrations of various trace elements, manganese, and iron in finer sediments presumably result from their association with the clay phases.

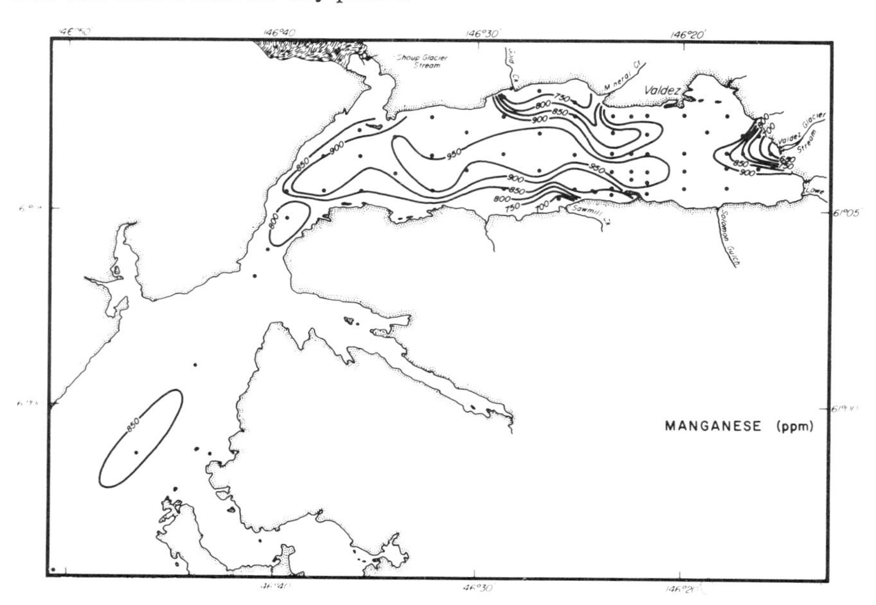

Figure 7-19. Manganese (ppm) in sediments, Port Valdez.

The degree of elemental differentiation resulting from size grading during sediment transport is fundamentally reflected in the differences in compositions of near-source and offshore sediments and is also manifested by differences among the various size fractions of the sediments. In this respect, comparison of the composition of silt fraction and bulk sediment from four stations in Port Valdez is of significant interest. Station PV-6 is located close to Valdez Valley on the alluvial fan, approximately 2 km west from the mouth of Valdez Glacier Stream (Fig. 7-10). PV-30 lies close to shore, approximately 2 km west from the mouth of Mineral Creek. PV-45 is located in deep water of the mid-fjord and is equidistant from Mineral Creek and Shoup Bay. PV-56 is situated between the two

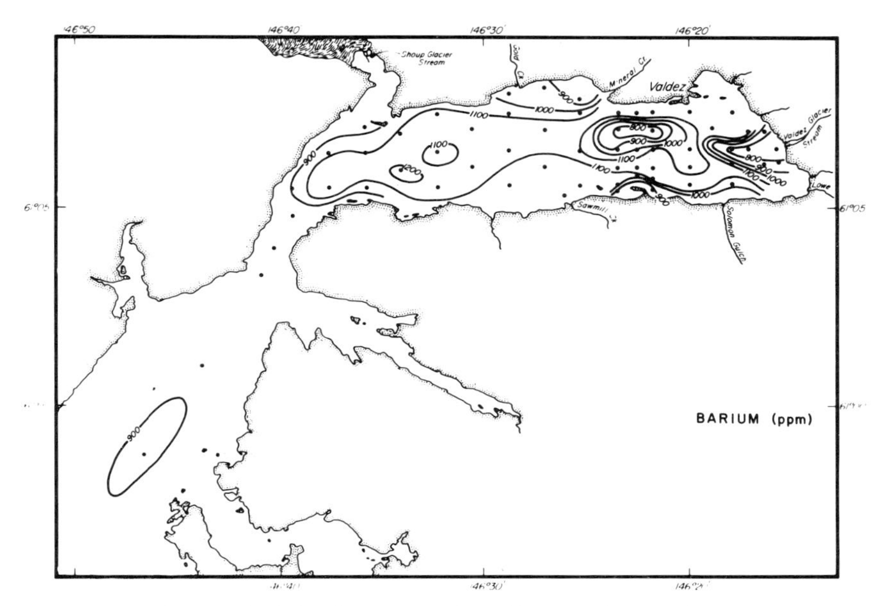

Figure 7-20. Barium (ppm) in sediments, Port Valdez.

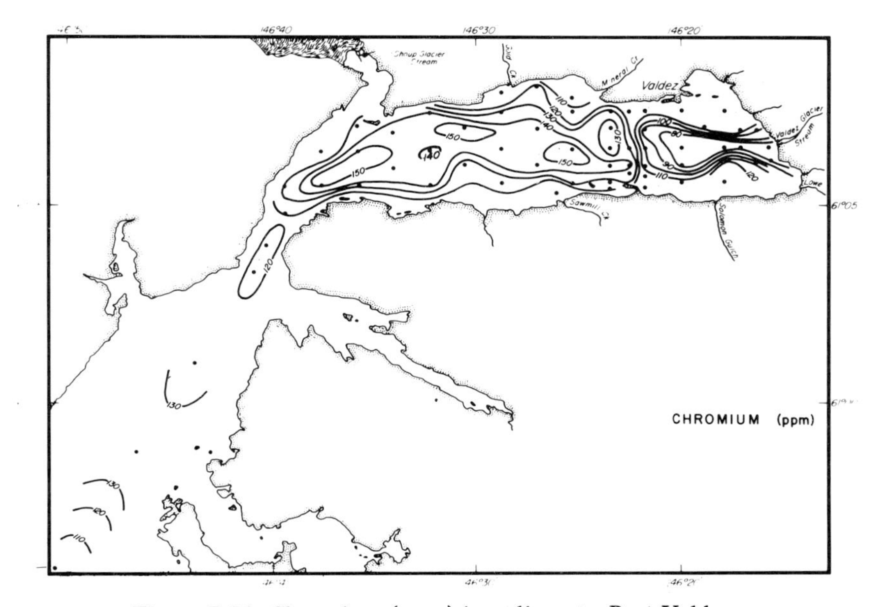

Figure 7-21. Chromium (ppm) in sediments, Port Valdez.

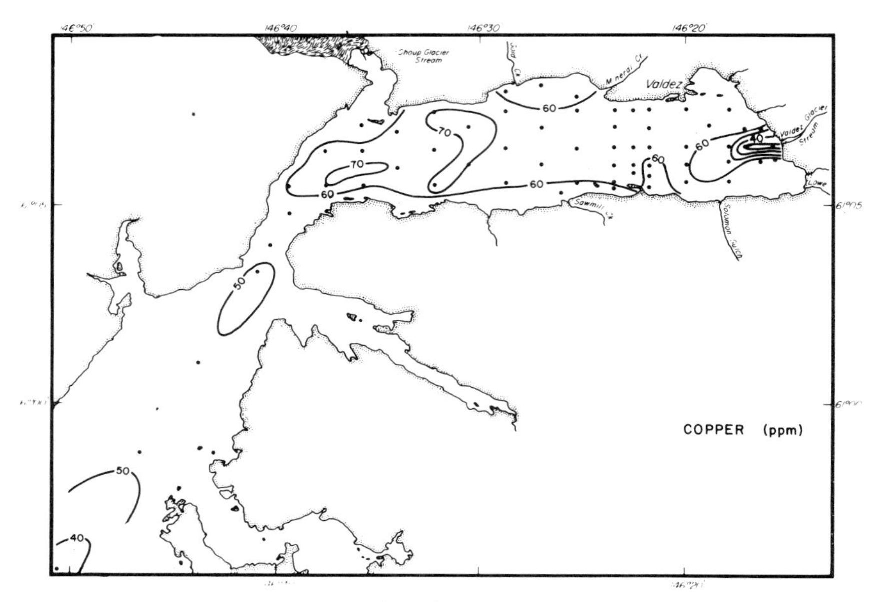

Figure 7-22. Copper (ppm) in sediments, Port Valdez.

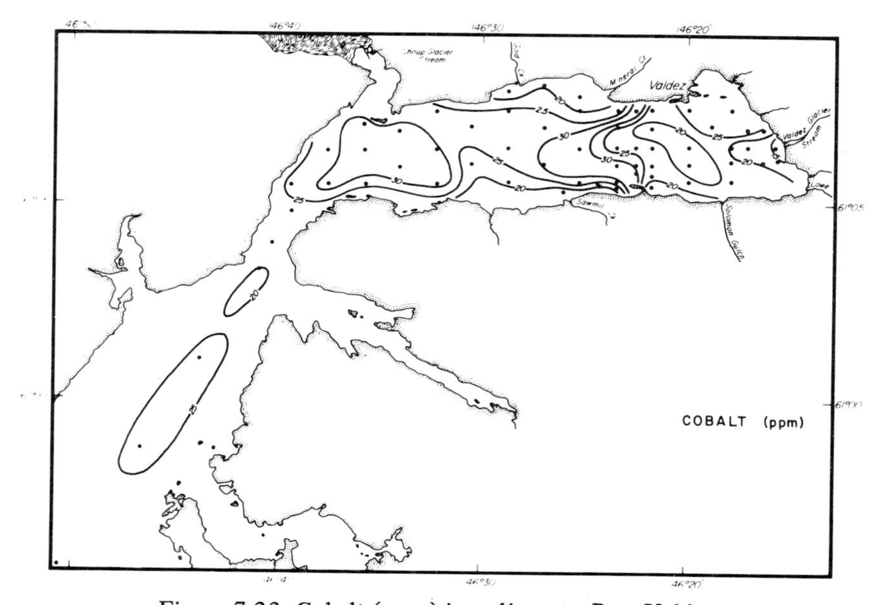

Figure 7-23. Cobalt (ppm) in sediments, Port Valdez.

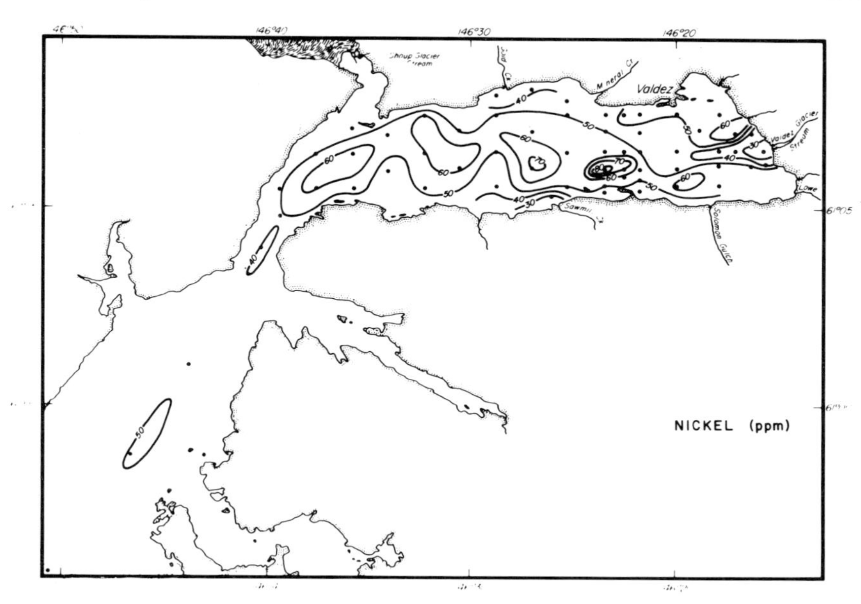

Figure 7-24. Nickel (ppm) in sediments, Port Valdez.

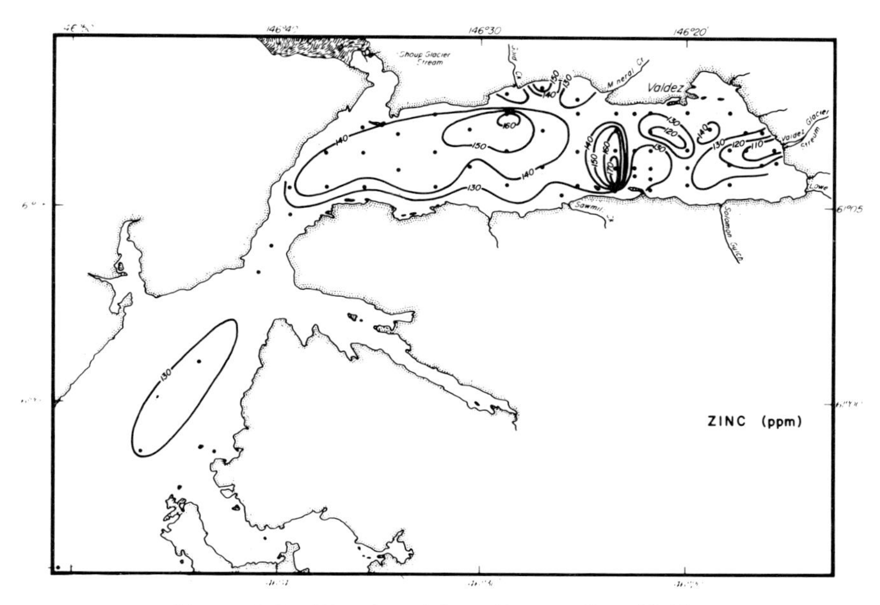

Figure 7-25. Zinc (ppm) in sediments, Port Valdez.

sills in the Narrows at the entrance. The silt fraction was separated from the bulk sediment using standard settling methods. A comparison of elemental distributions in the silt size fraction and the bulk sediment from the four stations, PV-6, PV-30, PV-45, and PV-56, clearly shows a progressive fining of sediments and a relative increase in the concentrations of various elements during sediment transport (Table 7-1). It should be noted that, contrary to this general trend, silicon, calcium and strontium decrease with decreasing sediment size.

Major elements, Al, K, and Mg, show noticeable differences in the bulk and silt size fraction of sediment. Average percent distribution for these three elements for all stations is as follows: aluminum 8.05 silt, 8.37 bulk; magnesium 1.78 silt, 1.90 bulk; potassium 1.58 silt, 2.05 bulk. The distribution of all elements is lower in the silt fraction. These relationships are most likely due primarily for separation of quartz from other minerals as the result of transportation and abrasion; quartz, with its inherently higher resistance to weathering, is deposited as coarser sediments near the source, while the lighter silicate minerals are carried further into the fjord. The particle morphology (shape) and the specific gravity of the various minerals would be expected to reinforce such separation.

Detailed analysis of Al, Mg, and K distributions shows two distinct environments, with a relatively larger concentration of these elements offshore in the western part of the port than shoreward in the eastern part. This distribution is similar to the mean size sediment distribution, which reflects two sedimentary facies.

SEDIMENT SOURCE AND TRANSPORT

The source for the sediments deposited in Port Valdez lies in glacial and glaciolacustrine deposits. These deposits have their principal origin in the Cretaceous phyllitic graywacke of the Valdez Group. This phyllitic graywacke, composed mainly of quartz, orthoclase and plagioclase feldspar, muscovite and chlorite, is the dominant outcrop material and is widely exposed in the area.

The sediment discharge into Port Valdez from various rivers and streams is mostly deposited near the shore, and a small fraction of it is transported offshore in suspension. The sediment input to the fjord is primarily dependent on the surface flow and, therefore, besides broad seasonal variations, many short-term extremes in sediment input can be expected due to heavy precipitation (Fig. 7-26). An estimated 26.35×10^{11} g of sediments in suspension were brought into Port Valdez during 1972 (Sharma and Burbank, 1973). Fine sediments are carried into the port in a distinct surface layer (sediment plume) of relatively low-salinity water overlying denser, higher salinity water. The thickness of this surface plume varies between 2 and 15 m, and the sediment load in suspension decreases seaward and with increasing water depth (Figs. 7-4 and 7-5). The heavy and well-defined plume originating from the Lowe River and Valdez Glacier Stream is slightly deflected to the north, extends westward, and

TABLE 7-1

Elemental Distribution in Silt Fractions and Bulk Sediment from Port Valdez

Element	PV - 6		PV - 30		PV - 45		PV - 56	
	Silt	Bulk	Silt	Bulk	Silt	Bulk	Silt	Bulk
Al %	8.38	8.83	7.58	8.68	8.36	9.36	8.06	8.60
Fe %	4.5	4.8	3.5	4.6	4.6	5.6	3.5	4.6
Mg %	1.93	2.04	1.28	1.71	1.76	2.14	1.37	1.73
Ca %	1.87	1.81	1.23	1.17	1.61	1.46	1.66	1.52
Na %	2.20	2.75	2.12	2.25	2.20	2.34	2.15	2.20
K %	1.64	1.83	1.55	2.11	1.76	2.32	1.39	1.95
Mn ppm	874	888	674	758	867	936	700	808
Ba ppm	815	800	740	886	897	1056		937
Co ppm	20	23	18	17	22	30	13	22
Cr ppm	99	98	96	106	104	119	80	111
Cu ppm	45	50	32	58	37	65	34	52
Ni ppm	32	32	25	45	30	60	30	40
Sr ppm	277	242	223	216	251	238	284	280
Zn ppm	98	142	106	128	119	144	109	125

Texture	PV - 6 %	PV - 30 %	PV - 45 %	PV - 56 %
Gravel	0.70	0.00	0.00	1.20
Sand	20.82	11.78	0.32	4.37
Silt	50.76	54.92	52.22	44.85
Clay	28.42	33.30	47.46	49.58

finally merges with the plume from Mineral Creek. The combined plume then proceeds west along the northern shore and dissipates past Gold Creek (Figs. 7-27 to 7-30). In the northwest, a small, narrow, and elongated sediment plume extends southward from Shoup Bay.

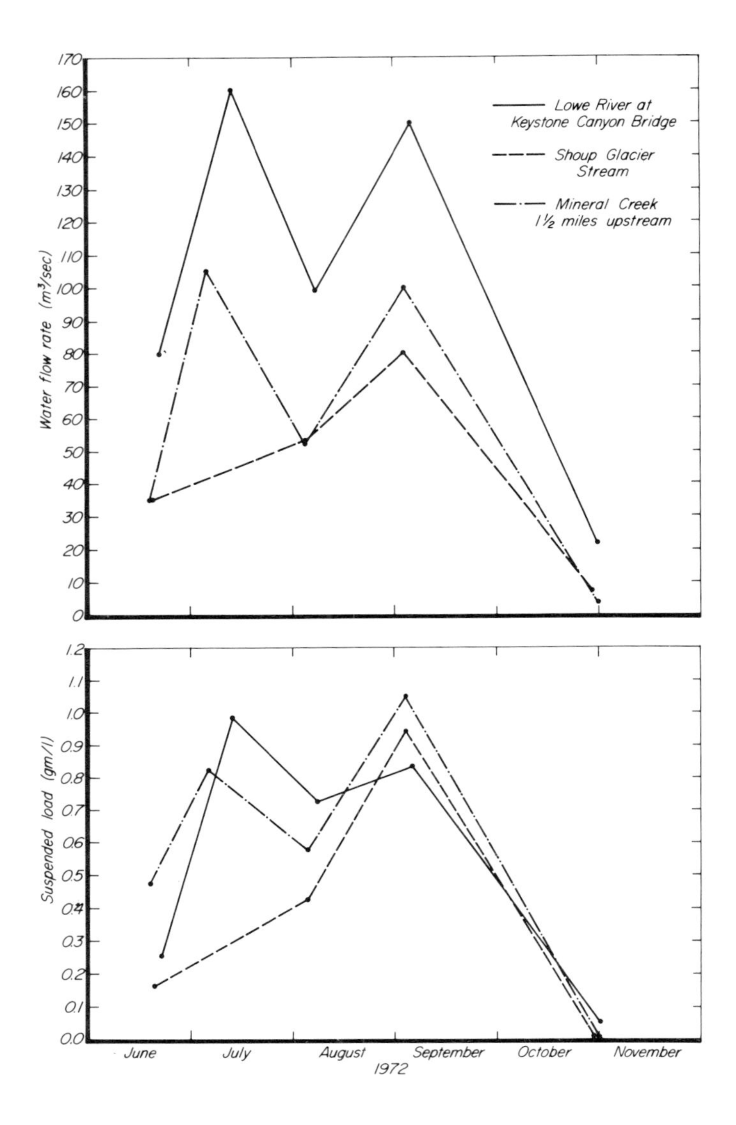

Figure 7-26. Discharge and suspended load of various rivers draining into Port Valdez.

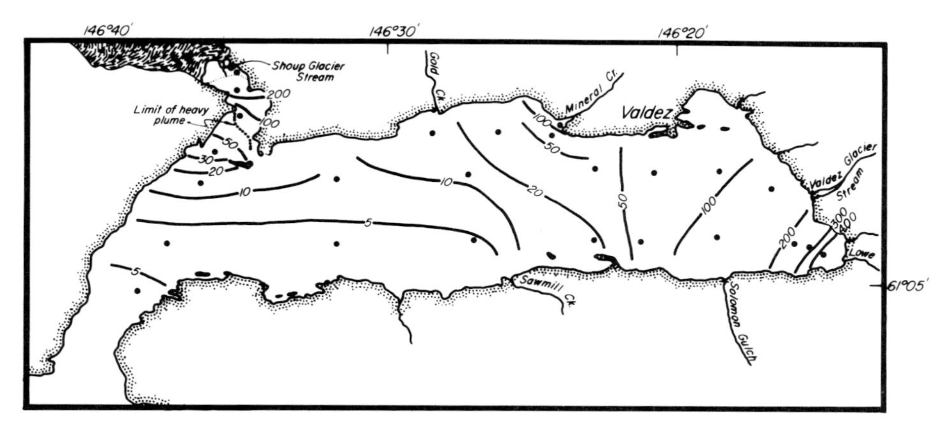

Figure 7-27. Distribution of the surface suspended load (mg/liter) on 2 August, 1972, Port Valdez.

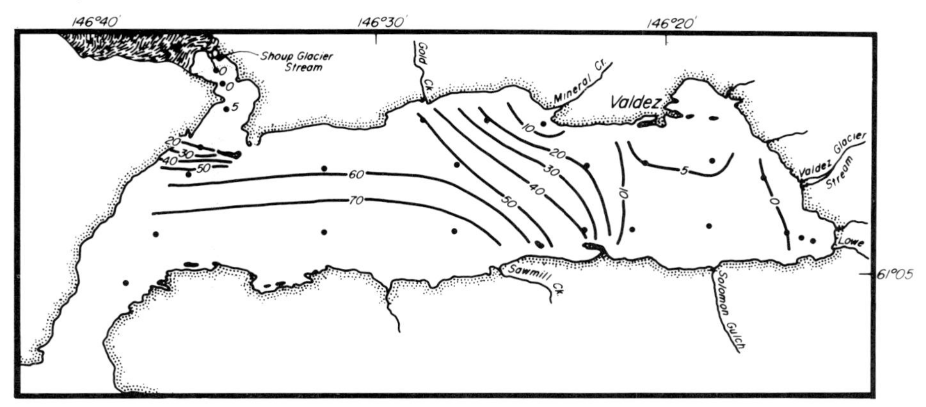

Figure 7-28. Variation of light transmissibility (%) in surface waters on 2 August, 1972, Port Valdez.

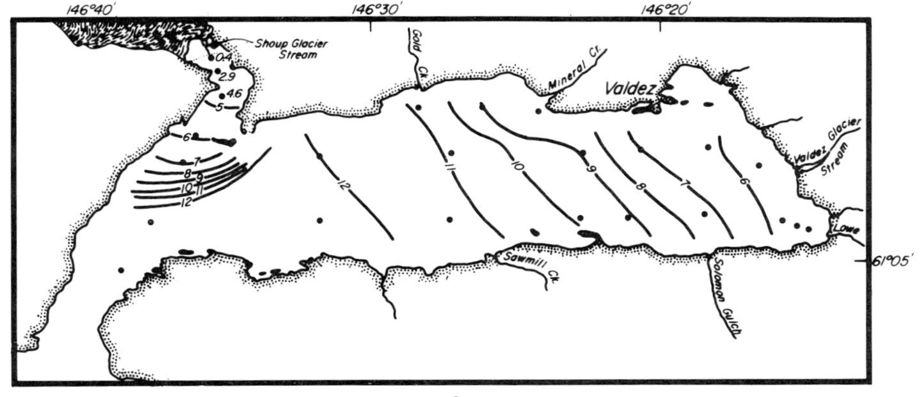

Figure 7-29. Surface water isotherms (°C) on 2 August, 1972, Port Valdez.

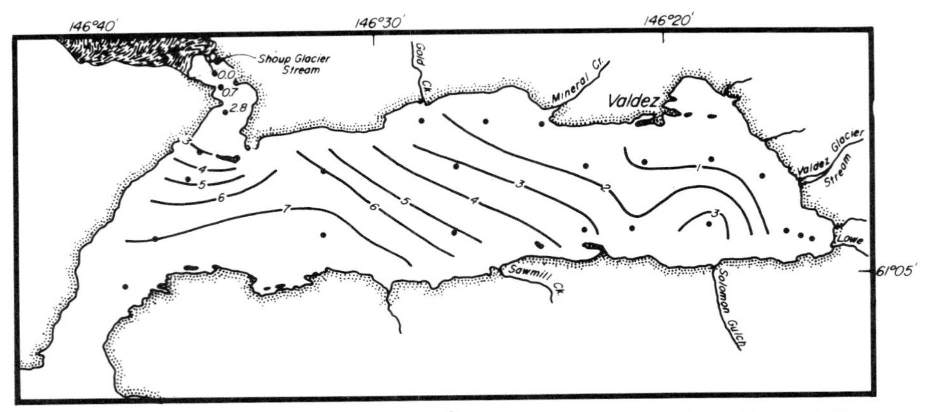

Figure 7-30. Surface water isohalines ($^{o}/_{oo}$) on 2 August, 1972, Port Valdez.

The ERTS-1 imagery of August 15, 1973, at approximately maximum flood tide, displays a surface suspended load distribution similar to that revealed by direct field measurement. This similarity in suspended load distribution as observed by these two independent methods emphasizes the usefulness of ERTS-1 imagery in other areas. The imagery not only delineates the major plumes in the east and in Shoup Bay but also distinctly shows smaller plumes along the southern shore (Fig. 7-31). There is a major influx of fresh sediment-laden water at the head of the fjord that flows to the northwest past the town of Valdez to form a counterclockwise gyre in the eastern half of the fjord. This westward moving plume is supplemented by input from Mineral and Gold creeks.

Shoup Bay, near the mouth of the fjord, is also a significant source for sediments. Most of the sediment discharge in this area, however, is deposited in the bay, and only a minor amount is carried into Port Valdez. The sediments in suspension that bypass the bay are carried into the Narrows along the northwest shores. An analysis of the imagery also indicates that during flood tide relatively clear sea water from Valdez Arm intrudes the port along the eastern shore of the Narrows. Coriolis force deflects the northerly moving incoming waters progressively to the right, forming a clockwise gyre in the western half of the fjord.

The maximum suspended load occurs within the plume (0 to 15 m), and the load below the plume decreases rapidly with increased salinity and depth—it generally ranges between 2 and 5 mg/ℓ down to a depth of about 100 m. The vertical distribution of the suspended load near the major sediment inputs and in the Narrows is complex and often displays numerous maxima and minima with increasing water depth. A maximum value of the subsurface suspended load in the Narrows usually occurs at about 20 m.

There are two major transport systems in Port Valdez, the river and episodic slide systems. The river system continually brings sediments to the port, and due to the gradient reduction and decrease of transport energy, the coarse river sediments are deposited near the river mouths, and thus build extensive deltas. The

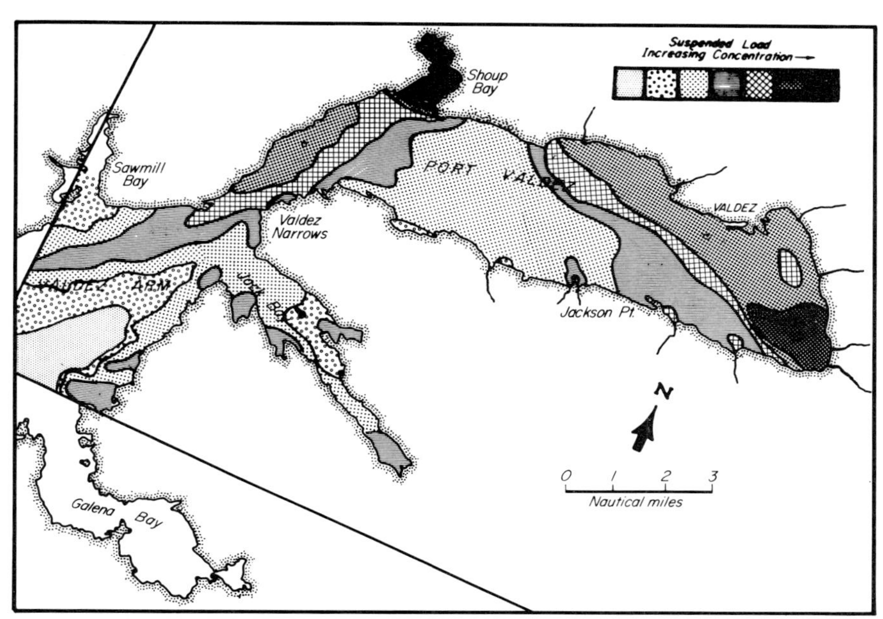

Figure 7-31. Isodensity distribution of reflectance in satellite imagery showing relative suspended loads in near-surface water on 15 August, 1973, Port Valdez.

bulk of the suspended sediments is carried seaward into the port in the surface layer, where it flocculates and settles out within a few kilometers from the river mouth.

The river detritus brought in suspension into the port during 1972 was estimated at 2.26×10^6 metric tons. It is believed that at least an equal amount of coarse detritus is carried into the fjord as bed load. This high amount of bed load is primarily due to glaciation, and to short streams with steep drainage basins. Most of the bed load, however, is deposited as alluvial fan material and is not carried farther into the fjord. An annual input of 2.26×10^6 metric tons distributed over an area of 105 km^2 gives a rate of sedimentation of 1.7 cm/yr. The sediment column deposited since the 1964 earthquake on the submarine fan near the head of the fjord gives a rate of 1.9 cm/yr. Because the sediments are not evenly distributed throughout the fjord, it is apparent that sedimentation in the eastern half is somewhat higher than that in the western half. The major sediment input pattern, together with the water dynamics in Port Valdez, gives rise to two sedimentary facies. The area near the head of the fjord consists of coarse sediments primarily brought as bed load, as well as in suspension, and is characterized by a higher rate of sedimentation. The western part of the port is covered by a thin, oxidized, and yellowish-brown clay sediment, with a slower rate of sedimentation. These two sedimentary facies also are distinctly reflected in the elemental distribution in sediments.

To some extent sediments are also carried in subaqueous layers generally termed "turbid layers" or "nepheloid layers." Discrete subaqueous horizons with higher concentrations of suspended sediments than adjacent waters above or below have been often observed in Port Valdez (Sharma and Burbank, 1973). The subsurface maxima (nepheloid layers) do not appear to be confined to any particular area in the port. The presence of an unusual nepheloid layer at station S-7 (Fig. 7-3) was observed during two consecutive low tides at depth of 100 m (Fig. 7-32). Surprisingly, the amount of suspended load observed at that depth was comparable to that of the surface. Furthermore, microscopic examination of the suspended sediments from 100 m depth revealed that the material consisted of well-sorted silt, while the sediments in the surface plume consisted typically of clay with some fine silt. The presence of occasional nepheloid layers suggests subaqueous transport in the inlet. The size and occurrence of these layers have not been monitored long enough to determine the magnitude of the subaqueous sediment transport.

Port Valdez is a typical fjordal basin, characterized by steep slopes and sharp constrictions. The rugged topography surrounding the fjord has been extensively glaciated. Recent glaciation has formed steep slopes with unstable deposits. Occasional snow and rainfall associated avalanches and earthquakes cause these unstable deposits to slide downslope toward the inlet. Large amounts of these sediments are mostly deposited in the near-shore region to form steeply inclined deltas. Additionally, the surf action along steep shores is generally very strong, thus undermining cliffs and leading to extensive cliff falls. A conglomeration of fragments comprising a wide range of sizes is often torn off the slope as a result of slumping, or due to shoreline scarps which slide downslope and fall into the basin. This sliding mass has tremendous kinetic energy, and during downslope movement the sediments are deformed in complex ways: They may be folded into blocks and at times form a homogeneous unstratified mass. In an extreme case of a very strongly disturbed zone, the primary sediments disintegrate completely. The disintegrated character of the sediments is indicated only by very small sporadic fragments which preserve the normal bedding of the mud, at times forming a distinctive, complex flow-like structure in clay. The shapes of mud fragments (clay balls observed in cores) in the disturbed layers vary from sharply angular to ellipsoidal. Sediments which have been in the path of the slide are generally intensely reworked, resulting in elutriation of fine sediments, as well as in graded bedding. The large and dense turbidity currents (due to downslope mass movement) also give rise to graded sedimentation. Downslope, at any given moment, grains of widely differing sizes are deposited, but despite the poor sorting laterally, excellent vertical sorting develops. The sediments from cores (Fig. 7-11) indicated that with increased distance from Fan I a decrease in grain size accompanied with thickening of their beds occurred horizontally, while the coarse material formed a prism with its base toward the head of the inlet. The average grain diameter on a graded bed appears to be proportional to the thickness of the bed.

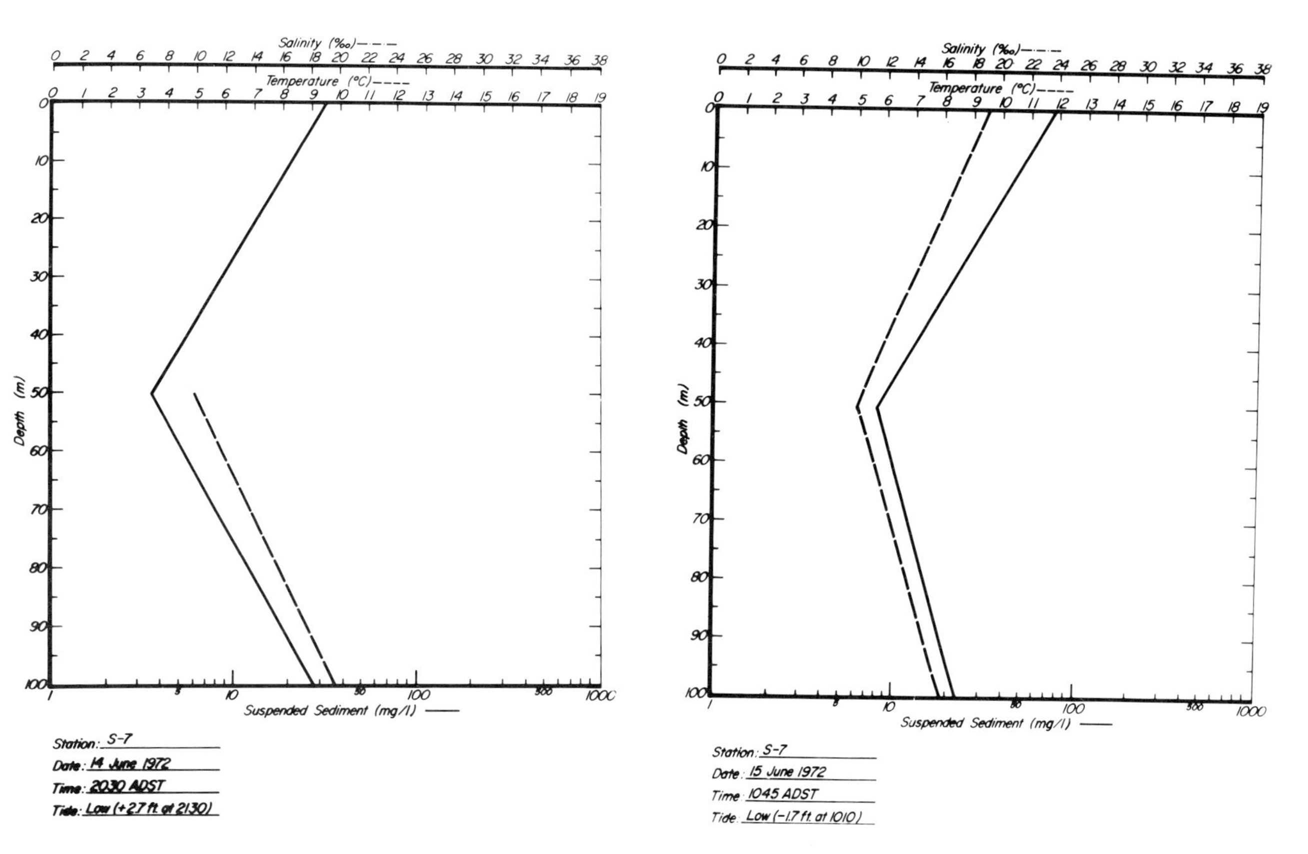

Figure 7-32. Depth profiles of suspended load and temperature, Port Valdez.

In summary, the sediment textural distribution in Port Valdez indicates two distinct depositional environments with characteristic sedimentary facies. The major input of sediments near the head and water circulation in the port result in coarser sediments being deposited in the eastern half and finer sediments in the west. The morphology of clay particles indicates minimal chemical alteration and diagenesis of sediments at the source and during transport. The clay size fraction entering Port Valdez contains approximately equal parts of mica and chlorite in the east. However, bottom sediments in the port itself become more chloritic westward. The textural and clay mineral variations are the result of a differential particle settling. The elemental distribution in sediments with Al, Mg, and K increasing westward conforms to the sedimentary facies pattern and, therefore, can be attributed to the sediment particle differentiation during transport and deposition.

REFERENCES

Carroll, D. (1970). Clay Minerals: A Guide to Their X-Ray Identification. Geol. Soc. Amer. Spec. Paper 126, 80 p.

Coulter, H. W., and R. R. Migliaccio. (1966). Effects of the Earthquake of March 27, 1964, at Valdez, Alaska. U.S. Geol. Survey Prof. Paper 542-c, p. 35.

Moffit, F. H. (1954). Geology of the Prince William Sound region, Alaska. *U.S.G.S. Bull.* 989-E:225-310.

Muench, R. D., and D. L. Nebert. (1973). Physical oceanography. In *Environmental Studies of Port Valdez.* D. W. Hood, W. E. Shiels, and E. J. Kelley (Eds.), Inst. of Mar. Sci. Occ. Publ. No. 3, Univ. of Alaska, Fairbanks, pp. 103-149.

Parham, W. E. (1966). Lateral variations of clay mineral assemblages in modern and ancient sediments. In *Proc. Intl. Clay Conf.* L. Heller and A. Weiss (Eds.), Israel Program for Scientific Translations, Jerusalem, pp. 135-145.

Plafker, G., and L. R. Mayo. (1965). Tectonic Deformation, Subaqueous Slides, and Destructive Waves Associated with the Alaskan Earthquake, March 27, 1964, and Interim Geologic Evaluation. U.S. Geol. Survey, Open-File Rept., p. 19.

Sharma, G. D., and D. C. Burbank. (1973). Geological oceanography. In *Environmental Studies of Port Valdez.* D. W. Hood, W. E. Shiels, and E. J. Kelley (Eds.), Inst. of Mar. Sci. Occ. Publ. No. 3, Univ. of Alaska, Fairbanks, pp. 15-99.

Tarr, R. S., and L. Martin. (1914). Alaskan Glacier Studies of the National Geographic Society in the Yakutat Bay, Prince William Sound and Lower Copper River Regions. Natl. Geog. Soc., Washington, D.C., p. 498.

von Huene, R., G. G. Shor, Jr., and E. Reimnitz. (1967). Geological interpretation of seismic profiles in Prince William Sound, Alaska. *Geol Soc. Amer. Bull.* 78:259-268.

CHAPTER 8
Northwestern Gulf of Alaska Shelf

INTRODUCTION

The northwestern Gulf of Alaska Shelf, as defined here, comprises the continental shelf lying between Resurrection Bay and Unimak Island (Fig. 8-1). The region lies between 53°40′ and 60°0′N latitude and 149°0′ and 165°30′W longitude. Landward, the northeastern part of the shelf is bordered by Kenai Peninsula. At the southern end of the Kenai Peninsula the shelf adjoins Cook Inlet. Near the mouth of Cook Inlet the shelf has a series of northeast-southwest oriented islands, the largest of which is Kodiak Island. The bathymetry surrounding these islands is complex because the islands bisect the shelf, forming 300-km-long and 40 to 65 km wide Shelikof Strait. South of Kodiak Island, the shelf is bordered by the Alaska Peninsula, a volcanic arc, to the north.

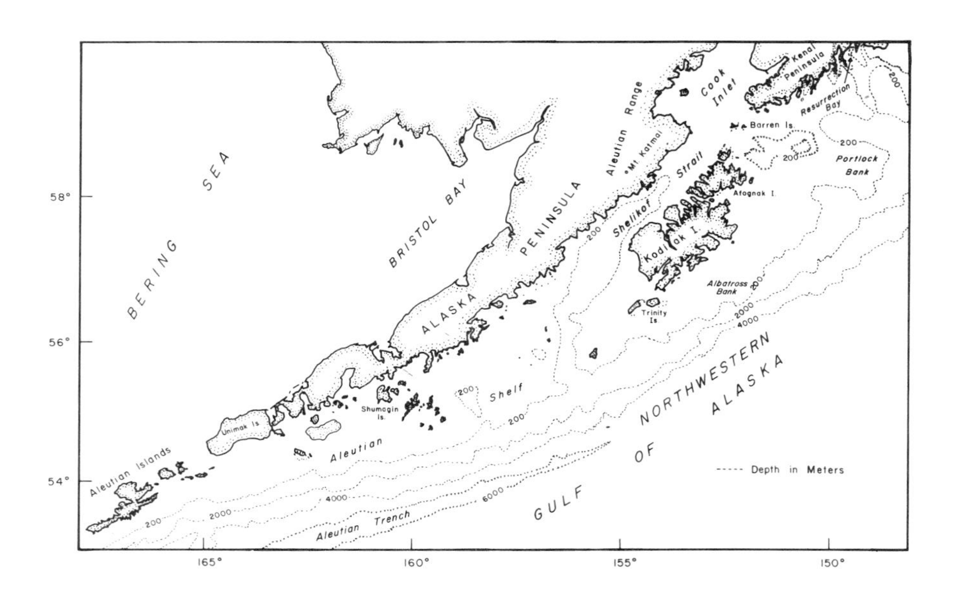

Figure 8-1. Index map with bathymetry (depth in meters), northwestern Gulf of Alaska.

Characteristically the shoreline of the shelf is young and rugged; its steep mountainous terrain and a highly irregular coastline indented by bays, inlets, lagoons, fjords, and channels are major features. Occasionally there are narrow beaches typical of young shoreline. The continental shelf is widest in the north, about 250 km; southward between Kodiak and Unimak islands it gradually becomes narrower, about 50 km near Unimak Island. Throughout its length, the shelf contains troughs and valleys and is dotted with islands of various sizes.

Seaward, the continental margin descends steeply to the Aleutian Trench to the southeast. The width of the rugged continental slope varies between 30 and 100 km; it is widest in the northeast and gradually narrows to the southwest. A unique feature of the slope is its steepness; from the edge of the shelf it drops more than 5,000 m to the trench. In general the slope varies between 3 and 10°. The upper part of the continental slope has a relatively smooth and gentler slope. The middle part is slightly steeper and contains few benches. The bottom section is very steep and rough (von Huene, 1972; Piper *et al.,* 1973).

GEOLOGY

Events of the geologic evolution of the region are well recorded in the bedrock exposed along the Kenai and Alaska peninsulas, and Kodiak Island. Some seismic data revealing the history of the adjacent shelf has also been obtained by various investigators (Piper *et al.,* 1973; von Huene *et al.,* 1971, and von Huene, 1972). The bedrock geology of the terrestrial region indicates that the shelf may contain two distinct and genetically different lithologic terrains, the Alaska Peninsula unit and the Kodiak-Shumagin Island unit. Each unit has its own lithologic characteristics, structural trend and geologic history. The boundary separating these two units, however, remains obscure. The basic difference in geologic character of these two regions may have been the result of a deep and extensive fault developed along the former boundary of the crustal plates. On the other hand these two regions may represent former sites of shallow and deep basins adjoined by a continental mass.

The Alaska Peninsula unit on the shelf consists of three major sequences. The basement rock is a partly metamorphosed sequence of clastic carbonate and volcanic rocks deposited between the Early Paleozoic and the Middle Jurassic, which is exposed on land and probably extends seaward. The middle sequence consists of deep marine clastic sedimentary rocks (flysch) deposited between the Middle Jurassic and Upper Cretaceous. The overlying Tertiary sequence is characterized by shallow marine and nonmarine clastics mixed with a variety of volcanic rocks. Distribution of these three units on the inner shelf is only vaguely known.

The Kodiak-Shumagin unit is comprised of two southwest-northeast trending narrow belts; the Kodiak Mesozoic province and the Kodiak Tertiary province (von Huene *et al.,* 1971). The Mesozoic belt, in part, is exposed on the Kenai

Peninsula and on the Kodiak-Shumagin Island system. The Tertiary province lies to the southeast and is mostly under the sea.

Mesozoic rocks exposed on the Kenai Peninsula, Kodiak Island and seaward of the Shumagin Islands are similar in lithology and structure. This similarity suggests that Mesozoic deposits form a continuous belt which extends along the entire shelf. The well-defined thick sequence observed in the outcrop area appears to extend offshore beneath the shelf. In the southeast outcrop areas, the upper boundary of the Mesozoic deposits is defined by a wide fault zone. Seaward, adjacent to this fault zone, lies the Kodiak Tertiary province and the subsurface seaward extension of Mesozoic rocks is masked by the overlying thick sequence of Tertiary materials.

The second main geologic province, the Kodiak Tertiary province, lies mostly under the sea. This is comprised of a thick sequence of Early to Late Tertiary, shallow to deep marine clastic sediments, of which only a small part is exposed along the southeast shore of Kodiak Island, or the Trinity and Shumagin islands. The offshore Tertiary sequence has been partially defined from geologic and geophysical surveys of the shelf and partly inferred from deap-sea drilling data obtained off-shelf (Shor, 1962; von Huene *et al.*, 1967, 1971, 1972; Kulm *et al.*, 1973). The major geologic elements of the Kodiak Tertiary province are a wide fault zone, a Late Tertiary Basin, and a broad arch along the edge of the continental shelf.

The Tertiary province is bordered on the northwest by a fault zone up to 30 km wide that runs parallel to the Mesozoic province. The fault separates the raised Kenai-Kodiak block from an offshore, subsiding, Late Tertiary Basin. The basinal structure, filled with 3 to 4 km of sediments, probably extends northeast into the central Gulf of Alaska region, as described previously. This basin also appears to extend southwestward along the outer shelf. Seaward, the basin is flanked by a broad arch that forms the margin of the continental shelf, but the surface expression of this anticlinal feature is not continuous. It appears to be a deep-seated structure along which frequent deformation, involving tilting, folding, faulting, and truncation, has occurred during periods of uplift and subsidence. Truncated strata suggest net uplift of the arch, and the arch was possibly exposed subaerially during glacial periods.

The advent of the Pleistocene epoch brought repeated glaciation of the adjacent mountain ranges and lowlands. Substantial glaciation of the region is obvious from the fjord-indented coastline. During major glacial advances the sea level receded toward the shelf margin, and ice covered much of the continental shelf. Most sediments eroded by the glaciers during low sea levels probably were deposited in the trench. Some of the sediments, carried as nepheloid layers, bypassed the trench and were deposited in the North Pacific Basin. Thick Pleistocene sediment deposits in the Aleutian Trench have been reported by Kulm *et al.* (1973) and Piper *et al.* (1973).

Contemporary sediments on the shelf form a thin, irregular veneer ranging to a maximum of tens of meters in thickness. Shallow banks generally either are devoid of or have a minimal amount of contemporary sediments, and the maximum thicknesses have accumulated in the depressions.

In summary, the region was a deep basin during the Paleozoic and Early Mesozoic eras, with widespread deposition of clastics, volcanics and carbonates. During the Jurassic, the sediments were intruded by large amounts of granitic materials, which formed the Alaska Peninsula, while deep water clastics were deposited on the shelf and slope to the southeast. During the Early Tertiary, the shelf region to the southeast was intruded by granitic materials, and uplift followed, forming the present continental shelf configuration. Although major deformations occurred during or after the intrusions, the area has been tectonically active throughout the Tertiary and has been undergoing continuing structural changes subsequently. Because of tilting and subsidence during the Tertiary, the shelf has accumulated a sediment thickness of over 4 km.

The Gulf of Alaska is well known for its high seismic and volcanic activity. One of the most powerful earthquakes ever recorded (1964 Alaska earthquake) and one of the greatest volcanic explosions (Mount Katmai, 1912) occurred in this region. The underthrusting of the Pacific crustal plate against the North American plate is not a continuous process but proceeds in a "jolty" fashion which causes tremors. Andesitic volcanisms, which often cause rather violent explosions, are typical for a region of plate convergence. About 12 volcanoes have been active in this region since 1760 and a few more probably erupted during post-glacial time.

Exposed coastal rocks are mainly of recent volcanic origin and are predominantly intermediate to basic in composition. They consist of andesites, andesitic basalts and tuffs. Mesozoic and Cenozoic sedimentary rocks are widespread on the Kenai Peninsula, Kodiak Island and the Alaska Peninsula. Some exposed metamorphic rocks have also been observed in the region.

BATHYMETRY

A detailed bathymetry of the region has been included in the "Bathymetric Atlas of the Northcentral Pacific Basin" prepared by the Scripps Institute of Oceanography (Menard and Chase, 1971). The shelf is characterized by numerous islands, plateau-like surfaces and sea valleys. The islands are generally found in the near-shore zone, which forms a narrow belt adjacent to the coastline, extending to the 30-50 m isobath. Numerous inlets and fjords are also found in this zone. The shallow zone is actively eroded by wave action.

Offshore, the main part of the shelf between the 50 and 220-m isobaths contains large, broad, plateau-like surfaces with banks. The plateau-like surfaces, have a very low gradient of 1-5′. These relatively smooth surfaces, however, are interrupted by many banks and shoals that often rise abruptly, some reaching above sea level to form islands. The banks and shoals found along the outer shelf are of particular interest. They show truncated beds and faulting, suggesting uplift during the recent past. Albatross Bank, off Kodiak Island, is a good example. The plateau-like shelf also includes a few closed depressions (Fig. 8-1).

The relatively flat shelf bottom is cut in many places by sea valleys, which separate the plateau-like surfaces. These sea valleys either are aligned essentially parallel to the shelf length or cut across the shelf. Those which run parallel or subparallel to the shelf are generally broad and flat-bottomed, with steep sides. The large size of these valleys suggests that they may have resulted from orogenic movement and may represent areas of rapid subsidence. A good example of such a sea valley is Shelikof Strait, which extends southwest for approximately 300 km and is 40 to 65 km wide and almost 300 m deep. At each end the valley has sills (about 100 m high). Shelikof Strait appears to be a continuation of Cook Inlet and may possibly be underlain by a thick prism of Tertiary sediments.

The smaller sea valleys traverse the shelf width. They generally extend seaward from bays and inlets along the near-shore zone. These U-shaped valleys are only 10 to 50 km wide and vary greatly in length. The longitudinal profile of the valley generally shows the deepest part in the middle, typical of a glacially formed valley.

The large sea valleys appear to be of tectonic origin, modified by glaciers during periods of low sea levels. The smaller valleys, on the other hand, may have been formed entirely by glaciers. Their origin is primarily inferred from the bathymetry of the northern shelf off Kenai Peninsula. It appears that during low sea levels, the glaciers in the vicinity of Resurrection Bay followed the fjords and converged into two major glaciers. One of these glaciers moved in a southeasterly direction to form a trough around the southwestern end of Montague Island. The other glacier continued on a southerly direction. Both channels have a rise in the seaward end of their profile inferred to be end moraines.

The region between Kenai Peninsula and the Kodiak group of islands is characterized by two straits separated by the Barren Islands. Off Kodiak Island, to the northeast and southeast lie two large banks; Portlock and Albatross, respectively. The Albatross Bank is divided by two troughs and has a complex floor with small steeply dipping scarps and depressions. The bank appears to be an eroded anticline with several fault scarps.

The major features of the shelf off Alaska Peninsula are the Shelikof Strait trough extending southward to the slope, various groups of islands, and banks.

HYDROLOGY

The coast adjacent to the northwestern Gulf of Alaska is rugged, steep and indented with fjords and embayments. Therefore most streams are short, with steep gradients, and are fed by small drainage basins. Because of the maritime climate the area is characterized by high precipitation and mild temperature. At the higher altitudes enough snow falls to form a few small glaciers.

The hydrologic data routinely obtained by the U.S. Geological Survey (1972) from this area has been published in "Water Resources Data for Alaska" (1971). An excellent summary of the hydrology of the region has been prepared by Feulner (1973a and 1973b). The runoff varies throughout the area; southeast Kodiak Island has the region's highest recorded mean annual runoff, approximately 1.1 $m^3/sec/km^2$, but in most of the region the mean annual runoff averages about 0.04 $m^3/sec/km^2$. The high precipitation, falling on bedrock with a thin soil cover, and the lack of aquifers result in little storage and high rates of runoff. The mean annual high monthly runoff occurs during the summer and abnormally high runoff follows storms.

HYDROGRAPHY

The main elements which influence the characteristics and dynamics of the shelf water are the offshore permanent counterclockwise gyre of the Alaska Current and the terrestrial freshwater input. The Alaska Current is the northern part of the gyre which occupies the northern Pacific Ocean, and its influence on shelf water is greatly regulated by the prevailing atmospheric conditions.

In the central Gulf of Alaska, the water carried by the broad and slowly northwestward-flowing Alaska Current is deflected to the southwest. As the current passes Kodiak Island it becomes narrower and more intense, forming the Alaska Stream, which continues along the Alaska Peninsula and the Aleutian Islands. The flow is most intense near the continental slope, and it is estimated that current speed there may vary between 25 and 50 cm/sec (McEwen *et al.,* 1930; Thompson *et al.,* 1936; Favorite, 1970).

The surface component of the southwest flowing current is greatly influenced by storm and seasonal winds. The extension of the winter Aleutian Low in the Gulf of Alaska increases the southwest transport of water and thus intensifies the current at depth as well as at the surface. This intensified circulation also induces a divergence in the Ekman transport of surface waters, causing upwelling. Winds associated with frequent winter storms cause local eddies.

The water characteristics of the Alaska Stream vary seasonally. In winter the surface layer is isohaline and isothermous to a depth of 100 m as a result of wind mixing and heat loss to the atmosphere. The temperature and salinity of the surface layer are $<3.5°C$ and $32.8°/_{oo}$, respectively.

With the onset of summer, the shelf water undergoes a very marked change, *i.e.,* seasonal variation. During this season, the freshwater discharge to the surface layer is maximal. This warm surface layer is further heated by solar radiation and often mixed vertically by storms. The surface water temperature may exceed 10°C but rapidly decreases with depth beneath the shallow thermocline at about 100 m.

The water dynamics on the shelf are not well known. However, it is certain that the Alaska Stream, flowing along the shelf break, should induce a general southwest drift in the shelf water. In winter, when there is seasonal wind stress, the drift should be maximum. In summer, however, the extension of the East Pacific High into the Gulf of Alaska should counter this drift. The water circulation on the shelf is also significantly influenced by tides, the fluvial runoff, and local winds. Tides are of a mixed type, with amplitudes varying between 2 and 4 m. Locally, the tidal currents and directions in the near-shore area are extremely variable. Seasonal surface outflow of fluvial waters, particularly during summer months, should significantly modify the circulation in the inshore waters. In near-shore areas, and specially in embayments, the surface runoff tends to counter the flood tide and accelerate the ebb tide.

The effect of winds on the shelf circulation is two-fold. First, the winds cause turbulence in surface waters, which may lower the pycnocline or destroy the water structure and could cause significant water movement on the shelf. Second, seasonal winds produce a surface wind stress which sets up a temporal circulation pattern. Locally, frequent storms may induce temporarily anomalous water characteristics and movements.

In general, the water masses formed on the shelf are more complex than offshore water. The complexity arises mainly from the seasonal variability of the surface runoff and mixing caused by erratic winds. Therefore the water structure varies locally and seasonally. The shelf water generally has a salinity between 32 and 33‰, with the temperature varying from 4 to 11°C. The outflowing fresh surface water causes intrusion of oceanic waters at depth, which appear to move along the depressions. Cold oceanic water is often observed in shelf depressions, forming ponds of waters distinctly different from normal shelf water. Whether this water has been brought to the shelf by outflowing water at the surface, or represents remnants of the preceding winter's shelf water at depth is not known.

SEDIMENTS

The textural distribution of bottom sediments from the northwestern Gulf of Alaska Shelf varies considerably, regionally as well as locally. The distribution is primarily affected by source, bottom topography, and hydrodynamics. Because of the major regional patterns in the textural parameters, the sediments are described in two groups:

1. Kodiak Shelf: the broad shelf surrounding Kodiak, Afognak, and Trinity islands, including Shelikof Strait
2. Alaska-Kenai Peninsula Shelf: the shelf off Kenai Peninsula between Resurrection Bay and Kodiak Island and the shelf off the Alaska Peninsula (Aleutian Shelf) between Kodiak Island and Unimak Pass.

Reconnaissance studies of shelf sediments from these regions have been reported by Gershanovich *et al.* (1964), Gershanovich (1968 and 1970) and Hampton *et al.* (1979). The mineral composition of the sediment fraction comprising the 0.01-0.1 mm size class (medium silt - fine sand) has been described by Bortnikov (1970).

The sediments from the shelf surrounding Kodiak Island primarily consist of gravelly sand and sandy gravel (Fig. 8-2 and Table 8-1). Occasionally, silt and clay are found in depressions. The sediment texture is primarily controlled by the bathymetry and hydrodynamics of the region. Coarse material is abundant on the banks and along the beaches, the rocky headlands, and small islands. The plateau-like areas, banks, and shoals are usually covered by poorly to extremely poorly sorted gravelly sand and sandy gravel. These sediments contain pebbles, cobbles, and skeletal carbonate fragments. Some areas of the banks are densely populated with organisms, which contribute significantly to the sediment deposits.

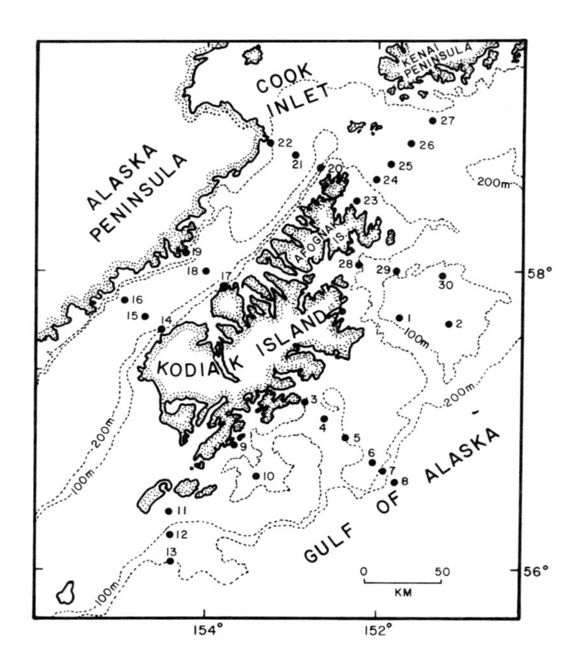

Figure 8-2. Bathymetry chart with station locations, Kodiak Island Shelf.

Sediments from the Peninsular Shelf show distinct regional textural characteristics and distributions. The shelf off the Kenai Peninsula consists of mainly fine sediments; the shelf adjacent to the Alaska Peninsula (Aleutian Shelf), on the other hand, is mostly covered with coarse sediments (Figs. 8-3 to 8-6). This distinctive diversity in texture may be the result of differences in provenance as well as sediment transport processes.

The sediment cover on the shelf off the Kenai Peninsula is mainly sand; gravelly, sandy and clayey silts; and silty clay (Figs. 8-3 to 8-7). They are extremely

TABLE 8-1

Textural Parameters of the Sediments from Kodiak Shelf

Sample Number	Gravel (%)	Sand (%)	Silt (%)	Clay (%)	Mean Size (M_Z)	Sorting (σI)	Skewness (Sk_I)	Kustosis (K_G)
KS-1		100.00			1.86	0.78	-0 24	1.33
KS-2	19.68	80.32			0.10	1.96	-0 29	0.99
KS-3	32.83	67.17			-0.35	2.30	-0.03	0.75
KS-4		13.01	64.89	22.10	0.55	2.39	0.28	1.42
KS-5		100.00			1.72	1.06	-0.20	0.92
KS-6		100.00			1.86	0.44	0.08	1.06
KS-7		93.71	3.92	2.37	0.88	2.85	0.30	0 96
KS-8	6.22	93.78			0.90	1.34	-0 37	0 89
KS-9		100.00			2.74	0.96	0.09	0.99
KS-10		20.74	53.91	25.35	6.39	2.75	0 09	1 00
KS-11		100.00			2.76	0.28	0.05	1.03
KS-12	57.86	30.22	9.09	2.83	-1.54	3 98	0.46	0 70
KS-13	14.71	64.72	13.20	7.37	1.99	3.36	-0.01	1 19
KS-14	44.57	55.43			-0.52	2.53	0 08	0.63
KS-15		1.01	45.48	53.51	8.5	2.34	0.18	0 97

TABLE 8-1 (Continued)

Sample Number	Gravel (%)	Sand (%)	Silt (%)	Clay (%)	Mean Size (M_Z)	Sorting (σI)	Skewness (Sk_I)	Kustosis (K_G)
KS-16		1.42	51.03	47.55	8.06	2.30	0.15	0.97
KS-17	19.19	80.81			0.29	2.81	-0.64	1.24
KS-18	33.45	53.38	7.66	5.51	0.56	3.81	-0.29	1.01
KS-19	62.43	20.24	10.75	6.58	-0.99	4.40	0.81	0.97
KS-20	35.56	54.24	4.71	5.49	-0.39	3.50	-0.09	1.24
KS-21	7.97	43.61	24.11	24.31	5.08	4.12	0.15	1.10
KS-22		74.95	15.20	9.85	4.07	2.10	0.59	1.74
KS-23	31.91	63.77	2.26	2.06	-0.03	2.94	-0.30	0.64
KS-24	18.03	75.94	2.09	3.94	0.73	2.88	-0.44	2.46
KS-25	7.57	84.28	2.57	5.53	1.05	2.39	-0.06	3.55
KS-26		95.84	1.28	2.88	1.56	1.11	-0.31	1.74
KS-27	20.60	57.94	5.96	15.50	1.42	5.79	0.0	2.83
KS-28	1.40	81.59	10.47	6.54	0.71	4.49	-0.27	1.43
KS-29		18.52	67.34	14.14	5.54	2.20	0.36	1.91
KS-30	36.07	59.40	2.50	1.33	-0.44	2.35	0.21	0.75

poorly sorted and their mean size ranges between 3.5 and 8.5ϕ (Figs. 8-8 and 8-9). Contrary to the usual offshore sediment grading, the sediment grain size in this region coarsens seaward. Due to a lack of sediment samples from the outer shelf, it is not certain whether this reverse grading extends to the shelf edge. In the central Gulf of Alaska, to the east, the outer shelf and slope indeed does consist of coarse glacial sediments. It is therefore conceivable that in this region the coarse relict glacial sediments in the near-shore area are covered with a thin layer of fine contemporary sediments. Seaward, the contemporary sediments taper off, leaving glacial relicts exposed. The seaward coarsening of sediments in this region, however, may be the result of elutriation of fines from contemporary sediments by the gradual seaward increase in water velocity.

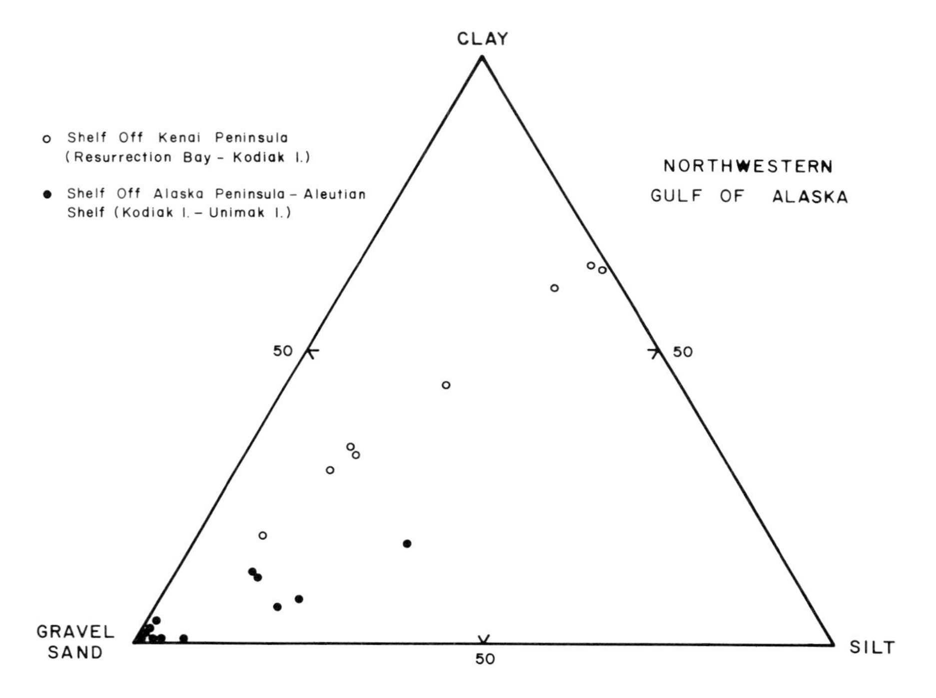

Figure 8-3. Percent gravel-sand, silt and clay in sediments, northwestern Gulf of Alaska.

The sediments from the shelf adjacent to the Alaska Peninsula are mostly clayey silty sands and locally sandy gravel (Figs. 8-3 to 8-7). The sediment mean size varies from 1 to 4ϕ, and sorting ranges between poor to extremely poor (Figs. 8-8 and 8-9). Fine sediments occur in the elongated depression of Shelikof Strait, and extend southwestward on the shelf. Clayey silt and silty clay mantle the floor of this broad sea valley. On the rest of the shelf mostly sands prevail and a general seaward fining of sediments occurs. Generally coarse sediments

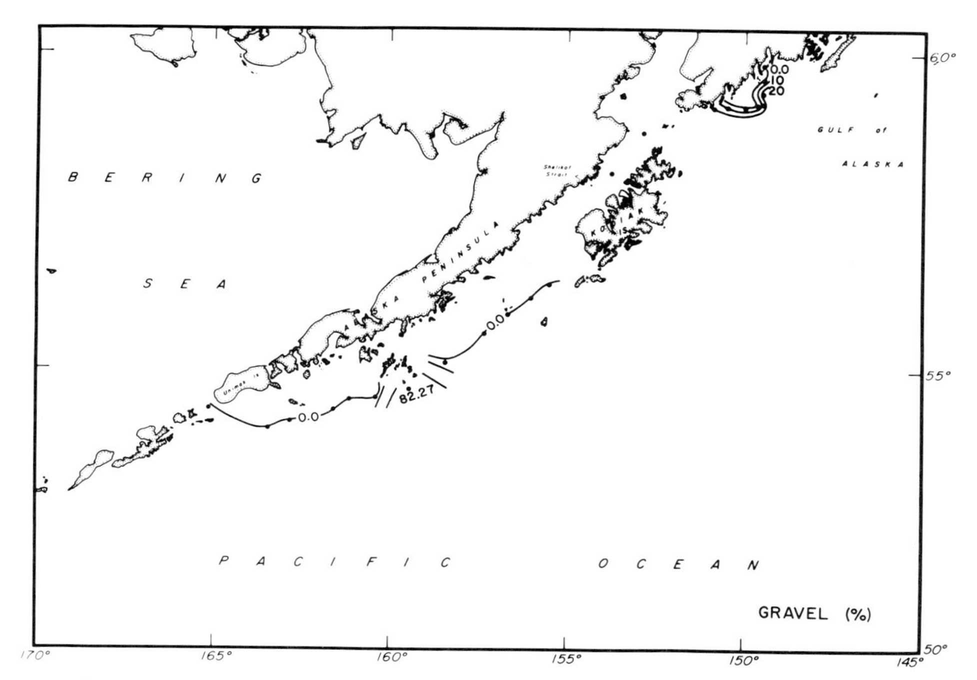

Figure 8-4. Weight percent gravel in sediments, northwestern Gulf of Alaska.

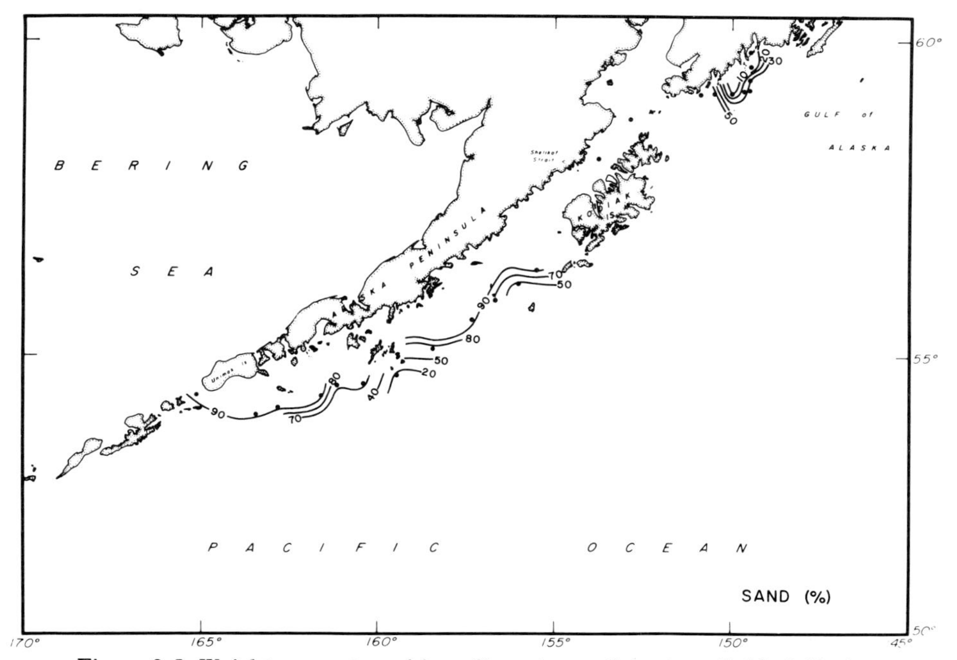

Figure 8-5. Weight percent sand in sediments, northwestern Gulf of Alaska.

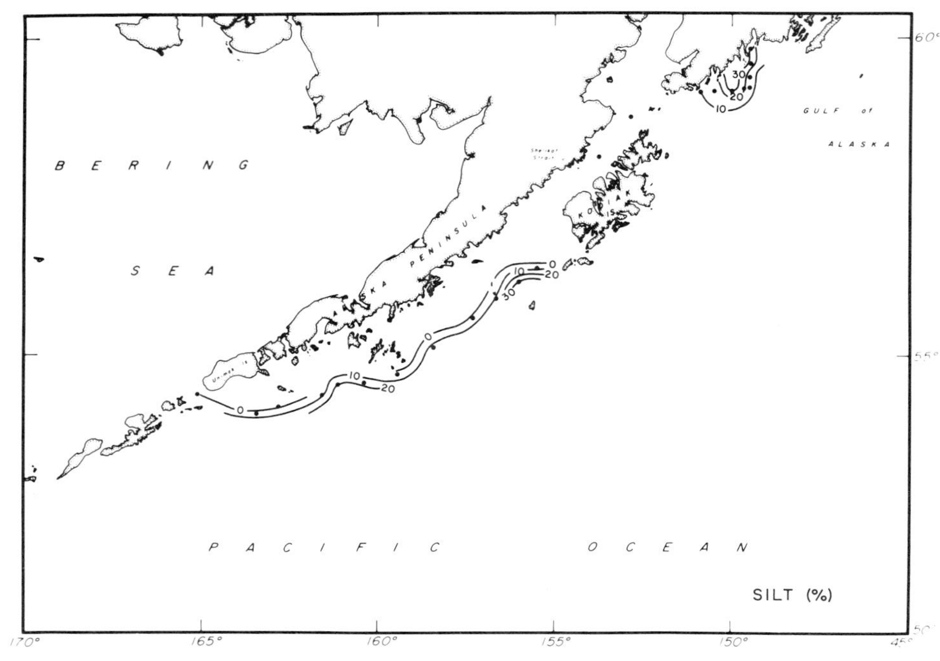

Figure 8-6. Weight percent silt in sediments, northwestern Gulf of Alaska.

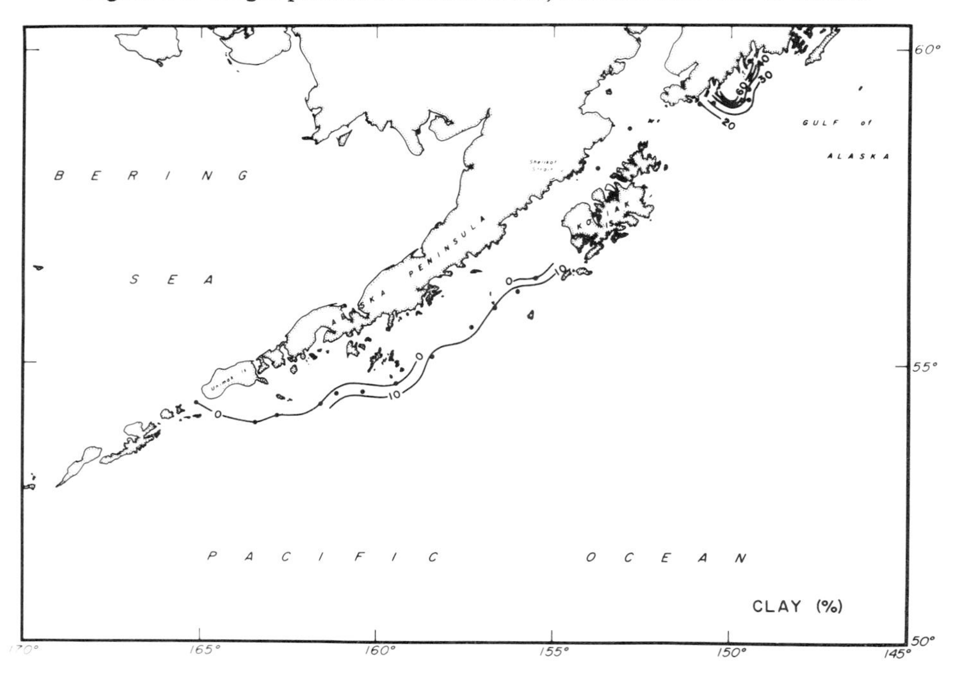

Figure 8-7. Weight percent clay in sediments, northwestern Gulf of Alaska.

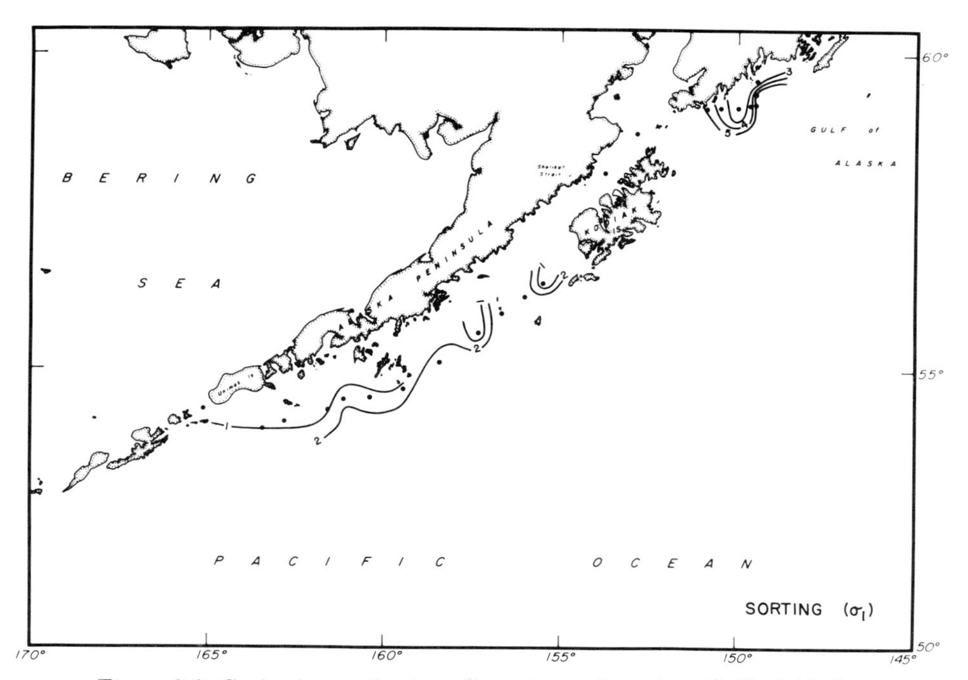

Figure 8-8. Grain size sorting in sediments, northwestern Gulf of Alaska.

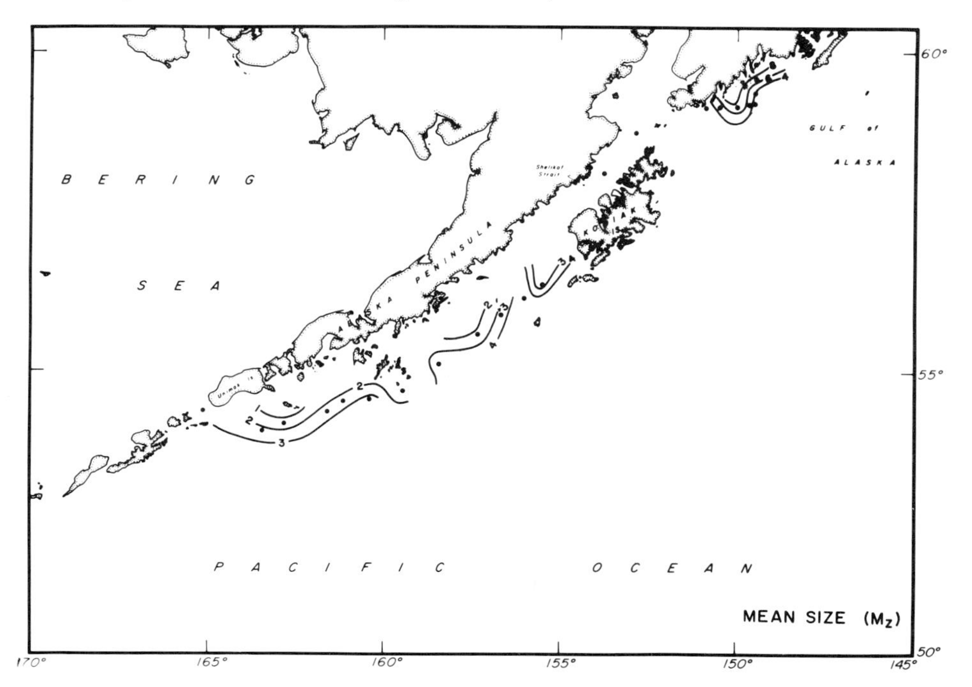

Figure 8-9. Sediment mean size distribution, northwestern Gulf of Alaska.

dominate in near-shore areas, with depths <80 m, while fine sand and silt mantle the middle and outer shelves, at depths varying between 80 and 150 m.

The mineralogy of the 0.05 to 0.1 mm (very coarse silt to fine sand) sediment fraction from the northwestern Gulf of Alaska Shelf has been described by Bortnikov (1970). The light mineral fraction consists mostly of quartz and orthoclase, with minor amounts of plagioclase and volcanic glass. In some samples, volcanic glass constituted as much as 90% of the total sample, probably Katmai ash ejected during 1912. Quartz, the most abundant constituent, generally varies between 5 and 30%, followed by orthoclase (2 to 25%) and plagioclase (1 to 9%).

The heavy mineral content (sp. gr. >2.9) in the bottom surface sediments varies from 2 to 10%, with local concentrations exceeding 50%. The most ubiquitous components in the heavy mineral suite are the opaque minerals (magnetite, ilmenite and titanomagnetite), amphiboles (mostly hornblende), pyroxenes (clino- and orthopyroxene), epidote and zoisite. Micas, garnet, zircon, tremolite and actinolite occur in small àmounts.

Organic carbon in sediments varies from 0.3 to 2.0% by weight. With the exception of one station most sediments contain approximately 0.5% organic carbon (Fig. 8-10). Interestingly, the organic carbon in the northeastern part of the shelf increases with increasing sediment grain size (Figs. 8-8 and 8-9). The sparsity of samples from the southwestern shelf does not show a specific relationship between organic carbon content and the sediment grain size distribution.

GEOCHEMISTRY

Chemical data from the northwestern Gulf of Alaska Shelf are rather scanty. This limitation precludes extensive discussion of the geochemistry, as well as the inferences which might be drawn therefrom. Nevertheless the data described here provides general information concerning the chemical composition of the sediments and elemental variations in this region. Distributions of most of the major as well as minor elements in the sediments display significant regional and local variations. These differences in distribution may be useful in delineating sources and movement of sediments on the narrow and complex shelf.

The distribution of aluminum in sediments shows a regional pattern. In the northeast (off the Kenai Peninsula), its abundance is about 8% and varies within a very narrow range. In the southwest (off the Alaska Peninsula) it varies between 5 and 9% (Fig. 8-11). In general, aluminum content increases offshore as well as southwestward. The distribution is similar to percent silt and the mean size distributions in the sediments; aluminum content increases with increasing silt content and decreasing sediment mean size. These similarities reflect the usual increase in aluminum content with increasing content of the silicates generally associated with fine (silts and clays) sediments.

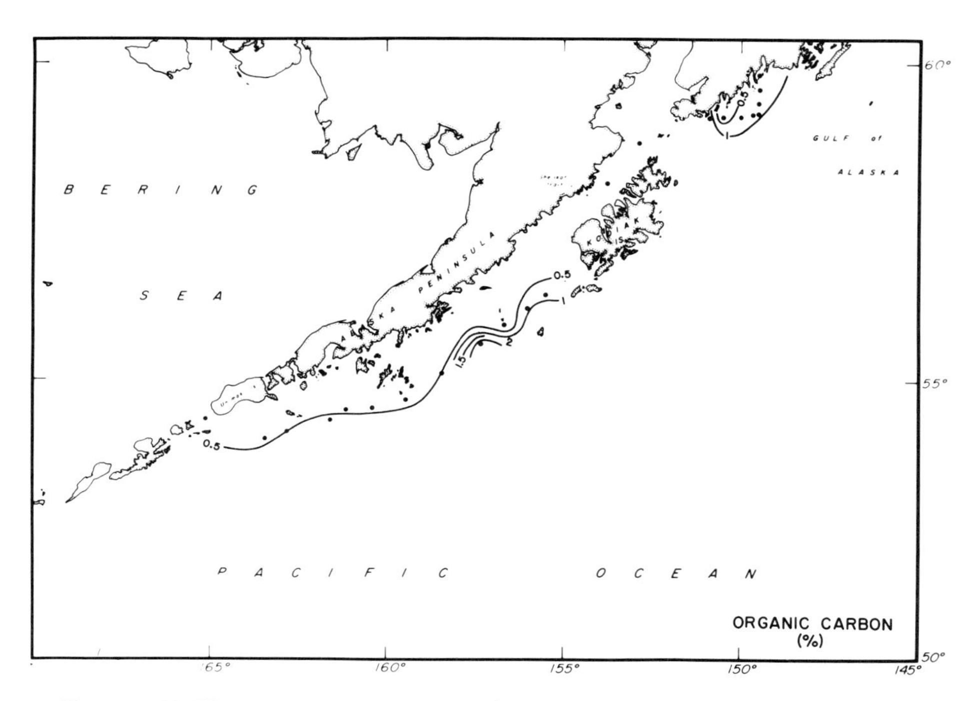

Figure 8-10. Weight percent organic carbon in sediments, northwestern Gulf of Alaska.

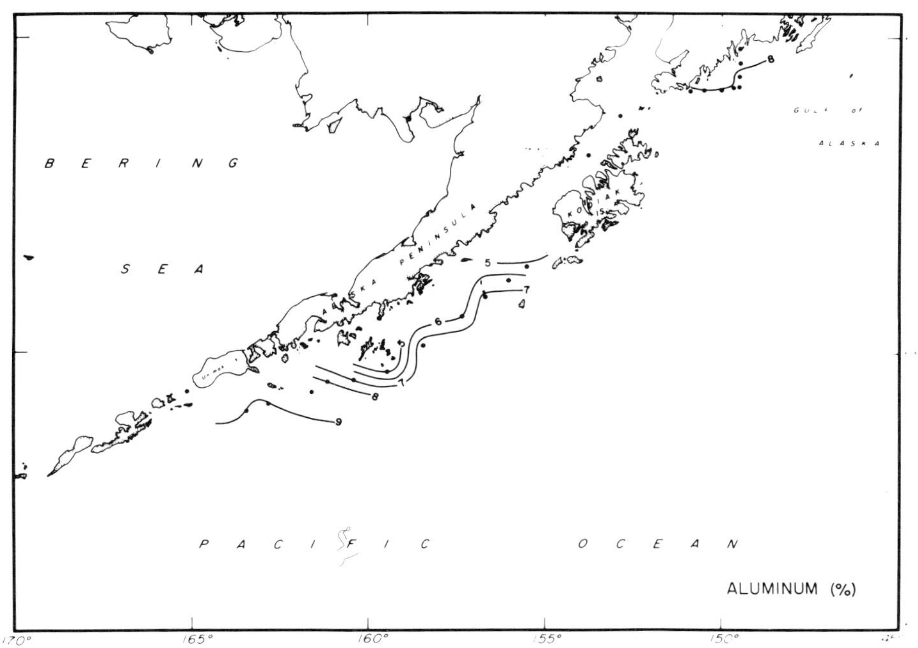

Figure 8-11. Percent aluminum in sediments, northwestern Gulf of Alaska.

Iron distribution in sediments is similar to that of aluminum. On the northeastern shelf, its abundance varies from 4.5 to 6.0% (Fig. 8-12), whereas in the southwestern shelf it varies over a wider range (2.5 to 6.5%). These variations appear to be related to the mean size and the calcareous biogenous content of the sediments. A decrease in sediment mean size is related to an increase of iron in sediments. Conversely, the iron content decreases with increased content of the calcareous biogenous component, as indicated by the distribution of calcium (Fig. 8-13). The regions of sediments with high calcium content (11.8%) are the result of considerable amounts of carbonate minerals contributed by calcareous shell material. Shell fragments are distributed throughout the silt, sand, and gravel sediment fraction.

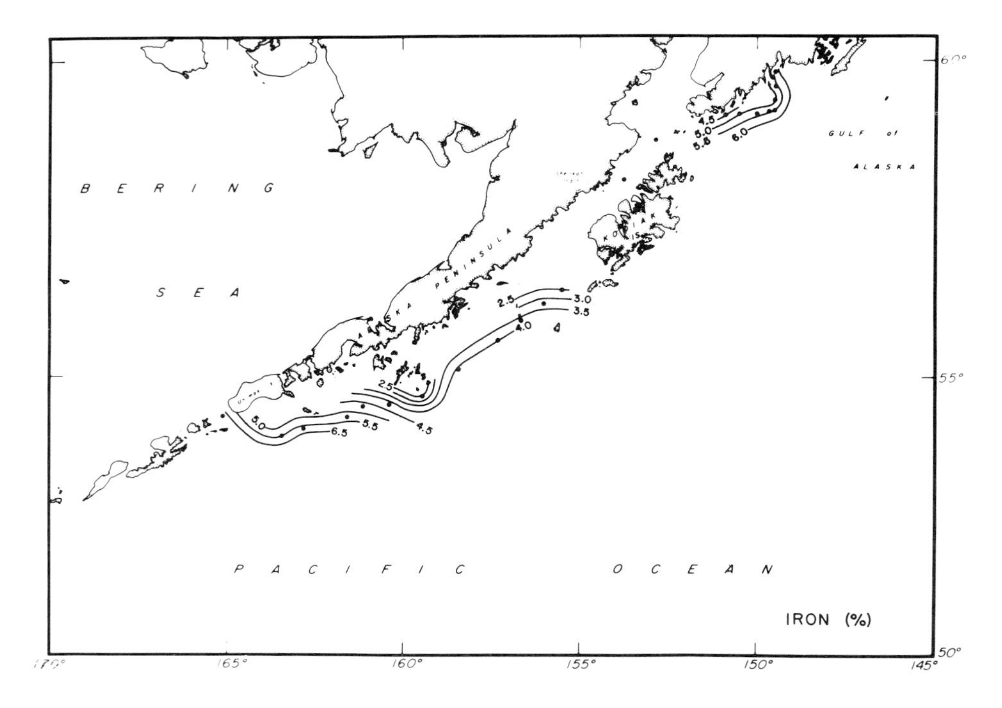

Figure 8-12. Percent iron in sediments, northwestern Gulf of Alaska.

The calcium distribution, like that of aluminum, also reflects the regional control on the elemental concentrations in the sediments. In the northeast part of the shelf, calcium content varies over a rather narrow range (2.0 to 4.0%), whereas to the southwest the concentration, on the average, is higher (3.0 to 6.0%) and the range is slightly broader (Fig. 8-13). The regional aspect and its influence on the elemental distributions is further manifested by the magnesium concentrations. Like calcium, magnesium varies in a narrower range (2.0 to

2.5%) in sediments from the northeastern area (Fig. 8-14), as contrasted with sediments from the southwestern shelf (1.0 to 2.5%).

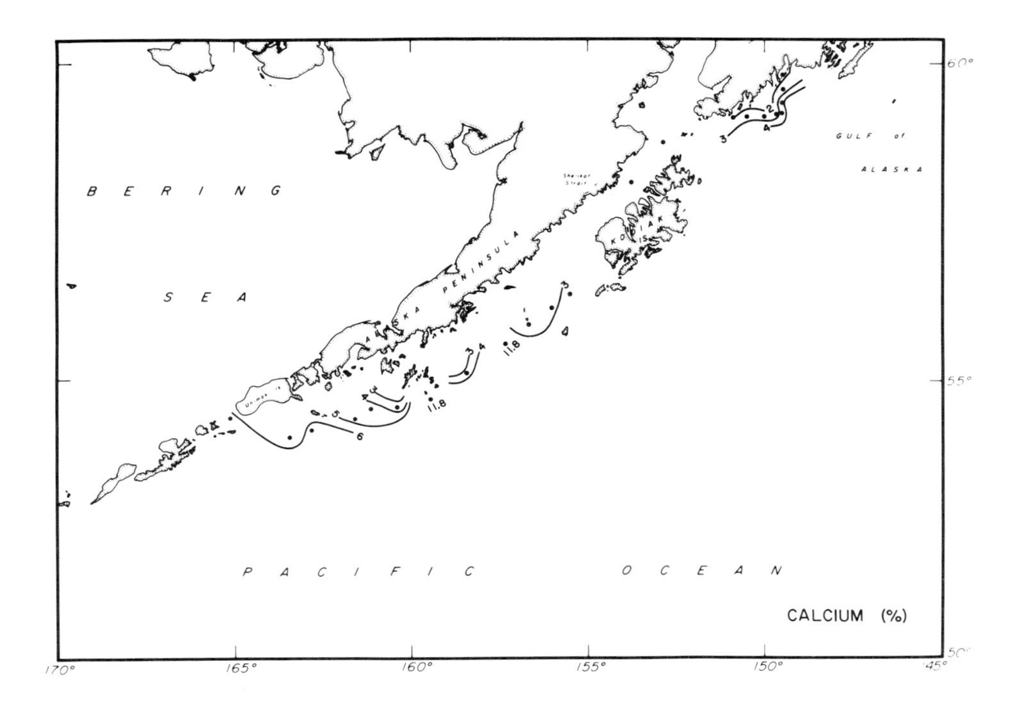

Figure 8-13. Percent calcium in sediments, northwestern Gulf of Alaska.

In general, both the calcium and magnesium distributions are related to sediment mean size; calcium and magnesium concentrations increase with decreasing grain size. Exceptions occur where the presence of calcareous shells not only increases calcium content abnormally but also increases the mean grain size of the sediments.

The distributions of sodium and potassium in these sediments are relatively simple. Both elemental concentrations increase seaward and show strong relationships with sediment mean size (Figs. 8-15 and 8-16), sodium and potassium contents increasing with decreasing grain size in sediments.

The manganese content (Fig. 8-17) varies over a considerable range (500 to 1,600 ppm). It increases seaward, and this appears to be related to decreasing sediment grain size. It is interesting to note that maximum manganese concentrations are found near the southwest tip of the Alaska Peninsula, off Unimak Island. The increased manganese in the sediment suggests a predominance of volcanic rocks at the source. Sediments from this area also contain relatively high amounts of aluminum, iron, and titanium.

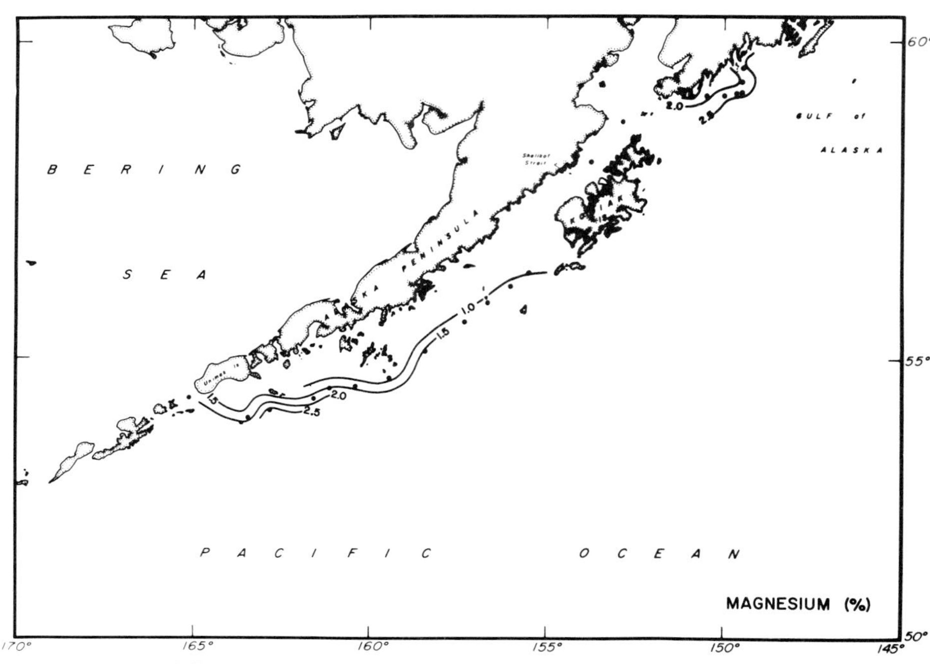

Figure 8-14. Percent magnesium in sediments, northwestern Gulf of Alaska.

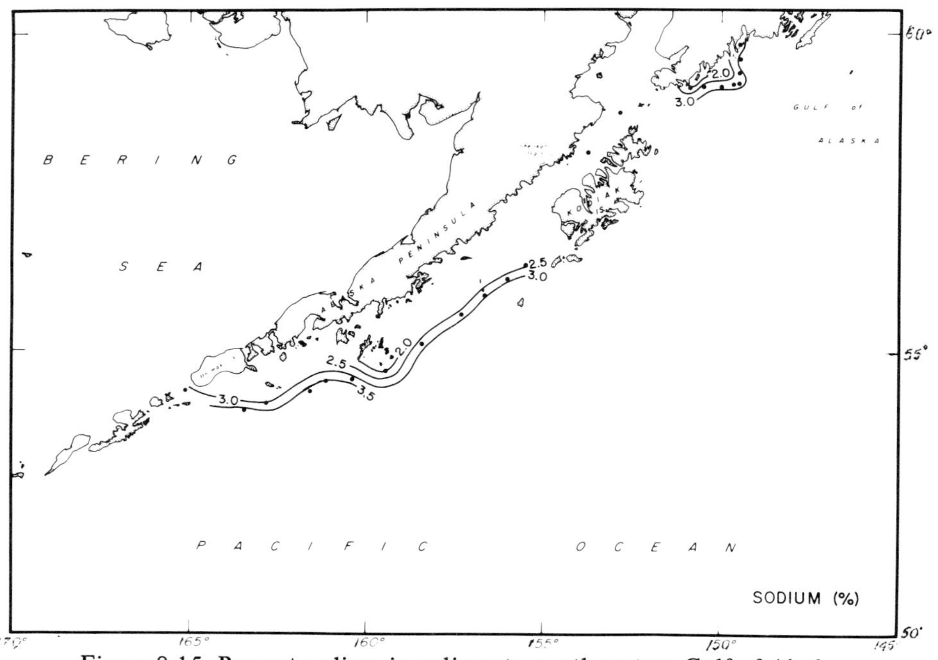

Figure 8-15. Percent sodium in sediments, northwestern Gulf of Alaska.

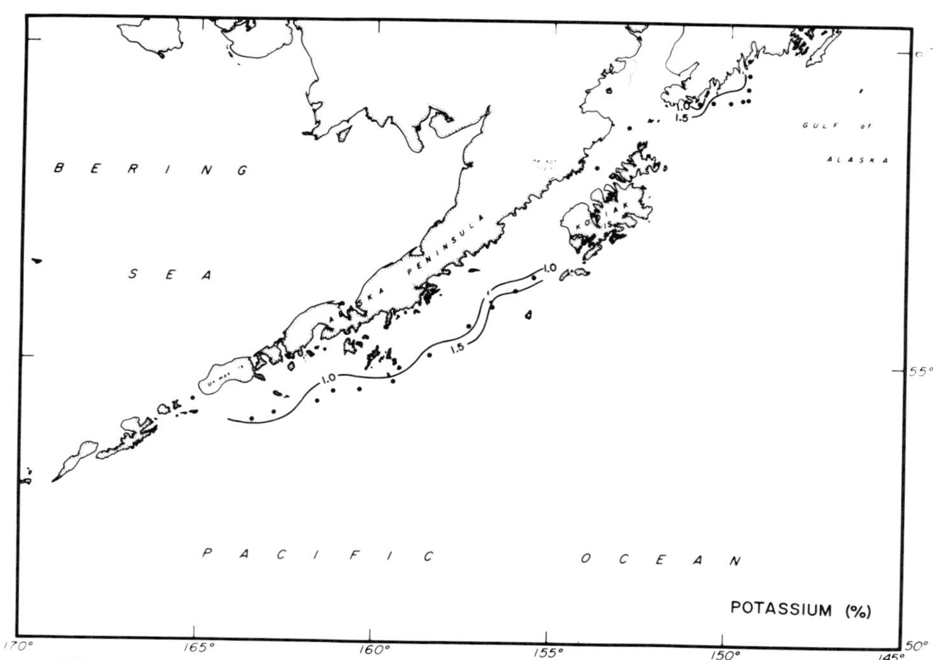

Figure 8-16. Percent potassium in sediments, northwestern Gulf of Alaska.

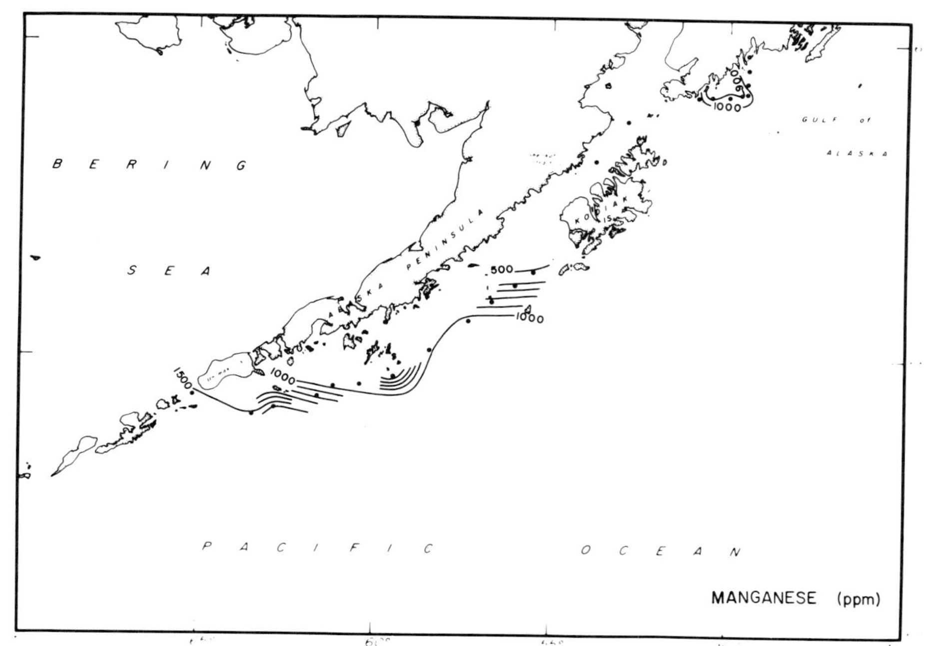

Figure 8-17. Manganese (ppm) in sediments, northwestern Gulf of Alaska.

The distribution of titanium (Fig. 8-18) is quite similar to that of other major elements. It increases with decreasing sediment mean size, and also with increasing distance from shore. Titaniferous minerals normally tend to aggregate in sand fractions, particularly the heavy sands, which are generally retained in the near-shore zone as lag deposits. The seaward increase of titanium in sediment is therefore somewhat unusual.

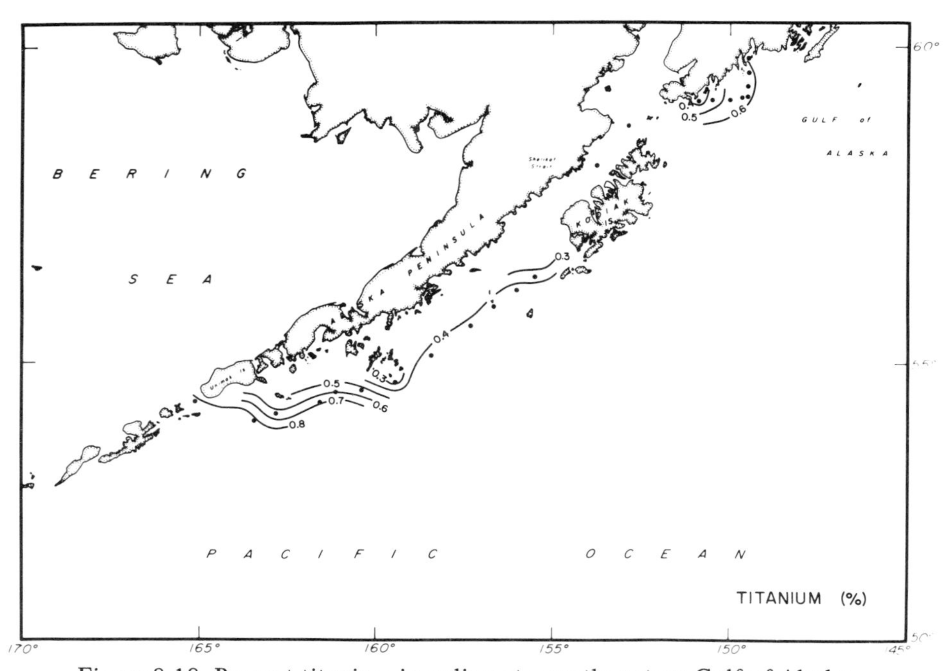

Figure 8-18. Percent titanium in sediments, northwestern Gulf of Alaska.

Among the minor elements cobalt, copper, and strontium show a normal seaward increase in sediments throughout the shelf (Figs. 8-19 to 8-21). Other minor elements, barium, chromium, nickel, and zinc, however, show some regional differences in concentrations as well as seaward variations. Sediments from the northeastern part of the shelf contain higher contents of barium, chromium, nickel, and zinc than those from the southwestern shelf (Figs. 8-22 to 8-25). These differences may be the result of differences in mineralogic composition of the source rocks, as well as of differences in sediment mean size from these regions, but all of these elements do show the usual overall progressive seaward increase in concentrations.

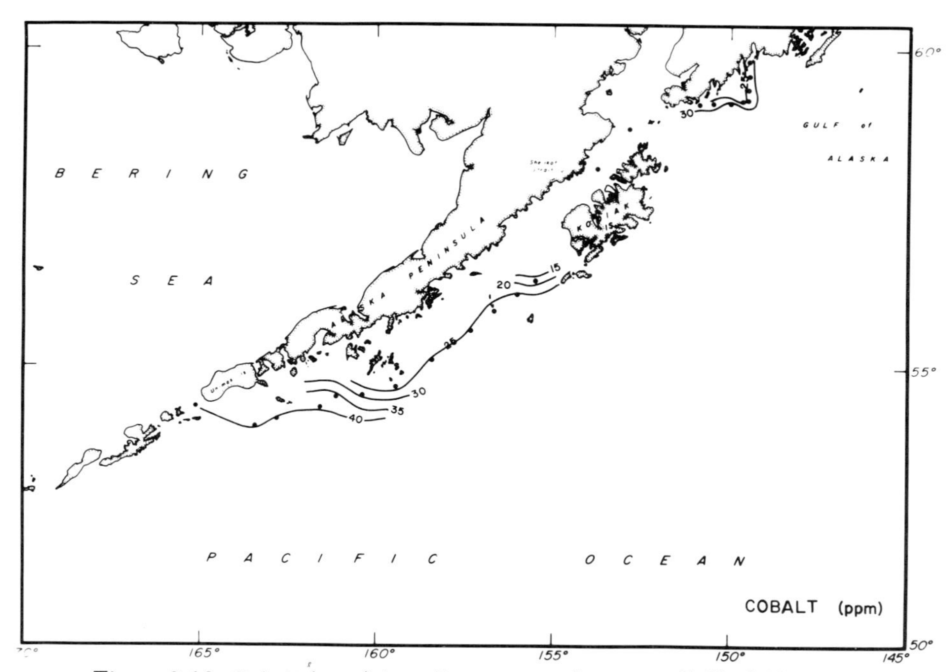

Figure 8-19. Cobalt (ppm) in sediments, northwestern Gulf of Alaska.

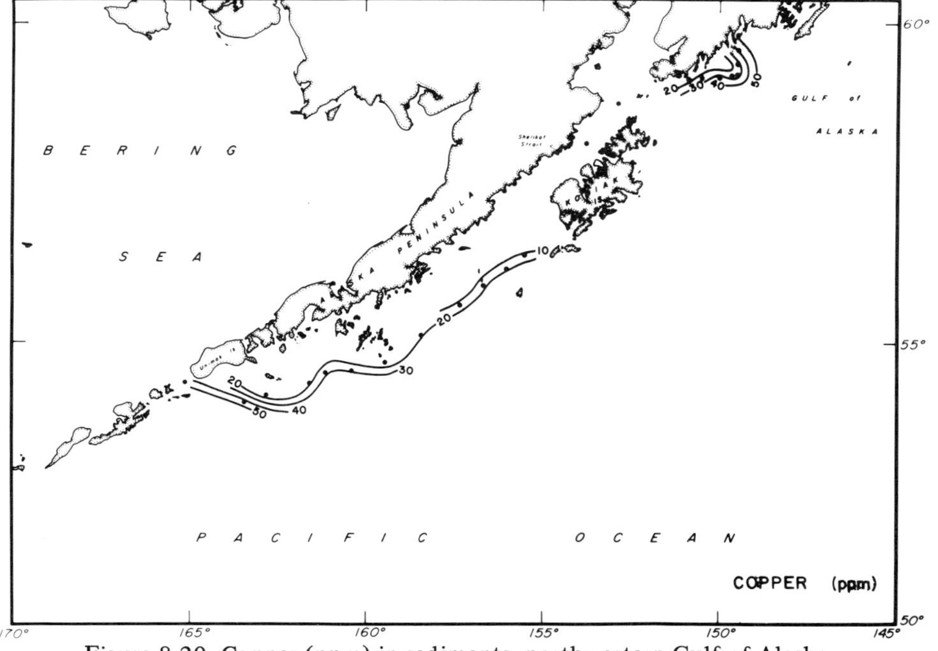

Figure 8-20. Copper (ppm) in sediments, northwestern Gulf of Alaska.

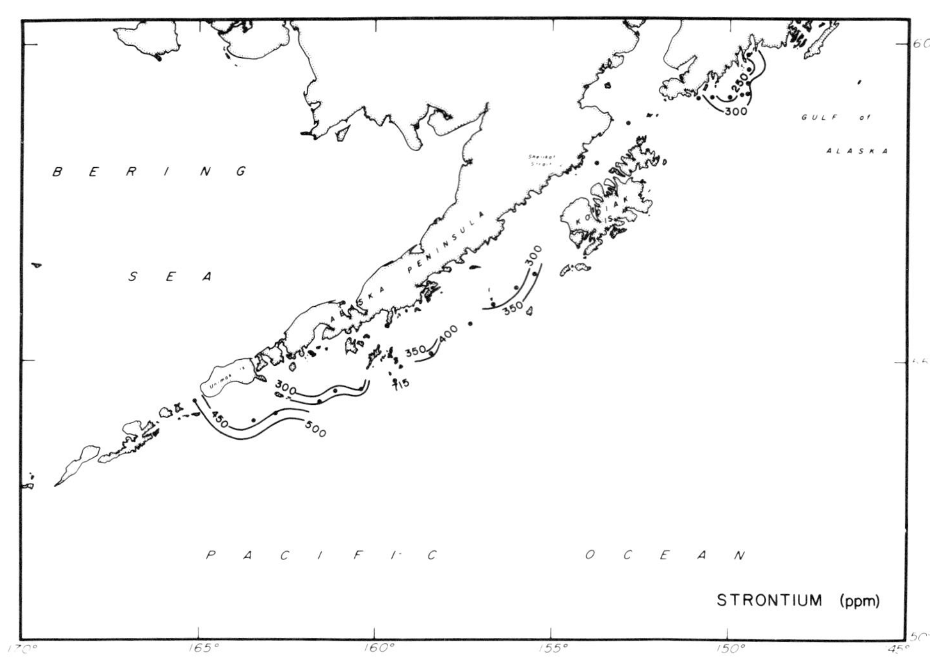

Figure 8-21. Strontium (ppm) in sediments, northwestern Gulf of Alaska.

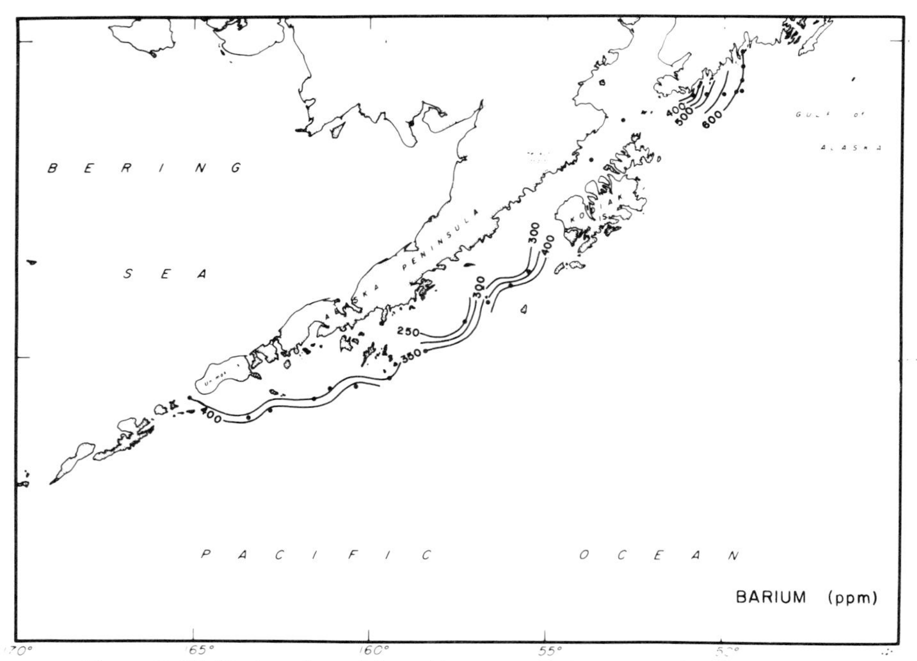

Figure 8-22. Barium (ppm) in sediments, northwestern Gulf of Alaska.

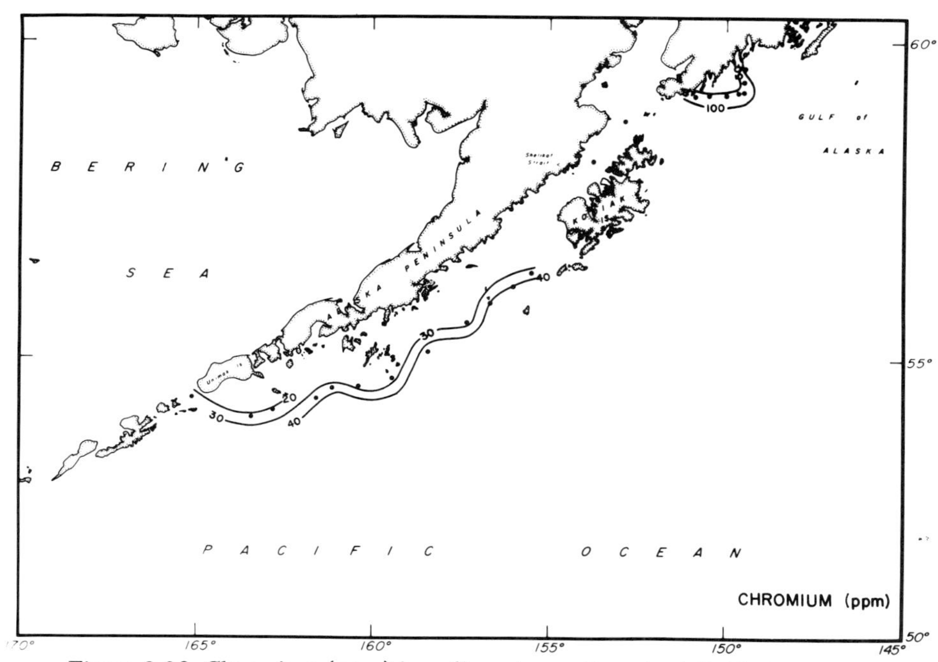

Figure 8-23. Chromium (ppm) in sediments, northwestern Gulf of Alaska.

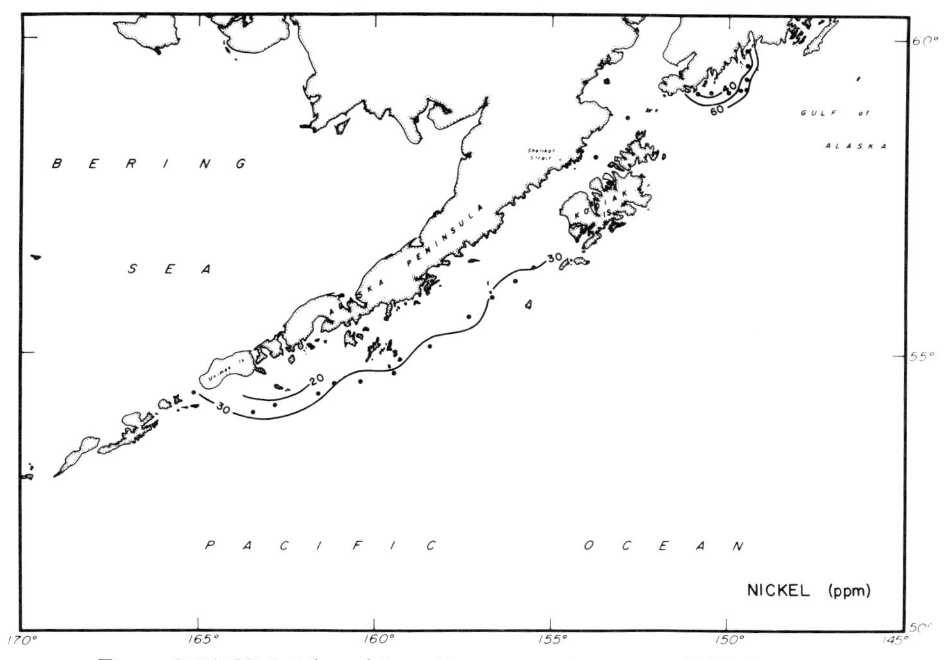

Figure 8-24. Nickel (ppm) in sediments, northwestern Gulf of Alaska.

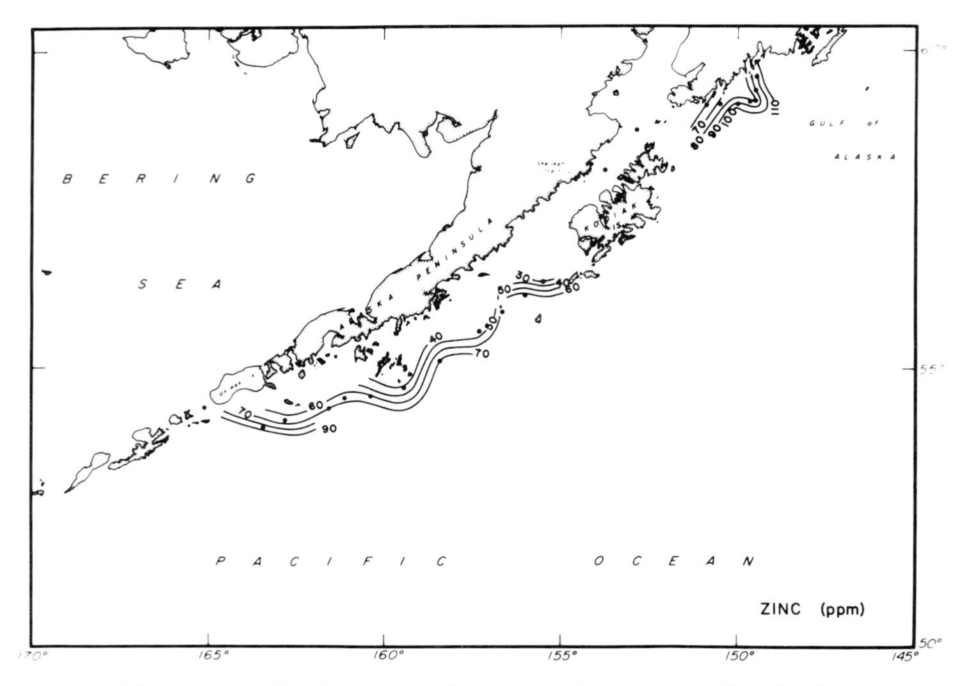

Figure 8-25. Zinc (ppm) in sediments, northwestern Gulf of Alaska.

SEDIMENT SOURCE AND TRANSPORT

The surface discharge along the Alaska Peninsula and Kodiak Island ranks among the highest in Alaska; but little is known concerning the nature and amounts of sediments these waters carry to the sea. During summer months the streams draining sizable ice fields of the Alaska Peninsula and the streams draining recent volcanic ash deposits often carry loads as high as 2,000 mg/ liter, while lowland streams generally carry <10 mg/ℓ (Feulner, 1973a and 1973b). The suspended load data described here is only an approximate average, based on a relatively few sporadic observations. Gauge records, continual flow rates, and suspended load measurements are not available for most streams from this region. It is difficult, therefore, to estimate the amount of detritus carried to the shelf.

Persistent cloud cover drastically reduces the availability of usable ERTS imagery to study sediment plumes in this region. Furthermore, the omnipresent haze precluded density slicing of the two available partly clouded images. Visual inspection of these images suggests that sediments from Cook Inlet are carried southwestward into Shelikof Strait in a relatively narrow band extending along the Alaska Peninsula. The sediment plume is well defined in the strait, but the sediments diffuse and disperse after leaving the strait, causing the plume to broaden and finally lose its identity.

The movement of sediment plumes in Shelikof Strait is complex and is mostly controlled by the water dynamics, dominated by tides. During the ebb tide, substantial amounts of sediments from Cook Inlet and the shallow shelf surrounding Kodiak Island move southwestward through Shelikof Strait. The southwestward movement of sediment is temporarily reversed during flood tide, and smaller sediment plumes in Shelikof Strait, extending northeastward, develop along the northwestern shores of Kodiak Island. Only small amounts of sediment are carried northward and the net sediment transport on the shelf is mostly to the south and southwest.

Measurements of the suspended load of surface shelf water between Resurrection Bay and Unimak Island during 8-9 July 1973 and 13-15 August 1973 suggest that the suspended load was low, with an irregular distribution (Fig. 8-26). The salinity-temperature distribution, however, shows seaward mixing of low-salinity warm waters originating as terrestrial discharge (Figs. 8-27 and 8-28).

The sources of the sediments deposited on the shelf between Resurrection Bay and Kodiak Island primarily lie to the northeast (Prince William Sound and Copper River), north, and northwest (Kenai Peninsula). Sediments from the northeast are brought to this region in suspension by the Alaska Current. Coastal detritus from the Kenai Peninsula is transported offshore by tides and deflected southwestward by the Alaska Current.

The geochemistry (elemental distribution) of the sediments from the shelf east of the Kenai Peninsula is signficantly different from that of the sediments to the southwest of Kodiak Island. This contrast in chemical composition of the sediments is due to the marked differences in the nature of the parent rocks in their provenances. The northern shelf mostly receives reworked glaciated sediments, with their origin in the metamorphosed rocks which dominate the Chugach Mountains to the northeast and the Kenai Mountains to the northwest. The sources for some of the very fine sediments deposited in this region may even lie as far south as the southeastern Alaskan Panhandle with subsequent westward transport by the omnipresent Alaska Current.

The southwestern shelf receives its sediments primarily from Cook Inlet and the adjacent landmasses, *i.e.,* Kodiak Island and the Alaska Peninsula. Sediments from Cook Inlet and Shelikof Strait are funneled offshore along the broad sea valley which cuts across the shelf south of Kodiak Island. Southwestward, the proportion of sediment contributed by the adjacent Alaska Peninsula increases progressively. Local biogenous contributions in shallow waters, particularly surrounding islands, are also important sources for sediments. The texture and geochemistry of sediments deposited on the southwestern proximity of the shelf (Figs. 8-29 and 8-30; samples 1 to 4), off Unimak Island, suggest an increasing

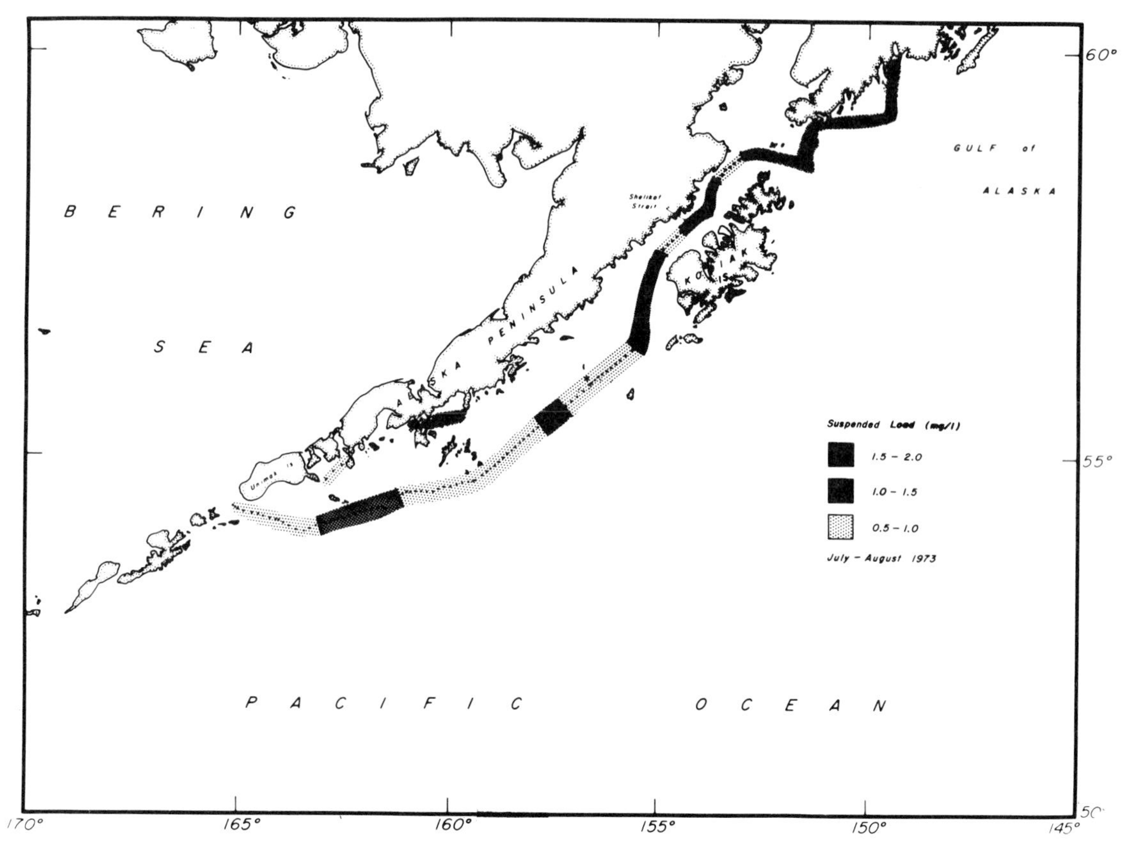

Figure 8-26. Distribution of the surface suspended sediment load during July-August, 1973, northwestern Gulf of Alaska.

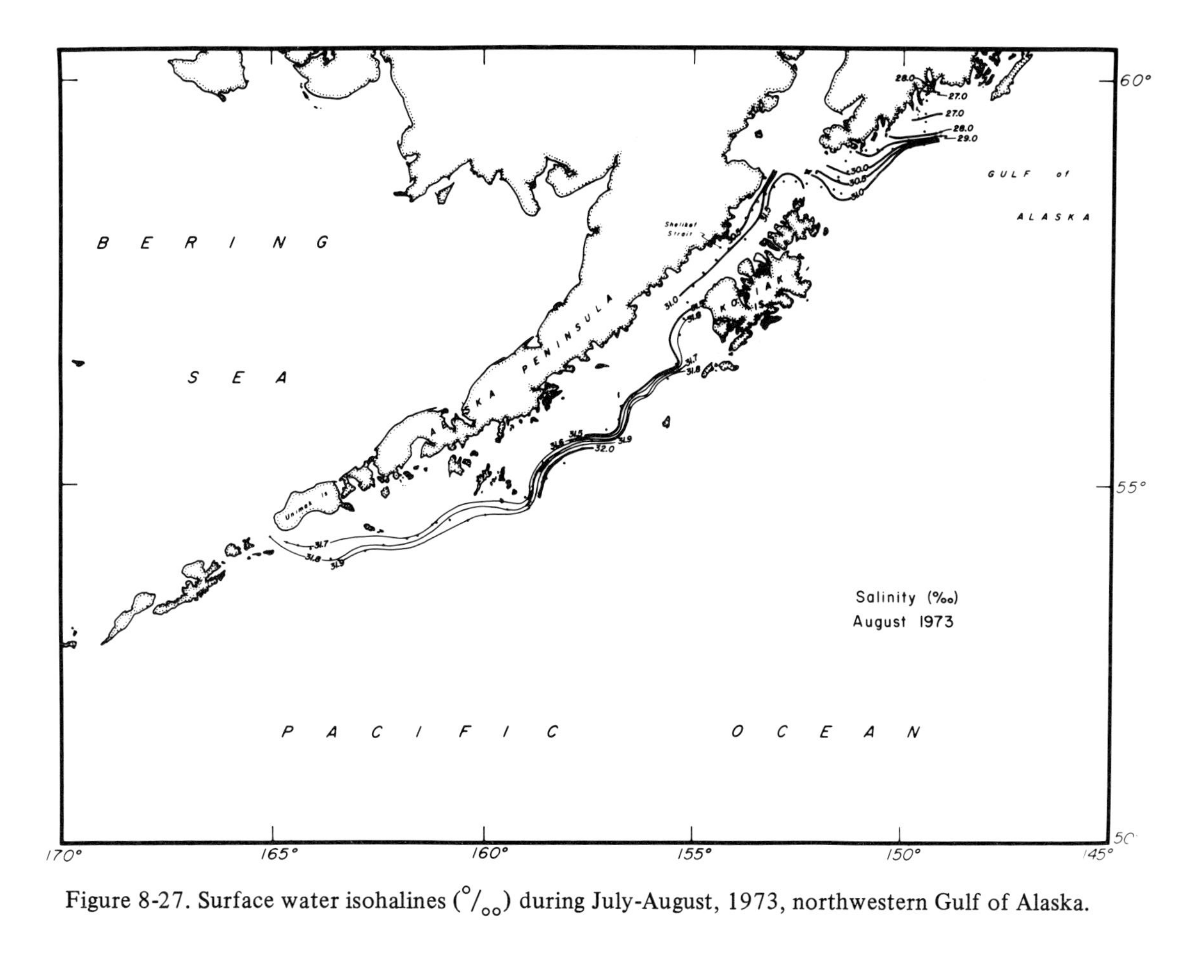

Figure 8-27. Surface water isohalines ($^{\circ}/_{\circ\circ}$) during July-August, 1973, northwestern Gulf of Alaska.

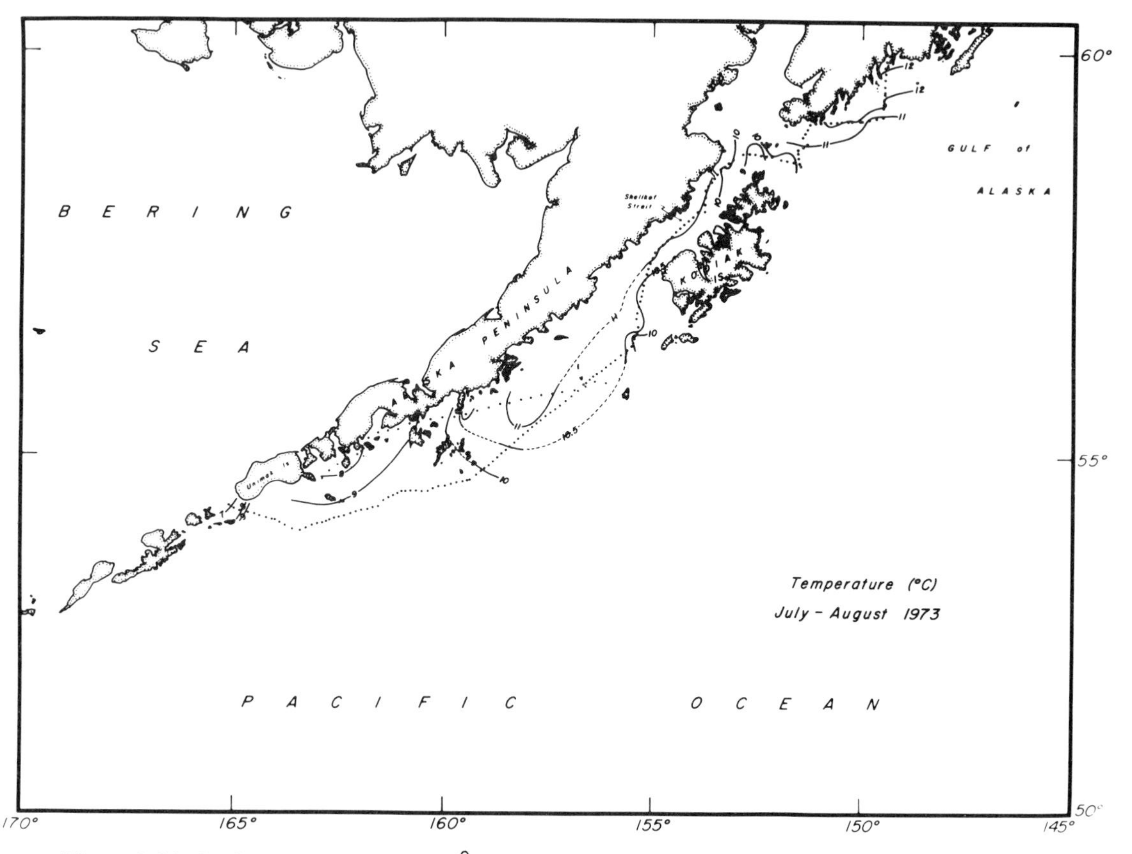

Figure 8-28. Surface water isotherms (°C) July-August, 1973, northwestern Gulf of Alaska.

contribution from basic rocks. Volcanic basic rocks exposed in this region probably provide significant amounts of sediments to the shelf. The mineral distribution in fine sand, reported by Bortnikov (1970), also suggests that the source for these sediments lies in the adjacent hinterland.

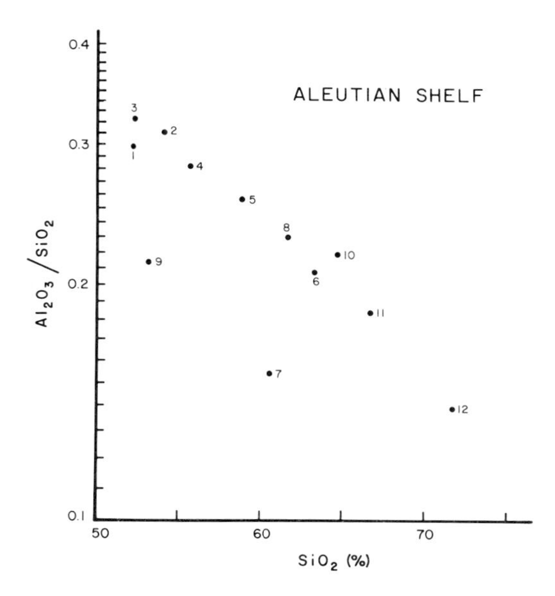

Figure 8-29. Al_2O_3/SiO_2 versus percent SiO_2 plot, northwestern Gulf of Alaska Shelf.

Sediments from the shelf off the Alaska Peninsula have some interesting elemental distributions. Except for Si and K, all major elements in sediments increase gradually southwestward. The variations in K content are slight but the increased content for other elements is noticeable. These increases are compensated by the decrease in Si (Fig. 8-30) and are primarily related to the composition of the provenance. Andesitic flows and pyroclastic rocks on the Alaska Pensinula are common. These rocks are basaltic in nature and, therefore, are poor in silica. Furthermore, the southwestward increase in alkalinity of basalts (R. B. Forbes, Prof. of Geology, Emeritus, University of Alaska, personal communication) is well reflected by the corresponding increase in calcium and other elemental concentrations.

The higher maturity factor for sediments from the northeastern part of the Aleutian Shelf (Fig. 8-29; samples 6 to 12) is caused by the addition of detritus from Cook Inlet and the central Gulf of Alaska. During transport the detritus from these sources is well differentiated and thus has a high maturity factor.

The other elemental ratios (Al_2O_3/Na_2O, K_2O/Na_2O, and CaO/MgO) shown in Figures 8-31 to 8-33 suggest southwestward migration of sediment. The southwestward drift of sediment on the Aleutian Shelf is caused by the Alaska Current, which has a similar flow near the shelf edge.

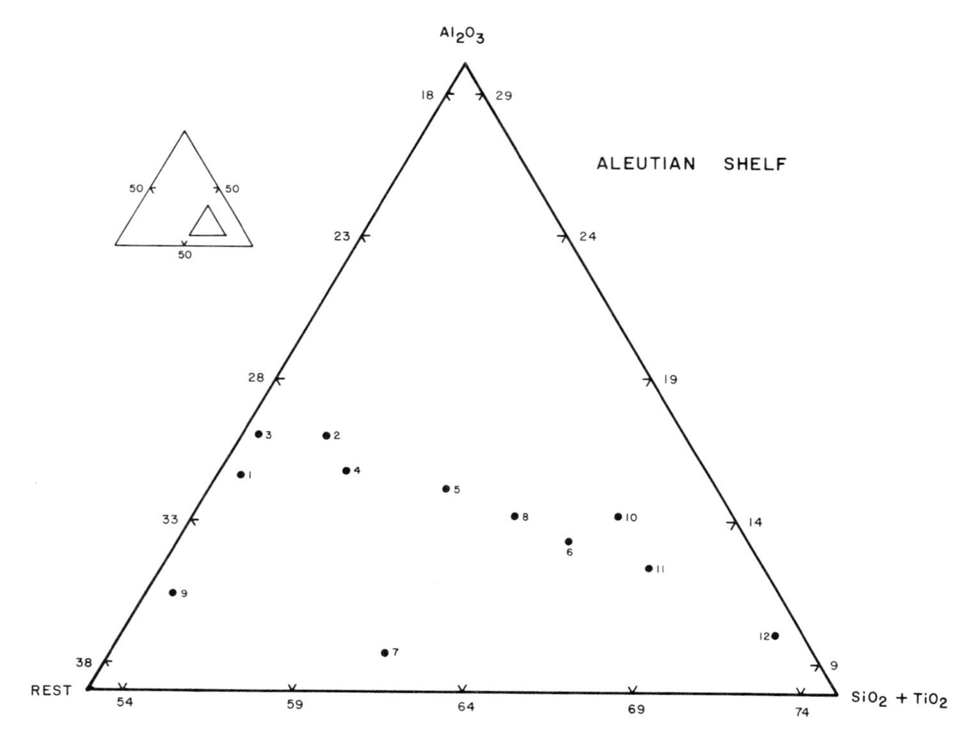

Figure 8-30. Triangular plot of percents Al_2O_3, $SiO_2 + TiO_2$, and Rest as end members, northwestern Gulf of Alaska Shelf.

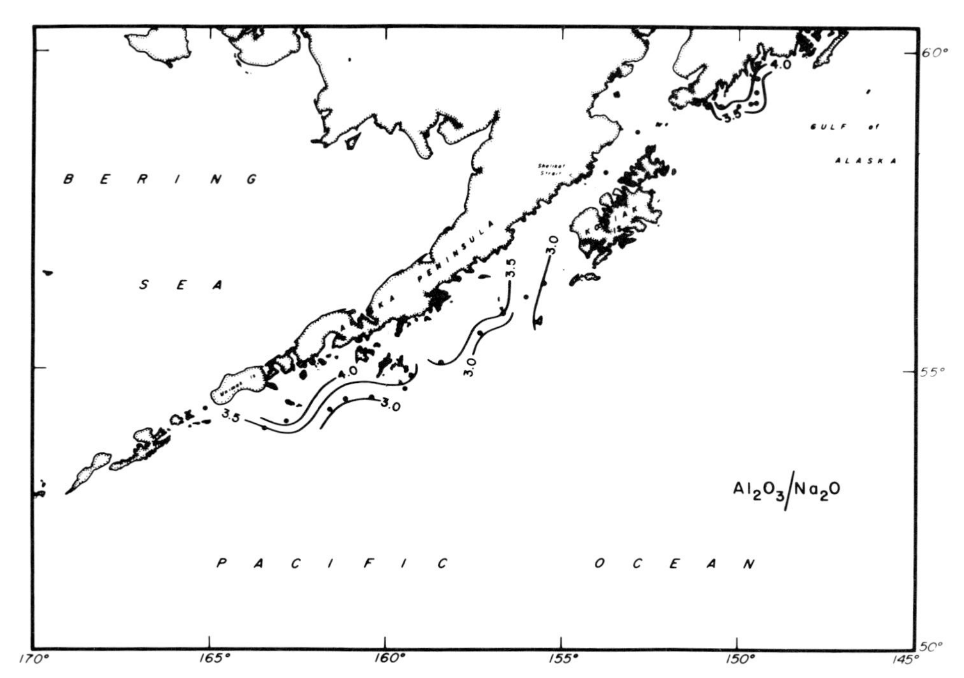

Figure 8-31. Al_2O_3/Na_2O variations in sediments, northwestern Gulf of Alaska.

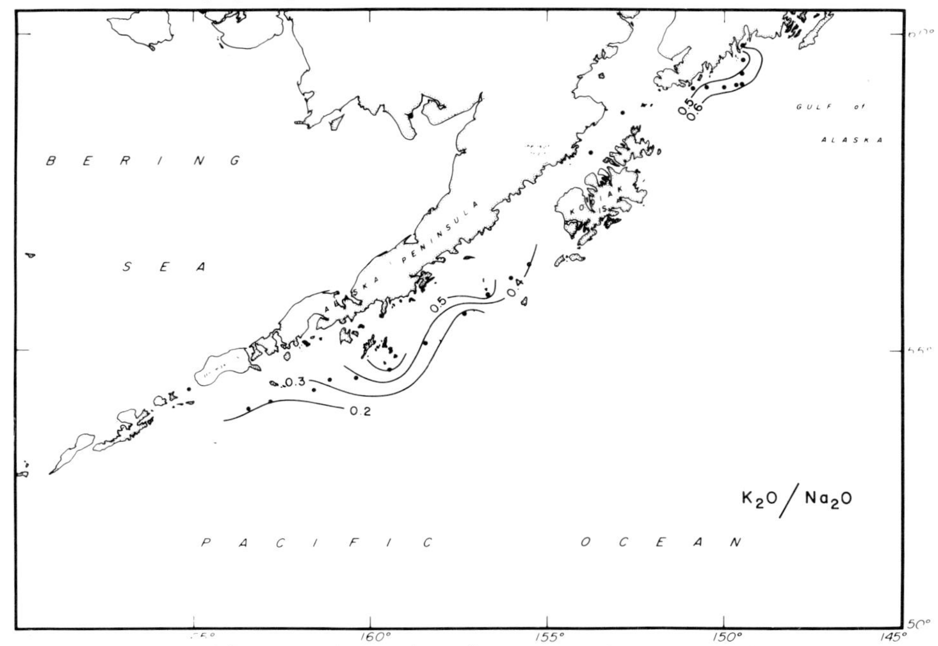

Figure 8-32. K_2O/Na_2O variations in sediments, northwestern Gulf of Alaska.

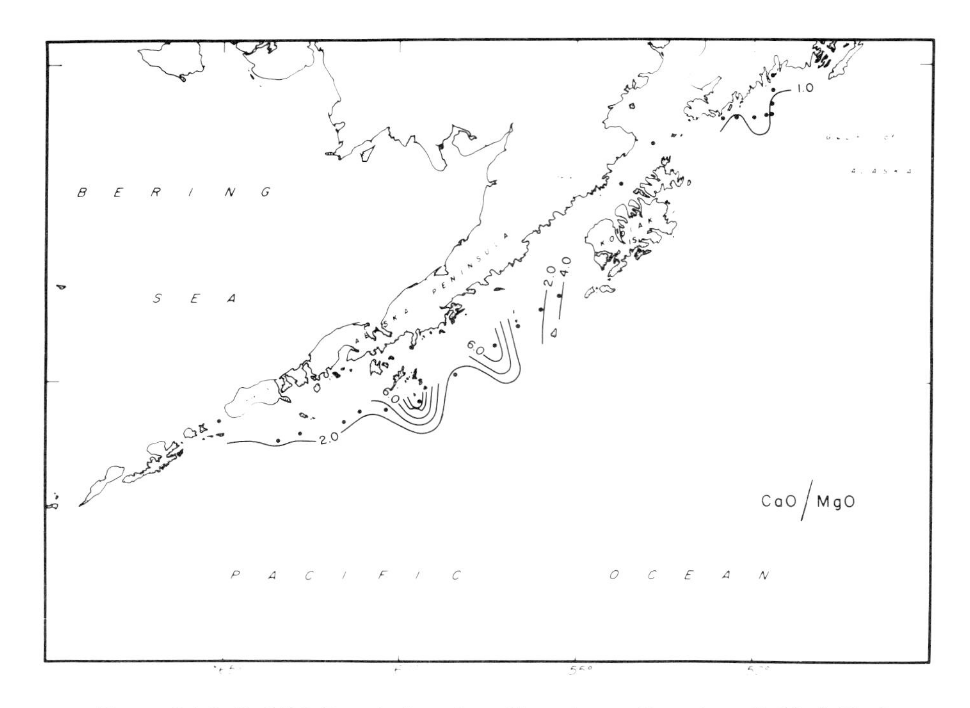

Figure 8-33. CaO/MgO variations in sediments, northwestern Gulf of Alaska.

REFERENCES

Bortnikov, V. S. (1970). Mineral composition of the coarse silt fraction of recent sediments in the Northwestern part of the Gulf of Alaska. In *Soviet Investigations in the Northeast Pacific, Part V.* P. A. Moiseev (Ed.), pp. 77-82. Transl. by Israel Program for Scientific Translations, Jerusalem.

Favorite, F. (1970). Fishery oceanography–VII. Estimation of flow in the Gulf of Alaska. *Commercial Fisheries Rev.* 32(7):23-29.

Feulner, A. J. (1973a). Summary of water supplies at Alaska communities, south-central region, Kodiak-Shelikof Subregion. Joint Federal-State Land Use Planning Commission, Resource Planning Team. Unpublished Rept. p. 24.

Feulner, A. J. (1973b). Summary of water supplies at Alaska communities, Southwest Region, Aleutian Subregion. Joint Federal-State Land Use Planning Commission, Resource Planning Team. Unpublished Rept. p. 13.

Gershanovich, D. E. (1968). New data on the geomorphology and recent sediments of the Bering Sea and the Gulf of Alaska. *Marine Geol.* 6:281-296.

Gershanovich, D. E. (1970). Principal result of latest investigations of bottom relief and sediments in fish grounds in the North Pacific Ocean. In *Soviet Fisheries Investigations in the Northeast Pacific Ocean, Part V.* All-Union Scientific Research Institute of Marine Fisheries and Oceanography (VNIRO) Trudy v. 70. P. A. Moiseev (Ed.), pp. 7-34. Transl. by Israel Program for Scientific Translations, Jerusalem.

Gershanovich, D. E., B. N. Kotenev, and V. N. Novihov. (1964). Relief and bottom sediments of the Gulf of Alaska. In *Soviet Fisheries Investigations in the Northeast Pacific Ocean, Part III.* All-Union Scientific Research Institute of Marine Fisheries and Oceanography (VNIRO) Trudy v. 53. P. A. Moiseev (Ed.), pp. 74-125. Transl. by Israel Program for Scientific Translations, Jerusalem.

Hampton, M. A., A. H. Bouma, T. P. Frost, and I. P. Colburn. (1979). Volcanic ash in surficial sediments of the Kodiak shelf–An indicator of sediment dispersal patterns. *Marine Geol.* 29:347-356.

Kulm, L. D., R. von Huene *et al.* (1973). Tectonic summary of Leg 18. In *Initial Reports of the Deep Sea Drilling Project. XVIII.* U.S. Govt. Printing Office, Washington, D.C.

McEwen, G. F., T. G. Thompson, and R. T. VanCleve. (1930). Hydrographic Sections and Calculated Currents in the Gulf of Alaska 1927 and 1928. Rept. Intl. Fish. Comm. 4. 36 p.

Menard, H. W., and T. E. Chase. (1971). Bathymetric Atlas of the Northcentral Pacific Ocean. H.O. Pub. 1302-S. Compiled by Scripps Institute of Oceanography. U.S. Naval Oceanographic Office, Washington, D.C.

Piper, D. J., R. von Huene, and J. R. Duncan. (1973). Late Quaternary sedimentation in the active eastern Aleutian Trench. *Geology* 1(1):19-22.

Shor, G. G., Jr. (1962). Seismic refraction studies off the coast of Alaska, 1956-1957. *Seismol. Soc. Amer. Bull.* 53:37-57.

Thompson,. T. G., G. F. McEwen, and R. Van Cleve. (1936). Hydrographic Sections and Calculated Currents in the Gulf of Alaska, 1929. Rept. Intl. Fish. Comm. 10, 32 p.

U.S. Geological Survey. (1972). *Water Resource Data for Alaska, 1971.* Water Resource Division, Anchorage, Alaska, p. 319.

von Huene, R. (1972). Structure of the continental margin and tectonism at the eastern Aleutian Trench. *Geol. Soc. Amer. Bull.* 83:1889-1902.

von Huene, R., E. H. Lathram, and E. Reimnitz. (1971). Possible petroleum resources of offshore Pacific-Margin Tertiary Basin, Alaska. *Amer. Assoc. Petrol. Geol. Mem.* 15(1):136-159.

von Huene, R., R. J. Malloy, G. G. Shor., Jr., and P. St. Amand. (1967). Geologic structures in the aftershock regions of the 1964 Alaskan Earthquake. *J. Geophys. Res.* 72:3649-3660.

von Huene, R., G. G. Shor, and R. J. Malloy. (1972). Offshore tectonic features in the affected region. In *The great Alaska Earthquake of 1964: Oceanography and Coastal Engineering.* National Academy of Sciences, Washington, D.C., pp. 266-289.

CHAPTER 9
Cook Inlet

INTRODUCTION

Cook Inlet, with its extensions, lies between 59° and 61°30′N latitude and 149° and 154°W longitude and covers more than 26×10^3 km^2. This large tidal estuary is a northeast-southwest oriented indentation into the southcentral Alaskan coastline (Fig. 9-1). It differs from the other indentations of the Pacific Coast of Alaska in that its head is well behind the coastal ranges and it has broad tributary valleys drained by large rivers. The estuary flows into the Gulf of Alaska, is 90 km wide at the entrance, and has an average depth of 100 m. From its entrance it extends northeast 280 km and, at the head, bifurcates into two arms; Turnagain Arm and Knik Arm, 80 km and 83 km long, respectively. Cook Inlet is divided into upper and lower inlets by a natural constriction near the East and West forelands.

The estuary is fed by the drainage from the surrounding steep mountains: the Aleutian Range and Alaska Range on the northwest, the Talkeetna Mountains to the northeast, and the Chugach-Kenai Mountains on the southwest. The topography is rugged, and numerous glaciers descend from these mountains, carrying large amounts of detritus. Major rivers, the Knik, Matanuska, Susitna, Little Susitna, and Beluga, discharge their loads near the head, in the upper inlet. Numerous smaller rivers and creeks also contribute their flow to the inlet.

Five active volcanoes, Mounts Augustine, Spurr, Redoubt, Iliamna, and Douglas, are located along the western margin. Eruptions from these volcanoes have been recorded during the recent past and have contributed volcanic ash to the adjacent marine environment.

GEOLOGY

Cook Inlet is the southwest flank of the arcuate Matanuska Geosyncline. Structurally, the sedimentary basin is bounded by the Talkeetna Geanticline to the northwest and by the Seldovia Geanticline to the southeast (Payne, 1955). The elongated basin is about 320 km long and 110 km wide and covers an area of about 40,000 km^2. In its simplest form, the basin is a graben, bounded by major fault zones on the north, west, and east that have been active since Eocene

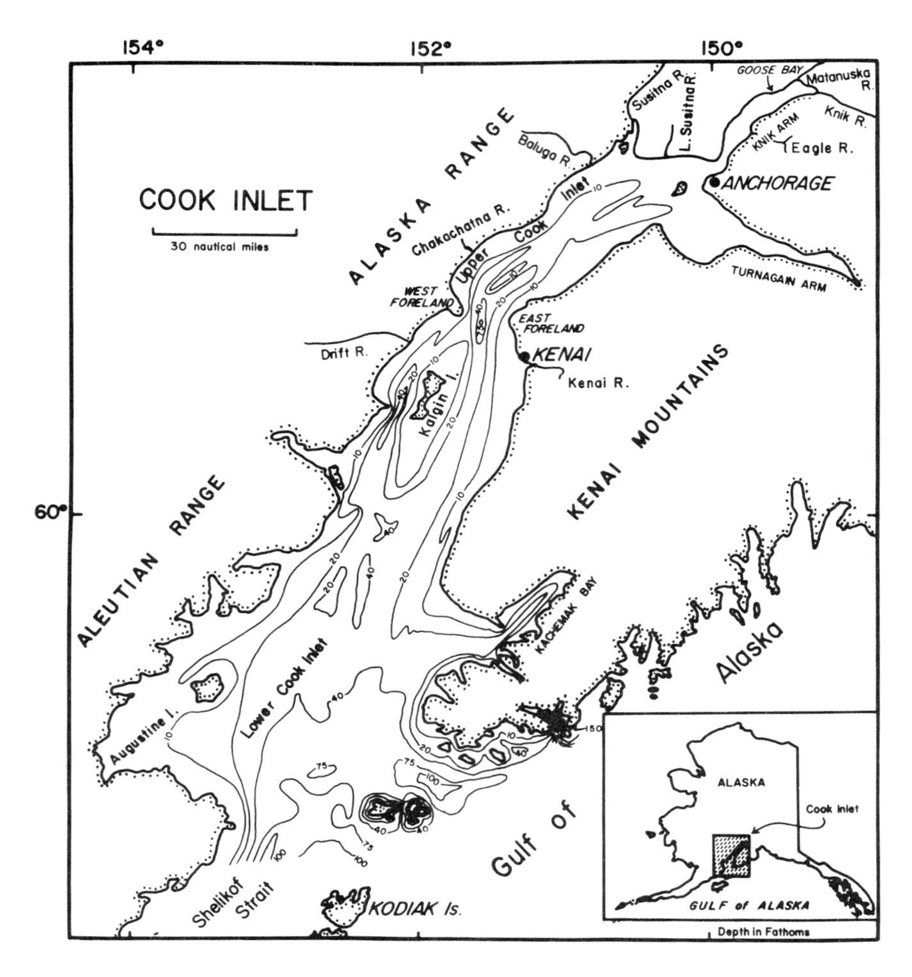

Figure 9-1. Bathymetric chart, Cook Inlet.

time. The basin is filled with a 20,000 to 25,000 m thick sequence comprised of marine and non-marine sediments as well as volcanic rocks.

The Cook Inlet region, a northeast-southwest oriented basin, was filled with geosynclinal sedimentary facies during the Paleozoic era, consisting of marine deposits and volcanic rocks. The Early Mesozoic era was a time of continued sedimentation, followed by a major orogeny at the close of the Early Jurassic or early Middle Jurassic that resulted in folding and uplift of the Talkeetna Geanticline to the north. Batholithic intrusion accompanied this orogeny. The elongated Matanuska Geosyncline (part of which is occupied by Cook Inlet), in the south of the Talkeetna Geanticline, was an area of the deformed lower Jurassic Talkeetna Formation which began to receive sediment from surrounding areas. The basin was an area of predominantly marine deposition from Middle Jurassic to Late Cretaceous time and of non-marine deposition during Tertiary time. The region is marked by Jurassic, Cretaceous, and Tertiary plutonic intrusions.

Cook Inlet lies in the trans-Pacific seismic zone and therefore in a region of continued tectonic activity. The orogenic movement generally occurs along the major Bruin Bay and Castle Mountain faults and numerous minor faults.

The stratigraphic sequence in Cook Inlet Basin includes over 12,000 m of Paleozoic and Early Mesozoic marine sediments and about 10,000 m of Tertiary nonmarine sediments. Mesozoic sediments consist of limestone, chert, volcanics, and clastic deposits. The strata in the inlet are divided into four units by three unconformities (Fisher and Magoon, 1978). At the beginning of the Tertiary, as a result of earlier orogeny, the basin took the form of a broad, linear, intermontane trough surrounded by low-relief topography (Kelly, 1963, 1968; Crick, 1971; Kirschner and Lyon, 1971). During the Tertiary, detritus typical of deltas was deposited by aggrading streams and other nonmarine sources. The early Tertiary sediments consist of conglomerate, sandstone, siltstone, and coal beds. During the later part of the Tertiary, the subsiding trough accumulated over 8,000 m of onshore sediments. These sediments contain hydrocarbons, which are commercially produced in Cook Inlet and from the Kenai Peninsula.

Structurally little change has occurred in Cook Inlet since the close of the Tertiary. Morphologically, however, the region has been repeatedly affected by glacial advances and retreats. At least five major Pleistocene glaciations have been observed in this region (Karlstrom, 1964). Glacier ice during the first three episodes ("Mount Susitna," "Caribou Hills," "Eklutna") filled Cook Inlet, and extended as far as Shelikof Strait. At the peak of the Mount Susitna glaciation, the ice elevation in the Cook Inlet region was 1,300 m. The contiguous sheet of ice during this time perhaps extended to the Copper River basin to the northeast, and into Bristol Bay to the southwest. The Caribou Hills glaciation was less extensive, with ice elevations between 750 and 1,000 m. At the close of glaciation, approximately 155,000 to 190,000 years B.P., the ice began to retreat. The Eklutna glaciation was less severe than preceding ones and closed about 90,000 to 110,000 years B.P.

During the succeeding "Knik" and "Naptowne" glacial episodes, ice covered only portions of Cook Inlet, with the Knik glaciation ice extending further than the Naptowne. Karlstrom (1964) reported that during these two glaciations upper Cook Inlet and parts of lower Cook Inlet were covered by a large glacial lake which was formed as a result of coalescence of ice lobes from Kachemak Bay in the southwest, and from the Aleutian Range in the west. The proglacial lake silts of Knik age are exposed to 300-m elevation. The final draining and subsequent encroachment of the sea in Cook Inlet began about 7,000 years B.P. at the close of the Naptowne glaciation. There have been two major ice advances since the last major glaciation; the "Tustumena" ice advance peaked between 5,500 and 3,200 years B.P., and the youngest, "Tunnel," climaxed between 1,500 and 500 years B.P.

BATHYMETRY

The bathymetry of Cook Inlet is relatively smooth, with three islands, and several shoals (Fig. 9-1). A maximum depth of over 100 m occurs near the entrance, and the inlet gradually becomes shallower toward its head, where the average depth is less than 20 m. There are numerous shoals throughout the inlet, some of which are located on the crests of folds. The upper inlet has extensive mud flats which, during low tide, cover an area in excess of 250 km^2.

HYDROLOGY

The Cook Inlet region lies in a transitional zone separating the continental and maritime climates and is characterized by cool summers, mild winters, heavy precipitation, and frequent storms. The average annual precipitation varies significantly from one place to another: It is over 7,000 mm at Susitna station, and less than 3,750 mm at Anchorage. The average annual precipitation over the entire area is about 5,000 mm, with the maximum precipitation occurring during the month of July and the minimum in January. The discharge pattern appears to be unimodal and is almost insignificant during the winter months (November through March) and the beginning of snow melting in April initiates the summer flow, which reaches its peak in July (Fig. 9-2). Discharge rates of various rivers into Cook Inlet suggests that the combined flow of the Susitna, Knik, Chakachatna, Matanuska, Eagle and Little Susitna rivers constitutes approximately 75% of the total freshwater inflow which is emptied in the upper inlet. The Kenai and Drift rivers are the major contributors of fresh water to the lower inlet. The total freshwater inflow during a half tidal cycle represents 1 to 3% of the total tidal volume (Rosenberg and Hood, 1967).

HYDROGRAPHY

A systematic hydrographic investigation in Cook Inlet was conducted during 1966-1967 by Matthews and Rosenberg (1969), followed in 1968 by Kinney *et al.* (1970). An extensive hydrographic survey was launched during 1972 and 1973 by Sharma *et al.* (1974). The latter investigators studied the seasonal variations of several water parameters throughout the inlet. These parameters were later correlated to ERTS-1 imagery in an attempt to delineate water and sediment movements in Cook Inlet. The salinity-temperature distribution in the inlet during September 1972 is shown in Figures 9-3 and 9-4. This distribution is typical of the ice-free season in Cook Inlet, and is similar to that observed by earlier investigators.

The Susitna, Knik, and Matanuska rivers constitute about 70% of the total fresh water entering the inlet. These rivers discharge along the northern shores of

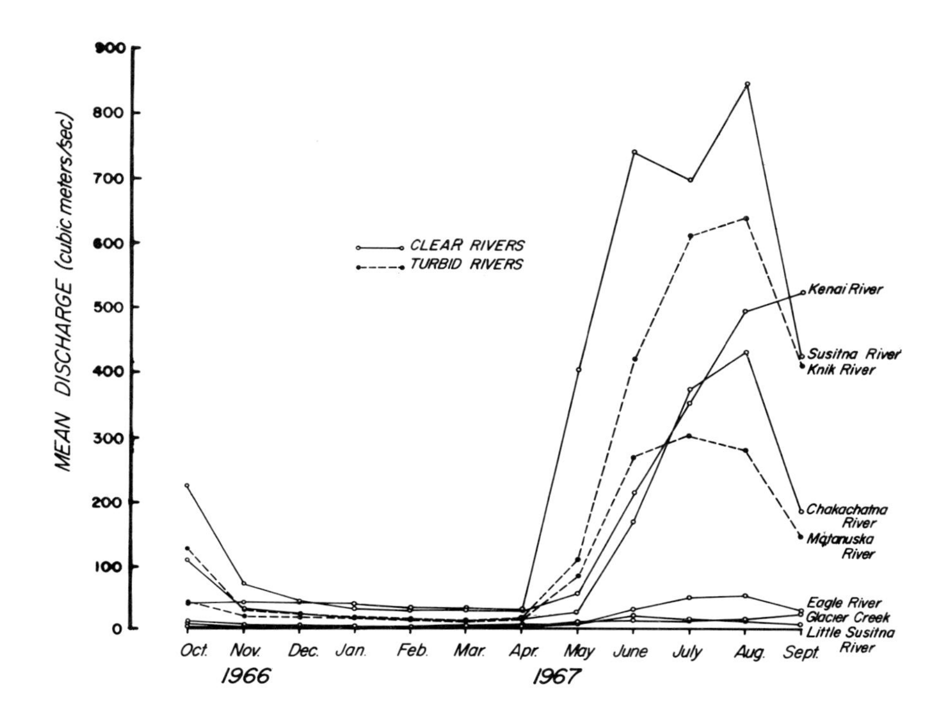

Figure 9-2. Monthly mean discharge of major rivers in Cook Inlet.

fresh water entering the inlet. These rivers discharge along the northern shores of the upper inlet. Salinity and temperature in upper Cook Inlet, therefore, change significantly seasonally and reflect the influence of freshwater input near the head. Broad seasonal variations in salinity-temperature are not unusual for high-latitude estuaries. The water parameters near the entrance of Cook Inlet, however, remain relatively unchanged during the ice-free season.

Cook Inlet has the tenth highest mean spring tidal range in the world, with a value of 10.0 m recorded for Turnagain Arm (U.S. Coast and Geodetic Survey, 1969). The tides are semi-diurnal with a marked inequality between successive low waters: The mean diurnal range in the inlet varies from 4.6 m at the mouth to 10.0 m in Knik Arm. The 2.2 times amplification of tidal range at the head is slightly less than 2.5 which occurs in the Bay of Fundy. There is, however, a 4.5 hour difference in tidal phase between the entrance and the head of the inlet. Such a large tidal range, coupled with the shallowness of the water, leads to high tidal current velocities throughout the inlet, as well as an occasional tidal bore in Turnagain Arm. Currents in excess of 3 m/sec in central Cook Inlet have been reported by National Oceanographic and Atmospheric Administration (NOAA) (1971).

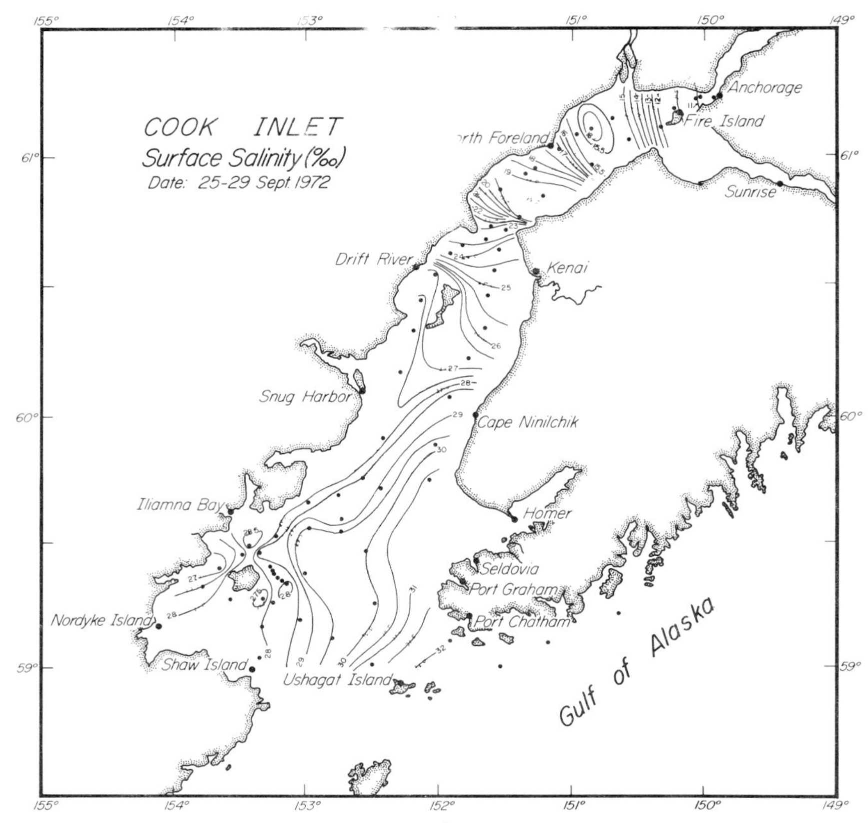

Figure 9-3. Surface water isohalines (°/oo) during 25-29 September, 1972, Cook Inlet.

Basically the hydrography of Cook Inlet is tidally driven and entails the mixing of two distinct water masses from the upper and lower inlets. Along the eastern shore of the inlet, the flood tide introduces saline water that moves as far north as Kenai, but freshwater at the head, contributed by glacial meltwater streams and rivers, completely dominates two-thirds of the upper Cook Inlet. These waters move outward as much as 1.5 km with each tidal cycle. An intense mixing of upper Cook Inlet water and northward moving saline water from the lower inlet occurs in the vicinity of the Forelands. This mixing is influenced by a multitude of factors, namely tidal currents; morphologic configuration, in particular, the narrow constriction near Forelands; coriolis force; and winds. In upper Cook Inlet, the turbulence caused by these factors generally renders the water vertically uniform, and stratified water masses occur only near the entrance of the inlet.

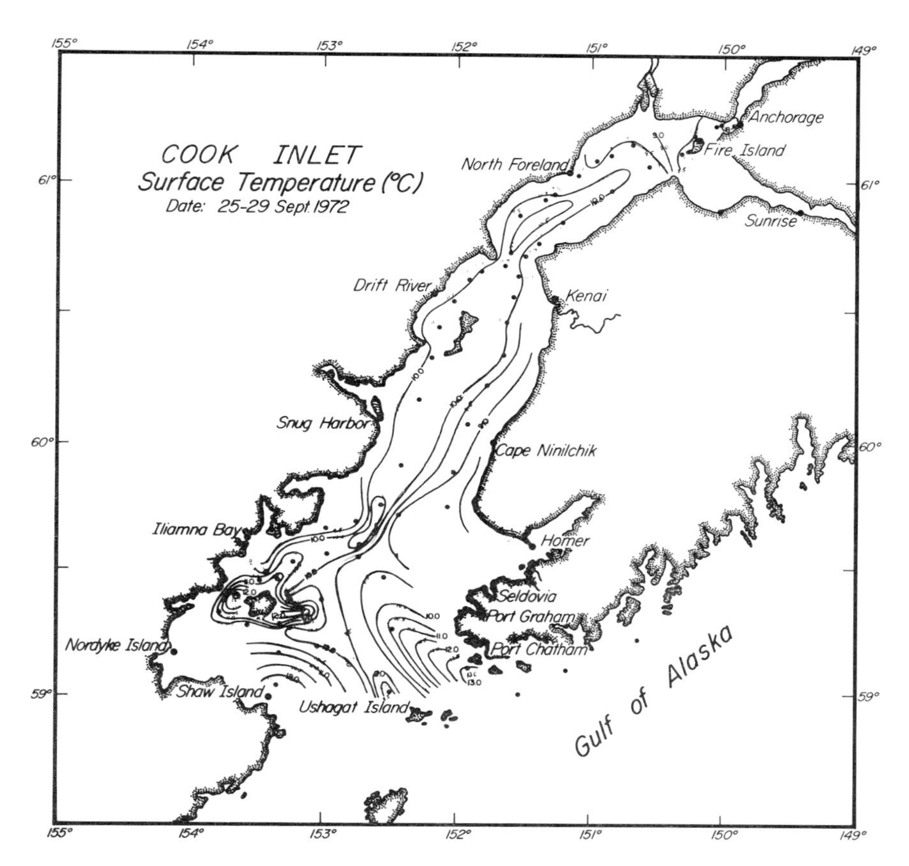

Figure 9-4. Surface water isotherms (°C) during 25-29 September, 1972, Cook Inlet.

Incoming marine water during flood tide is deflected to the right because of the coriolis force and sets up a counterclockwise circulation in lower Cook Inlet. The northward flow of the water is significantly restricted by the narrow constriction near the Forelands. This damming of water produces a jet effect, which in turn causes turbulence and mixing.

Upper Cook Inlet is generally covered by heavy ice during four months of the year. The ice cover in Cook Inlet begins with the freezing of river discharge covering large tidal flats during low tides. Relatively fresh and shallow water on tidal flats forms a thin layer of sheet ice. Flood tide picks this ice up, breaks it into small pieces, and redeposits it upon the flats, with some pieces stacked on top of others. This cycle is repeated many times, and alternating layers of thin ice and frozen sediments form ice blocks of up to 5 to 7 m thick, generally called "stamukhi." Occasionally, high tides carry these stamukhi to mid-stream where they are broken up and reconsolidated to form floe or float ice. Ebbing tides carry this floe ice to the Foreland constriction with an average net seaward movement of about 4 km/day. The thickness of ice in the narrows reaches an excess of 2 m. The ice floes during the passage through the Forelands are broken into smaller pieces which flow into the lower inlet. The retention of ice floes in the upper inlet is found to be approximately 28 days.

SEDIMENTS

Sediments in Cook Inlet consist mainly of gravel and sand, with minor proportions of silt and clay. These sediments do not reflect any textural gradation in relation to distance from source or to the water depth (Sharma and Burrell. 1970). The sediment distribution in Cook Inlet suggests three distinct sedimentary environments.

In the upper inlet, east of the Susitna River (Fig. 9-5), the sediments are composed of sand (Facies I). Facies I mostly contains unimodal, well-sorted fine sand. The sands in mid-channel are negatively skewed, whereas most samples are positively skewed. Sorting and removal of silt and clay in this region are a result of the high-energy environment.

The middle inlet is mantled with sandy gravel and gravel (Facies II). Sediments of Facies II consist of 50 to 100% gravel. The gravel mainly occurs in mid-channel and near the Forelands, becoming sandy gravel in shallow areas.

Southward, in the lower inlet, the gravel grades into silty sand (Facies III). The region surrounding Mount Augustine consists of coarse sediments with carbonate skeletal fragments. The medium and fine sand in the lower inlet also consist of biogenous calcareous clam, coral, sponge, and byrozoan fragments. The shallow bays and numerous tidal flats are covered with muddy sand and mud.

GEOCHEMISTRY

Although sediments from Cook Inlet proper have not been analyzed, geochemical data for sediments from various streams draining into the inlet have been collected by the State of Alaska, and the data are available as file reports (Jasper, 1965).

SEDIMENT SOURCE AND TRANSPORT

The suspended sediment load in the waters of Cook Inlet have been measured sporadically since 1966. A comprehensive seasonal sampling program was conducted during 1972-1973. Data collected from various cruises and oil platforms

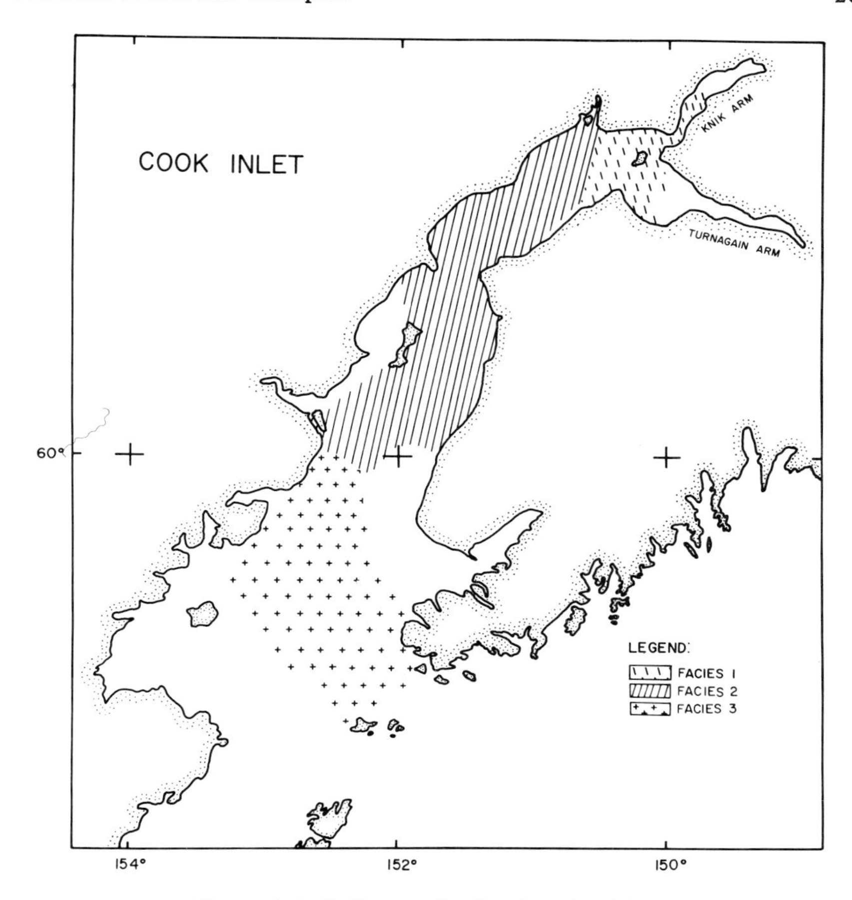

Figure 9-5. Sediment distribution, Cook Inlet.

have been reported by Sharma *et al.* (1974). During a July 1966 cruise, the measured sediment load exceeded 2 g/ℓ near the head of the inlet and decreased rapidly southward (Sharma and Burrell, 1970). The waters in the upper inlet are vertically mixed, but maximum values for suspended load were generally observed at a depth of about 10 m. The sediment concentration in the waters of the upper inlet is dependent on the river discharge: An increased river flow carries a heavier load (Fig. 9-6).

The suspended sediment load in the lower inlet varies between 1 mg/ℓ and over 100 mg/ℓ. The saline waters near the entrance generally contain between 1 and 2 mg/ℓ throughout the year. The waters in this part of the inlet are stratified, and suspended load invariably increases with increasing depth. As the oceanic water mass is carried toward the head, it mixes with turbid water, and thus the suspended load gradually increases (Fig. 9-7) and becomes more vertically uniform.

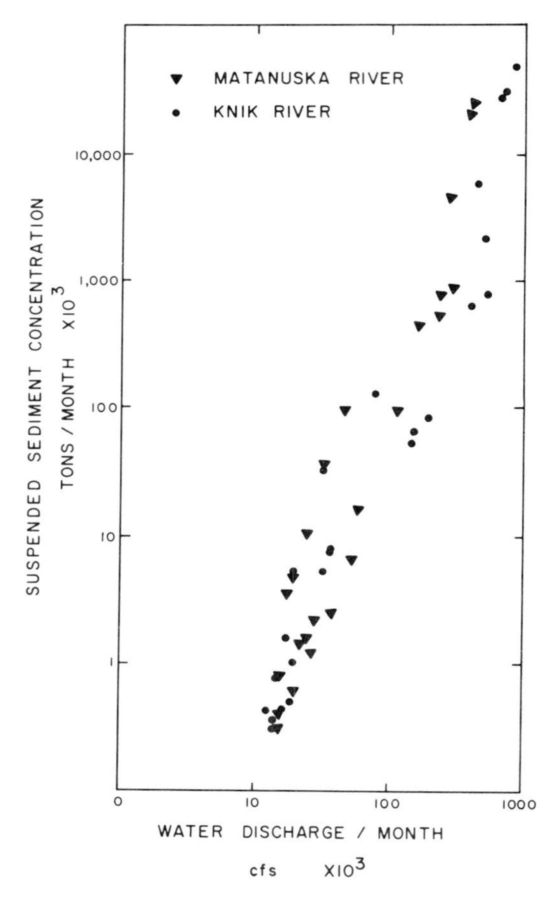

Figure 9-6. Suspended load concentration versus water discharge plot for Matanuska and Knik rivers.

The relative distribution of sediments in near-surface layers of Cook Inlet waters has also been studied using ERTS-1 imagery. An overwhelming advantage of ERTS-1 imagery over the data collected using conventional onboard ship methods is that it provides synoptic data over a large area. In Cook Inlet, the ERTS-1 imagery analysis is extremely helpful because swift currents and rapidly mixing water masses continually alter water properties throughout the inlet. The isodensity analysis of various images (Figs. 9-8 and 9-9) shows, in general, a suspended sediment distribution similar to that observed by direct measurements. The density-sliced images, however, show details of circulation in various regions of the inlet which were not apparent from the onboard measurements (Sharma *et al.*, 1974; Burbank, 1974).

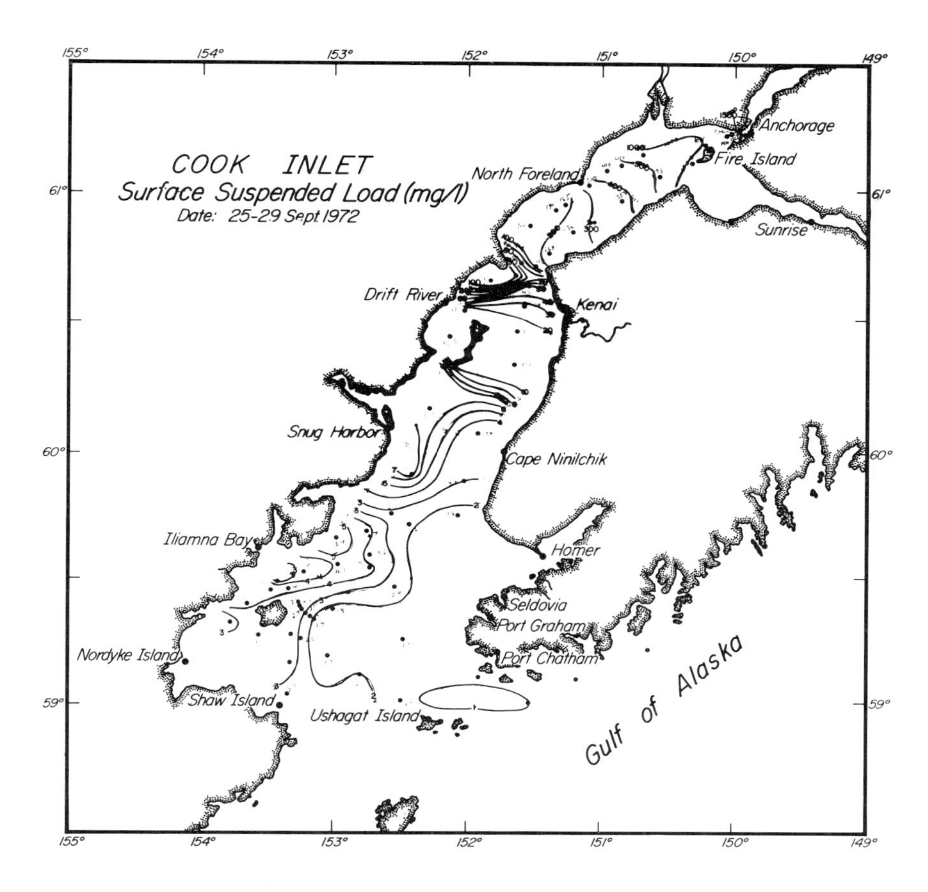

Figure 9-7. Distribution of the surface suspended sediment load during 25-29 September, 1972, Cook Inlet.

Water circulation in Cook Inlet, as interpreted from onboard measurements and ERTS-1 imagery, is counterclockwise in the lower inlet and quite variable in the upper inlet. The water moving northward through the narrows, with its high current velocity, is not significantly affected by coriolis force and therefore is not deflected to the right. This water flows in a northerly direction and moves along the western side of Middle Ground Shoal. This upward moving water loses its momentum and becomes diffused with the discharge of the Susitna and Knik rivers. The analyses of various ERTS-1 images also shows some intricate mixing and water movements in the vicinity of the Forelands. This complexity is the result of the difference in tidal phase at the head and the entrance of the inlet. During some stages of the tides, for example, the waters at the entrance as well as at the head flow toward the Forelands. The

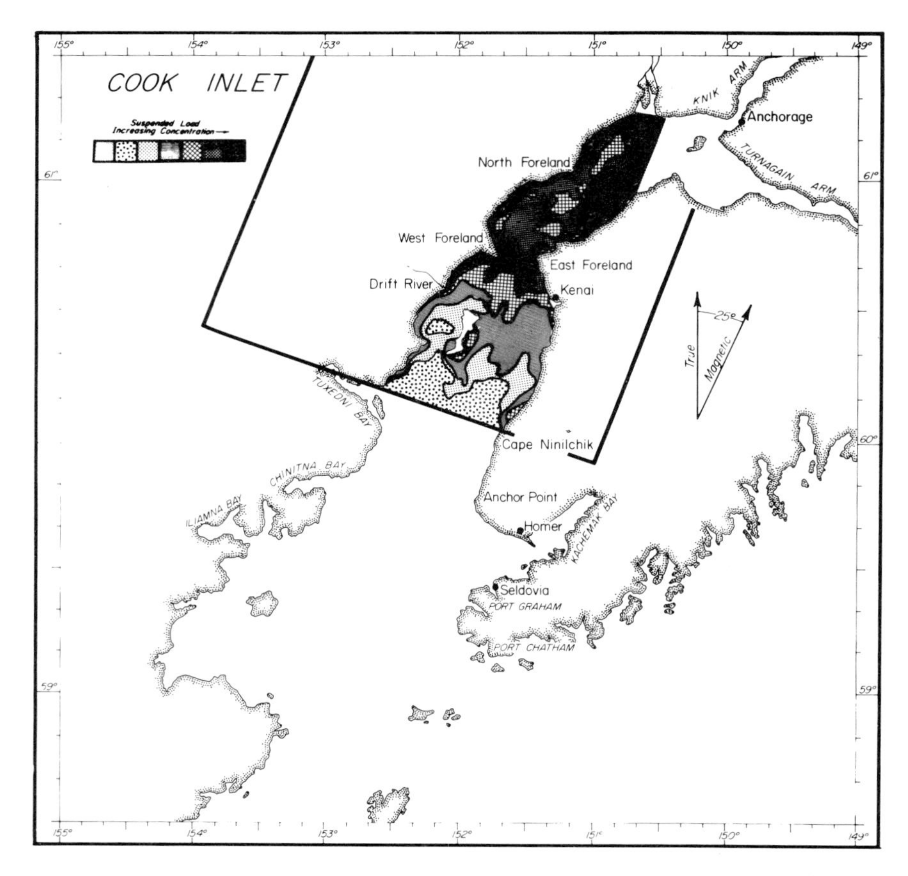

Figure 9-8. Isodensity distribution of reflectance in satellite imagery showing relative suspended loads in near-surface water of upper Cook Inlet on 4 November, 1972.

meeting of these two water masses is not only complex but continually changing as the tidal cycle progresses.

Intrusion of saline water along the southern shore sets up a counterclockwise circulation in Kachemak Bay (Fig. 9-10) due to the combined effects of the tidal and coriolis forces. The relatively turbid discharge originating near the head of the bay is carried along the northern shore. Most of the detritus is deposited in the near-shore area to form large mudflats. A small portion of the detritus is carried into Cook Inlet.

The distribution of sediments in Cook Inlet is complex. Particularly noteworthy are the absence of coarse and medium sands in Facies I and the abrupt change into Facies II, which consists of gravel (Fig. 9-5). This unique distribution is a result of the combined factors of water movement, to and fro tidal pulsation, basin morphology, ice and wind.

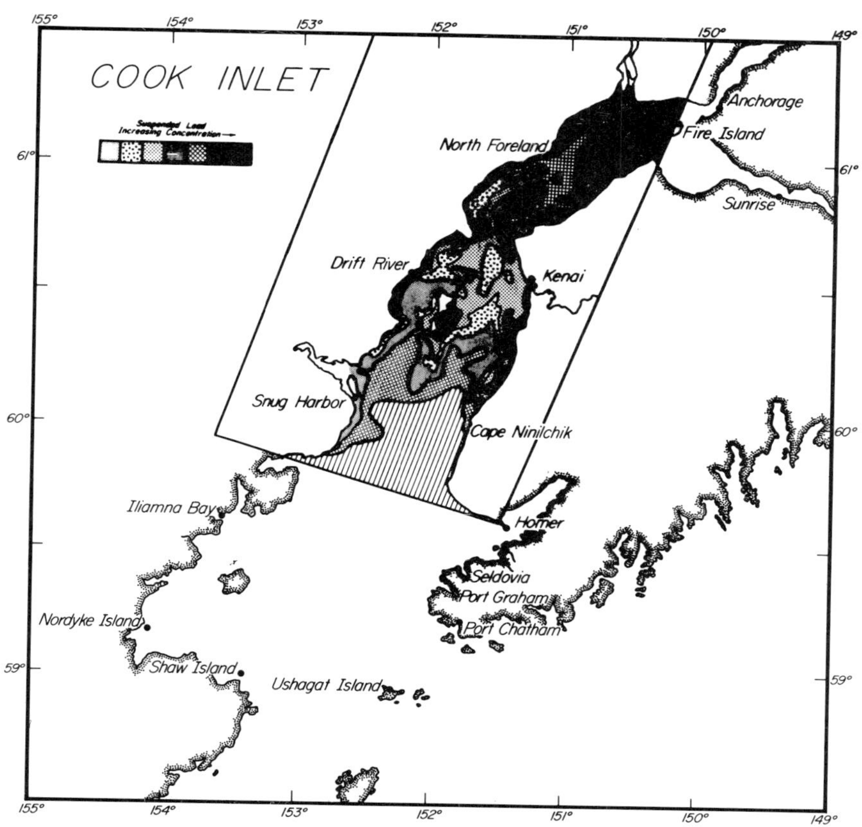

Figure 9-9. Isodensity distribution of reflectance in satellite imagery showing relative suspended loads in near-surface water on 24 September, 1973, Cook Inlet.

The suspended sediment concentration near Anchorage is much higher than that observed southwest of the Susitna River. Therefore, it appears that most sediments contributed by the rivers are deposited in upper Cook Inlet, particularly the sand deposits east of the Susitna River, designated as Facies I. With the exception of a few samples from the channel, the sands are positively skewed, which are typical of an environment of sediment accumulation. The sorting of the sand is a result of shifting of sediments swept by rotary tidal currents. Frequent shifting of sand is also suggested by the series of parallel submerged sand ridges, about 10 m apart and a few meters high, observed in this area. The tidal currents in the shallow waters of upper Cook Inlet are swift and quite capable of carrying fine sand in suspension. The current velocities in this area, measured and computed by Matthews and Mungall (1972), are comparable to those in

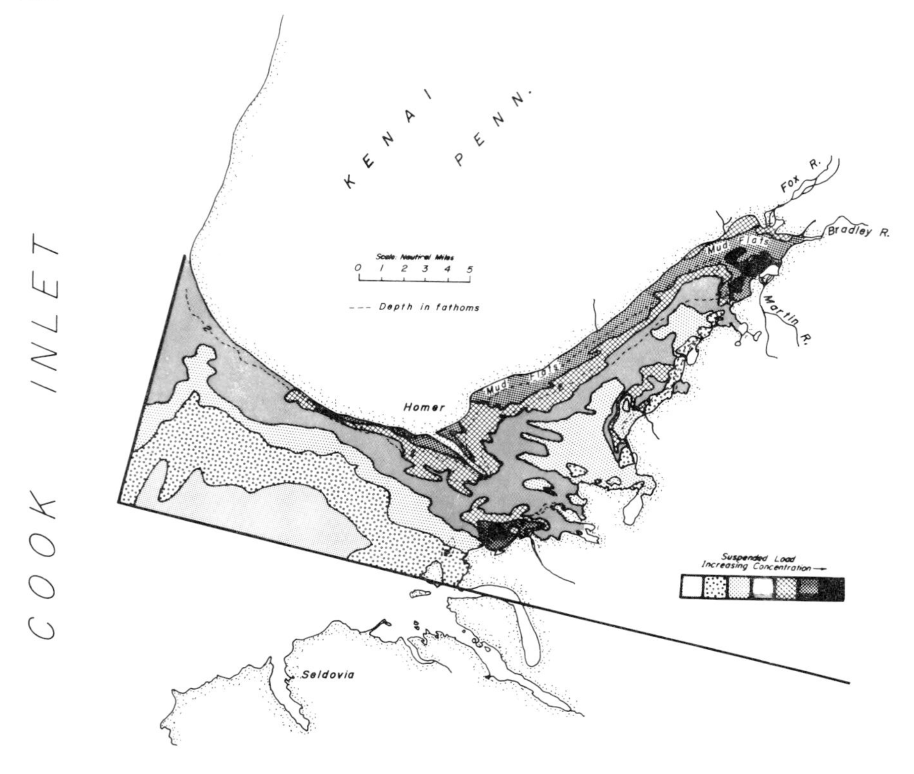

Figure 9-10. Isodensity distribution of reflectance in satellite imagery showing relative suspended loads in near-surface water on 24 September, 1973, Kachemak Bay.

areas farther south, where sand is absent and only gravel is retained on the bottom. Although the currents and turbulence in waters of the upper inlet are sufficiently high to retain sediments in suspension, the capacity of the water to carry them offshore is limited. The water primarily moves to and fro along the longitudinal axis of the inlet, with a net offshore movement of only about 1 to 2 km during each tidal cycle. During the high-discharge season, the sediment input near the head exceeds the rate of removal by the seaward movement of the water. Movement of sediments during this period is mostly through mixing of upper and lower inlet waters near the Forelands rather than as discrete plumes.

An unusual sediment distribution, *i.e.*, the preponderance of gravel in Facies II and the sharply defined boundaries between each facies, is related to the water dynamics in the inlet. The gravel in the Forelands, for instance, is attributed to the strong tidal currents. As the tide comes in, the constriction in the Forelands restricts the flow and thus causes a damming of water in the lower inlet. This rise of water in the lower inlet causes a jet effect north of the Forelands. Swift currents and turbulence in this area churn up the bottom sediments and move them toward the head of the inlet. Thus northward grading in sediment mean size relates to decreasing water energy with increasing distance from the Forelands. The competency of water to carry sediment is maximum near the Forelands, and decreases with increasing distance from the Forelands. The ebb tibe has a similar effect on the sediments deposited south of the Forelands.

Sedimentation south of the Forelands and in the lower end of the inlet is typical of an estuary as well as of a bay. The southward flowing detritus mixes with marine waters and settles out. The fine particles are generally deposited near the mouth of the inlet, in shallow bays along the shore, and in areas with weak tidal currents. Some sediments bypass the inlet and are transported in suspension into Shelikof Strait.

The effect of tidal currents on sediment transport and deposition in the vicinity of the Forelands is further illustrated by the distribution of sediments in suspension during a tidal cycle. At maximum flood and ebb stages of tides, the suspended load in water was maximum, while during slack tides it was significantly lower (Sharma and Burrell, 1970). The increase in suspension undoubtedly reflects increased turbulence in the vicinity of the constriction.

The amount of suspended sediment varies considerably throughout the Foreland area during each tidal cycle, as well as during each season (Fig. 9-11). The amount of sediment movement through the Forelands during each tidal cycle has been estimated by dividing the area into five sections and computing the volume of water flow through each section. These estimates are derived by dividing the distance between the East and West forelands into five equal parts. The current speed in this region has been computed from the tidal heights (Matthews and Mungall, 1972). The flow rates are then obtained by multiplying the cross-sectional area with the maximum current speed in each section; they are given in Table 9-1. The computed data indicate that although the current velocities in section 2 are not the highest, over 60% of the total volume of water passes through this section because of its greater depth. Even though stronger flood currents occur in sections 3, 4 and 5 (station No. CCC-7 East Foreland, Fig. 9-11), the stronger ebb currents occur in sections 1 and 2 (station No. CCC-10 West Foreland, Fig. 9-11), and the latter carry higher loads of sediment in suspension. Suspended load measurements and the computed flow data suggest that a significant amount of sediment is carried south of the Forelands. The southward movement of sediments occurs along the West Forelands as evidenced in the ERTS-1 images.

Winter ice formed in upper Cook Inlet also carries significant quantities of clay, silt and sand, which are incorporated during the formation of ice sheets on the tidal flats. Ice sheets with thin layers of sediments are stacked and carried to mid-channel, where they form floes. These ice floes with incorporated sediments then move southward through the Forelands. The passage through the constriction breaks the floes into smaller pieces, which disintegrate rapidly in the lower inlet and thereby unload the frozen sediment. The ice-rafted sediments are

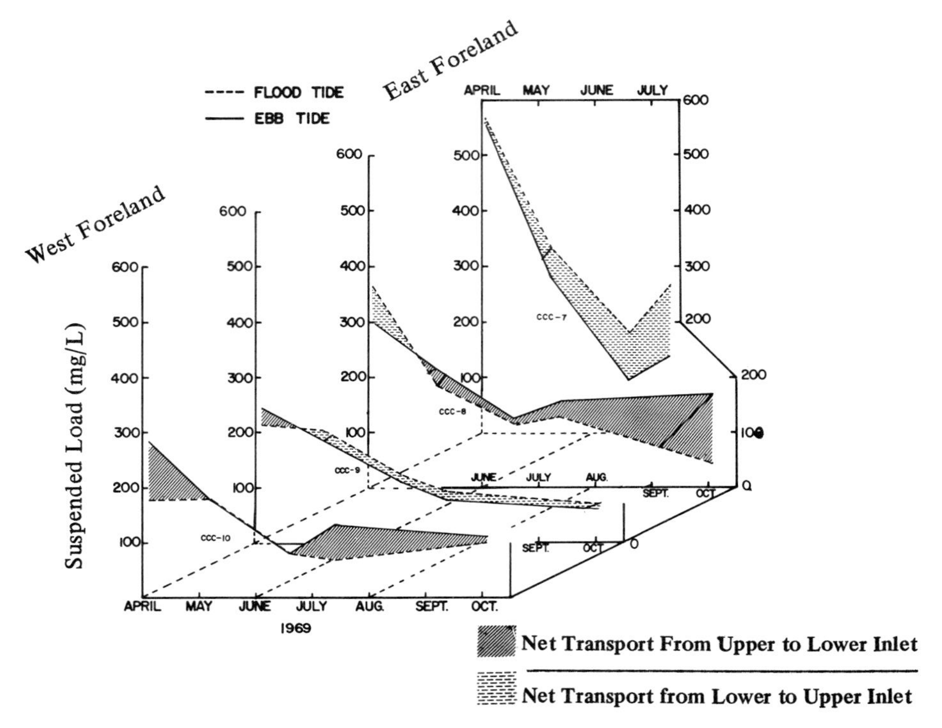

Figure 9-11. East-west profile between the Forelands, Cook Inlet, showing the net transport of sediments during flood and ebb tides.

TABLE 9-1

Computed Flow Rates Through the Forelands

	Sections					
		West		East		
	1	2	3	4	5	Total
Flow Rate x 10^4 m^3/sec	5.1	48.1	14.2	7.9	1.0	76.3
Percent Flow	6.7	63.0	18.6	10.3	1.4	100.0
Area of Section x 10^4 m^2	3.5	26.2	7.5	5.2	8.6	51.0
Percent of Area	6.9	51.4	14.6	10.2	17.0	100.0

probably largely put into suspension, to be redistributed and redeposited by tidal currents.

River detritus brought into the inlet is predominantly clay, silt, and sand. Occasionally, during flood conditions, large amounts of gravel are also transported by the rivers, For example, volcanic activity in the winter of 1965 caused an unusual annual break-up which flooded many river banks along the western shores of the inlet. The Drift River, in particular, transported large quantities of gravel into the inlet during this breakup. Occasional brief large floods resulting from the sudden draining of the ice-dammed lake in the Aleutian Range may also contribute gravel to the inlet.

The sediment texture distribution in upper Cook Inlet indicates that gravel dominates along the western shore between West Foreland and the Susitna River, while sand and sandy gravel are dominant along the eastern shore. The absence of sand in the west is caused by the swift current which accompanies the flood tide. This current entrains sediments finer than gravel and carries them to the northeast. The incoming water sets up a clockwise current north of the Forelands, in contrast to the counterclockwise circulation in the lower inlet. This clockwise circulation was also noticed in the ERTS-1 imagery discussed earlier.

Sediment discharge from the rivers (Table 9-2), together with the suspended load distribution in Cook Inlet, clearly point out that the drainage area to the north is the major source for sediments in the inlet. Over 70% of the river sediments are discharged near the head of the inlet. In the lower inlet, the Kenai River, which empties its discharge just south of East Foreland, is an important source of sediments. Numerous other small streams and rivers entering the inlet from both sides also contribute sediments to the inlet.

Waters of the lower inlet are bordered by high cliffs composed of poorly indurated to unconsolidated sediments. Frequent storms in the region develop waves which undercut the cliffs and thus introduce sediments into the inlet. Corals, clams, sponges, and other skeleton-building organisms are also important contributors of biogenous sediments to lower Cook Inlet.

The major sources of sediments near the head and the southward movement of sediments are further suggested by the southward increase in total cation-exchange capacity of sediment in suspension (Sharma and Burrell, 1970). The total cation-exchange capacity of suspended sediment material increases from 3.0 mEq/100 g north of Anchorage to between 20.0 and 90.0 mEq/100 g south of the Forelands.

Table 9-2

Suspension Loads of Rivers Draining into Cook Inlet

River	Date	Suspension (mg/L)	Remarks
Kasilof	October 31, 1968	51.6	Kasilof River bridge on Homer Highway
	July 12, 1969	42.2	
Kenai	September 27, 1968	98.6	Kenai River bridge in Soldotna
	October 31, 1968	22.1	
	July 12, 1969	26.5	
	September 30, 1969	10.9	
Knik	September 26, 1968	74.2	Knik River bridge near Palmer
	July 9, 1969	942.9	
	October 1, 1969	320.0	
Matanuska	September 26, 1968	92.4	Matanuska River bridge near Palmer
	June 9, 1969	3773.4	
	October 1, 1969	23.0	

REFERENCES

Burbank, D. (1974). Suspended sediment transport in Alaskan coastal waters. Unpublished Masters Thesis, Inst. of Mar. Sci., Univ. of Alaska, Fairbanks, p. 222.

Crick, R. W. (1971). Potential petroleum reserves, Cook Inlet, Alaska. In: Future petroleum provinces, North America. *Amer. Assoc. Petrol. Geol. Mem.* 15(1):109-119.

Fisher, M. A. and L. B. Morgan (1978). Geologic framework of lower Cook Inlet, Alaska. *Amer. Assoc. Petrol. Geol. Bull.* 62(3):373-402.

Jasper, M. W. (1965). Geochemical Investigations of Selected Areas in South-central Alaska, 1964. Division of Mines and Minerals-Geochemical Rept. No. 4. Dept. of Natural Resources, State of Alaska, 31 p.

Karlstrom, T. N. V. (1964). Quaternary Geology of the Kenai Lowland and Glacial History of the Cook Inlet Region, Alaska. U.S. Geol. Survey Prof. Paper 443.

Kelly, T. E. (1963). Geology and hydrocarbons in Cook Inlet basin, Alaska. Backbone of the Americas. *Amer. Assoc. Petrol. Geol. Mem.* 2:278-296.

Kelly, T. E. (1968). Gas accumulations in nonmarine strata, Cook Inlet basin, Alaska: Natural gases of North America. *Amer. Assoc. Petrol. Geol. Mem.* 9(1):49-64.

Kinney, P. J., J. Groves, and D. K. Button. (1970). Cook Inlet Environmental Data. R/V *Acona* cruise 065-May 21-28, 1968. Inst. of Mar. Sci., Univ. of Alaska, College, Alaska, Rept. No. R70-2, p. 120.

Kirschner, C. E., and C. A. Lyon. (1971). Stratigraphic and tectonic development of the Cook Inlet petroleum province. Unpublished manuscript.

Matthews, J. B., and J. C. H. Mungall. (1972). A numerical tidal model and its application to Cook Inlet, Alaska. Sears Foundation: *J. Marine Res.* 30(1):27-38.

Matthews, J. B., and D. H. Rosenberg. (1969). Numeric Modeling of a Fjord Estuary. Inst. of Mar. Sci., Univ. of Alaska, Fairbanks, Rept. No. 69-4, p. 77.

National Oceanographic and Atmospheric Administration (NOAA). (1971). Tidal Current Tables, Pacific Coast of North America and Asia. U.S. Dept. of Commerce, National Ocean Survey, Washington, D.C., p. 254.

Payne, T. G. (1955). Mesozoic and Cenozoic Tectonic Elements of Alaska. U.S. Geol. Survey, Misc. Geol. Inv. Map. 1-84, scale 1:5,000,000.

Rosenberg, D. H., and D. W. Hood. (1967). Descriptive oceanography of Cook Inlet, Alaska (abs): *Amer. Geophys. Union Trans.* 48(1):132.

Sharma, G. D., and D. C. Burrell. (1970). Sedimentary environment and sediments of Cook Inlet, Alaska. *Amer. Assoc. Petrol. Geol. Bull.* 54(4):647-654.

Sharma, G. D., F. F. Wright, J. J. Burns, and D. C. Burbank. (1974). Sea-surface Circulation Sediment Transport and Marine Mammal Distribution, Alaska Continental Shelf. Rept. NAS5-21833. Geophysical Institute, Univ. of Alaska, Fairbanks, Alaska, p. 77.

U.S. Coast and Geodetic Survey. (1969). Tide tables, West Coast of North and South America. U.S. Dept. of Commerce, Washington, D.C., p. 224.

CHAPTER 10
Bering Shelf

INTRODUCTION

The Bering Sea is a unique subarctic water body that lies between 52° and 66°N and 162°E and 157°W (Fig. 10-1). A relatively small sea (1% of the world oceans), it contains approximately 3.7×10^6 km^3 of water. It is bounded on the north by the Chukotka and Seward peninsulas, with a narrow (85 km) shallow passage, the Bering Strait, connecting it to the Chukchi Sea and the Arctic Ocean. In the south, it is partly separated from the Pacific Ocean by a 1900-km-long ridge and a chain of islands forming the Aleutians. Hydrographically, the sea is an immense bay in the northern part of the Pacific Basin and exchanges water with the Pacific and Arctic oceans. Geologically it is a merging site for two gigantic structures: the Alaska Orocline in the east and the Chukotka Orocline in the west.

The sea covers an area of 2.25×10^6 km2 and has an unusual bathymetry. More than half (53%) of the Bering Sea floor consists of a gentle, uniformly sloped continental shelf and a very steep continental margin. Approximately 80% of the shallow shelf lies adjacent to Alaska and the Gulf of Anadyr. With the except of several islands (Diomede, King, St. Lawrence, Sledge, Stuart, Hall, St. Matthew, Nunivak, the Pribilofs, Hagemeister, High, Crooked, Summit, and Round), the shelf floor is essentially featureless, in this respect displaying a degree of leveling and a slope uniformity that are extremely rare on other shelves of the world oceans.

The unusual features of this shelf are its width and the gradient. Compared to the average world shelf width of 65 km (Shepard, 1963), the Bering Shelf is about 500 km wide in the southeast and increases to over 800 km in the north. The average gradient (0.24 m/km) of the Bering Shelf is also markedly less than the average gradient of the world shelves reported by Shepard (1963).

In comparison with the other parts of the Alaskan Shelf, the Bering Shelf has received the most attention. The earliest recorded mapping of the region was accomplished during two monumental expeditions of the famed Danish explorer Vitus Bering between 1725 and 1743. The region was further explored by the equally well-known British navigator, Captain James Cook, in 1778 and by his successor in 1779. Numerous scientific expeditions conducted in the Bering Sea between 1850 and 1950 have been chronicled and reviewed by

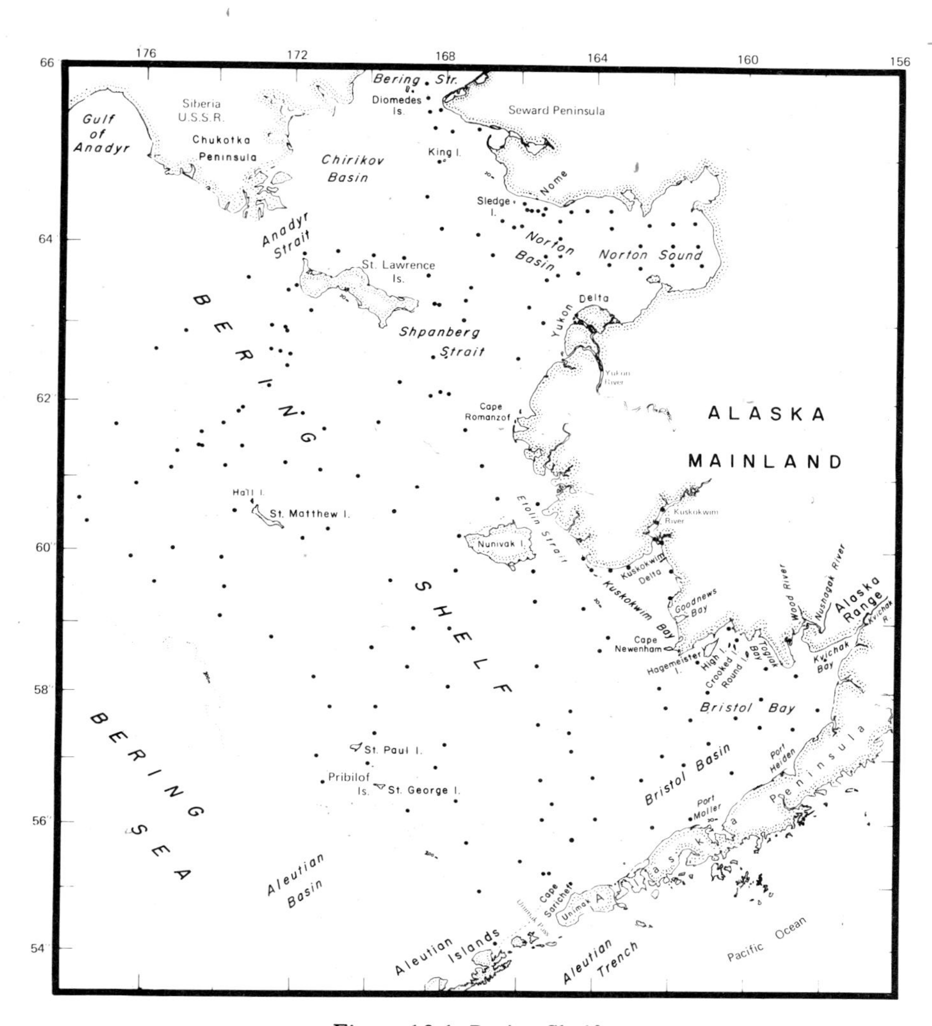

Figure 10-1. Bering Shelf.

Lisitsyn (1966). Since 1950, the tempo of scientific investigations in the Bering Sea has increased enormously. Various American, Canadian, Japanese, and Russian organizations have obtained significant geologic, biologic, meteorologic, and hydrographic information from this region. Prior to the 1960s the reason for the interest in this region was the exploitation of its biologic resources and its strategic location. Exploration for minerals during the next two decades (gold and platinum in the 1960s and petroleum in the 1970s) has caused much of the recent activity in the Bering Sea.

GEOLOGY

The origin and early history of the Bering Sea remain obscure, and the available data lead to many theories as to origin. The most widely accepted theory relates the origin of the Bering Sea and Aleutian Arc to subcrustal convection currents, ocean floor spreading, and continental drift. The convection cell origin, based on interpretation of gravity data, has been proposed by Menard (1967). In grossly simplified terms, the convection currents arising under the mid-Arctic Ocean Rise are deemed to have caused southward movement of Alaska and the Bering Sea, thus forcing the northern plate to ride bodily over the Pacific floor.

Churkin (1972) has suggested that the continental shelf of the Bering Sea is a submerged part of the larger American-Siberian continental plate, and that the abyssal basin of the Bering Sea is the northern extension of the Pacific Ocean. The subduction zone between these plates was located along the present Bering Slope through most of the Mesozoic era. Evidence for a number of faults along the slope has been found in Scholl *et al.* (1966). The continental margin has an unusually steep slope of about 20° and appears to be of a tectonic origin. Seismic reflection data obtained along the continental margin also indicate that the continental basement is downflexed and extends seaward into the Aleutian Basin. However, the plate contact moved southward sometime during Late Cretaceous-Late Eocene time and formed the Aleutian ridge and trench system (Scholl and Buffington, 1970; Scholl *et al.*, 1970, 1975). The southward movement of the subduction zone may have been the result of the bending of North America and Siberia due to rifting in the Atlantic and Arctic basins, respectively (Hopkins and Scholl, 1970).

Major structures on the eastern Bering Sea Shelf are dominated by the Alaskan Orocline (Scholl *et al.*, 1968; King, 1969; Scholl and Hopkins, 1969; Grantz *et al.*, 1970a and 1970b; Scholl and Marlow, 1970; Verba *et al.*, 1971; Kummer and Creager, 1971). The broad, concave, faulted flexures buried under the shelf were formed at the close of the Cretaceous. During the Early Tertiary, the area was exposed and eroded, and faulting, folding and development of the basin continued throughout the Tertiary.

The nature and sequence of the rocks deposited on the shelf during the Paleozoic era has not been studied in detail. The Paleozoic rocks surrounding the northern Bering Shelf, when extrapolated to the subsurface, should form the basement rock for the Norton and Chirikov basins. Southward, in the offshore region between the Yukon and Kuskokwim rivers, the shelf is perhaps underlain by Mesozoic rocks of the Koyukuk Geosyncline (Gates and Gryc, 1963). Between the Kuskokwim River and Togiak Bay the Kuskokwim Geosyncline probably extends offshore under the shelf. The Bristol Bay shelf appears to be the extension of the Alaska Range Geosyncline.

Seismic records obtained by Moore (1964) and Scholl *et al.* (1966 and 1968) indicate that the shelf subsurface deposits can be divided into three general groups. The lowest unit is an "acoustic basement" consisting of folded rock below a strongly reflecting horizon. This acoustic basement has been related to pre-Cambrian and Paleozoic sedimentary and metamorphic rocks in the northern shelf, to Cretaceous and Early Tertiary volcanic rocks in the central shelf, and to Jurassic and Cretaceous flysch-type sediments in the south (Nelson *et al.*, 1974).

Overlying the acoustic basement is a thick sequence of gently deformed marine and non-marine sediments of Middle to Late Tertiary age. This sequence is termed the "main layered sequence" and reaches a thickness of over 3 km in the northeast-trending Bristol Basin near Kvichak Bay (Scholl *et al.*, 1966). Tertiary sediments in cores obtained from this region consist of shale, siltstone, sandstone, and conglomerate (Hatten, 1971).

A thin layer of Quaternary sediments overlies the main layered sequence generally unconformably (Grim and McManus, 1970; Nelson, 1970; Kummer and Creager, 1971). These sediments consist of silt, sand, and gravel which were deposited by streams and glaciers during the Pleistocene, and their thickness rarely exceeds 100 m.

The Precambrian and Early Paleozoic history of the Bering Sea shelf is virtually unknown, and the data on events leading to the evolution of the region during the Late Paleozoic is scanty at best. During the Late Paleozoic and throughout the Mesozoic eras, the shelf was a site of deposition for various geosynclinal sediments, including siliceous clastics, cherts, carbonates and volcanic rocks. Sedimentation was temporarily interrupted during the Middle Jurassic because of the uplift associated with intrusion of the Alaska-Aleutian Range batholith, which provided a source of sediments during the Middle Jurassic and Cretaceous. Sediments added during the Mesozoic extended the shelf westward to its present position. During Late Cretaceous and Early Paleocene time these rocks were intruded by plutons. Volcanic material associated with this episode formed a volcanic belt which extended from Siberia across the Bering Shelf, between St. Lawrence and St. Matthews islands, to the Alaska mainland. The Mesozoic era concluded with uplift and erosion, which resulted in the low-relief surface that is now so distinctly apparent as the top of the acoustical basement rock.

The Cenozoic era began with continued erosion, as well as with the ejection of volcanic material along the southern edge of the shelf which led to the formation of the present-day Aleutian Ridge and Aleutian Islands. Volcanic activity has been persistent in this area since that time. The uplift of the Alaska Range during the Late Miocene diverted a considerable portion of the flow of detritus that was previously carried into the Gulf of Alaska to the Bering Shelf. Terrigenous clastic sediments derived from the Alaska mainland formed thick deposits on the shelf as well as along the continental margin.

During Quaternary time, most of the adjacent land bordering the Bering Shelf was intermittently glaciated, and a thin layer of glacial sediments was deposited on some parts of the shelf. Kummer and Creager (1971) and Grim and McManus (1970) reported glacial sediments in the Gulf of Anadyr and north of St. Lawrence Island. These sediments had their source in Siberia to the east. Glaciers from the Seward Peninsula carried sediments southward a few kilometers seaward of the present coastline. Two stages of glaciation have been identified in the northern region.

Evidence for at least four episodes of glaciation on the southern shelf have been reported (Muller, 1952; Péwé, 1953; Karlstrom, 1957, 1964; Fernald, 1960; Coulter, 1962; and Porter, 1967). Glaciers extended south from the Kuskokwim Mountains and to the north and west, respectively, from the Aleutian and Alaska ranges. Maximum ice cover occurred during the Illinoian glaciation, when ice extended westward on the continental shelf.

On the central shelf, buried channels and river silt reported by various investigators suggest flow of the Yukon River seaward on the shelf during periods of low sea level. Yukon River sediments observed between St. Lawrence and St. Matthew islands (Knebel and Creager, 1973) were deposited between 11,000 and 16,000 years B.P. Moore (1964) found buried channels between the present Yukon Delta and St. Lawrence Island and north of the delta. The sea valley in the Bering Strait can be traced to smaller rivers that drained Seward and Chukotka peninsulas. The other rivers, the Anadyr and Kuskokwim, had steadier courses: the Anadyr flowed southward through the Gulf of Anadyr and the discharge from the Kuskokwim was carried southward into Bristol Bay and then westward to the continental margin.

Some of the major islands of the Bering Sea are of volcanic origin. Smaller islands (Diomedes, King, and Sledge islands) are granite. The largest, St. Lawrence Island, is on an upwarp of Late Paleozoic or Mesozoic sedimentary and plutonic rocks and folded volcanic rocks of the Okhotsk volcanic belt. A complex of Late Tertiary and Quaternary volcanics locally overlies the older sequence. The next largest, Nunivak Island, to the southeast, is mostly composed of basalt. A sedimentary sequence containing graywacke and siltstone is exposed along its northern shore (Coonrad, 1957).

The small islands on the central shelf, St. Matthew and Hall, are part of the Okhotsk volcanic belt and consist of deformed and uplifted rocks of Cretaceous age. St. George and St. Paul, along the outer edge of the southwestern shelf, are composed of basalt and tuffaceous material (Lisitsyn, 1966).

BATHMETRY

The regional bathymetry of the large Bering Shelf has been described by Gershanovich (1963), Grim and McManus (1970), Kummer and Creager (1971), Sharma *et al.* (1972), Sharma (1972; 1974a and 1974b), Askren (1972), Knebel (1972), and McManus *et al.* (1974). For the most part the shelf is extremely flat with an average gradient of about 0.2 m/km (Fig. 10-2). The northern shelf displays a few basins and small banks. One large elongate and two small circular depressions are conspicuous in Norton Sound. The region between the Yukon River and St. Lawrence Island, the Shpanberg Strait, has two linear depressions, with an intervening northwest trending submarine ridge. Two large banks, one

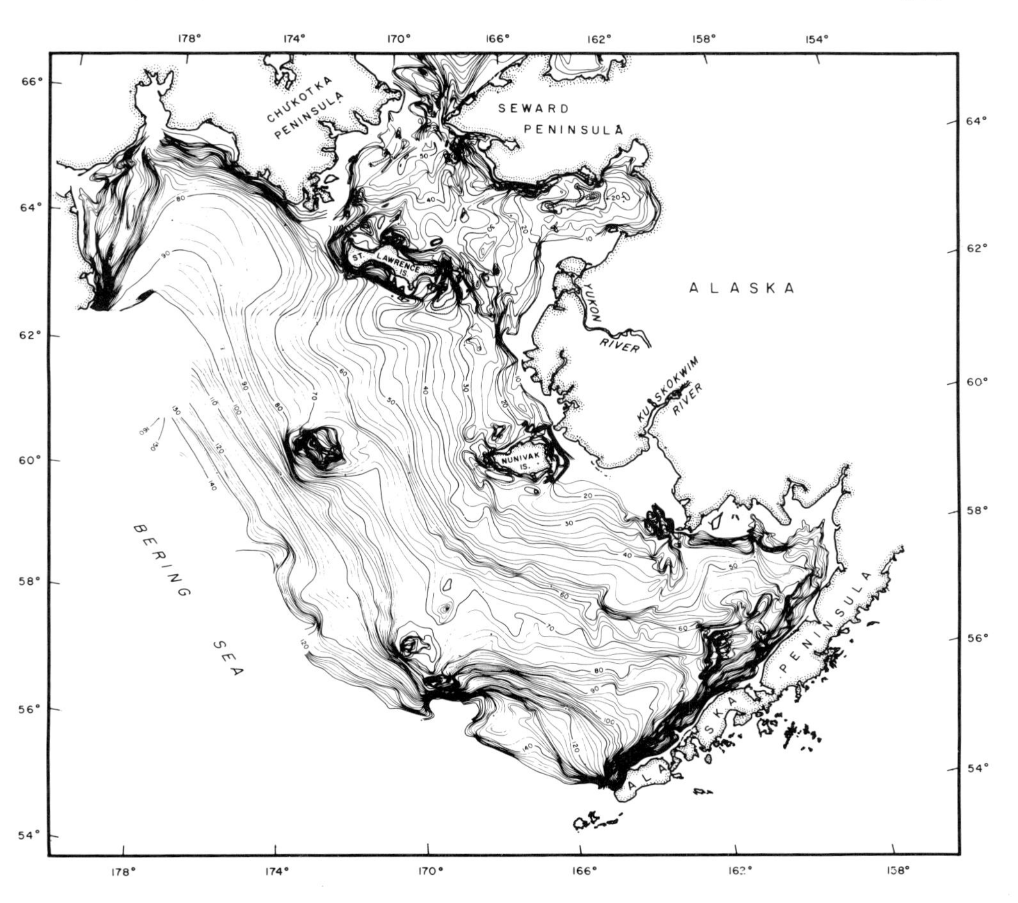

Figure 10-2. Bathymetry chart, Bering Shelf. (Courtesy of Dr. J. S. Creager, University of Washington, Seattle.) Depth in meters.

south and the other northeast of St. Lawrence Island, are important features of the central shelf. The Bristol Bay–Pribilof Islands region has the most salient bottom relief irregularities, including a distinct northeast trending trough along the Alaska Peninsula. The bottom topography in this region may be in part a result of structural features characteristic of the transition between the epicontinental shelf and the geosynclinal zone, and partly a result of superimposition of younger Cenozoic rocks on older geologic structures in this region.

Transitory bottom forms include channels, swales and ridges, and small closed depressions. Channels are common in Kvichak and Kuskokwim bays. The narrow troughs and ridges, which have about 10 m of relief, are the salient features of the shallow waters near the head of Bristol Bay. Northward, near Nunivak Island, a closed depression and channels are the conspicuous bathymetric features. A prominent channel lies east of Nunivak Island and extends northward along the shoal adjacent to the Alaska mainland.

HYDROLOGY

Most of the Bering Shelf lies in the subarctic latitudes, and therefore a cyclonic atmospheric circulation predominates in this region. The annual weather patterns are largely controlled by the Honolulu, Arctic, and Siberian highs, and the Aleutian Low. During summer, the Honolulu High occupies the northern Pacific Ocean, and its intensification generates southerly and southwesterly winds in the eastern Bering Sea. With the onset of winter, the summer Honolulu High moves to the southwest and is replaced by a large intense Aleutian Low. This shift permits the movement of the Arctic High farther southward and results in predominant northeasterly winds on the shelf. The seasonal winds significantly influence the currents on the shelf. The direction, intensity, and duration of winds influence the shelf water exchange between the Pacific Ocean to the south and the Arctic Ocean to the north.

The Bering Shelf lies in the paths of both extratropical cyclonic and Asiatic anticyclonic storms. The storms occur so frequently that sometimes several are present in the region. Frequent summer and winter storms intensify currents and sometimes reverse the general flow. Furthermore these storms often destroy water stratification in shallow regions.

The shelf climate is influenced by the mild North Pacific Ocean and the Bering Sea maritime climate to the south and west, as well as by the cold continental subarctic climate to the east. The southern shelf is strongly affected by meteorologic conditions prevailing along the Aleutians, and the climate is therefore milder than that prevailing to the north. Cloudy skies, moderately heavy precipitation, and strong surface winds characterize the shelf weather. Average summer temperatures vary from 10°C in Bristol Bay to about 8°C in Norton Sound; the average winter minimum ranges from -14 to -18°C (Environmental Data Service, 1968). The annual precipitation is 600 mm at the Pribilof Islands and 400 mm near the northwest tip of St. Lawrence Island.

Major rivers discharging fresh water and sediments into the eastern and northwestern shelf are the Yukon, Anadyr, Kuskokwim, Wood, Nuyakuk–Nushagak, and Kvichak. Mean annual discharges of the Yukon, Anadyr, Kuskokwim, Nuyakuk, and Wood rivers have been described by Roden (1967) and are shown in Table 10-1. The high-latitude drainage area that feeds these rivers has a typical unimodal discharge pattern, which reaches its peak in June.

Ninety percent of the annual flow occurs between May and October and approximately 60% of the mean annual discharge takes place during the months of June, July and August. According to Roden (1967), it is reasonable to assume that the total mean annual freshwater discharge into the Bering Sea exceeds the discharge into the Pacific Ocean of the combined states of California, Oregon,

TABLE 10-1

Drainage Area and Discharge of Major Rivers in the Bering Sea (Roden, 1967)

River	Drainage Area (km^2)	Mean Annual Discharge (m^3/sec)
Yukon River at Kaltag	766,600	6,220
Anadyr River at mouth	200,000	1,660
Kuskokwim River at Crooked Creek	80,550	1,270
Nushagak River near Dillingham	3,860	164
Wood River at Aleknagik	2,870	140

and Washington. The significance of such a large discharge becomes even more important when it is considered that this discharge is added during a 6-month period. Because of this large input of fresh water along the Alaska mainland, a distinct near-shore Alaskan coastal water is formed during the summer.

HYDROGRAPHY

Waters of the Bering Shelf are complex and extremely dynamic. A large part of the shelf can be regarded as an immense, shallow, high-latitude estuary. The exceptional characteristic of the shelf waters during summer is their erratic variability. The waters of this immense subarctic shelf are continually influenced by the intrusions of Pacific and Arctic waters, river discharge, wind stress, and variations of air temperature. The frequent brief but violent storms, in particular, alter water density structure and cause upwelling and downwelling of large volumes of water. Because of the significant dominance of the relatively warm Pacific water intruding through the southern passes and the influence of cold polar waters from the Anadyr Gulf and Chukchi Sea in the north, the shelf waters can be broadly divided into two major hydrographic regimes: (1) the northern region between the Bering Strait and St. Matthew Island is dominated during the summer by cold Gulf of Anadyr water, and during winter it is covered with ice; (2) the southern region between St. Matthew Island and the Alaska Peninsula is mostly ice free throughout the year; but occasionally the severe winter conditions may cause ice to form further south. The waters of both regions are continually modified near the surface by river discharge and by shoreward movement of saline waters at depth. The formation and distribution of various water masses on the shelf have been described by LaFond and

Pritchard (1952), Arsen'ev (1967), Ohtani (1969), Kitano (1970), and Takenouti and Ohtani (1974).

Water regimes on the southern and northern Bering Shelf are different and are discussed separately. There are four dominant water masses on the southern Bering Shelf; oceanic water, outer shelf water, mid-shelf water, and coastal water. The oceanic waters are the Pacific and Bering Sea waters. The characteristic Pacific water has temperatures between 4 and 8°C, and a salinity range of from 32.0 to $33.2^\circ/_{oo}$. Bering Sea water, on the other hand, is slightly colder (2 to 8°C), with salinity varying between 32.6 and $33.2^\circ/_{oo}$. The Pacific water entering the shelf through Unimak Pass is brought by the Alaska Current, part of a counterclockwise gyre formed in the northern Pacific Ocean. By the time the waters of the Alaska Current reach Unimak Pass, they have been well mixed with the surface flow from southeast Alaska, the central Gulf of Alaska, and the Aleutians. This well-mixed water, with relatively stable salinity, enters the southern shelf over a sill depth of 150 m.

The outer shelf water of the southern regime forms a discontinuous zone of temperature (0.5°-9°C) and salinity (32.7-$33.0^\circ/_{oo}$) with the oceanic water along the edge of the shelf. This water mass is formed as a result of mixing of the Pacific, Bering Sea, and shelf waters and is found along the outer edges of the continental margin.

The mid-shelf water covers a broad zone extending from the outer margins of the shelf shoreward to the isobaths of 40 to 50 m. This water is characterized by salinity between 31.0 and $32.6^\circ/_{oo}$ and temperature between 1.8 and 9.0°C. The isohalines are generally parallel to the isobaths but the stratification of water in summers is often destroyed by severe storms, while during winter rapid cooling and the presence of drift sea ice render the water vertically uniform.

The fourth water mass, coastal water, is formed as a result of mixing of surface runoff and shelf water in the near-shore areas and extends to the 40 to 50 m isobaths. The salinity of the water is less than $31.0^\circ/_{oo}$, with temperatures ranging between 1.3 and 18.2°C (Figs. 10-3 and 10-4). During winter the surface coastal water freezes and forms an ice cover. In shallow coastal areas with high freshwater input the freezing results in shorefast ice.

The upper reaches of shelf waters on the southern regime are partly covered by sea ice and the region between the Pribilofs and St. Matthew Island during December to April has a 10 to 70% ice cover. The southward extension of winter ice has a strong influence on the water characteristics of the shelf waters during the following summer.

The northern shelf has three types of waters: the outer shelf water, the inner shelf water and the coastal water. The outer shelf water is formed at depths greater than 100 m by the mixing of inner shelf water with eastward intruding Bering Sea water of 2 to 4°C temperatures and 33-$34^\circ/_{oo}$ salinity. The resultant water mass is well stratified and extends along the outer shelf, northward past St. Lawrence Island and into part of the Bering Strait.

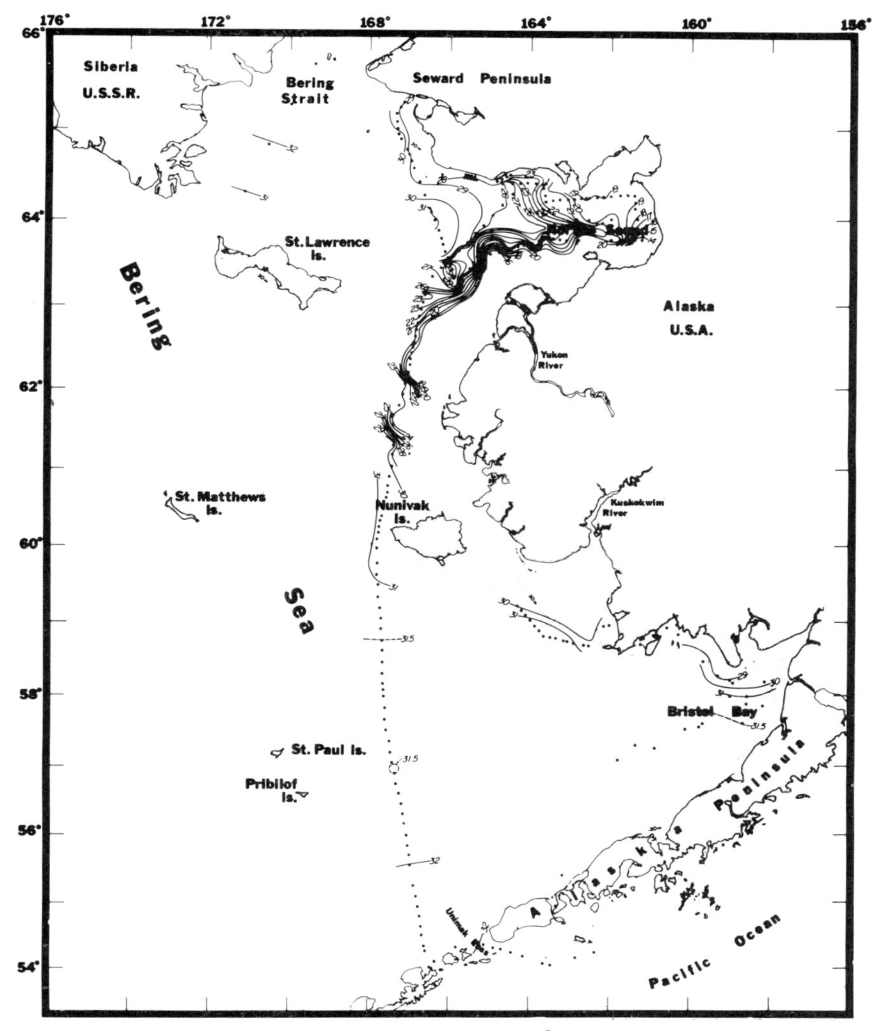

Figure 10-3. Surface water isohalines (°/$_{oo}$), Bering Shelf.

The inner shelf water of the northern regime is formed as a result of freezing of the surface layer and sea ice cover during winter. In winter the entire region is covered with ice, and the water becomes nearly isothermal and isohaline. The water has a distinctive character, with -1.7 to 2.0°C temperature and 32-33°/$_{oo}$ salinity. During summer months, when the region is ice free, this water retains its identity and is commonly observed between the 20- and 40-m isobaths.

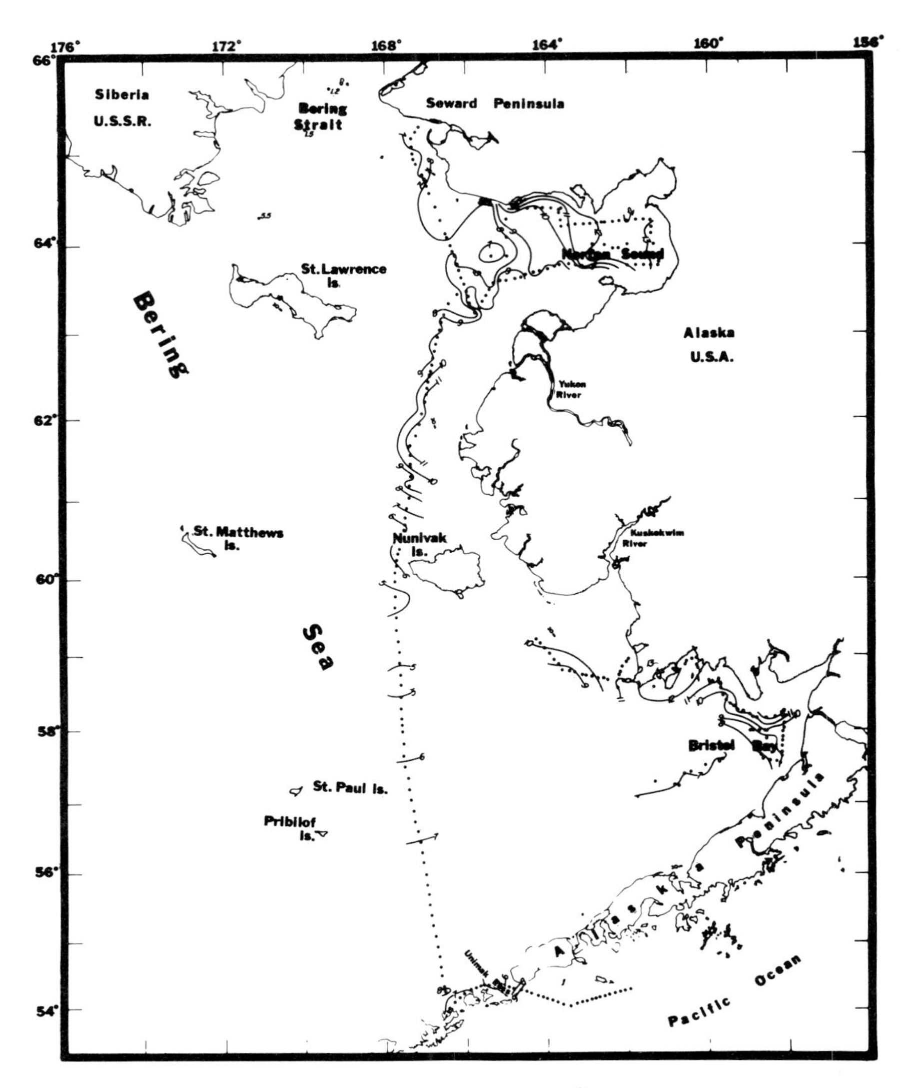

Figure 10-4. Surface water isotherms (°C), Bering Shelf.

A narrow band of coastal water along the Alaska mainland is characterized by 8-11°C temperature and salinity of less than 31°/₀₀ (Figs. 10-3 and 10-4). This water is formed as a result of the mixing of inner shelf water and the coastal runoff. During summer months, when the Yukon River discharge is high, the Norton Sound is entirely covered with coastal water. With the onset of colder weather during autumn, this water freezes to form shorefast ice, which initiates the beginning of the winter ice cover in the northern Bering Shelf.

Ice conditions in winter in the northern regime are severe. Observations recorded over 30 years (U.S. Navy Hydrographic Office, 1961) indicate that generally the ice cover begins in November and reaches its maximum extent in March. The southward extension of sea ice on the Bering Shelf is highly dependent upon the atmospheric pressure systems and the prevailing winds (Konishi and Saito, 1974). During January through April the ice coverage in the region lying north of Nunivak Island is between 80 and 90% of the sea surface. The ice cover begins to recede northward in May, and by early July the ice has generally retreated northward of Bering Strait.

Only fragmentary knowledge of the movement of waters at the surface and at depth on the Bering Shelf exists. The paucity of direct current measurements in the region permits one to predict, with considerable uncertainties, only a generalized circulation pattern. The general circulation on the shelf and the basin has been described by Ratmanof (1937), Barnes and Thompson (1938), Goodman *et al.* (1942), Saur *et al.* (1952, 1954), Dobrovol'skii and Arsen'ev (1959), Favorite and Pederson (1959), Favorite *et al.* (1961), Hebard (1961), Dodimead *et al.* (1963), Sharma *et al.* (1972), and Favorite (1974).

The driving force for the water movements on the shelf is a combined effect of wind, tide, and surface runoff. The tidal range on the shelf is moderately low and ranges from 2.4 m at the head of Bristol Bay in the south to 0.6 m near the Seaward Peninsula in the north. Winds and storms develop short-term local and regional circulation patterns. Current reversals in response to changes in the wind direction have been reported at many locations by the U.S. Coast and Geodetic Survey (1964).

The incoming tidal currents (175 cm/sec) off Scotch Cap in Unimak Pass exceed the outgoing tidal currents (150 cm/sec) by 25 cm/sec (U.S. National Ocean Survey (1973b). The northward current through Unimak Pass may be considerably accelerated by the influence of an atmospheric depression north of the chain (U.S. Coast and Geodetic Survey, 1964). Under such meteorologic conditions the current velocity in the pass may exceed 300 cm/sec, resulting in the transfer of large amounts of Pacific water into the Bering Sea. Part of the water passing through Unimak Pass is deflected to the east and continues north of the Alaska Peninsula, while the rest flows northward along the outer shelf.

The tides in Bristol Bay are amplified near the heads of shallow embayments. Mean ranges are at Cape Sarichef, Unimak Island, 1.0 m; Port Moller, 2.3 m; Kvichak Bay, 4.6 m; Nushagak Bay, 4.7 m; Kuskokwim Bay, 4.1 m; Goodnews Bay, 1.9 m; and St. Paul Island (the Pribilofs), 0.6 m (U.S. National Ocean Survey, 1973). Hebard (1961) reported near-shore tidal currents of 40 to 85 cm/sec along the northern Alaska Peninsula and 50-75 cm/s in central Bristol Bay.

In addition to the tidal influence, semi-permanent currents form a counterclockwise circulation on the southern shelf. The currents forming the gyre have been measured by Hebard (1961) and by Natarov and Novikov (1970), who show them to vary seasonally and to be significantly influenced by changes in wind direction.

Farther north, the tidal currents are 40 cm/sec off the west coast of Nunivak Island; 40 cm/sec off Northeast Cape, St. Lawrence Island; and 50 cm/sec at Sledge Island, west of Nome. Near-bottom current speeds of 30 to 40 cm/sec in the northern Bering Shelf have been estimated by McManus and Smyth (1970). Surface and near-bottom currents of 15 to 72 cm/sec and 15 to 34 cm/sec, respectively, have been reported in the Bering Strait by Creager and McManus (1967).

The flow pattern of northward moving Pacific water on the shelf is somewhat unclear. Many investigators have stipulated formation of various smaller gyres, particularly in the regions lying north and south of St. Lawrence Island. These gyres may be seasonal. No concerted attempts have been made to determine the intensity and the configuration of these currents, or whether these currents prevail throughout the year.

The northward flow of water through the Bering Strait has been well documented by various investigators. The current velocity in the strait was obtained by direct current measurements and computations from the density distribution (Fujii, 1973). The mean speed of the northward flow was 0.95 knots and the volume of transporting entering the Chukchi Sea was estimated at 1.97 SV. The north-setting currents persist throughout the year with only temporary reversals.

SEDIMENTS

The surficial sediments from the Bering Shelf consist of a varying mixture of clay, silt, sand, and gravel. Most of the shelf is covered by either sand or silt. Gravel and clay components are generally absent or constitute only a minor proportion of these sediments. The textural distribution of sediments on the Bering Shelf is complex because of the extremely variable and localized source input (rivers) and the variance in sediment transport energy (currents). The distribution is also influenced locally by the action of wind wave, tidal, and permanent water circulation, as well as the effects of ice. Regionally, semi-enclosed Bristol Bay and Norton Sound display a sediment texture which is slightly different from that observed on the open shelf (Figs. 10-5 and 10-6). Local textural anomalies in near-shore zones may result from river detritus input and/or exposed relict glacial deposits.

The shelf deposits of Bristol Bay have been described in detail by Sharma (1972, 1974a, and 1974b, and Sharma *et al.,* 1972). These studies show that in Bristol Bay proper, near-shore sediments consist of gravel and coarse sand, with the greater part of the central shelf being covered with fine to medium sands, while farther offshore the sediments become progressively finer (Fig. 10-7). The floors of the shore indentations and some bays are covered with yellowish-brown clayey silt and clayey silty sand, whereas the open shore is generally mantled with pebbly sediments. The sediment mean size decreases with increasing distance from shore and with water depth (Fig. 10-8). Sorting in sediments is

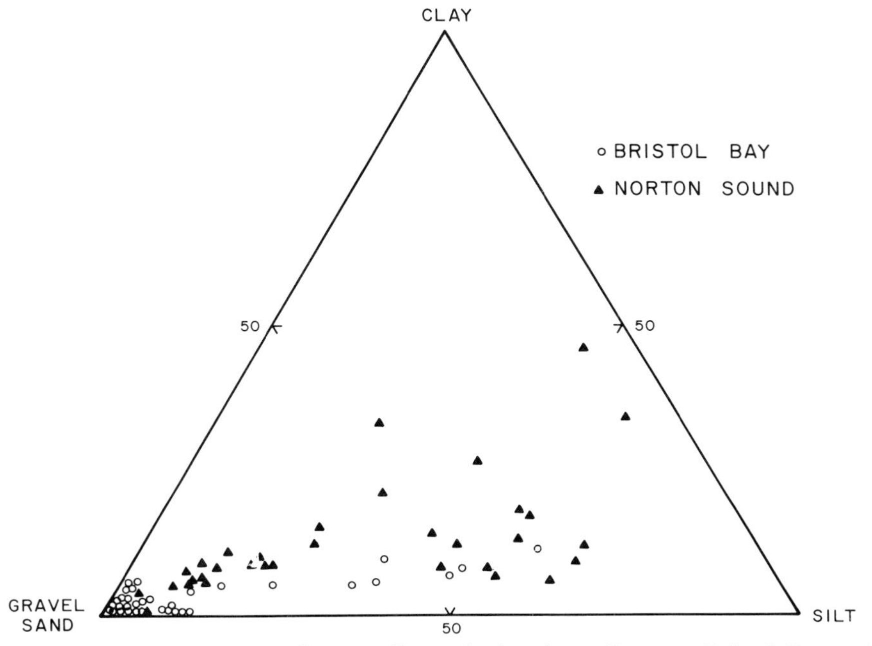

Figure 10-5. Percent gravel-sand, silt, and clay in sediments, Bristol Bay and Norton Sound.

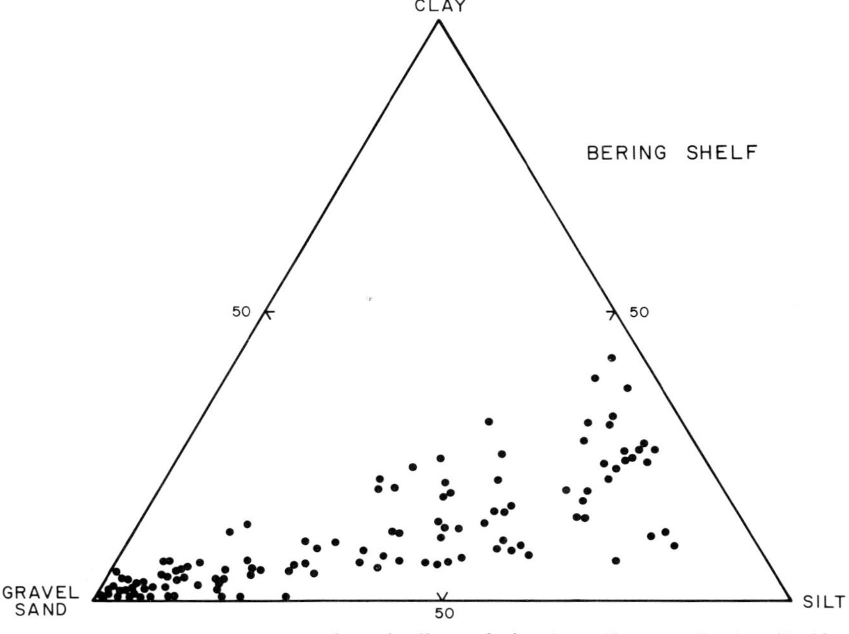

Figure 10-6. Percent gravel-sand, silt, and clay in sediments, Bering Shelf.

related to the sediment mean size: Near-shore coarse sediments are extremely poorly sorted, medium and fine sands deposited on the mid-shelf are moderately well sorted, and offshore the sorting deteriorates with increasing amounts of silt and clay components (Figs. 10-9, 10-10, and 10-11). The medium and fine, moderately well-sorted sands on the mid-shelf have nearly symmetrical size distributions but progressively grade into strongly coarse-skewed sediments shoreward, and strongly fine-skewed sediments toward the continental margin. Most sediments are leptokurtic to extremely leptokurtic.

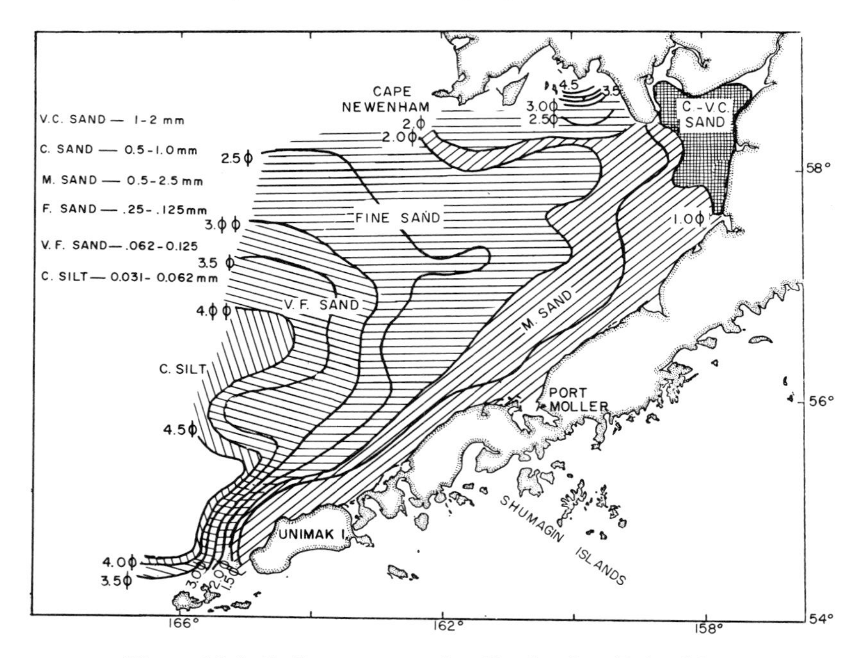

Figure 10-7. Sediment mean size distribution, Bristol Bay.

The distribution (by weight percent) of organic carbon in the surface sediments of the Bristol Bay region varies between 0.05 and 0.55 (Fig. 10-12). The organic carbon increases with increasing clay content in the sediments (Figs. 10-11 and 10-12). Therefore, higher organic carbon contents occur in sediments deposited near the continental margin, as well as in Togiak Bay.

The sediment cover of the shelf area between Unimak Pass, Nunivak Island, and St. Matthew Island has been described by Askren (1972). This triangular region of the shelf consists of sand, silty sand, sandy silt, and sandy clayey silt. The eastern shallow portion, with depths of less than 60 m, is covered with sediments containing 75% or more sand. Westward, a narrow zone which lies between the 60 and 75-m isobaths consists of sediments comprised of a range of mixtures of sand and silt. The outer shelf, with depths greater than 75 m, is mantled with clayey silt and some sand (Figs. 10-13 and 10-14). The shallow

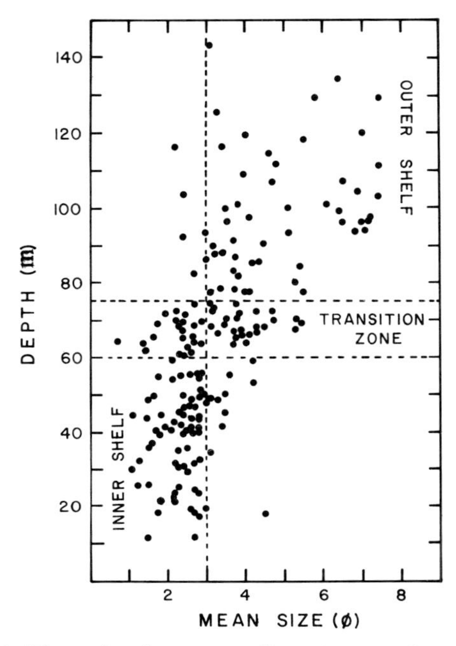

Figure 10-8. Water depth versus sediment mean size plot, Bering Shelf.

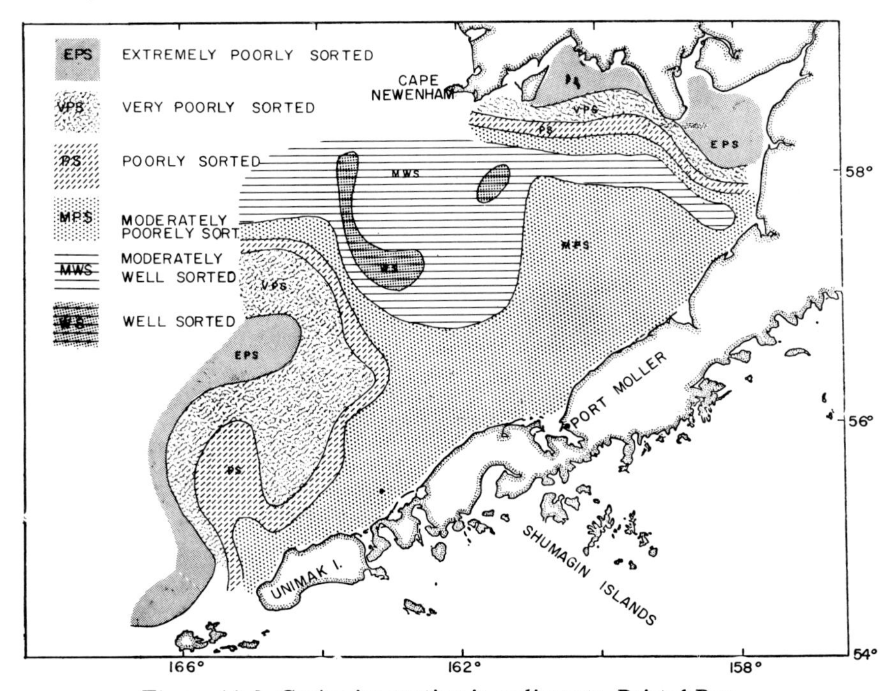

Figure 10-9. Grain size sorting in sediments, Bristol Bay.

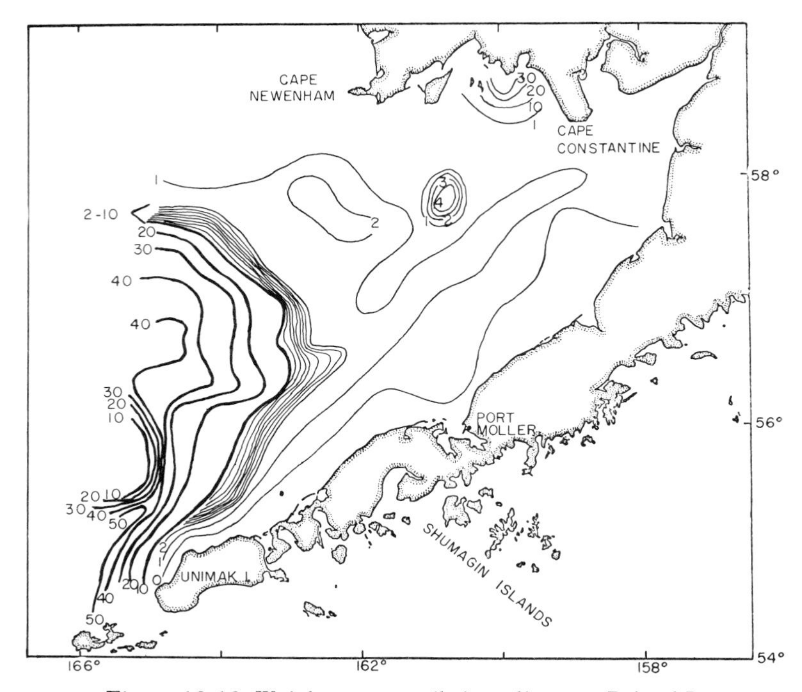

Figure 10-10. Weight percent silt in sediments, Bristol Bay.

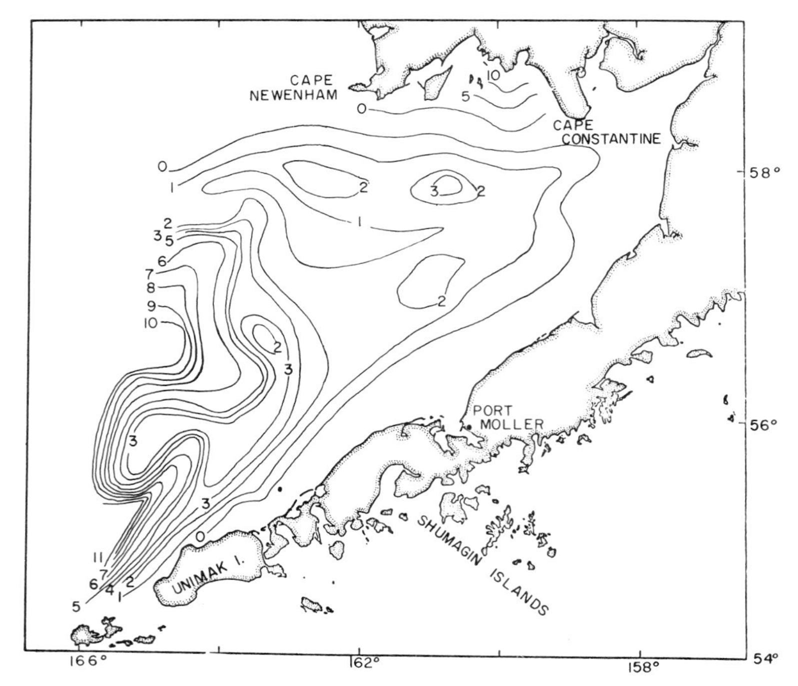

Figure 10-11. Weight percent clay in sediments, Bristol Bay.

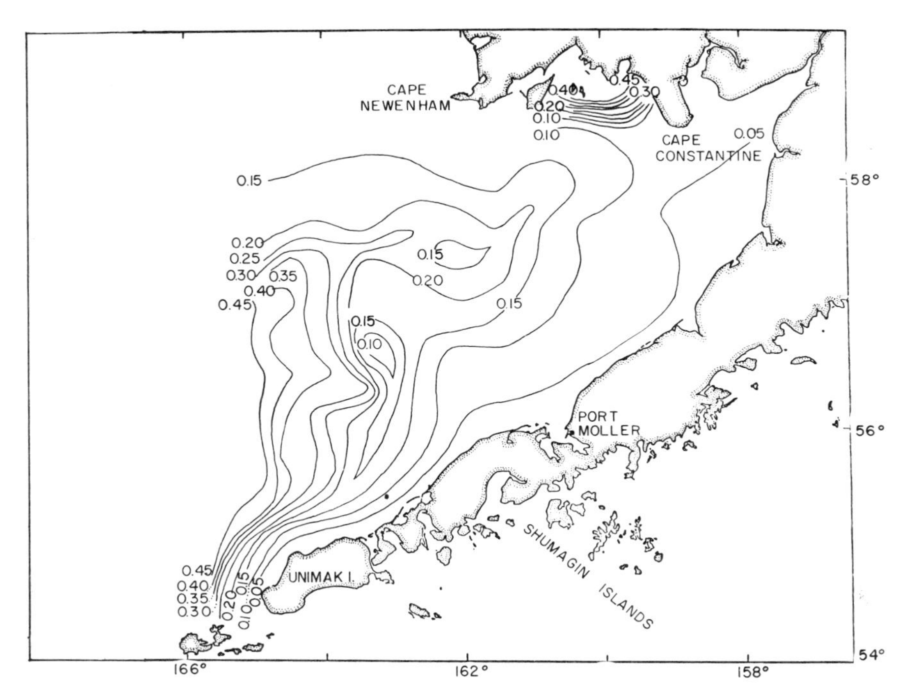

Figure 10-12. Weight percent organic carbon in sediments, Bristol Bay.

shelf sands are moderately well sorted, but sorting deteriorates with increasing depth and increasing silt-clay fraction (Figs. 10-15, 10-16 and 10-17). Most sediments are fine to strongly fine skewed and show platykurtic to extremely leptokurtic size distributions (Fig. 10-18).

The interrelation between the water depth and the sediment mean size distribution on the southern Bering Shelf deserves closer scrutiny. The combined plots of sediment mean size versus water depth (Fig. 10-8) obtained from Askren (1972) and Sharma (1972) distinctly show that in depths less than 60 m, the shelf is covered with sediments coarser than 3 ϕ. On the other hand, the outer shelf (depths greater than 75 m), is mantled only with sediments finer than 3 ϕ. The intermediate zone (depths between 60 and 75 m) has sediments of a wide range of grain size, coarser as well as finer than 3 ϕ. Significant changes in sorting and skewness of the sediments also occur in this transition zone between the 60- and 75-m isobaths. The mean size, sorting, and skewness isopleths run almost parallel to the isobaths, particularly in the area just west and south of Nunivak Island (Figs. 10-15, 10-16, and 10-17).

The textural characteristics of sediments deposited on the shelf between St. Matthew and St. Lawrence islands have been described by Knebel (1972). He observed sands to be the most ubiquitous components in the sediments. However, the sediments size distribution on the central Bering Shelf is complex, and the seaward size grading with water depth is not obvious.

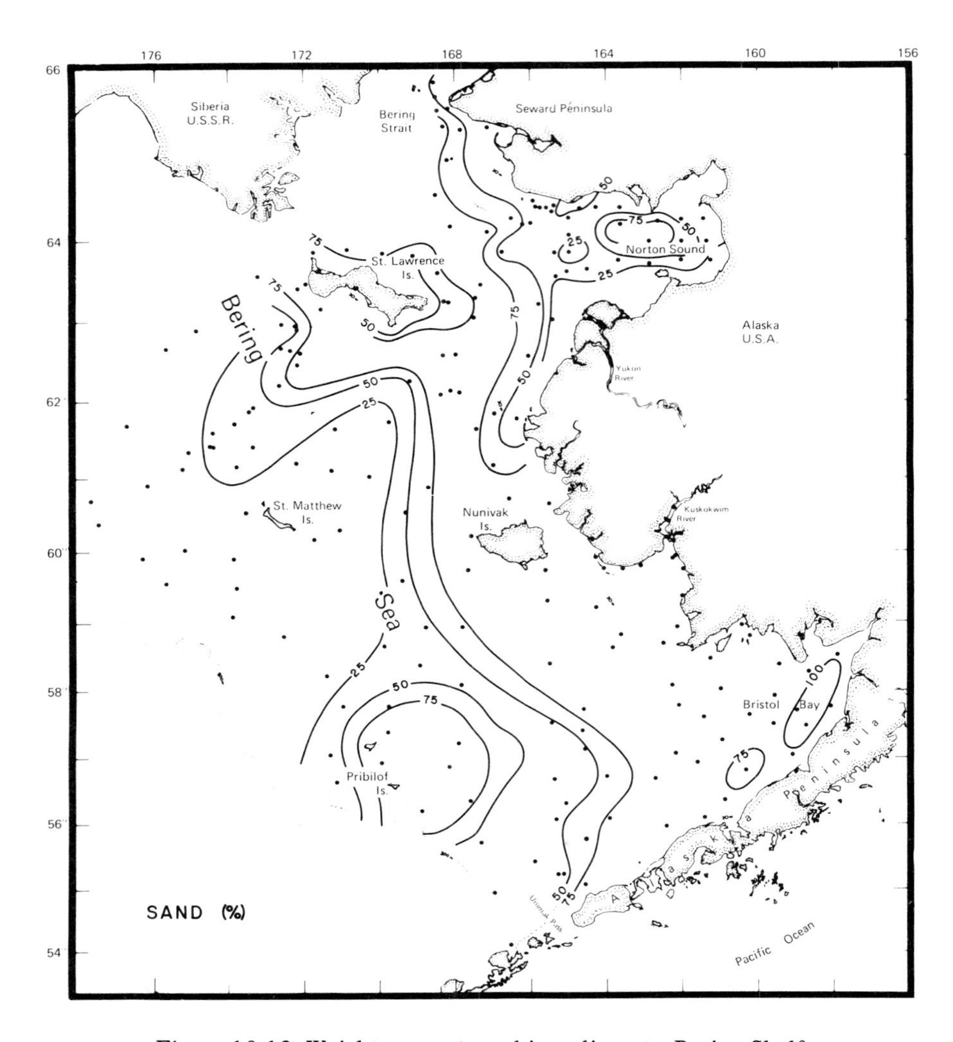

Figure 10-13. Weight percent sand in sediments, Bering Shelf.

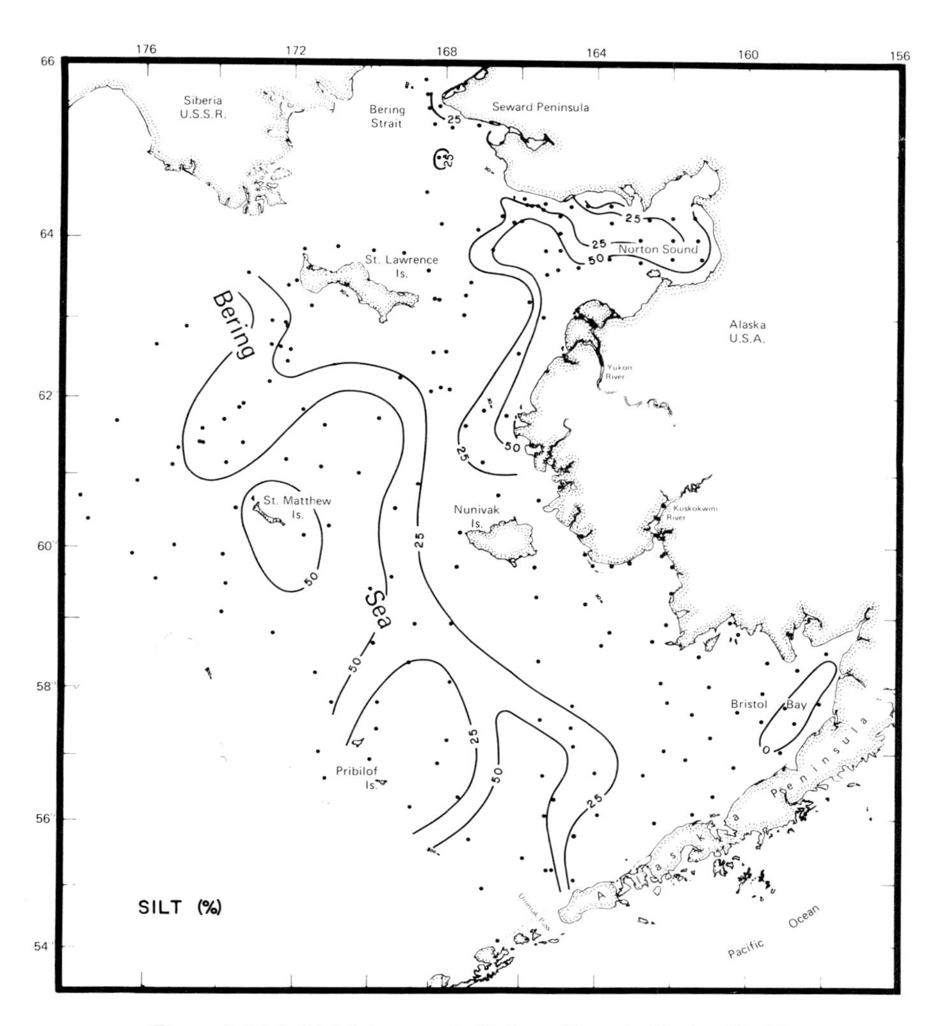

Figure 10-14. Weight percent silt in sediments, Bering Shelf.

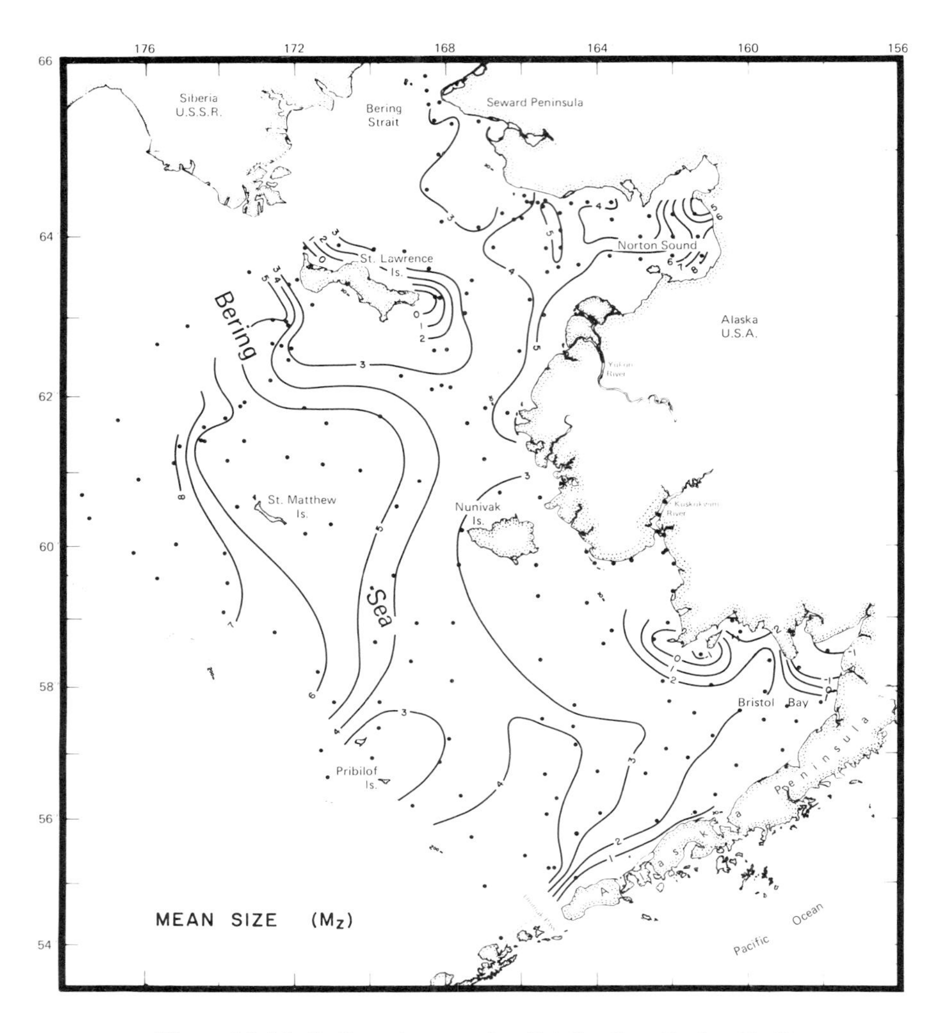

Figure 10-15. Sediment mean size distribution, Bering Shelf.

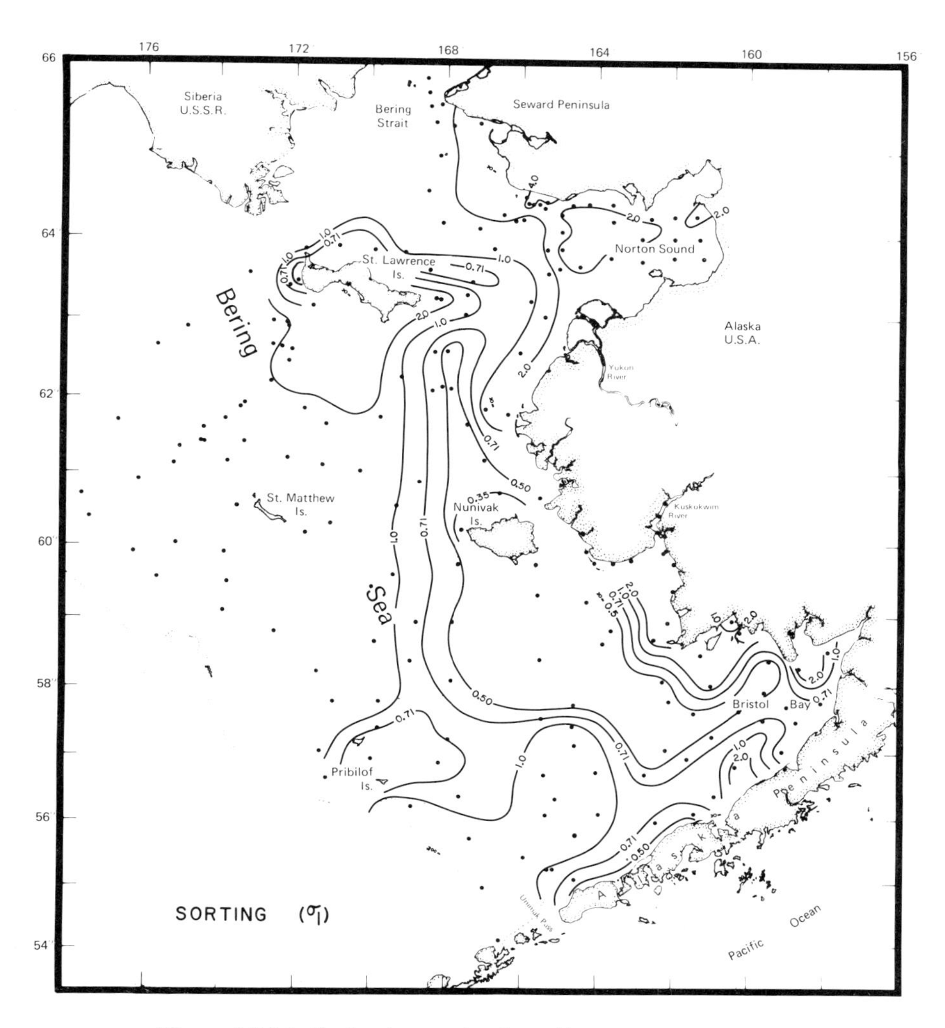

Figure 10-16. Grain size sorting in sediments, Bering Shelf.

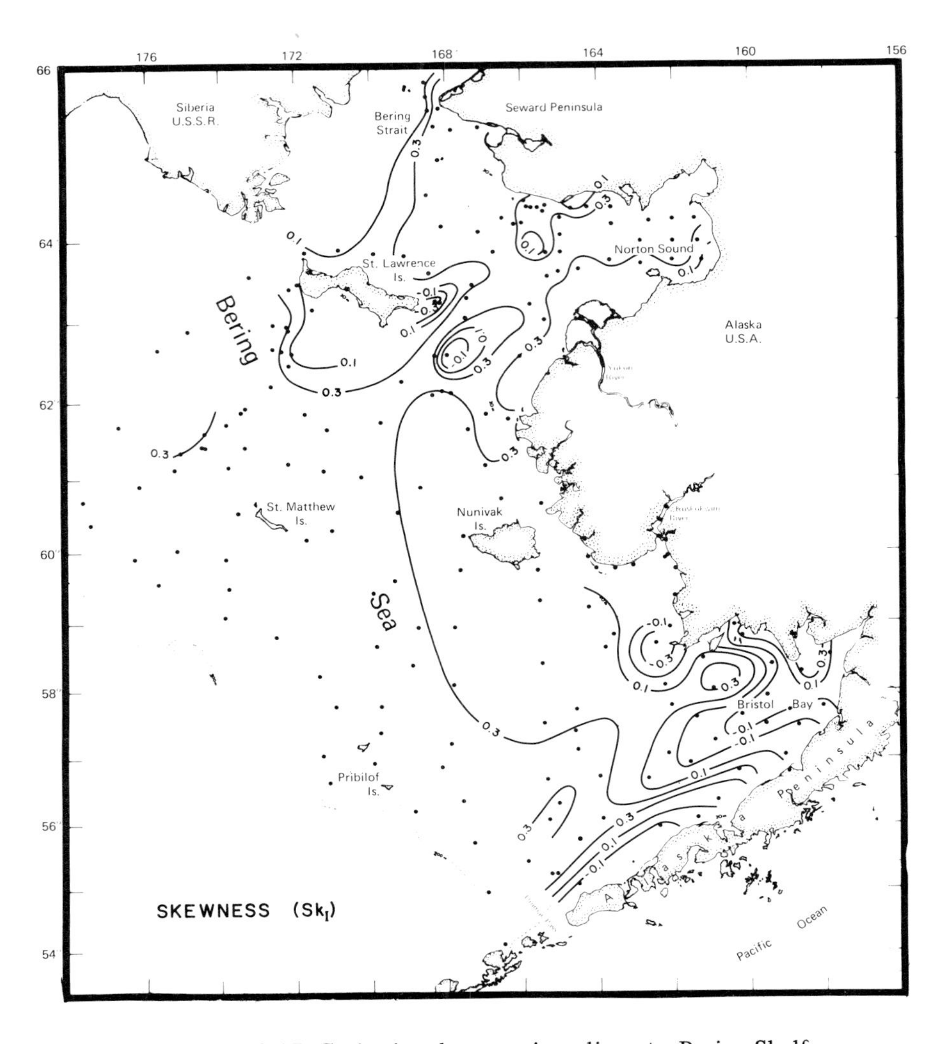

Figure 10-17. Grain size skewness in sediments, Bering Shelf.

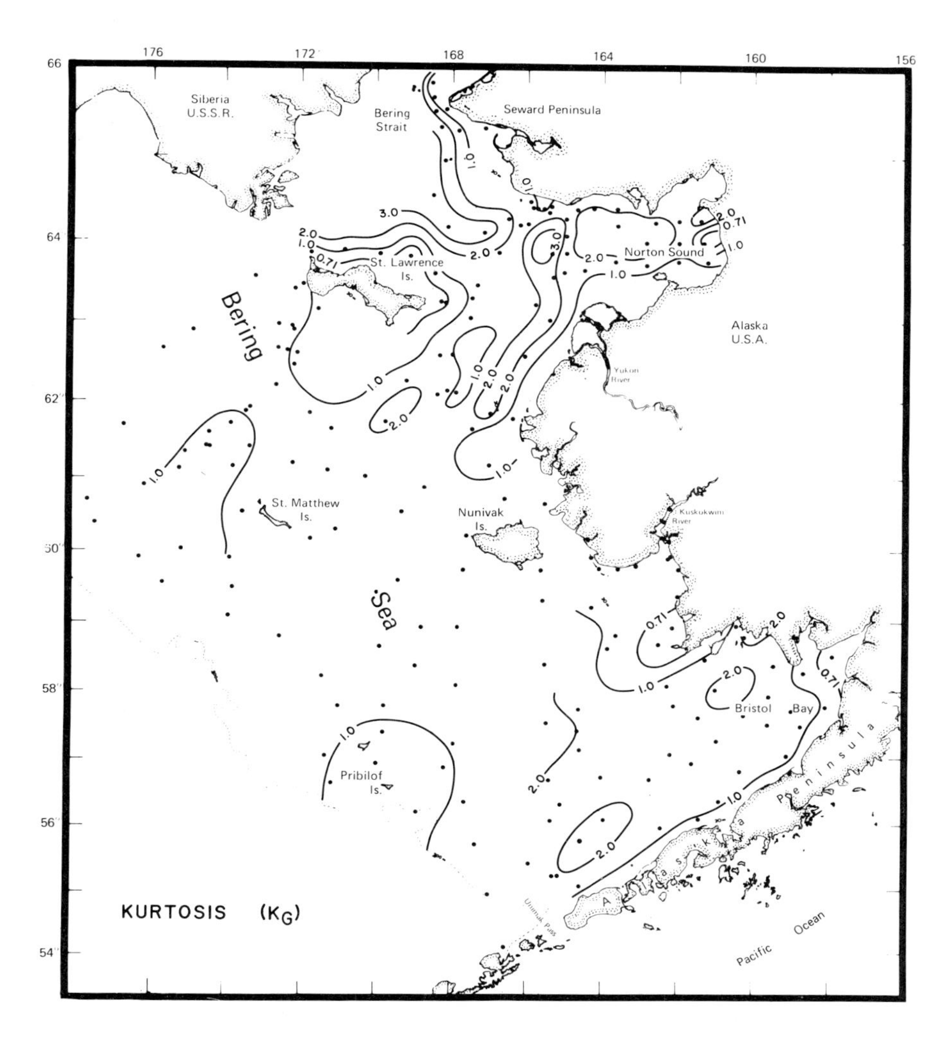

Figure 10-18. Grain size kurtosis in sediments, Bering Shelf.

The easternmost area of the central shelf, a narrow elongated belt adjacent to the Alaska mainland, running parallel to the shore, and shallower than the 10-m isobath, consists of silty sands. Shoreward of this belt the silt content of the sediments increases rapidly. Offshore, however, the amount of sand increases and reaches a maximum (approximately 90%) between the 20- and 40-m isobaths. The isopleths of sediment mean size in the near-shore region are parallel to the isobaths, but they also show an unusual relationship in that the sediment grain size decreases with decreasing water depth (Fig. 10-15). This unusual shoreward decrease in mean size is primarily due to an increase in the silt component, contributed by the discharge of the Yukon River and other coastal streams. This increasing silt fraction also attributes to the poorer sorting in the sediments.

Sediments with predominant sand fractions also cover a large shallow bank south of St. Lawrence Island (Fig. 10-13). The shelf areas with water depths of >40 m along the western periphery, and north of St. Matthew Island, consist primarily of silt with minor sand and clay components. Locally, isolated patches of gravel and gravelly sediments are found in the near-shore regions east and northwest of St. Lawrence Island. In general, however, the gravel component is insignificant and of limited lateral distribution, and the more prevalent sand and silt components complement each other in these sediments. The sediment mean size distribution on the bank south of St. Lawrence Island, in general, conforms to the bathymetry in a usual manner, such that the sediment mean size decreases with increasing water depth. North of St. Matthew Island the region does not show any definitive textural distribution which could be related to either water depth or the source. The sediments are coarsely to very finely skewed and have platykurtic to extremely leptokurtic size distributions.

The northern Alaskan Bering Shelf is defined by the Bering Strait to the north, and by a 46 m deep sill across Shpanberg Strait, between the Alaska mainland and St. Lawrence Island, to the south. This shallow shelf, with the exception of three passages, is surrounded by landmasses. The eastern part, Norton Sound, is a semi-enclosed embayment, less than 30 m deep. The slightly deeper (~50 m) area to the west and north of St. Lawrence Island, usually called the Chirikov Basin, has a complex bathymetry. The sediments from the Chirikov Basin and the regions offshore of Nome have been studied in detail by Creager and McManus (1967), McManus *et al.* (1969), Venkatarathnam (1969), Nelson and Hopkins (1972), McManus *et al.* (1974), and Sharma (1974a and 1974b).

The sediments on the northern shelf consist of gravel, sand, and sandy to clayey silts. Sand is ubiquitous and is widely distributed over more than 50% of the shelf. Gravel and gravelly sand occur in the passages of Bering, Anadyr and Shpanberg straits. Gravel patches also lie along the coast between Nome and the Bering Strait, as well as along the northern coast of St. Lawrence Island. A narrow belt of gravel protrudes approximately 60 km northward from St. Lawrence Island (McManus *et al.*, 1969). Because of the glacial origin of the gravel, its distribution is complex (Nelson and Hopkins, 1972; Sharma, 1974a and 1974b).

The Chirikov Basin, with the exception of few small anomalous areas, is covered with sand of very coarse to very fine size range. The sand component in sediments from this region usually exceeds 75% (McManus *et al.*, 1969). To the east, a northwest oriented narrow belt extending from the Yukon River delta to Nome has a low sand content (Fig. 10-13). In Norton Sound, the sediments are mostly very fine to medium sands. It is interesting to note that gravel complements sand in the Chirikov Basin, while silt complements sand in Norton Sound, perhaps suggesting erosional and depositional environments, respectively.

The shallow region (<10 m depth) extending along the Alaska mainland from Cape Romanzof to eastern Norton Sound is mantled with silt. The Yukon River silt also extends from its delta northwest toward Nome (Fig. 10-14). The clay content in the sediments is generally less than 10 percent, but in some areas of Norton Sound clay constitutes as much as 15% of the bulk sediment.

The sediments of the northern shelf are poorly to extremely poorly sorted, with finely skewed to nearly symmetrical size distributions, which range from leptokurtic to extremely leptokurtic.

The amount of organic carbon in sediments is associated with the distribution of Yukon River silt. The higher concentrations of organic carbon (>1.0% by weight) occur near the Yukon River mouth and decrease offshore to <0.5% (Fig. 10-19).

The heavy mineral distribution of Bristol Bay sediments has been described by Sharma *et al.* (1972) and Sharma (1974a,b). The contents of heavy and light minerals (weight percent) in the 1.5 to 2.0 ϕ and the 2.5 to 3.0 ϕ fractions, respectively, are shown in Figure 10-20. The heavy minerals percentages for discrete size fractions (½-ϕ intervals) of the sediments are shown in Figures 10-21 and 10-22 for the sample locations, given on Figure 10-20. Higher concentrations of heavy minerals in sediments along the Alaska Peninsula are related to the adjacent beaches, which are enriched with heavy minerals. The beach deposits and their heavy mineral distributions have been studied by Berryhill (1963). Heavy minerals in Bristol Bay are predominantly hypersthene (strongly pleochroic in green and brown), some actinolite, brown hornblende ("basaltine"), and opaques (magnetite and ilmenite) (Fig. 10-23). Hypersthene grains are mostly euhedral, and magnetite octahedrons are common. Diopside is found in minor amounts, and chlorite, epidote, garnet, sillimanite, sphene, staurolite, tremolite, and zircon are present as traces (<1.0%). Sediments directly north of Unimak Island contained above average amounts of pleochroic dark brown hornblende ("basaltine"). The light mineral assemblage contains mainly feldspars and quartz, with a predominance of feldspar.

Factor analysis of heavy mineral characteristics in two size fractions of sediments from the Bristol Bay Shelf identified two factors that accounted for 92% of the variations. The characteristics employed were the relative abundances of 16 major heavy minerals. Factor I sediments are enriched with opaques, hypersthene, diopside, and hornblende, however, factor II sediments are depleted in opaques, hypersthene and diopside. The distribution of these factors suggests that fractionation of the minerals is largely dependent upon the sediment mean size. The factor I end member predominates in sediments of 2 ϕ mean size, and factor II represents sediments of 3 ϕ mean size. Thus, the differentiation of the heavy minerals, it appears, is the result of hydraulic sorting in the sediments, and therefore should be indicative of the sources and migratory paths for these sediments.

The composition and concentration of heavy minerals from the central Bering Shelf have been studied by Knebel and Creager (1974), who reported that the

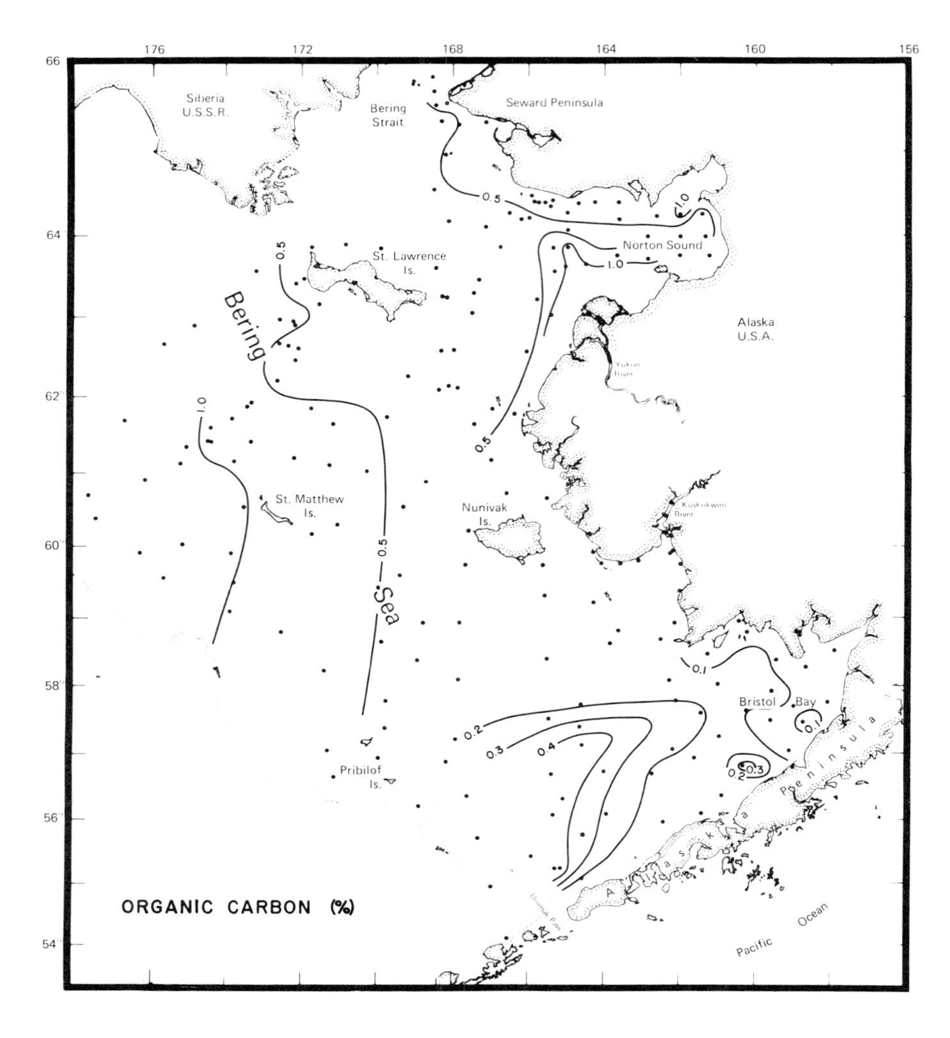

Figure 10-19. Weight percent organic carbon in sediments, Bering Shelf.

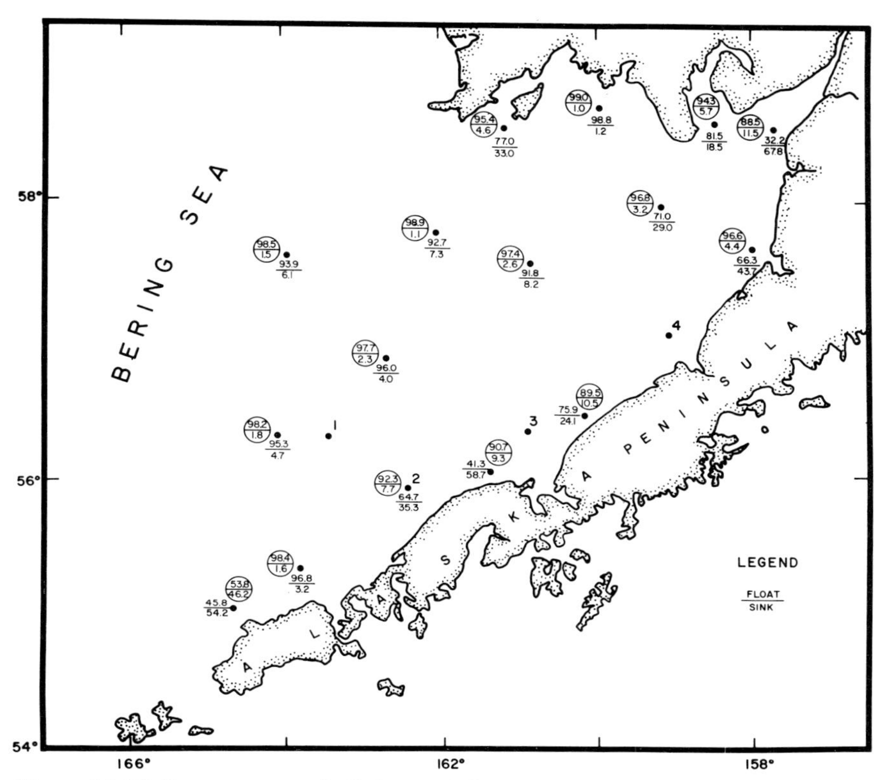

Figure 10-20. Percentages of light and heavy minerals in 0.125-0.177 mm (circle) and 0.25-0.35 mm size fractions of sediments, Bristol Bay.

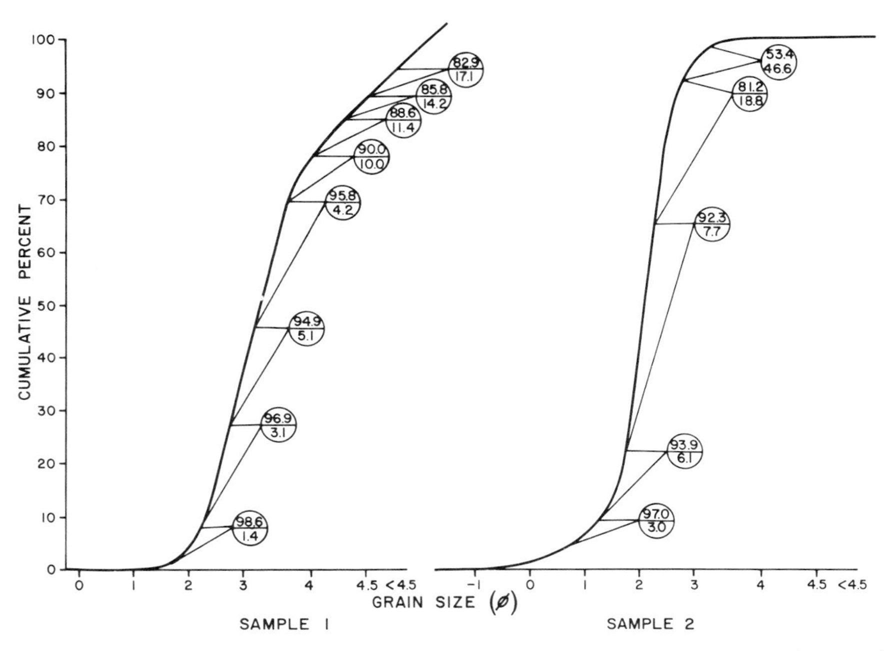

Figure 10-21. Cumulative plots and percentages of light (top) and heavy (bottom) minerals in various size fractions of sediments from stations 1 and 2, Bristol Bay.

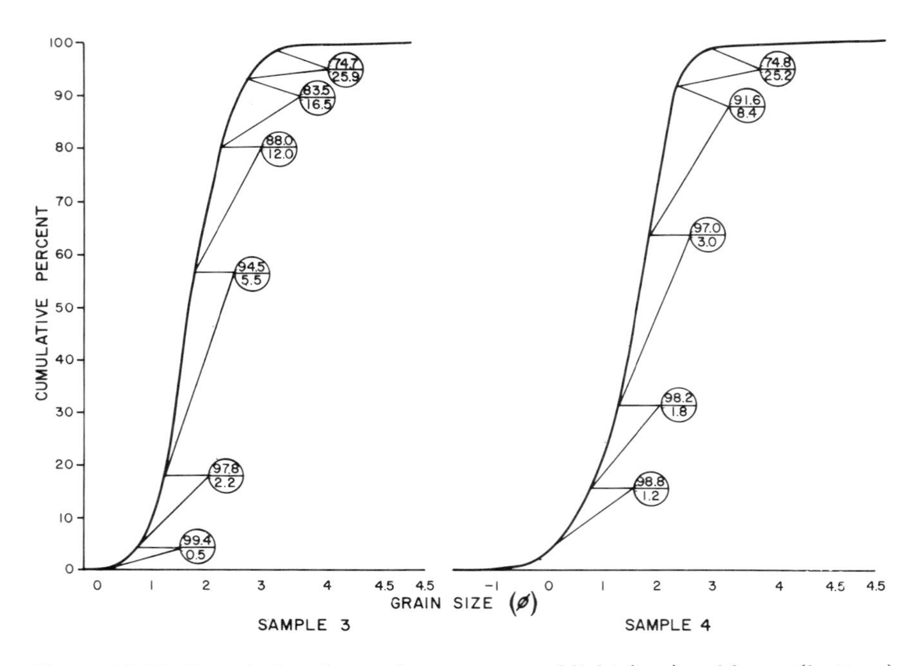

Figure 10-22. Cumulative plots and percentages of light (top) and heavy (bottom) minerals in various size fractions of sediments from stations 3 and 4, Bristol Bay.

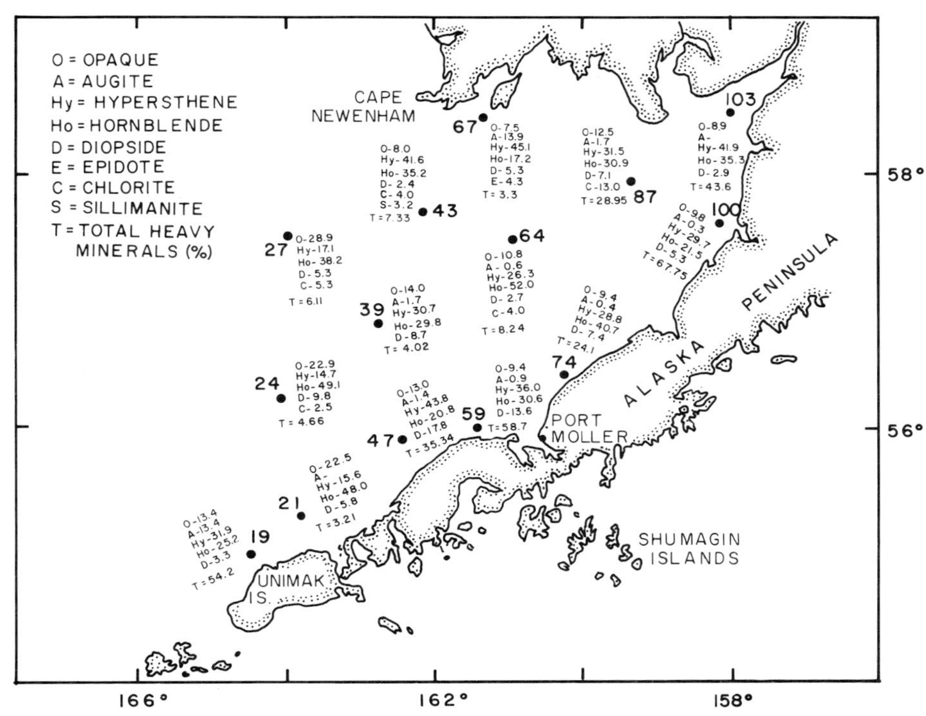

Figure 10-23. Percent minerals in heavy fractions of sediments, Bristol Bay.

major constituents of the heavy minerals in sediments from this region are clinopyroxene (mostly augite), epidote, amphibole (mostly hornblende), opaques and orthopyroxene (mostly hypersthene). The minor constituents consist of garnet, rock fragments, weathered grains, and sphene, while apatite, chloritoid, pyrite, tourmaline and zircon occur in trace amounts. Because of the homogeneous distribution of the heavy minerals, Knebel and Creager (1974) subjected their data to a Q-mode analysis in an attempt to determine the sources and the environments of deposition of the sediments in this region. Three factors, accounting for 89 percent of the variability, were chosen. Factor I is characterized by amphibole, factor II is associated with sphene and orthopyroxene, and factor III is related to rock fragments; the other components are common in all three factors.

The factor I heavy mineral assemblage is found along the southern and eastern parts of St. Lawrence Island and Shpanberg Strait. Sediments with factor II heavy mineral suites are confined to insular near-shore zones of southern St. Lawrence and northern Nunivak islands. The factor III distribution is prevalent in most sediments and occurs around St. Matthew Island, offshore southwest of St. Lawrence Island, and along the Alaska mainland.

Knebel and Creager (1974) concluded that the relatively homogeneous distribution of heavy minerals in these sediments is controlled primarily by sorting (by size, shape, or specific gravity). The heavy mineral composition and content in sediments from the central Bering Shelf, however, differ significantly from those of modern Yukon River sediments. Although the region is mostly mantled with modern, or a mixture of modern and relict sediments (palimpsest), the provenance for these sediments could not be delineated by the factor analysis.

Light and heavy mineral assemblages of the northern Bering Shelf sediments have been studied by Venkatarathnam (1971), Sharma (1974b), and McManus *et al.* (1974). The light mineral suite consists mainly of quartz and feldspar. In the 2.75 to 4.00 ϕ class, quartz constitutes more than 50% of the light mineral fraction on the east-west shoal in northern Norton Sound and the northern Chirikov Basin. In other areas, feldspar content exceeds that of quartz (McManus *et al.*, 1974). The distribution pattern of plagioclase and K-feldspar ratios seems clear, however; near the Yukon River mouth it is high (2.0 to 4.0), and it gradually decreases northwest into the Chirikov Basin, where it is only 0.5 to 1.0.

Nine heavy mineral assemblages and their areal distributions in the northern Bering Shelf have been described by Venkatarathnam (1971). The heavy mineral content in sand (1 to 4 ϕ) rarely exceeds 10%. Higher concentrations are found in beach and near-shore sediments, and in brownish sands from elongated regions between the 34- and 36-m isobaths west of St. Lawrence Island. On the basis of the heavy mineral content in sand (1 to 4 ϕ), the northern shelf can be broadly divided into three regions: An area of low heavy mineral content extends northwest-southeast along the broad shallow valley in the central part of the northern shelf, separating two areas of relatively high concentrations on either sides; Norton Sound in the east and Chirikov Basin in the west.

McManus *et al.* (1974), in order to better define heavy mineral provinces of this region, subjected the heavy mineral distribution in the most common modal size fine sands (2.75 to 4 ϕ class) of 100 samples to *Q*-mode factor analysis. Four factors explained 90 percent of the variability in the samples. They suggested that the concentrations of heavy minerals in the sediments are controlled by the source rocks in this region, and the areal extent of each factor corresponds to the adjacent sediments provenance. Clinopyroxenes, rock fragments, orthopyroxenes and amphiboles dominate in their factor I, which has high values near the Yukon Delta. Factor I sand, it appears, is introduced into this region by the Yukon River. Factor II sands are characterized primarily by garnet, and to a lesser extent by epidote, staurolite, chloritoid and sphene. The mineral assemblage of this factor is characteristic of metamorphic rocks, and the distribution pattern of this factor relates its origin to the metamorphic terrane of the adjacent southwestern Seward Peninsula. A small near-shore area north of western St. Lawrence Island, representing factor III, is correlated with the adjacent Quaternary olivine basalts.

The significance minerals of factor IV are clinopyroxene, amphibole, rock fragments and epidote. Factor IV sands are widely distributed in Chirikov Basin and northeastern Norton Sound. The similarity of mineral assemblages in sediments from south of Anadyr Strait, the southern coast of Chukotka Peninsula, and the

Chirikov Basin suggests that these sands originate from the acidic plutons and Cretaceous-Jurassic volcanics of the southern Chukotka Peninsula. A similar source for the factor IV sands of northeastern Norton Sound can be located in the acidic plutons and Cretaceous-Jurassic volcanics of the southeastern Seward Peninsula.

GEOCHEMISTRY

The large Bering Shelf has a complex distribution of sediments. The texture and the mineralogy of these sediments, discussed earlier, are generally related to regional provenance. The influence of the Yukon and Kuskokwim rivers on the texture of sediments in the northern and southern Bering Shelf is apparent. Sediment grading in Bristol Bay strongly indicates the control of bathymetry on the sediment texture. Of further interest is the extent to which the distribution of major and minor elements on the shelf is similarly influenced by the factors which control the sediment texture. However, the variations in elemental concentration in the sediments from the shelf should be somewhat more complex than the textural variations. In addition to the local and regional distributions, the shelf may display a general south-north variation in some elements due to differences in climatic regimes.

The aluminum content in sediments from the shelf varies significantly: On the southern shelf near Unimak Pass the clayey sediments contain up to 7% aluminum, while the gravelly sands from Bering Strait contain less than 3% aluminum. The higher distribution of aluminum (9%) in the clays deposited off the shelf north of the Aleutian Pass is characteristic of deep-water sediments (Fig. 10-24). Fine sediments (clay minerals accompanied by plagioclase from andesitic basalt) rich in aluminum are also carried to the shelf through upwelling and in suspension by the north- and eastward moving Pacific water. The medium sands of the midshelf generally contain less aluminum than the sandy clayey silts of the outer shelf.

The coarse near-shore deposits north of the Alaska Peninsula have relatively high contents of aluminum, which rapidly decrease offshore in Bristol Bay. The higher aluminum content may be the result of greater amounts of heavy minerals in these sediments. The high aluminum concentrations in the Togiak Bay region are attributable to the large amounts of fine clayey sediments deposited in the quiescent bay. A large region between Kuskokwim Bay and the Pribilof Islands is covered with sand containing between 5.5 and 6.0% aluminum. Northward, however, a small region adjacent to the southern shore of Nunivak Island has lower aluminum contents.

Sediments from the offshore region lying between the Pribilof and St. Lawrence islands display a close relationship between the sediment mean size and aluminum concentrations. The coarse sands from St. Lawrence Bank have lower aluminum contents, while the clay-silts deposited offshore contain relatively

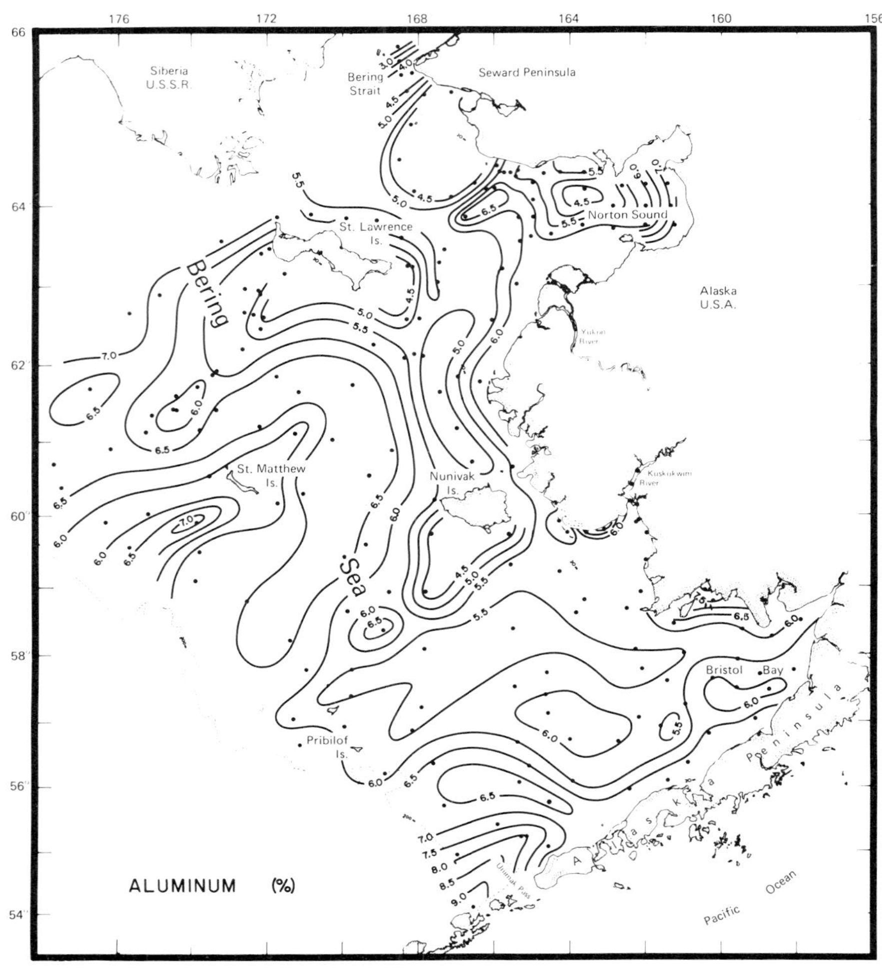

Figure 10-24. Percent aluminum in sediments, Bering Shelf.

more aluminum. The silts and sands from the shallow area lying adjacent to the Alaska mainland contain between 5 and 6.5% aluminum. In eastern Norton Sound the occurrence of high aluminum content is related to greater amounts of the fine fraction (silt and clay) sediments, while the lower aluminum distributions in the central sound are related to well-sorted sands. Sediments from northeastern Chirikov Basin, between St. Lawrence Island and the Bering Strait, have the lowest aluminum contents.

The distribution of iron in sediments from the Bering Shelf is somewhat analogous to that of aluminum. The iron ranges from slightly less than 1.5% to over 6.0% (Fig. 10-25). There are two near-shore regions of iron-rich sediments. The high content of iron in the coarse sediments deposited adjacent to the Alaska Peninsula, and in the near-shore coarse sediments in the vicinity of Nome is the result of greater amounts of heavy minerals in these sediments. Generally, the sand deposits of the mid-shelf contain between 2 and 2.5% iron, while the silt deposits of the Yukon Delta and offshore contain higher (3.0%-4.5%) concentrations.

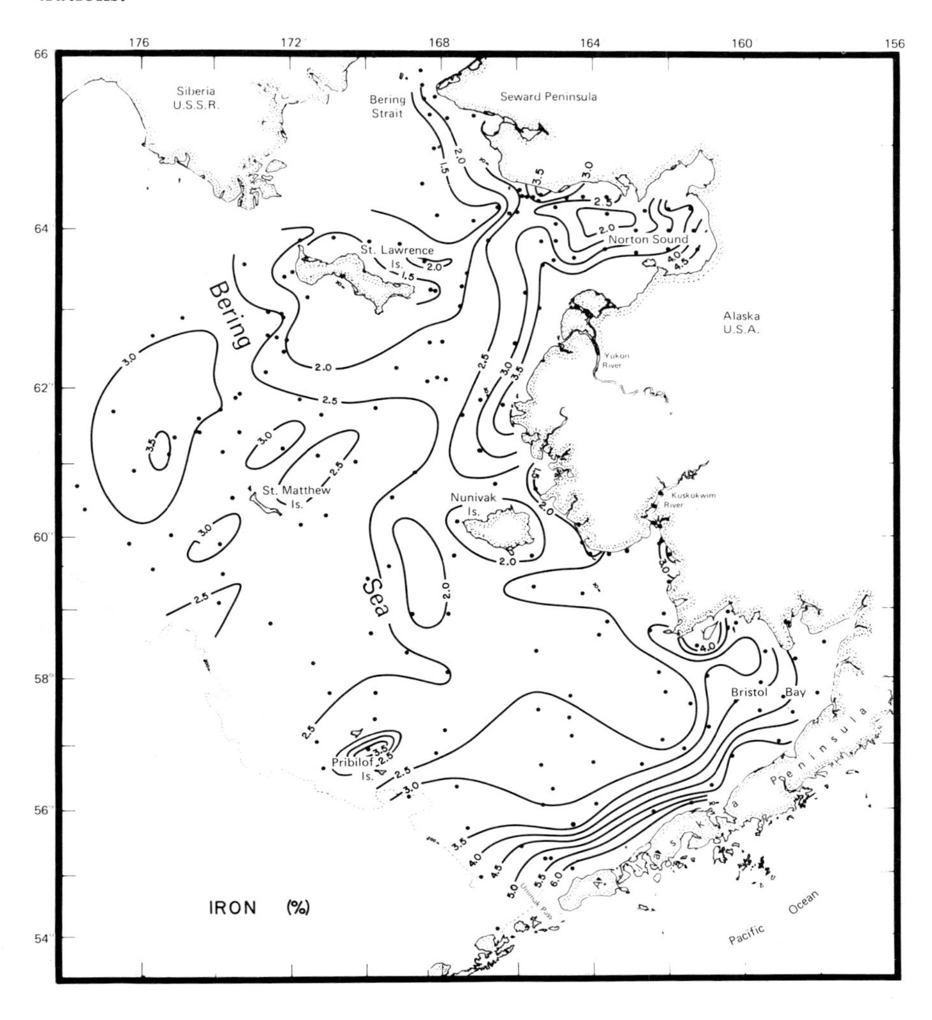

Figure 10-25. Percent iron in sediments, Bering Shelf.

It is interesting to note that the iron concentration is higher in the near-shore sediments and decreases rapidly offshore to a depth of approximately 50 to 60 m. The low concentration of iron in near-shore sediments from the vicinity of Nunivak and St. Lawrence islands is, however, an exception to the general distribution pattern. There is also a decrease in the iron content in sediments, from south to north. The highest content, over 6.0%, occurs near Unimak Pass, and the content decreases northward to about 1.5% in Bering Strait.

Except in the near-shore region, the calcium content in sediments from the Bering Shelf varies within a narrow range. The distribution of calcium does not appear to be related to either the bathymetry or the sediment texture. The maximum concentrations, as well as variations, of calcium occur along the coastal regions of the Alaska Peninsula and the Pribilof Islands (Fig. 10-26). The high calcium observed locally in near-shore sediments is attributed to the presence of calcareous shell fragments. The remainder of the shelf is covered with sediments containing about 2% calcium. The lowest concentrations of calcium are observed in the region lying between St. Lawrence Island and the Bering Strait.

Magnesium in sediments from the Bering Shelf varies between 0.5 and 2.5%. However, most of the shelf is mantled with sediments containing magnesium within a comparatively small range of 0.75 to 1.0% (Fig. 10-27). The maximum values are observed along the near-shore region of the southwestern Alaska Peninsula. The coastal sediments generally contain more magnesium than those deposited farther offshore. Sediments deposited near the mouths of the Yukon and Kuskokwim rivers also contain slightly more magnesium than sediments on the adjacent shelf. Significantly lower concentrations are observed in the near-shore sediments along the southwestern Seward Peninsula, and in the insular sediments surrounding St. Lawrence Island. The distributions of magnesium in sediments correlate neither with the bathymetry nor with the sediment size distribution on the shelf.

A marked difference in the distribution pattern of sodium, relative to those of other elements, is apparent in Figure 10-28. This divergent distribution primarily reflects the higher sodium content of the sediments on the inner and outer shelves, compared with sediments from the central shelf. The sodium is mostly contributed by Na-feldspars. The sediments on the central shelf are continually subjected to hydraulic action by currents and often by storm waves. This hydraulic action causes Na-feldspar to break into smaller particles, which are removed from this region. Maximum concentrations of sodium are observed north of Unimak Pass and in the offshore region northwest of St. Matthew Island. The remainder of the shelf shows a relatively uniform distribution of sodium. This pattern of sodium occurrence in sediments indicates that the sodium is influenced neither by the sediment mean size nor the water depth. It appears that sodium decreases with increasing distance from the shore, but this general trend reverses itself along the outer shelf.

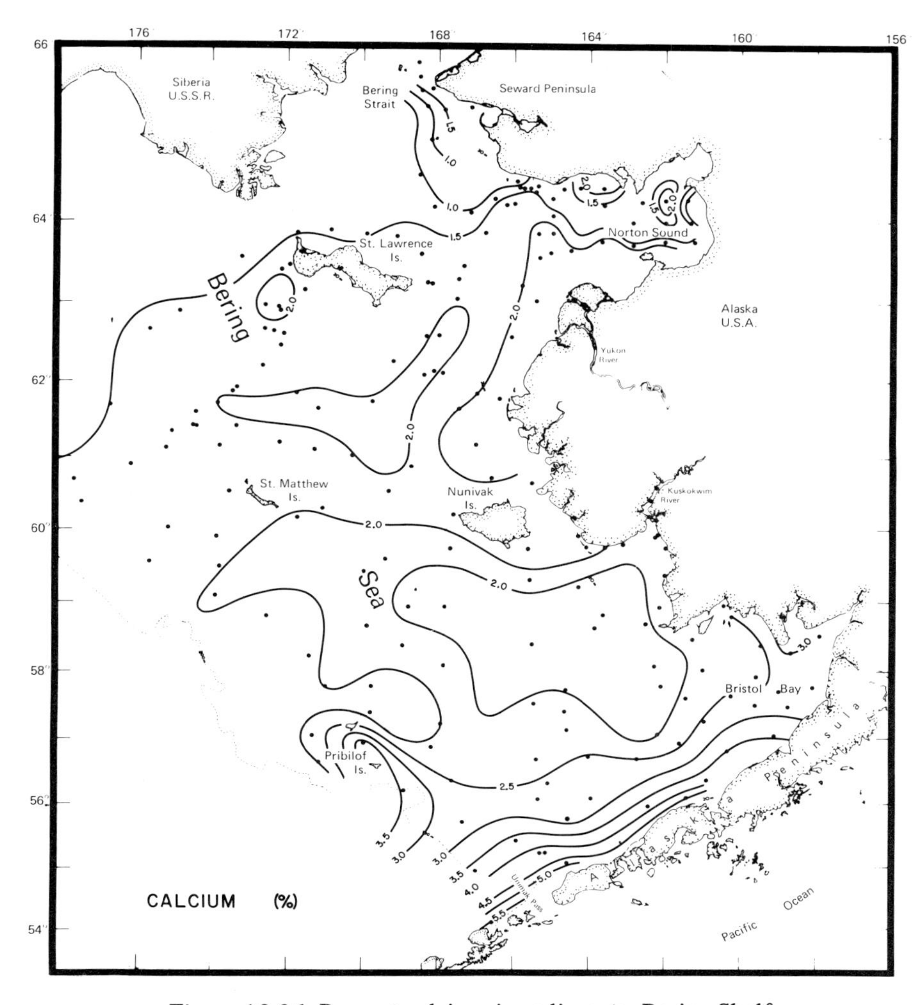

Figure 10-26. Percent calcium in sediments, Bering Shelf.

The concentration of potassium in Bering Shelf sediments (Fig. 10-29) is strongly influenced by the particle size distribution (Fig. 10-15). The coarse near-shore sediments along the Alaska Peninsula have lower potassium, while offshore clayey sandy silts deposited northwest of St. Matthew Island have higher potassium contents. The Yukon River silts deposited off the delta and in southeastern Norton Sound also have a relatively higher content of potassium. Low concentrations are observed near Unimak Pass, Kvichak Bay, and in the narrow elongated belt extending north and southeast of Nunivak Island. The general northward concentration decrease exhibited by other elements in the Bering Shelf sediments is not discernible in the potassium distribution.

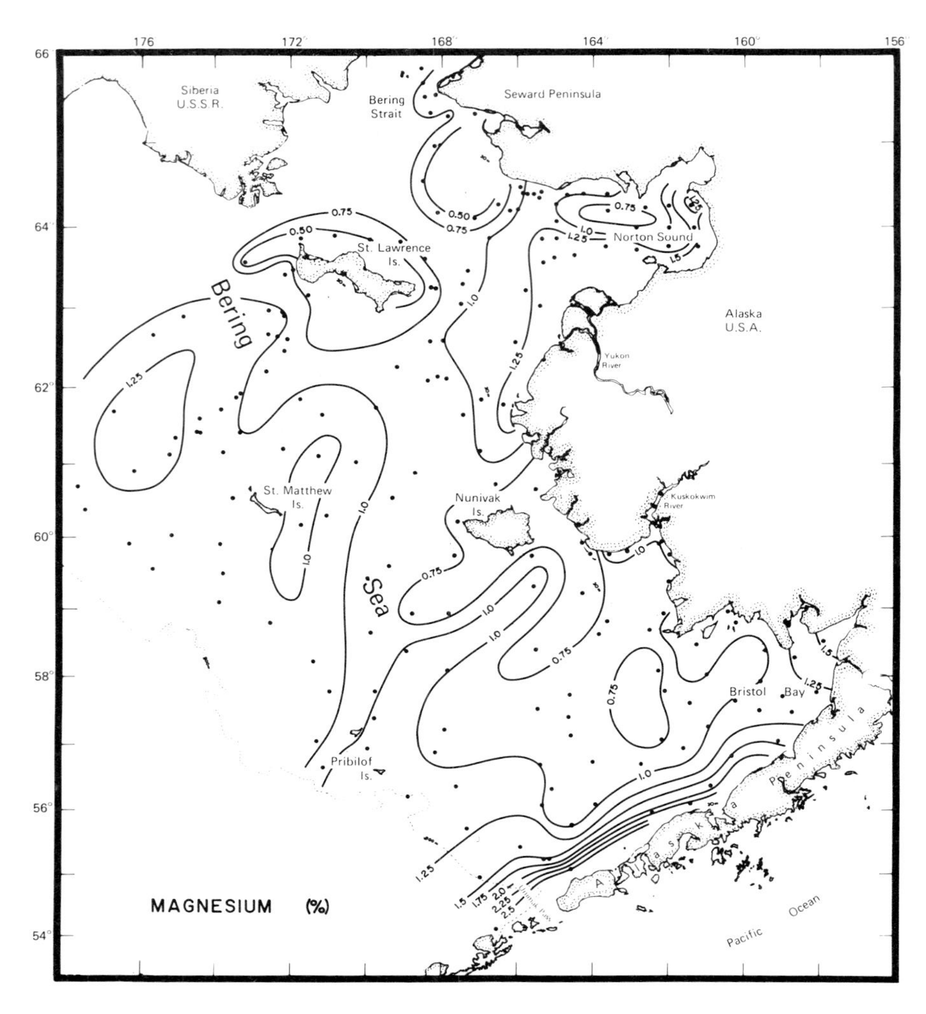

Figure 10-27. Percent magnesium in sediments, Bering Shelf.

Of all the major elements in the sediments from the Bering Shelf, manganese shows the greatest variations (Fig. 10-30): differences over one order of magnitude are observed. High manganese contents (1,200 ppm) are observed in sediments deposited north of Unimak Pass and along the shallow region lying north of the Alaska Peninsula, while offshore the concentration decreases rapidly. The variance of manganese in sediments deposited on the open shelf and Norton Sound is not as striking as it is along the Alaska Peninsula. Here, over most of the shelf, the manganese content varies only between 400 and 500 ppm.

Sediments in the proximity of river mouths and deltas have slightly higher manganese contents than adjacent offshore sediments. The apparent differences

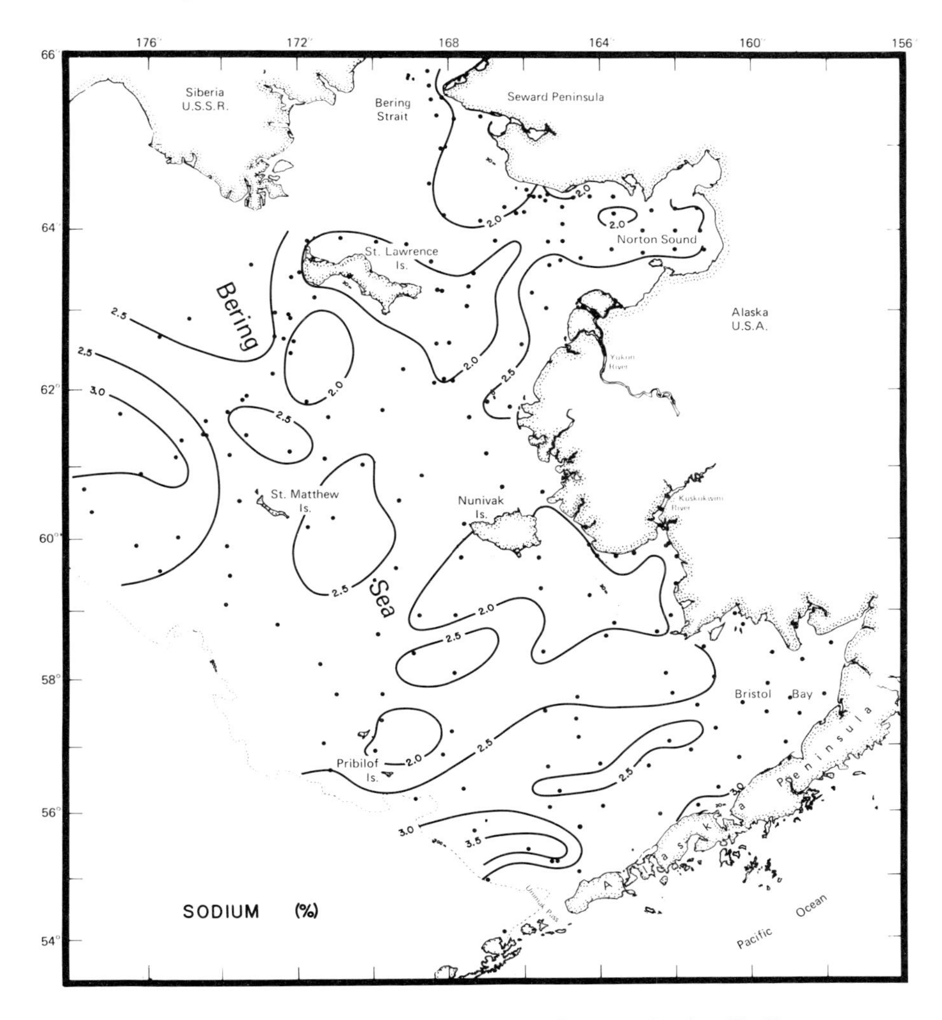

Figure 10-28. Percent sodium in sediments, Bering Shelf.

in manganese in sediments from the Kuskokwim and Yukon deltas may be a result of differences in the sediment texture rather than the provenance. The Yukon Delta consists predominantly of silt, whereas the Kuskokwim Delta contains mostly sand. The influence of source rock on the amounts of manganese in sediments, however, is apparent in the near-shore deposits of the Alaska Peninsula. Manganese and iron in sediments from the Bering Shelf show similar distribution.

The titanium content is generally higher in near-shore sediments than in those deposited offshore (Fig. 10-31). Relatively high concentrations of titanium occur in sediments deposited near Unimak Pass, whereas sediments significantly

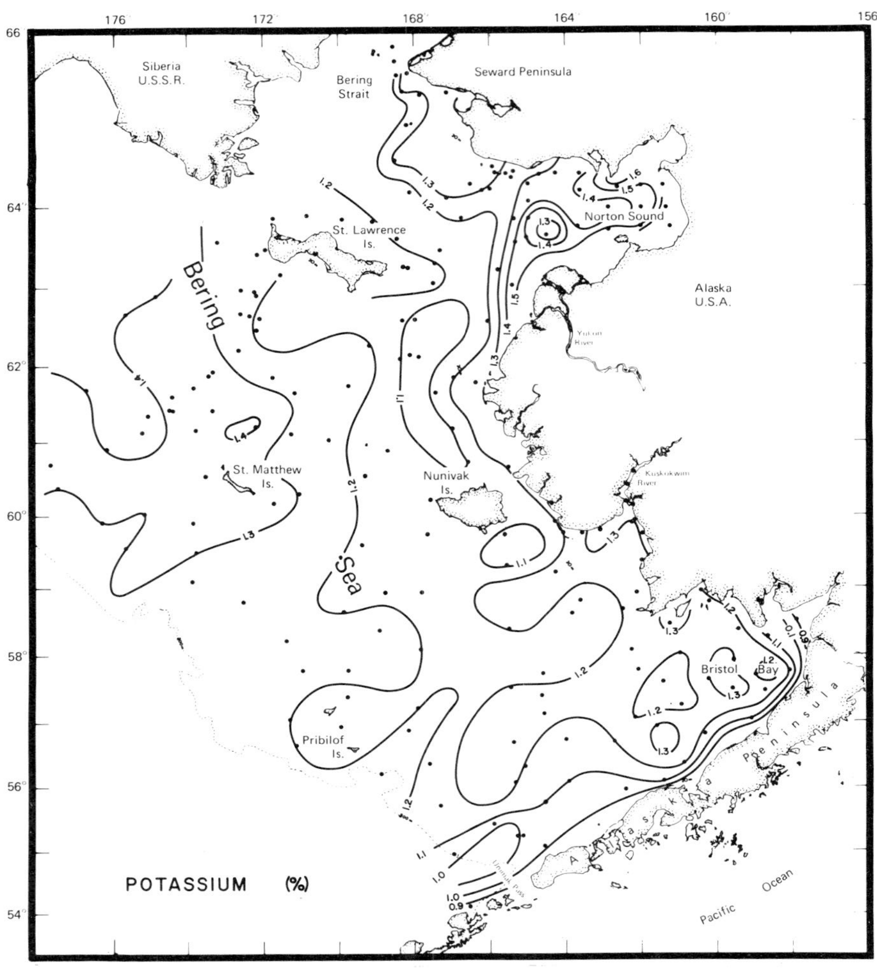

Figure 10-29. Percent potassium in sediments, Bering Shelf.

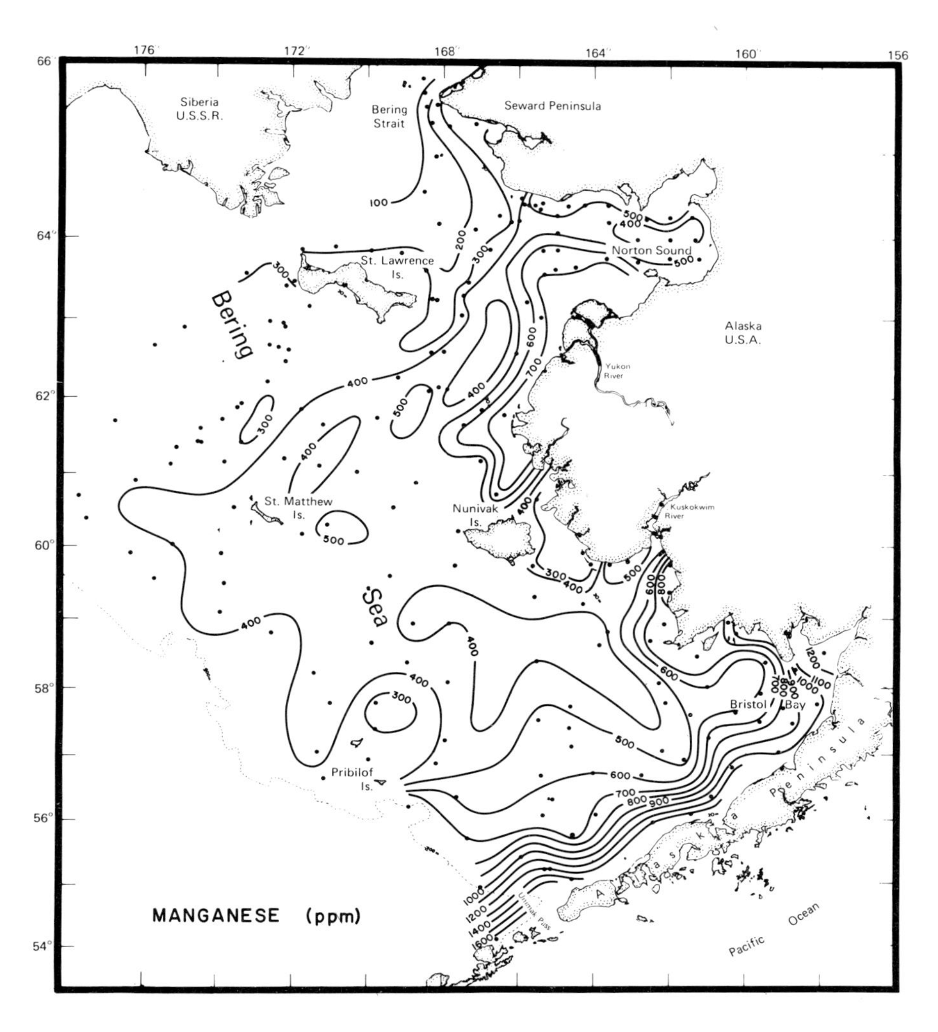

Figure 10-30. Manganese (ppm) in sediments, Bering Shelf.

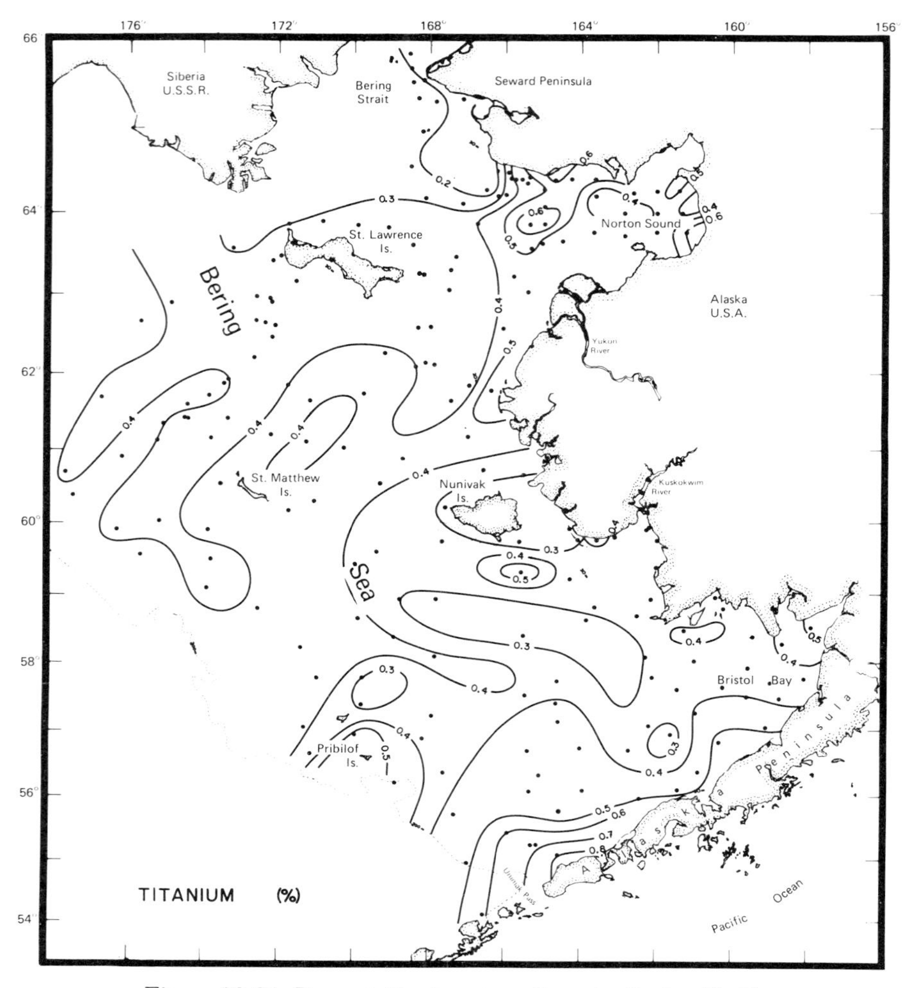

Figure 10-31. Percent titanium in sediments, Bering Shelf.

low in titanium lie along the southwest shores of Seward Peninsula. In most areas, the titanium content usually varies between 0.3 and 0.5% and shows no apparent relation with the grain size of the sediments. However, the titanium distribution in sediments appears to be quite closely related to the source for the sediments on the shelf. For example, the high content of titanium in sediments deposited along Unimak Island, with a gradual decrease offshore, indicates that Unimak Island is an important source for the sediments in this region. The mineral distribution, discussed earlier, led to a similar conclusion.

Among the minor elements of the Bering Shelf sediments, barium displays the greatest variations in distribution, ranging from a low of 150 ppm near Unimak Pass to a high of 550 ppm in southeastern Norton Sound and Shpanberg Strait. There are three large regions with high barium content on the shelf (Fig. 10-32). The first, a region of complex shape bounded by the 450-ppm isopleth, lies southwest of Kuskokwim Bay. The second, a region near the Yukon River delta extending northwest into Norton Sound, contains between 400 and 500 ppm barium. Although the sediments from the third region, east and southeast of St. Lawrence Island, are not located near the Yukon River mouth, they contain 480 to 550 ppm barium. In general, the near-shore sediments contain concentrations of barium similar to those deposited farther offshore.

Chromium content in the sediments from the Bering Shelf varies from 10 to 85 ppm (Fig. 10-33). Chromium content is quite low in the near-shore sediments deposited along the Alaska Peninsula, but it gradually increases offshore. A narrow belt of sediments extending from Cape Romanzof to southeast Norton Sound along the Alaska mainland contains a higher content of chromium. On the mid-shelf, three anomalous regions of sediments with <30 ppm chromium lie southwest of Kuskokwim Bay, south of St. Lawrence Island, and in Norton Sound. Offshore, along the outer shelf, the sediments generally contain between 40 and 50 ppm chromium. Chromium content does not appear to be related to sediments mean size, although increased clay and silt fractions do substantially enrich sediments in chromium. The distribution of chromium in sediment deposited near the Yukon Delta may reflect the effect of the provenance.

The distribution of cobalt in the sediments is relatively simple and mostly varies between 10 and 20 ppm (Fig. 10-34). Slightly higher contents of cobalt in the near-shore sediments along the Alaska Peninsula and in southeast Norton Sound perhaps result from increased heavy mineral concentrations in sediments, as well as being a reflection of the provenance. The cobalt distribution does not appear to be related to the sediments texture.

The Bering Shelf sediments contain about 30 ppm copper (Fig. 10-35). Relatively higher contents of copper are found in clayey silt deposited between Unimak Pass and the Pribilof Islands, as well as in near-shore sandy silts deposited between Cape Romanzof and southeast Norton Sound along the Alaskan mainland. Sands from Bristol and Kuskokwim bays have low copper contents (<10 ppm). With the exception of a few local anomalies, such as near Port Moller, Platinum, and a large area southeast of the Seward Peninsula, where

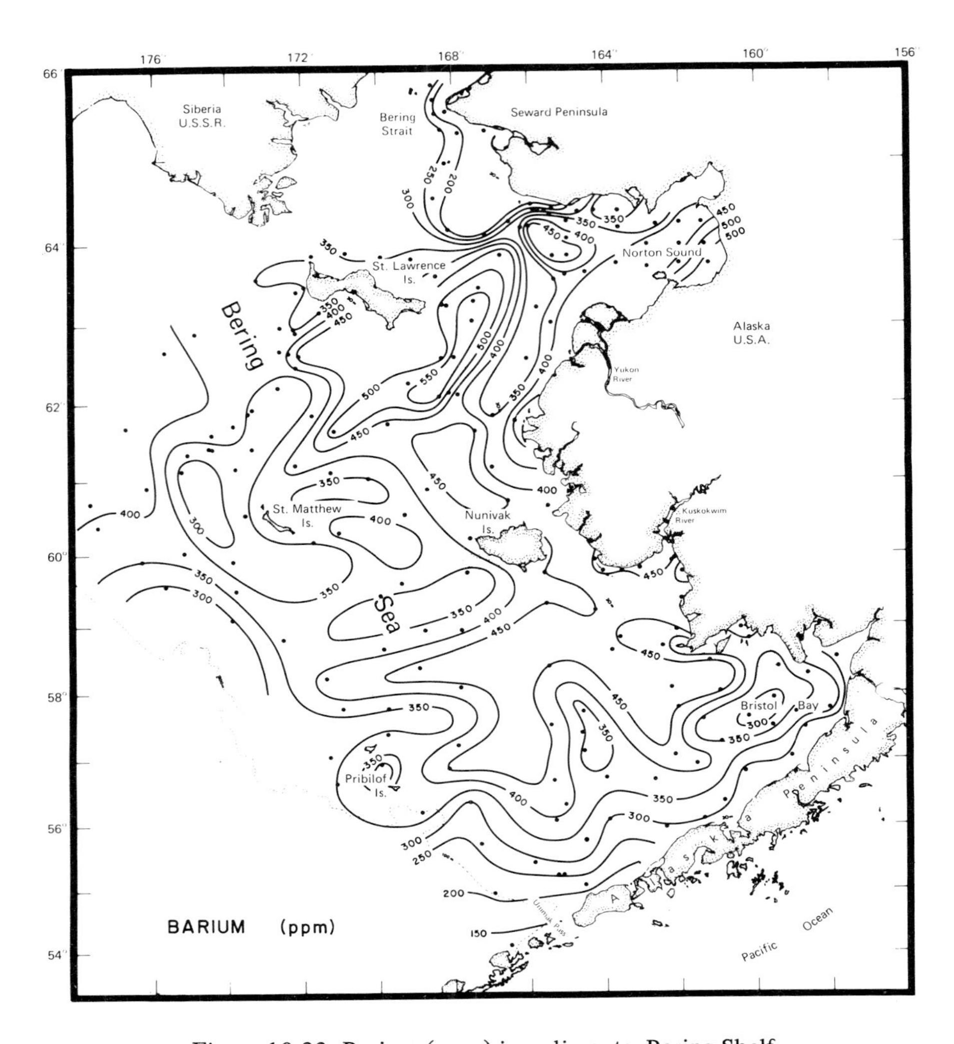

Figure 10-32. Barium (ppm) in sediments, Bering Shelf.

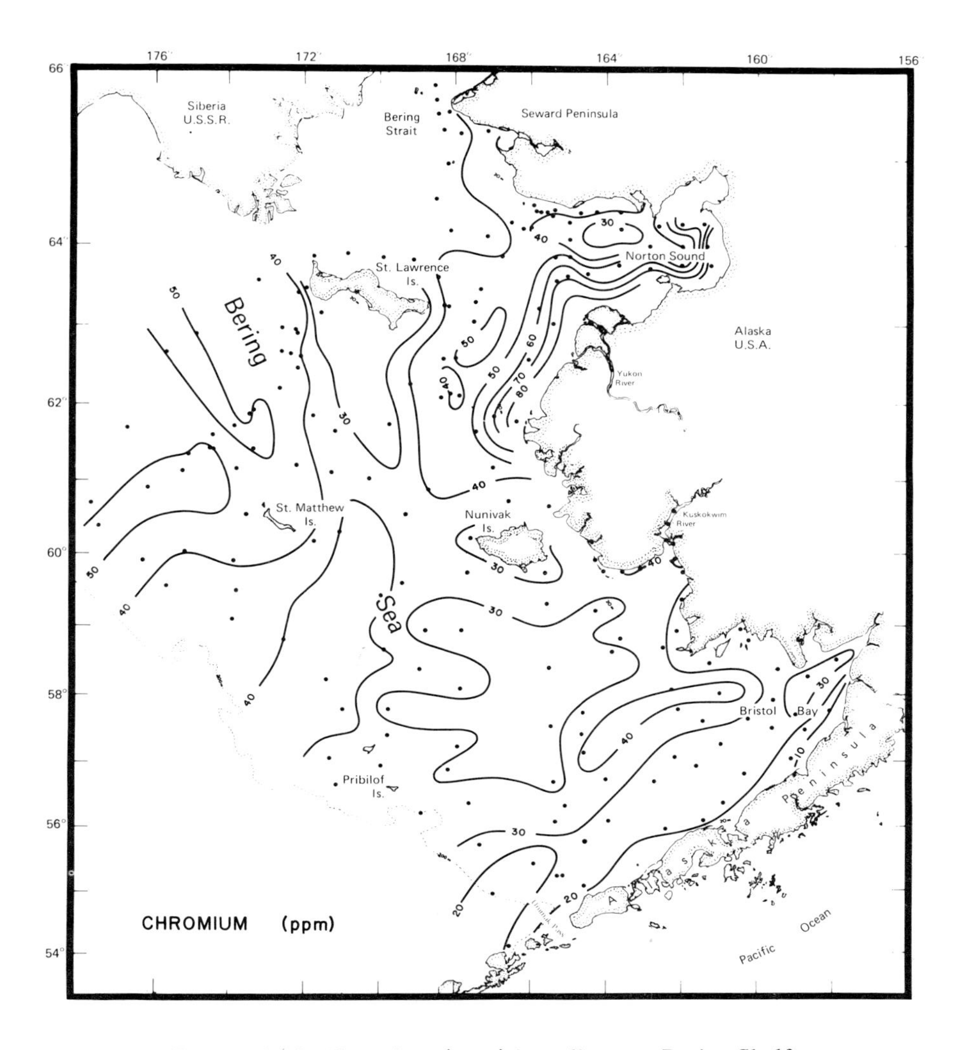

Figure 10-33. Chromium (ppm) in sediments, Bering Shelf.

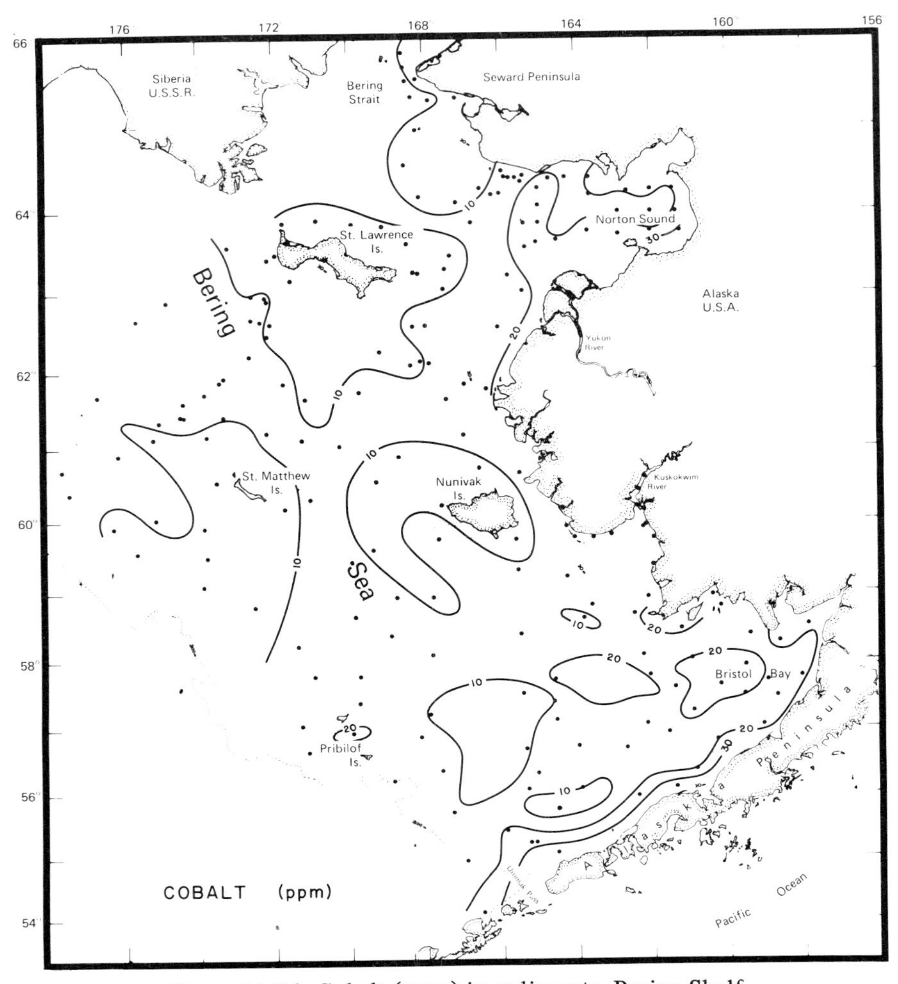

Figure 10-34. Cobalt (ppm) in sediments, Bering Shelf.

coarse sands contain 30 ppm, copper content is generally related to the percentage of silt and clay fractions in sediments (Figs. 10-14 and 10-35).

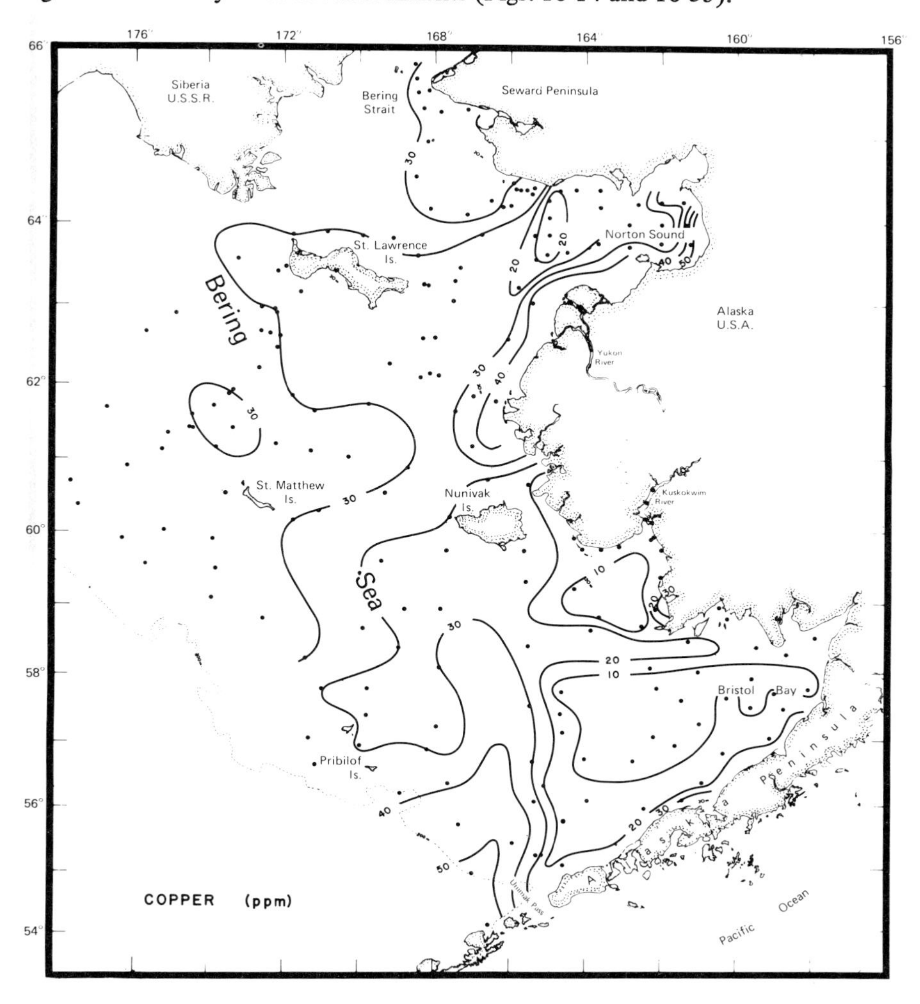

Figure 10-35. Copper (ppm) in sediments, Bering Shelf.

The nickel content in sediment varies between 10 and 45 ppm, and its distribution is extremely complex (Fig. 10-36). High concentrations are observed in regions south of Kuskokwim Bay, near the Yukon Delta, in Norton Sound, and offshore southwest of St. Lawrence Island. It is interesting to note that sediments from most of the Bering Shelf south of Cape Romanzof contain less nickel (<25 ppm) than sediments from the northern shelf (>25 ppm). In part, the higher concentrations appear to be related to greater amounts of silt and

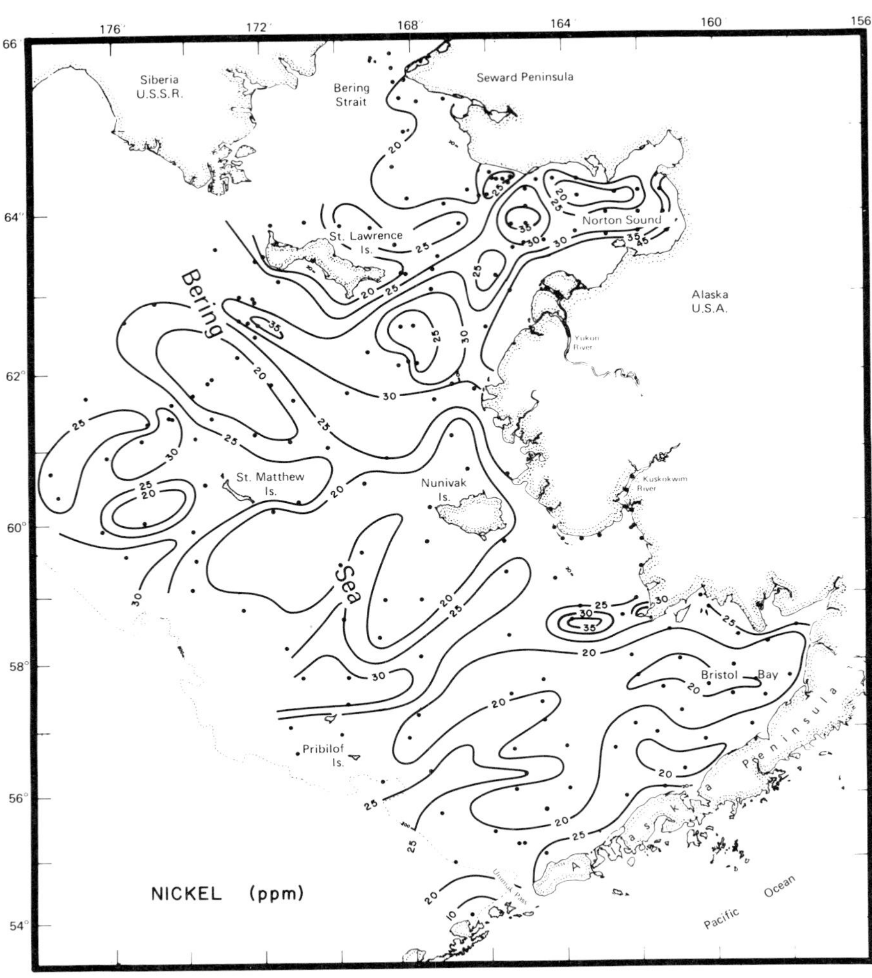

Figure 10-36. Nickel (ppm) in sediments, Bering Shelf.

clay, although this relationship does not prevail throughout the region. For example, fine sediments north of Unimak Pass contain less than 20 ppm nickel.

The distribution of nickel is more strongly related to the provenance than to texture. The Yukon River detritus contains significantly high nickel concentrations, which are manifested in the sediments deposited in adjacent areas. The observed nickel distribution suggests that Yukon River sediments migrate to the east into Norton Sound, northwest toward Nome, and offshore south of St. Lawrence Island.

Maximum strontium contents occur in near-shore sediments along the Alaska Peninsula with decreases offshore as well as northward. There is a gradual and distinct south to north decrease in strontium contents of sediments from the

Bering Shelf. In the southern shelf and in the eastern Bristol Bay area, the strontium concentration is 300 to 500 ppm; in the central shelf between the Pribilof and St. Lawrence islands, levels decrease to the 250 and 300 ppm range; north of St. Lawrence Island strontium contents are less than 250 ppm (Fig. 10-37). The distribution of strontium is not related to sediment texture.

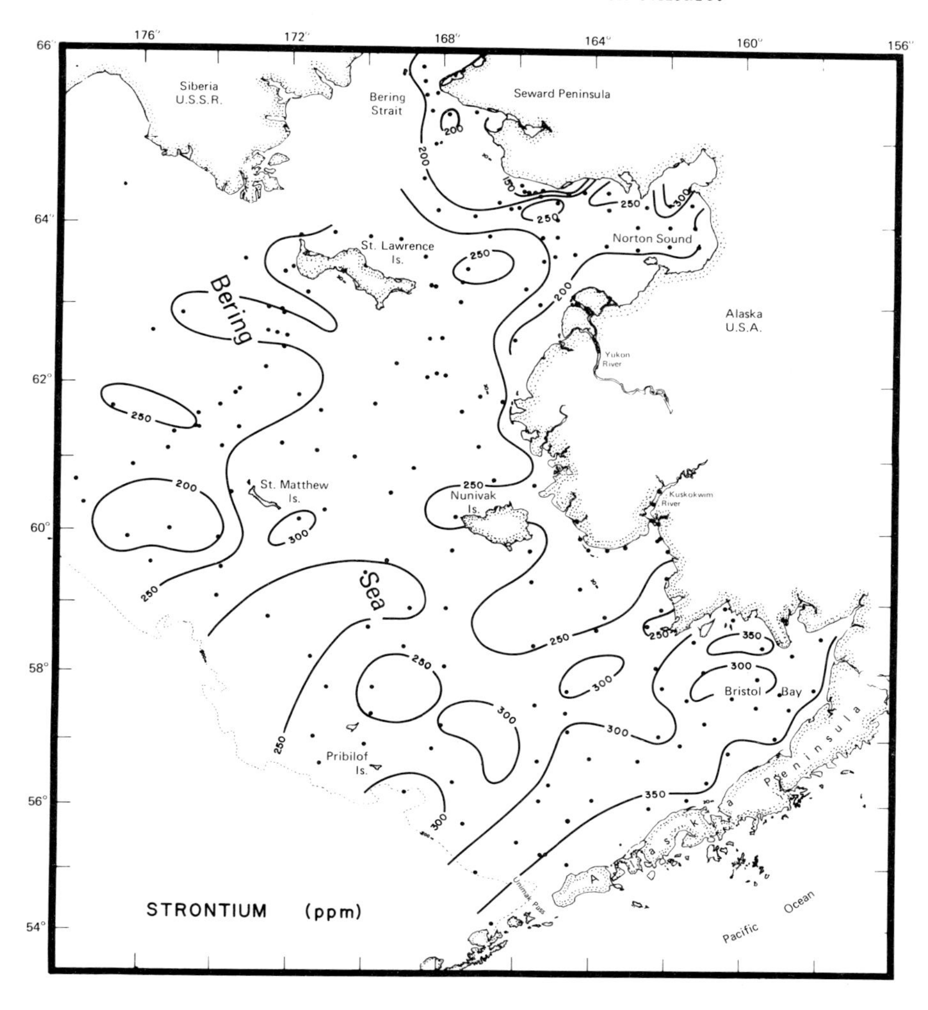

Figure 10-37. Strontium (ppm) in sediments, Bering Shelf.

The distribution pattern of zinc in the Bering Shelf sediments is complex. Low concentrations are observed southwest of Seward Peninsula as well as from a large region extending from southwest of Kuskokwim Bay northward to the southeast corner of St. Lawrence Island (Fig. 10-38). Higher contents of zinc are observed in near-shore sediments north of the Alaska Peninsula, along the Alaska mainland in the proximity of the Yukon Delta, and in the offshore region west of the St. Matthew and St. Lawrence islands. Two small isolated regions of high zinc occur west of St. Lawrence Island and between the Pribilof and Nunivak islands. The distribution of zinc appears to be related to sediment provenance and to texture.

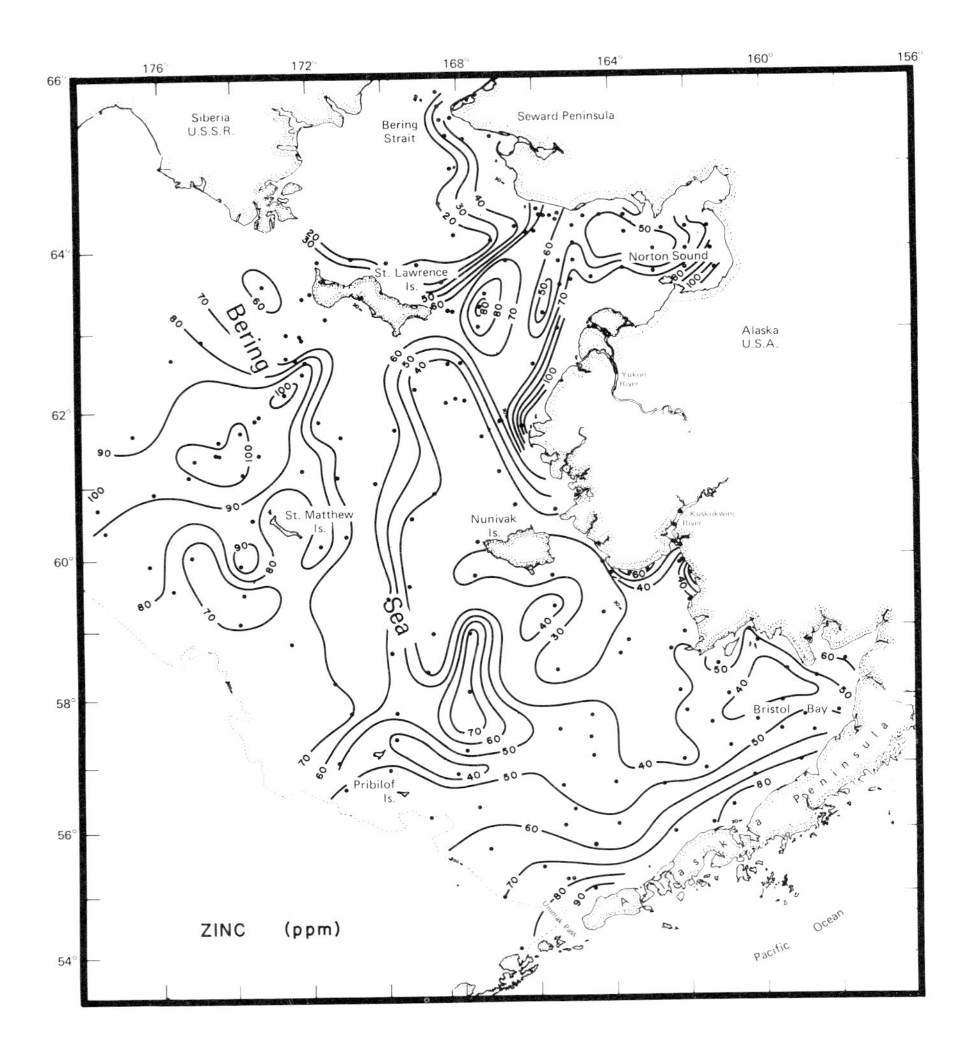

Figure 10-38. Zinc (ppm) in sediments, Bering Shelf.

SEDIMENT SOURCE AND TRANSPORT

Most sediments deposited on the eastern Bering Shelf originate along the eastern coast. The magnitudes of sediment discharge from the two major rivers, the Yukon and Kuskokwim, have been computed by various investigators. The estimates of suspended load input vary from 88 to 100 million metric tons per year for the Yukon River, and approximately four million metric tons per year for the Kuskokwim River. The sediment contributions from the many rivers (Kvichak, Nushagak, Wood) and streams draining the Alaska Peninsula remain totally unknown. There are numerous streams and rivers which drain the Alaska mainland and debouch into the shelf. These streams each may carry as much as 500-2000 mg/ℓ of sediment in suspension. Furthermore, no estimates of the sediments eroded from the coastal regions have been made, although it is believed that the sediment contribution from coastal zones during periodic storms and sea surges can be significant.

The Yukon River has been suggested as the major source for sediments in the Bering Sea by Lisitsyn (1966), who estimates that this river supplies 90 percent of the fluvial-borne sediments which enter the Bering Shelf. Although the annual sediment input from the Yukon River ranks 19th among rivers of the world (Inman and Nordstrom, 1971), its contribution in relation to the total sediment input onto the Bering Shelf appears to be lower than that estimated by Lisitsyn (1966).

The importance of various streams and rivers as sediment contributors to the shelf can be evaluated to some extent by the distribution of the respective surface sediment plumes observed on ERTS-1 imagery. The density of sediment plumes is then related to the *in situ* measurements of the suspended load. Because of the immense size of the shelf the distribution of suspended load is described in two sections: the southern and northern shelves.

Sediment Transport in Suspension

Southern Bering Shelf. In Bristol Bay various loci of sediment input, as well as the counterclockwise movement of surface sediments in the near-shore zone can be easily seen in the density-sliced ERTS-1 images (Figs. 10-39 to 10-41). A significant local sediment source in Port Heiden is indicated by a dense surface plume, which covers the bay and extends northeastward along the shore. Further up the coast at the head of the bay, sediments brought by the Kvichak River are carried westward as a surface plume into Kvichak Bay (Fig. 10-39). As this sediment-laden water moves westward along the northern coast, the Kvichak River sediment plume merges with the Nushagak River plume, which dominates the Nushagak Bay (Figs. 10-39 to 10-41). The combined waters continue to move westward around the Nushagak Peninsula, and the plume subsequently disperses in Togiak Bay.

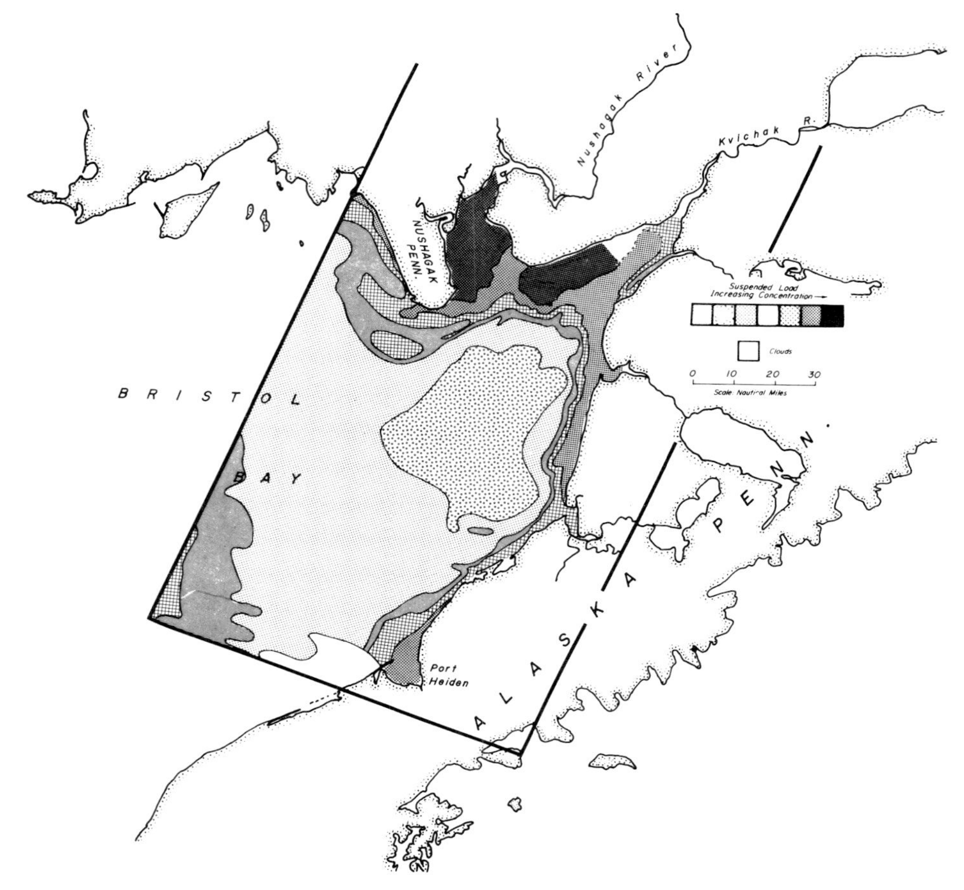

Figure 10-39. Isodensity distribution of reflectance in satellite imagery showing relative suspended load in near-surface water on 15 October 1973, Bristol Bay.

Kuskokwim River detritus forms a dense plume and covers most of Kuskokwim Bay (Fig. 10-42). Because of the cloud cover, the offshore movement of the river sediments unfortunately could not be traced in the available imagery. The configuration of the plume, however, indicates a southward movement of surface water along the shores of the bay. The ERTS-1 image of 31 August 1972 indicates that the plume along the western shore extends seaward of the bay and continues southwestward, retaining its identity for a short distance along the shores of the Alaska mainland west of Nunivak Island in Etolin Strait. The movement of sediments along the eastern shore is to the south (Fig. 10-42). A small but distinct westward moving near-shore plume, originating in Togiak Bay to the east, passes Cape Newenham, and is deflected northward into Kuskokwim Bay. This northward deflection of the turbid water may be the result of coriolis force.

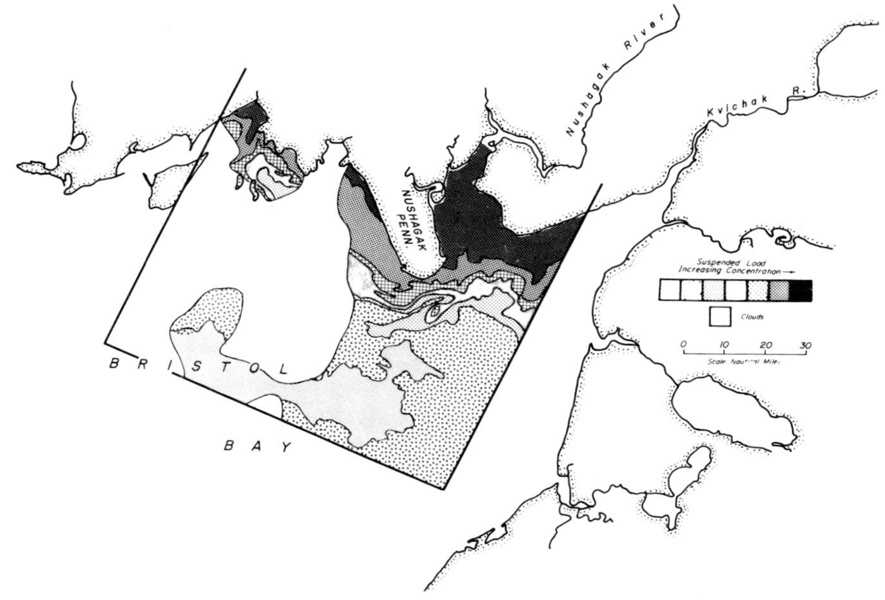

Figure 10-40. Isodensity distribution of reflectance in satellite imagery showing relative suspended load in near-surface water on 15 September, 1972.

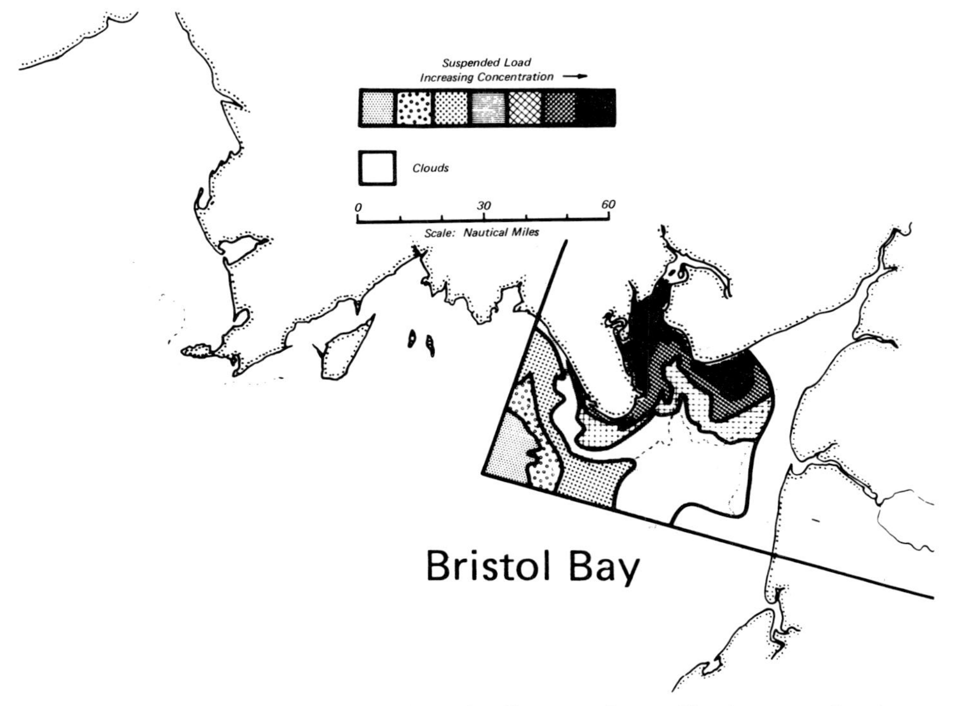

Figure 10-41. Isodensity distribution of reflectance in satellite imagery showing relative suspended load in near-surface water on 2 October, 1972, Bristol Bay.

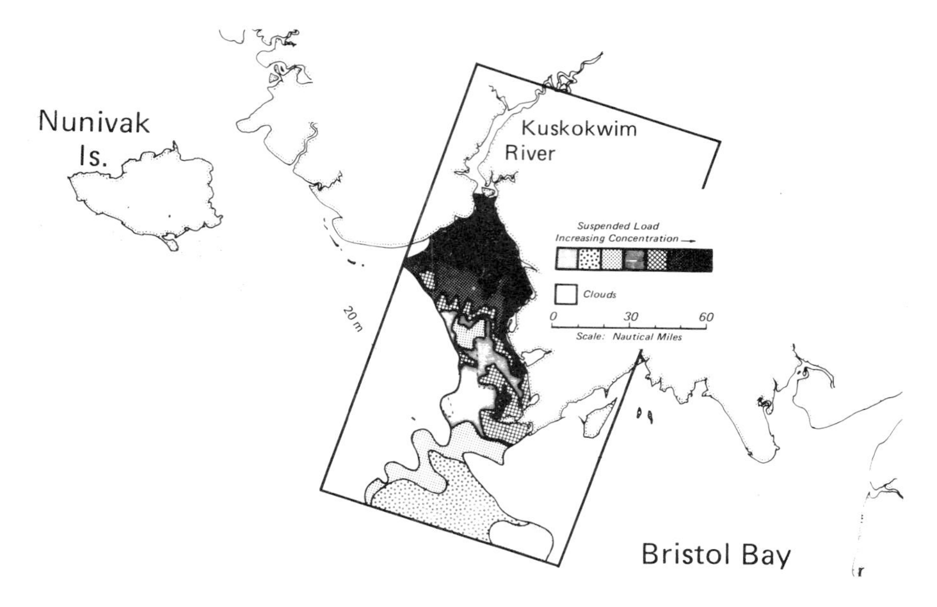

Figure 10-42. Isodensity distribution of reflectance in satellite imagery showing relative suspended load in near-surface water on 5 November, 1973, Kuskokwim Bay.

North of various Aleutian passes, and near the Pribilofs, large concentrations of biologic material were observed on filter papers, and high suspended loads may well be indicative of increased biologic productivity as a result of upwelling in these regions. The suspended load on the shelf waters, however, consisted primarily of sediment detritus. Of particular interest is the tongue of relatively turbid water which apparently originates in Kuskokwim Bay and diffuses to the south and west. At depth (Figs. 10-43 to 10-46) this turbid layer extends farther south and west. High suspended sediment concentrations south of Kuskokwim Bay near Cape Newenham were also noted by Neiman (1961). The subsurface extension and the increase in sediment concentration in the near-bottom waters suggests sediment movement from the bay to the south and the west, which are apparently carried by the southwestward moving Bristol Bay gyre. Evidence for such a water movement has been presented by Hebard (1961) and Natarov and Novikov (1970). The latter authors concluded that southwestward moving currents should be more intense in this region during winter and early summer than later in the summer. Intensity of water movement may be partly enhanced by the prevailing wind. During winter, the wind is northeasterly and in July reverses to a southerly direction. It should be noted that the suspended load data described was obtained during early summer and the due southward movement of sediments may only be seasonal.

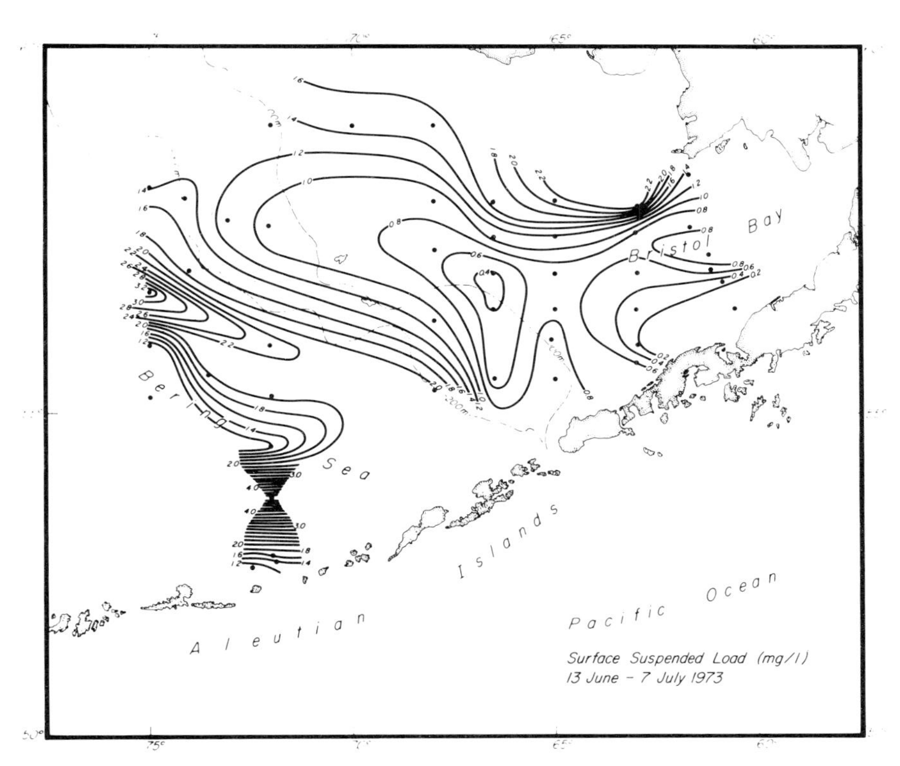

Figure 10-43. Distribution of the surtace suspended load during 13 June-7 July, 1973. Kuskokwim Bay.

During July 11 to August 11, 1973, however, samples for suspended load determinations were collected under moderate storm conditions (Fig. 10-47). Even under these stormy conditions, the sediment plume near the head of Bristol Bay is primarily restricted to the warm and low-salinity coastal waters moving westward along the northern shore. More importantly, the suspended load in the near-bottom waters increased 2 to 3 fold, suggesting roiling of bottom sediments during storms and their resultant near-bottom transport westward.

Northern Bering Shelf. Suspended load measurements of central Bering Shelf water were made on a north-south track and the load distribution is shown in Figure 10-47. The sediment loads in the surface waters are typical of marine water and vary from 0.5 to 2.0 mg/ℓ. Measurements of suspended sediments in near-bottom waters on the central shelf have been made by Neiman (1961) and McManus and Smyth (1970). Neiman (1961) reported suspended sediment concentrations between 7.5 and 17.5 mg/ℓ with high concentrations near Cape Romanzof and north of St. Matthew Island.

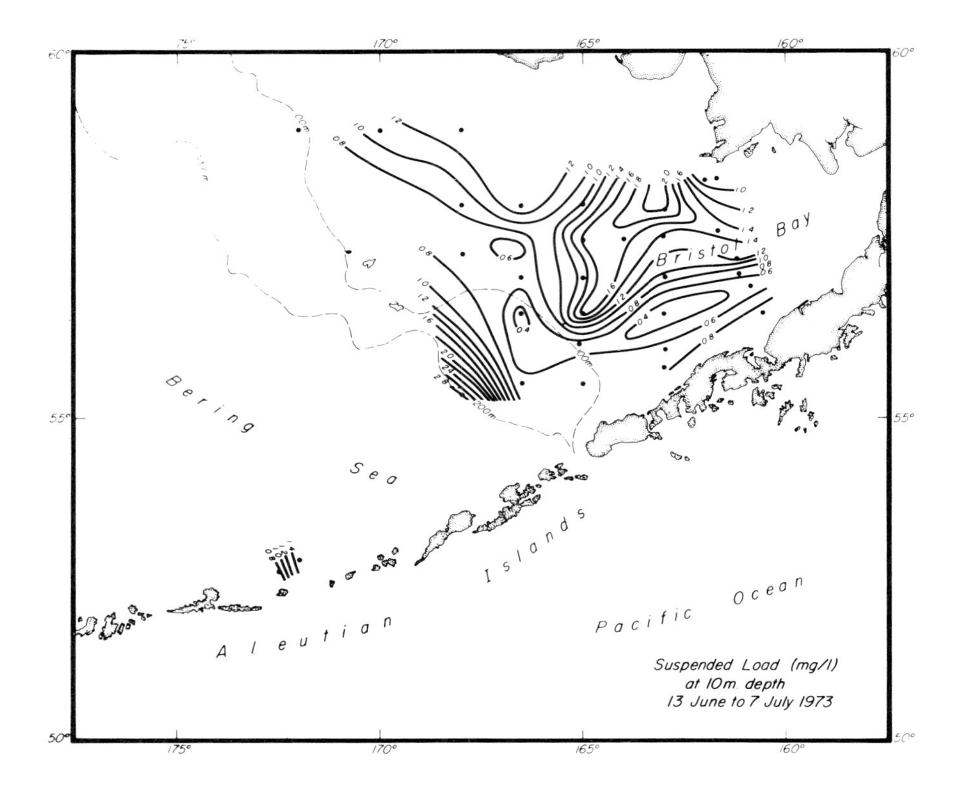

Figure 10-44. Distribution of the suspended load at 10 m water depth, during 13 June-7 July, 1973, southern Bering Shelf.

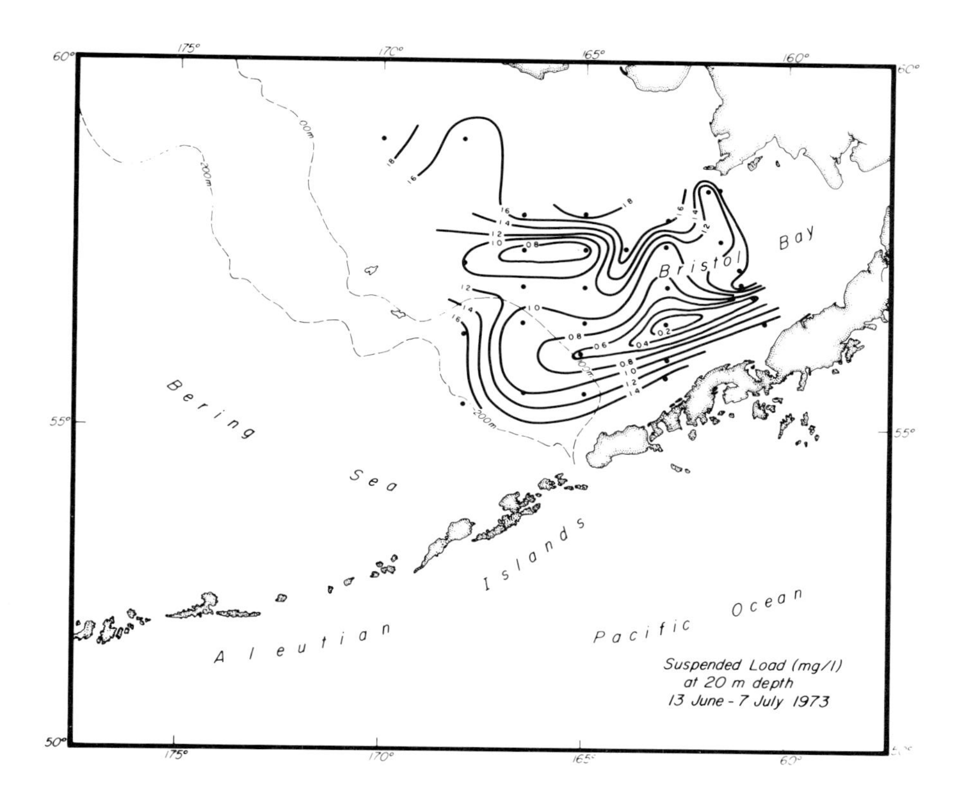

Figure 10-45. Distribution of the suspended load at 20 m water depth during 13 June-7 July, 1973.

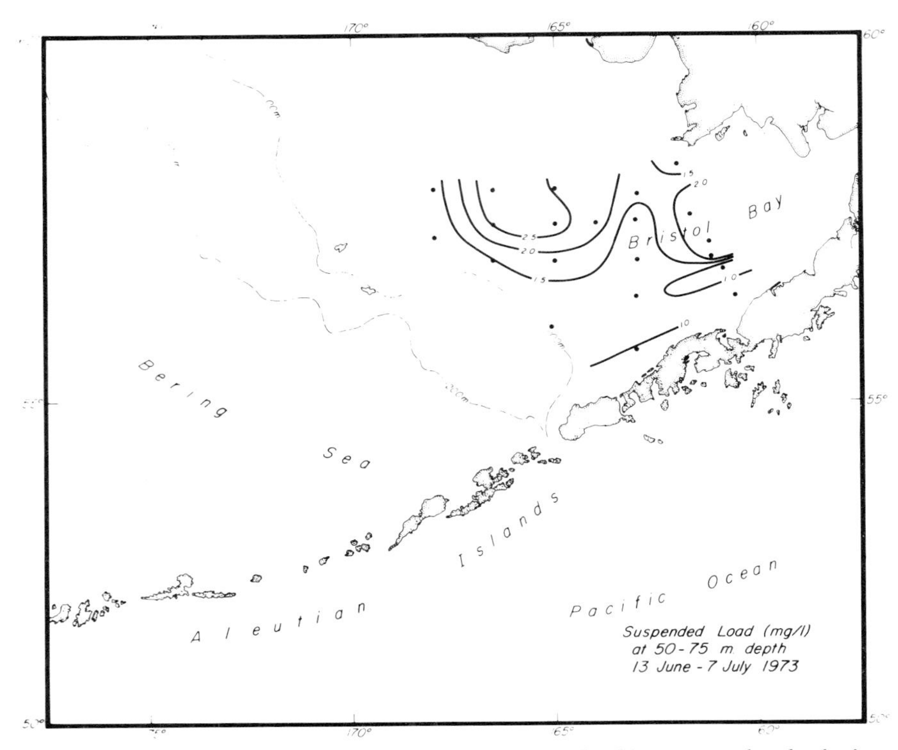

Figure 10-46. Distribution of the suspended load at 50-75 m water depth, during 13 June-7 July, 1973, Southern Bering Shelf.

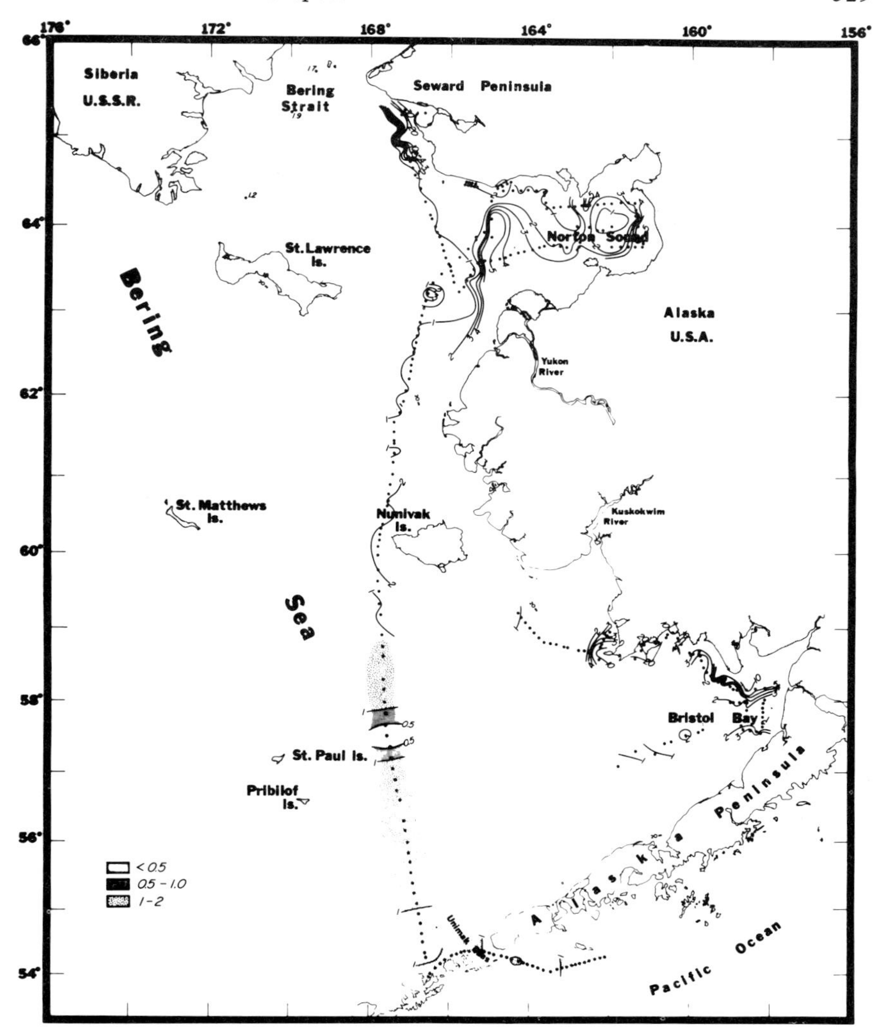

Figure 10-47. Distribution of surface suspended load (mg/ℓ) during 11 July-11 August, 1973, Bering Shelf.

The near-shore region along the Alaska mainland north of Nunivak Island is dominated by the large Yukon River plume (Figs. 10-48 and 10-49), which covers an area over 100,000 km². Its southern boundary extends as far as Cape Romanzof, and to the north the plume surrounds the entire subdelta and extends eastward into Norton Sound. The eastward extension of the plume in Norton Sound forms a counterclockwise gyre and disperses as complex eddies. Much of the fine sediment carried in suspension settles out in the eastern portion of the sound. A small but distinct sediment plume, with its origin in Norton

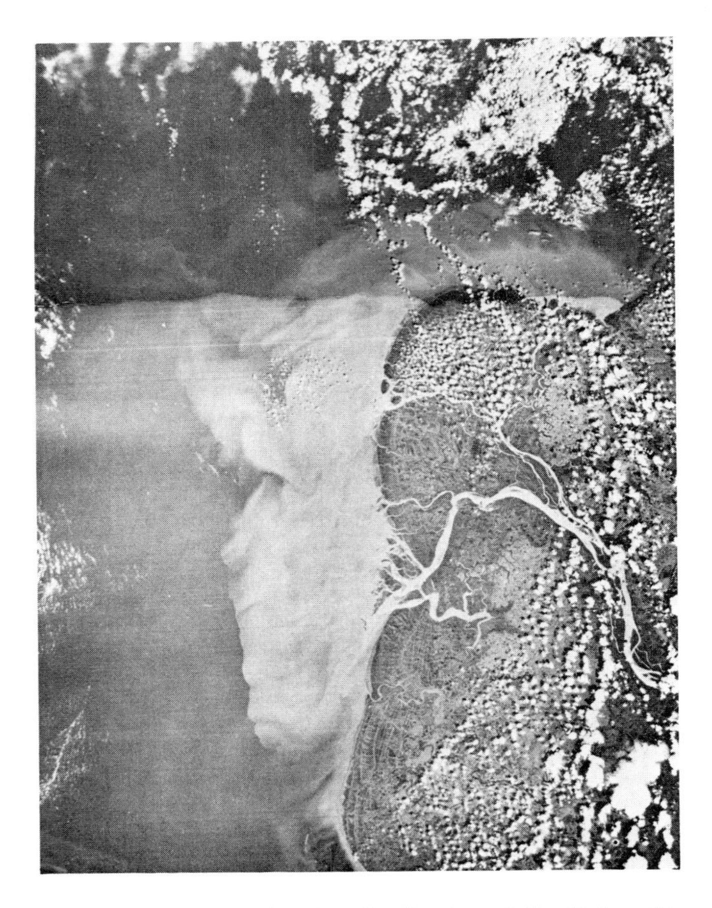

Figure 10-48. ERTS-1 imagery showing distribution of the Yukon River plume on 11 August, 1973, Bering Shelf.

Bay, extends southward out of the bay, turns slightly to the west and blocks the northward advance of the Yukon River plume. The distribution of water properties and the dense plume suggest that the Yukon River sediments are mostly retained in Alaska coastal waters. Some diffusion of sediments from the coastal waters into the offshore northward moving currents is also apparent from the ERTS-1 imagery.

The isodensity distribution in ERTS-1 imagery (Fig. 10-50) indicates a northward extension of the Yukon River plume from its subdelta. This northward offshoot of the plume is a combined effect of tide, wind, and the offshore northward moving currents. The plume is dense near the subdelta and retains its identity northwestward for about 100 km, where it disperses rapidly. Maximum northward extension of the plume should occur under ebb tide and southerly

Figure 10-49. ERTS-1 imagery showing distribution of sediment plumes on 19 September, 1972, eastern Norton Sound.

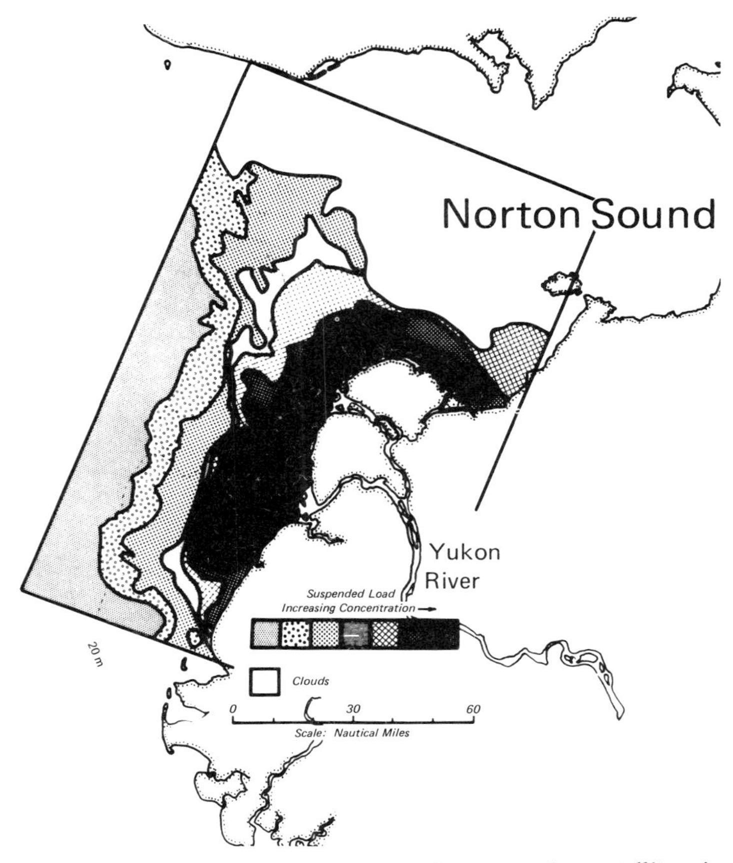

Figure 10-50. Isodensity distribution of reflectance in satellite imagery (Fig. 10-48) showing relative suspended load in the near-surface water of northeastern Bering Shelf on 11 August, 1973.

wind conditions. Isolated but distinct plumes emerging from the southern and southwestern coast of the Seward Peninsula were observed in various ERTS-1 images. The orientation and distribution of these plumes suggested a general movement of sediments toward the Bering Strait.

The suspended load distribution in surface water in the vicinity of the Yukon River delta and in Norton Sound during July 1973 is shown in Figure 10-47. The sediment load in surface water in this region ranged from 1 mg/ℓ to 5 mg/ℓ. The variations of sediment concentrations in suspension and in the plume density conform well and suggest that the surface suspended sediments move directly north-northwest from the river mouth toward Nome, as well as into Norton Sound along the southwest shore.

The distribution of suspended sediments in Norton Sound also conforms to the density distribution of the plume observed in ERTS-1 imagery (Fig. 10-49).

Maximum concentrations of sediments occur along the eastern and southern shores of Norton Sound, suggesting sediment movement in a counterclockwise gyre (Figs. 10-49 and 10-51). Suspended load distribution at depth (Figs. 10-52 and 10-53) in Norton Sound is similar to that observed in the surface waters. Therefore, the major movement of Yukon sediments in bottom water, as in surface water, is to the east along the southern coast, as well as north-northwest toward Nome. It is interesting to note that the highest sediment concentrations in surface water are found in eastern Norton Sound, whereas in the near-bottom water the maxima are observed between the mouth of the Yukon River and Nome.

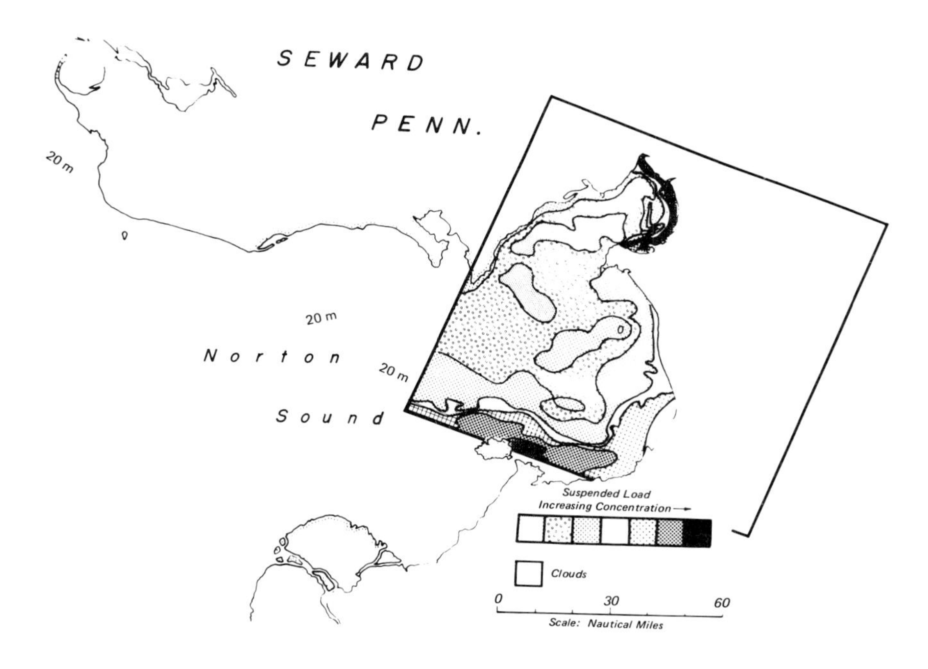

Figure 10-51. Isodensity distribution of reflectance in satellite imagery (Fig. 10-49) showing relative suspended load in near-surface water of eastern Norton Sound, on 19 September, 1972.

A detailed survey of the distribution of suspended sediments in surface and near-bottom water off Nome was conducted by Sharma (1974c) and is shown in Figure 10-54. In the surface waters, the near-shore sediment load (in excess of 6 mg/ℓ) decreased seaward (2 mg/ℓ). In the water column, the sediment concentration generally increased with depth. A maximum suspended load of 17.2 mg/ℓ at 10-m depth was observed in the near-shore water west of Nome. High sediment concentration (>10 mg/ℓ) at 10- and 15-m depths also occurred near the

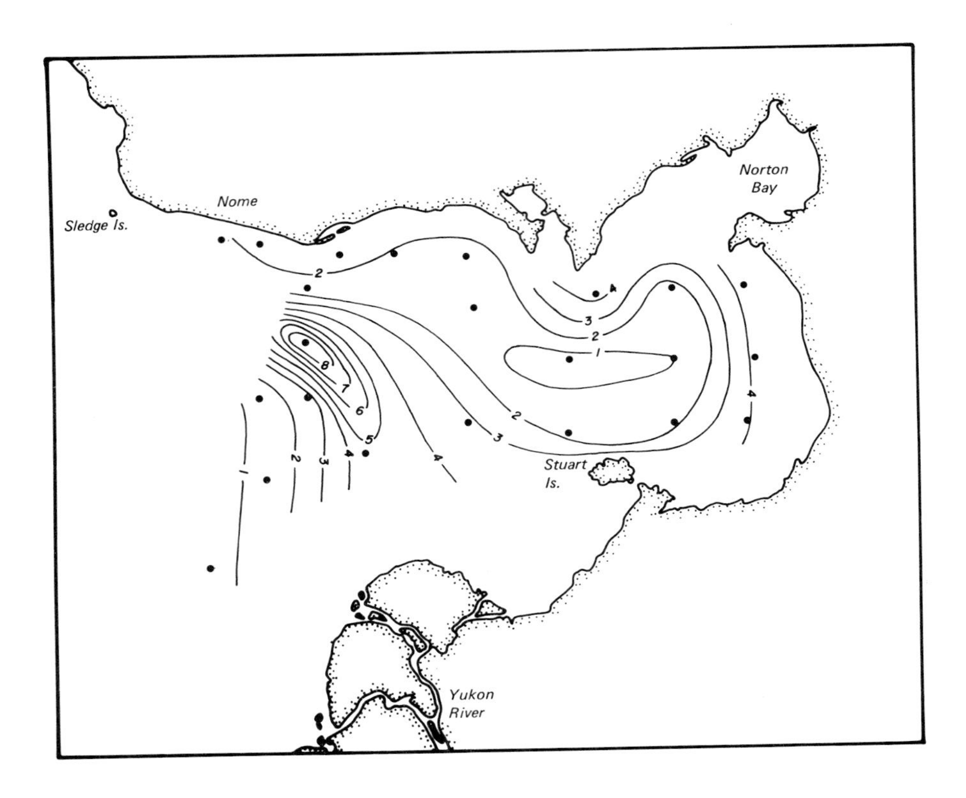

Figure 10-52. Distribution of suspended load (mg/ℓ) at 5 m water depth during 20-24 July, 1973, Norton Sound.

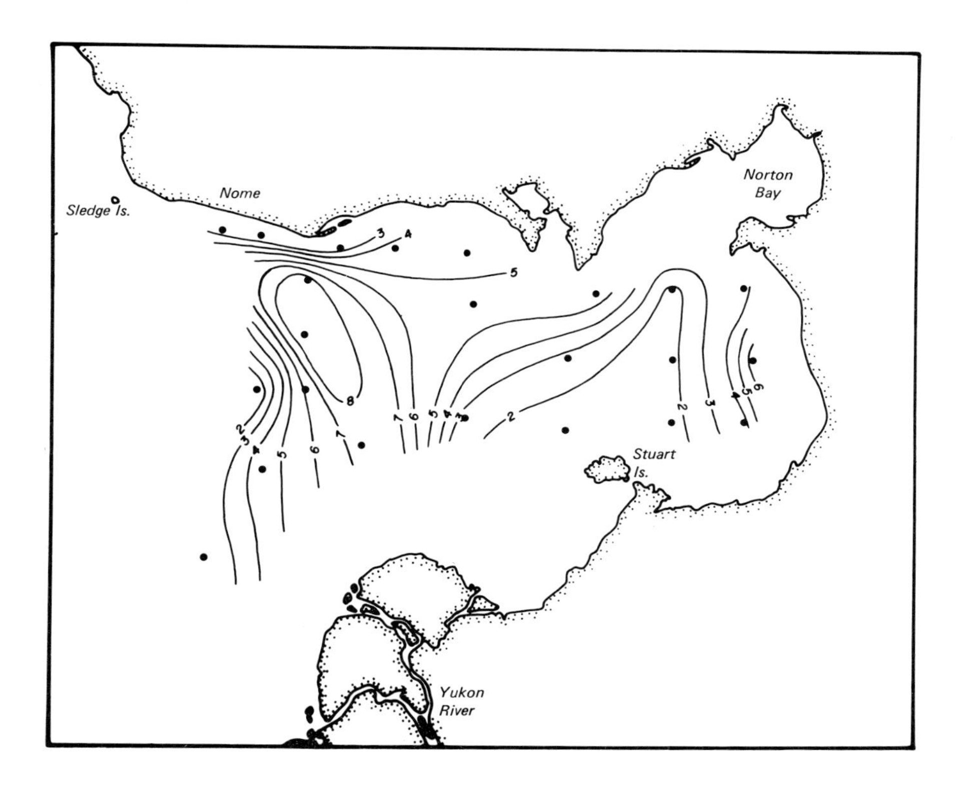

Figure 10-53. Distribution of suspended load (mg/ℓ) at 10 m water depth during 20-24 July, 1973, Norton Sound.

mouth of the Snake River. McManus and Smyth (1970) reported that the bottom water from the vicinity of Nome contained sediments in excess of 20 mg/ℓ, whereas the concentrations offshore decreased to 5 mg/ℓ.

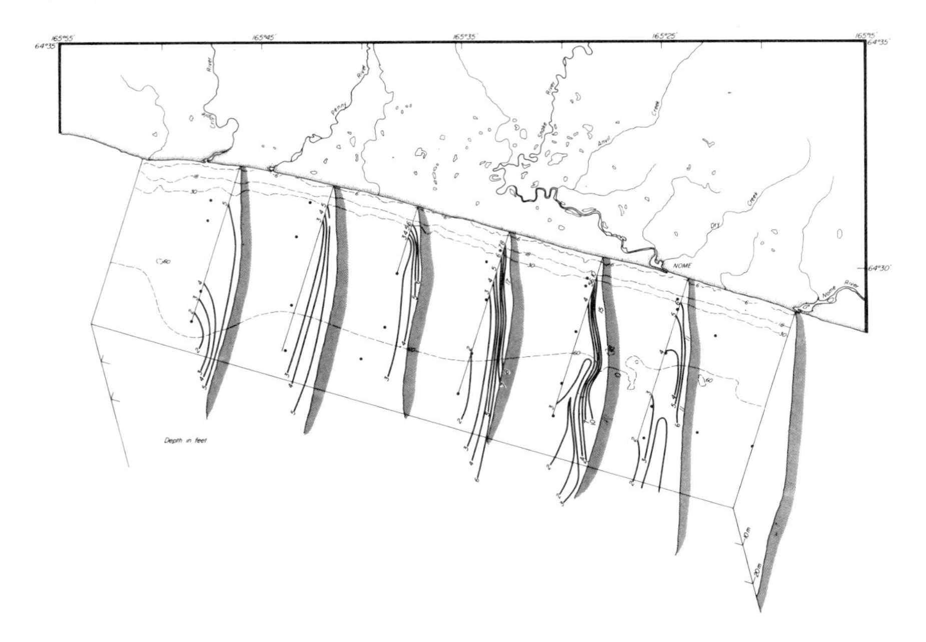

Figure 10-54. Vertical and spatial distributions of suspended load in waters off Nome, 20-23, July, 1973.

The transects of suspended load depth profiles near the entrance of the Snake River indicate an anomalous distribution in comparison with other transects to the west. Saline waters with low sediment loads appear to intrude shoreward at depth (10 and 20 m) near the mouth of the river. This subsurface shoreward movement of water may be caused by the Snake River water outflow at the surface, thus setting up a typical estuarine circulation.

The water movement and its effect on the suspended sediment distribution in Norton Sound and adjacent areas perhaps can be best explained by the salinity and temperature distributions in this region. The isotherms and isohalines of surface waters are shown in Figures 10-3 and 10-4, and those at 5- and 10-m depths are given in Figures 10-55 to 10-58. The distribution of water properties clearly suggests that relatively warm and fresh Yukon River runoff flows northward and eastward and forms a prism with its base along the delta, and the eastward moving near-surface water forms a counterclockwise gyre. The water properties at 10-m depth, however, suggest an intrusion of colder, more saline water toward Norton Bay (Fig. 10-58). The source for this water may lie in the Gulf of Anadyr, with subsequent transportation to the north of St. Lawrence

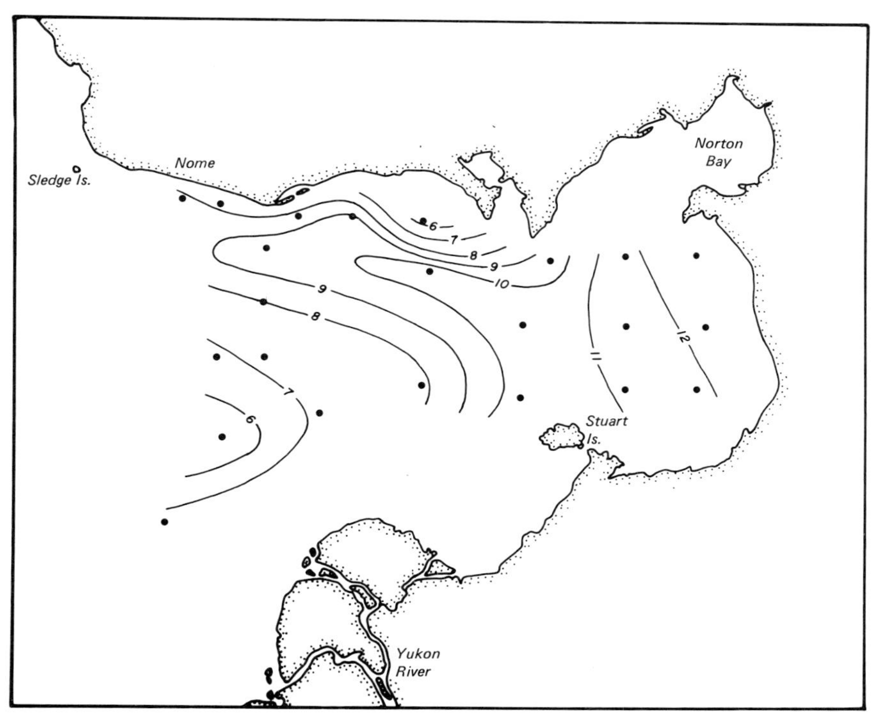

Figure 10-55. Isotherms (°C) at 5 m water depth during 20-24 July, 1973, Norton Sound.

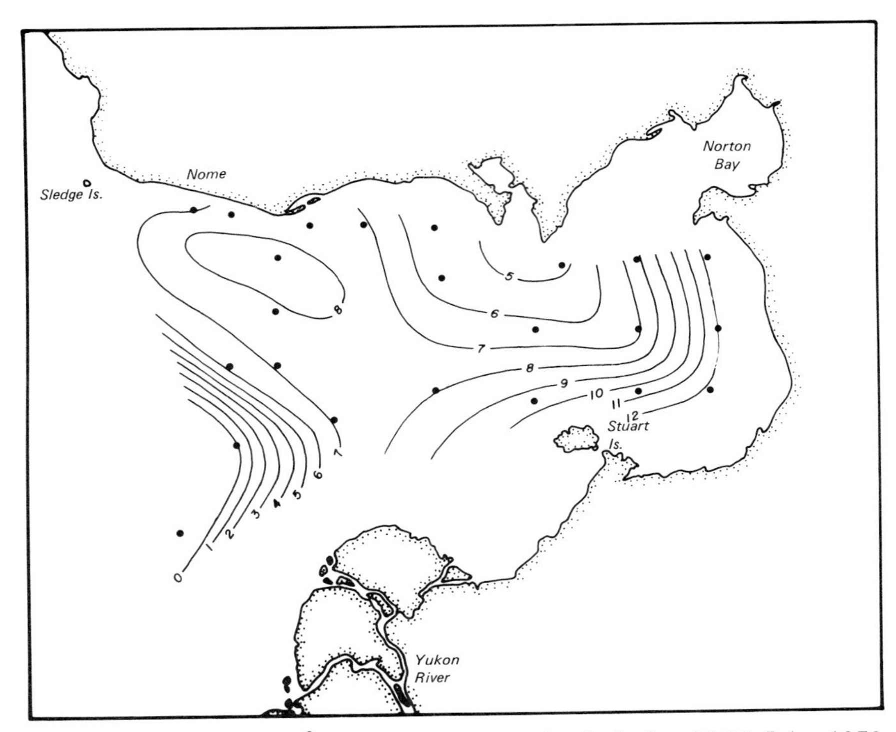

Figure 10-56. Isotherms (°C) at 10 m water depth during 20-24 July, 1973, Norton Sound.

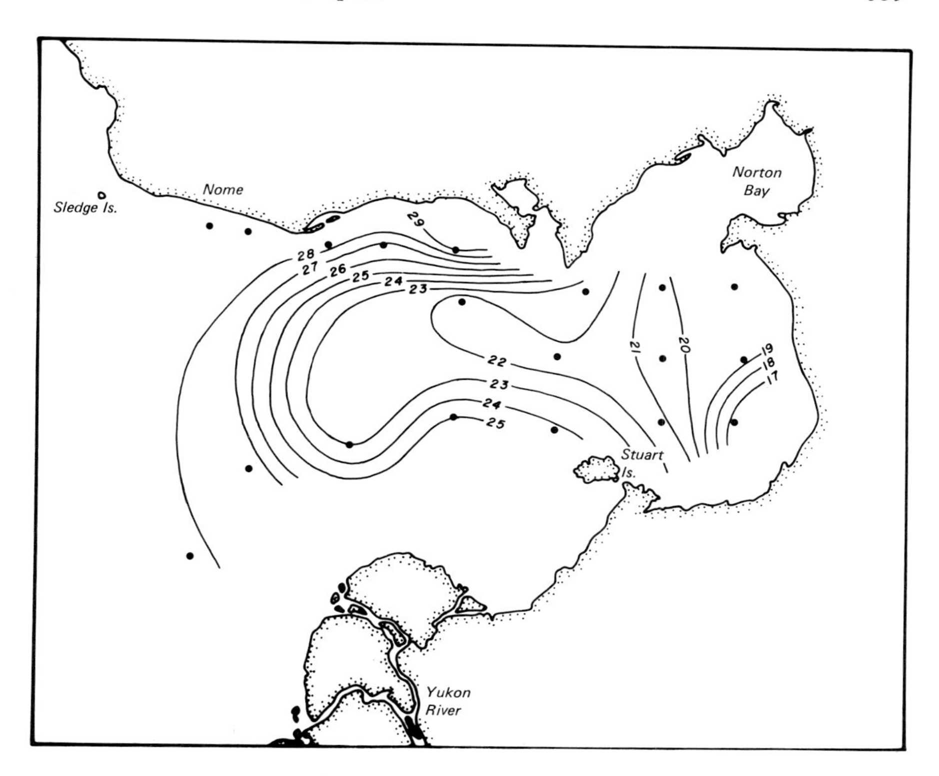

Figure 10-57. Isohalines (°/₀₀) at 5 m water depth during 20-24 July, 1973, Norton Sound.

Island by the permanent currents which transfer water from the Bering Sea into the Chukchi Sea. The intrusion of cold water near the bottom, with upwelling along the southern shore of the Seward Peninsula, is also suggested by the temperature-salinity distributions at 5-m depth, as well as by the suspended load relationships discussed earlier. Concurrent with the movement at depth, the mixed water formed between 0 and 5 m in northeastern Norton Sound appears to move westward out of the sound. This outflow is channeled through the center of the sound and is bounded by marine water on the north and south.

The relative sediment distribution in the surface waters of the northern Chirikov Basin and Bering Strait regions inferred from density slicing of ERTS-1 images are shown in Figures 10-59 and 10-60. A dense plume appears to develop in the Bering Strait along the northwestern shore of the Seward Peninsula and extends northeast along the Prince of Wales Shoal (Fig. 10-59). The Prince of Wales Shoal is shallow, with an average water depth of about 7 m. The absence of detrital sources in this area indicates that this plume may be the result of resuspension of bottom sediment by a turbulent, northward moving water current.

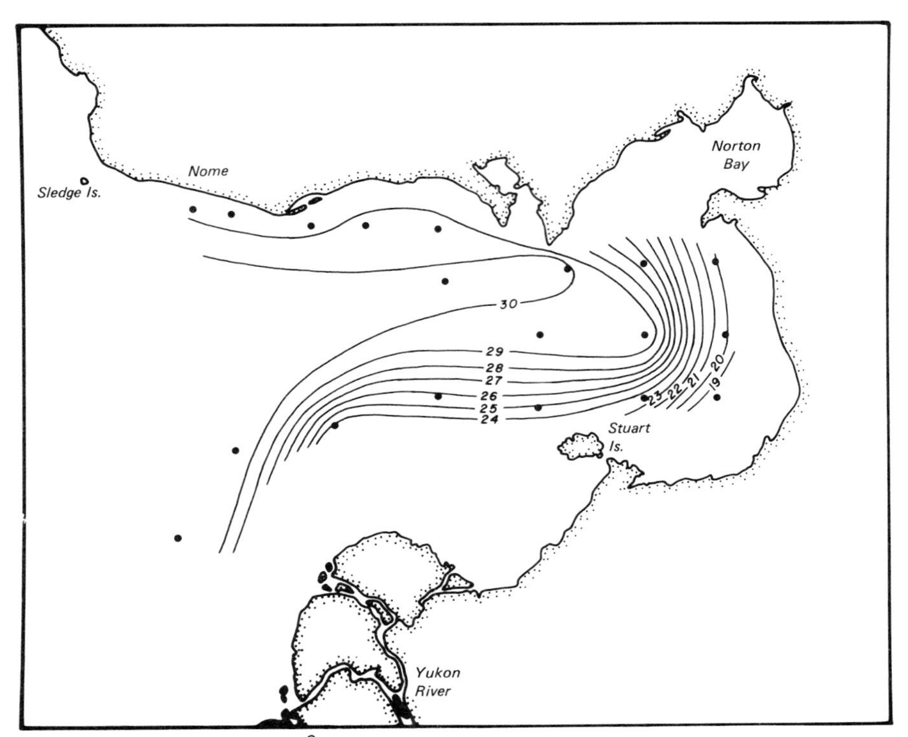

Figure 10-58. Isohalines ($^{\circ}/_{\circ\circ}$) at 10 m water depth during 20-24 July, 1973, Norton Sound.

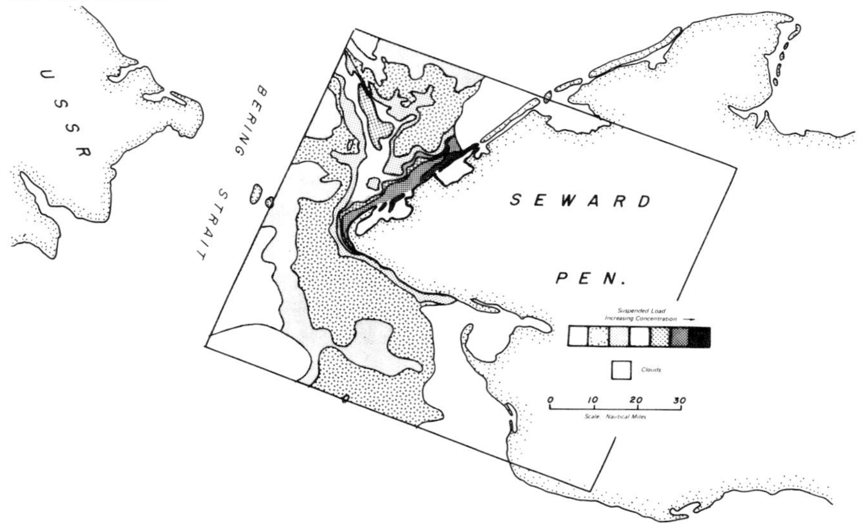

Figure 10-59. Isodensity distribution of reflectance in satellite imagery showing relative suspended load in near-surface water of eastern Bering Strait on 9 July, 1973.

Dense sediment plumes are generally observed primarily along the ice edge, suggesting melting sea ice as a major source for surface suspended sediment. High sediment concentrations appear to follow the northward receding ice. North of Bering Strait sediments rapidly settle out, presumably because of a decrease in turbulence as well as the relatively coarse nature of the sediment carried by the ice. The analysis of ERTS-1 images (Fig. 10-60) demonstrates that winter ice can be an important agent for sediment transport on the Bering Shelf.

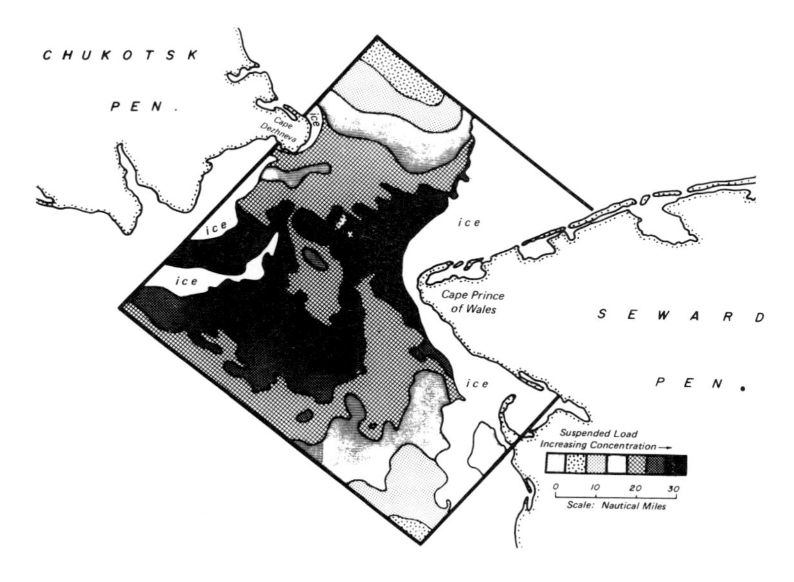

Figure 10-60. Isodensity distribution of reflectance in satellite imagery showing relative suspended load in near-surface water of eastern Bering Strait on 22 June, 1973.

Ice-Rafted Sediments

During the winter months, part of the Bering Shelf is covered with ice, which begins forming near the shore in December, and often lasts until late in April. By mid-winter the pack ice covers the northern shelf and often extends as far as the Pribilof Islands. During May, the ice begins to break up along the coast, and by late June the ice recedes northward beyond the Bering Strait. The significance of ice as an agent of weathering erosion and sediment transport in the central and southeastern Alaska shelves is quite apparent. But the formation of sea ice and its role as a sediment carrier in the Bering Sea is somewhat obscure. Visual examination of the winter shelf ice invariably shows that ice usually contains biogenic as well as detrital material. The Bering Shelf ice and the sediments therein have been described by Sharma *et al.* (1971). Reconnaissance surveys, microscopic examination of a few sea ice samples, and measurements of sea ice sediments from the shelf suggested that the nature of sea ice, and the amount of associated sediments, varies regionally as well as temporally (Figs. 10-61 to 10-63).

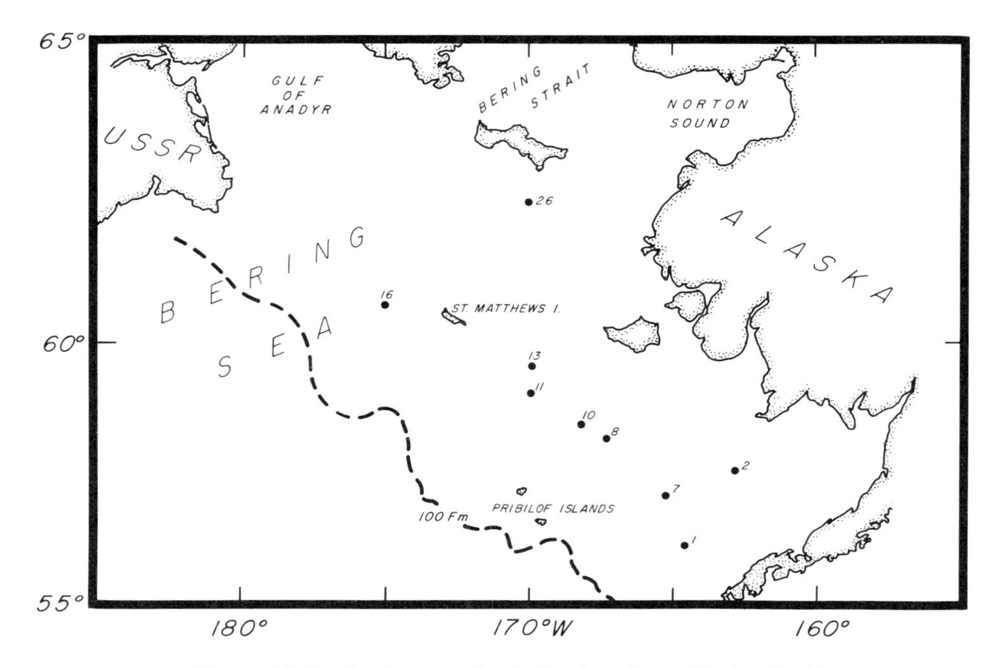

Figure 10-61. Sea ice sample station locations, Bering Shelf.

Layered sea ice of the Bering Shelf displayed a consistent crystallographic structure. Usually three layers were observed: a top layer of milky, fine-grained ice with air pockets, and vertically oriented C-axes; an intermediate layer of thick, clear, dense ice consisting of long, tapered crystals with horizontal C-axes, similar to common lake and stream ice; and a bottom layer of clear, very fine-grained, dense and milky ice, with air pockets (Figs. 10-64 and 10-65). Variations in ice crystal size with depth, particularly in the intermediate layer, also enhance the layered ice appearance (Fig. 10-65).

The maximum concentration of sediments in the sea ice was found at the lower boundary of the intermediate ice layer with horizontal C-axes. The sediments in this zone were distributed irregularly. Sea ice sediments collected from a few locations consisted mostly of silt (35 to 80%) and clay (12 to 60%), with minor amounts of sand (1 to 8%). The bottom sediments from the same locations, however, contained mostly medium to fine sands (Fig. 10-66). The nature of the clay minerals and their distribution in sea ice sediments throughout the shelf were quite uniform. The fine particle size, uniformity of texture, and, to some extent, the clay minerals distribution suggest that these sediments may have been incorporated in the ice while in suspension. The clay and silt observed in

Figure 10-62. Sediment distribution in the Bering Shelf ice.

Figure 10-63. Sediment-laden sea ice, Bering Shelf.

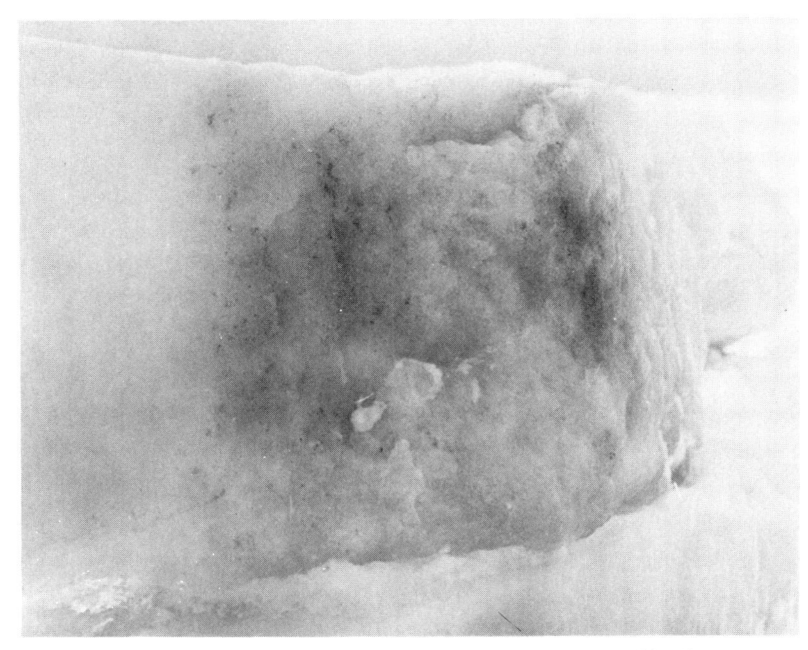

Figure 10-64. Layering of sediments in the Bering Shelf ice.

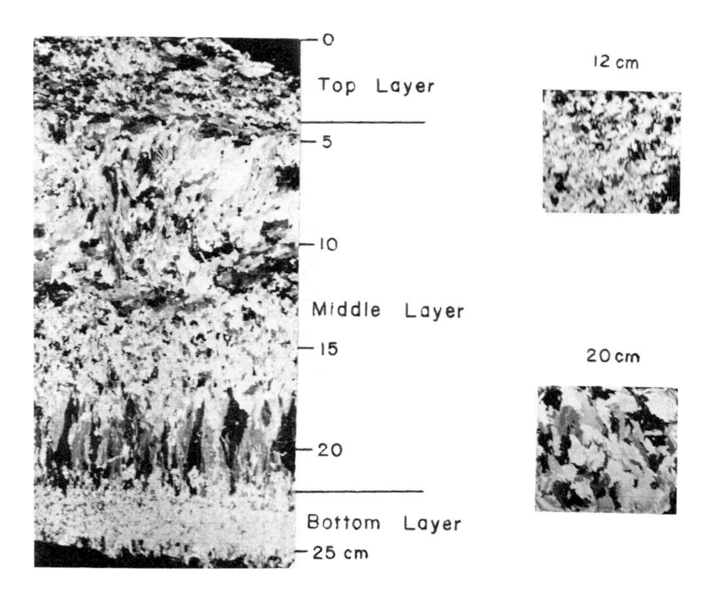

Figure 10-65. Crystal structure along vertical and horizontal sections of Bering Shelf ice.

sea ice easily could have been resuspended in shallow waters by storm waves, which roil the bottom and should cause vertical mixing in the water column. Mixing in the water column and roiling and resuspension of bottom sediments in Norton Sound and Bristol Bay under moderate storm conditions was indicated by salinity-temperature gradients and by increased concentration of sediments in the bottom waters during 1973. Extensive mixing may ultimately destroy the water structure, leading to rapid accretion of sea ice. The simultaneous destruction of vertical water structure, resuspension of bottom sediments, and rapid accretion of sea ice by winter storms may be the prime mechanism for inclusion of sediments in sea ice.

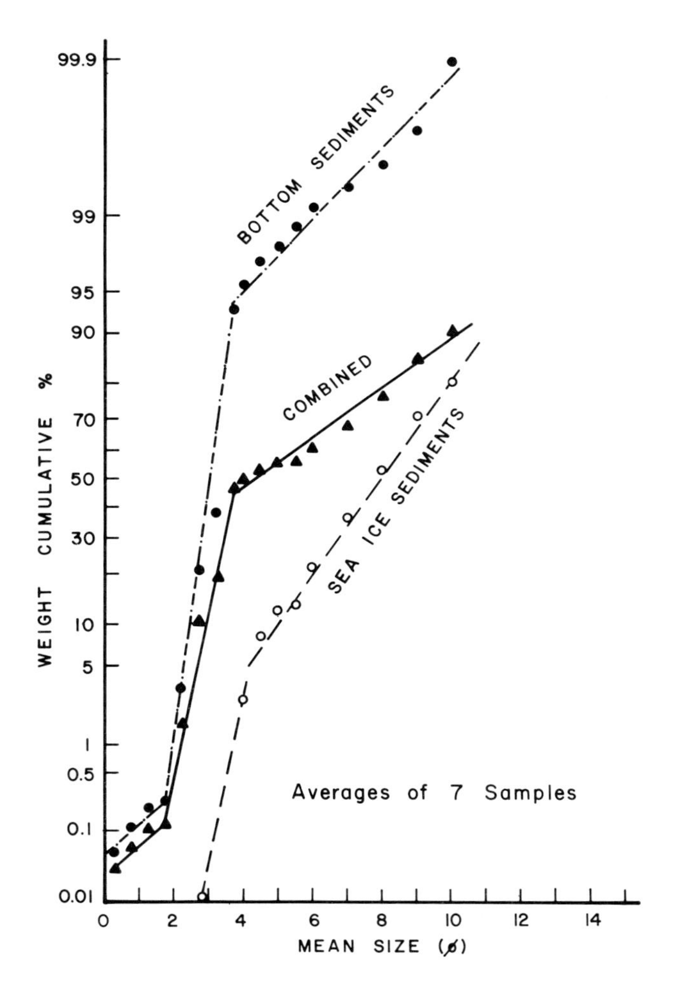

Figure 10-66. Size distributions of the bottom and sea ice sediments, Bering Shelf.

Ice with vertically oriented C-axes is generally formed under turbulent conditions, whereas ice with horizontally oriented C-axes is formed in a quiet environment. The concentration of sediments near the top of the bottom ice layer (with vertically oriented C-axes) further supports the suggestion of resuspension and inclusion of sediment, particularly the fine sand fraction, in sea ice during storms.

Sediment transport on the Bering Shelf by sea ice varies regionally as well as seasonally. Sea ice in the northern Bering Sea generally moves southward from the Bering Strait, and the rate of movement may be as high as 1.25 km/hr (Sharma *et al.,* 1974). The irregular distribution of sediment in sea ice, together with the erratic movement of ice on the Bering Shelf, makes it impossible to estimate the amount of sediment transported annually by sea ice. Nevertheless, the large stretches of sea ice containing significant amounts of sediment observed during the January-February 1970 cruise suggest that sea ice may be an important agent for sediment transport on the Bering Shelf.

From the preceding discussion, it is apparent that the sediments on the Bering Shelf are carried in suspension, along the bottom, and to some extent in ice as well. The sediment transport, therefore, varies seasonally. Because of the paucity of *in situ* measurements, neither the permanent currents nor the more ephemeral circulation created by storms, tides, and other causes are clearly defined. Except for a few measurements of the intensity and direction of water movement on some parts of the shelf, the water circulation and its effect on sediment distribution on most of the shelf can only be deduced from the general oceanographic setting described earlier, as well as from the sediment properties, which are primarily based on granulometry, heavy mineral distribution, and geochemical characteristics.

Bottom Sediment Transport

On the southern and southeastern shelves, the sediment mean size is closely related to the water depth (Fig. 10-8). Besides the usual decreasing sediment grain size with increasing water depth relationship, there are three distinct sediment textural regimes with are closely associated with water depth. Sediments coarser than 3 ϕ (mean size) are mostly restricted to shelf depth of less than 60 m; sediments varying in size from 1.5 ϕ to 5.5 ϕ are deposited between the 60- and 75-m isobaths; and sediments finer than 3 ϕ are deposited offshore with water depths $>$75 m (Fig. 10-8). The sediment mean size is dependent on the distribution of fine sand on the shallow shelf ($<$75 m) and on the distribution of silt on the offshore shelf ($>$75 m) (Figs. 10-7 and 10-14). On the basis of sand and silt distributions and other sedimentary parameters, Sharma *et al.* (1972) broadly divided Bristol Bay into inner and outer shelves. Askren *et al.* (in preparation), based on sediment factor analysis using grain size distribution as variables, classify sediments into five factors: The distribution of sediment factors I and II and the distribution of factors III and IV correspond to the inner and outer shelves, respectively, defined by Sharma *et al.* (1972), and factor V sediments cover the offshore area lying west of St. Matthew Island.

The grain size statistical parameters of 105 sediment samples from Bristol Bay, when plotted, revealed some interesting textural relationships. Sediment mean size versus skewness plots, and mean size kurtosis plots display the conventional sinusoidal interrelationships between the parameters (Figs. 10-67 and 10-68). A gradual offshore decrease in the sediment grain size of the predominant modes and the mixing of various modes are well illustrated in Figure 10-68. The composite diagram of Figure 10-69 also displays the interrelations between sediment mean size, skewness, and kurtosis, and their distribution in Bristol Bay. In this diagram, various sediment size classes (3.0 to 4.5 ϕ, 2.0 to 3.0 ϕ, 1.0 to 2.0 ϕ, and -1.0 to +1.0 ϕ) were plotted on a plane with kurtosis as the abscissa and skewness as the positive and negative ordinates. Each of these plots then was used to construct a three-dimensional graphic representation of the data, by placing the baseline (abscissa) of each plot along a common line in space, each plot being rotated 90° in space with respect to the preceding one, in sequence of increasing particle size range, from left to right (Fig. 10-69). Interestingly, the resultant data configuration shows a conceptual sinusoidal band, which seems to relate the sediment mean size to the water energy near the bottom. Since the sediment mean size is closely related to water depth, the energy imparted at the bottom and its capacity to transport sediment become directly related to the water depth. Maximum energy is imparted in shallow water, and the sediment transport capacity decreases with increasing water depth. The energy required to achieve textural gradation in waters up to 100 m deep, and over such a large area, is enormous. Such amounts of energy seem only to be available from the severe fall and winter storms which often ravage large areas of the southern Bering Sea. Sediment transport by storm waves in Bristol Bay has been described by Sharma *et al.* (1972; Sharma, 1972).

The first percentile versus median grain size (CM patterns) relations of sediments from the southeastern shelf (Sharma *et al.*, 1972) and from the southern shelf (Askren *et al.*, in preparation) reveal that the sediments are predominantly transported either by rolling or in suspension. The sediment texture distribution also indicates that large areas of the shelf are covered with moderately to well-sorted fine sand. Extensive distribution and sorting of fine sands are perhaps attributable to the relative mobility of this material since the competent velocity for roiling of sediments either coarser or finer than fine sand is considerably greater.

As indicated by CM patterns, as well as by other considerations, in the absence of strong currents the movement of sand in suspension on a significantly large shelf can be achieved only by large wind waves generated during storms. The frequent mild storms produce waves which usually are capable of roiling the bottom sediments in waters up to 60 m deep. As indicated by increased suspended loads during moderate storm activity, storm waves remove most of the silt and clay particles from the shallow-depth areas and carry them offshore.

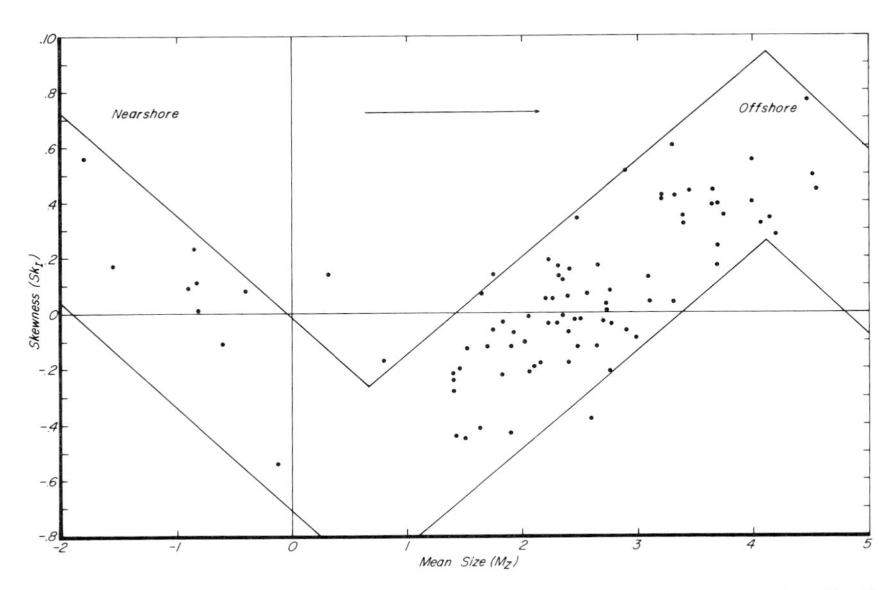

Figure 10-67. Mean size versus skewness plot of sediment grain size, Bering Shelf.

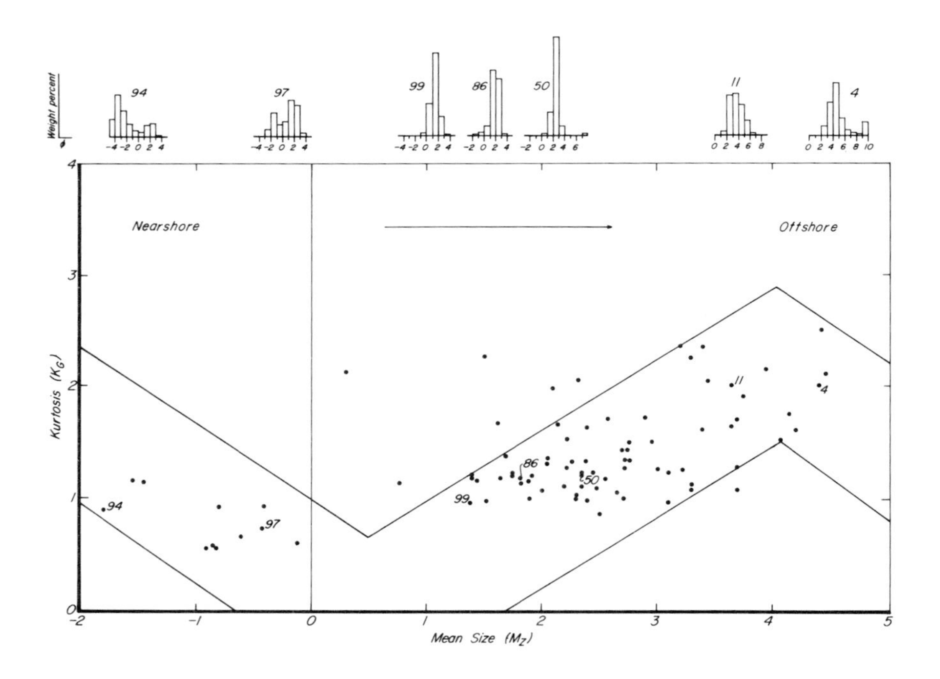

Figure 10-68. Mean size versus kurtosis plot of sediment grain size, Bering Shelf.

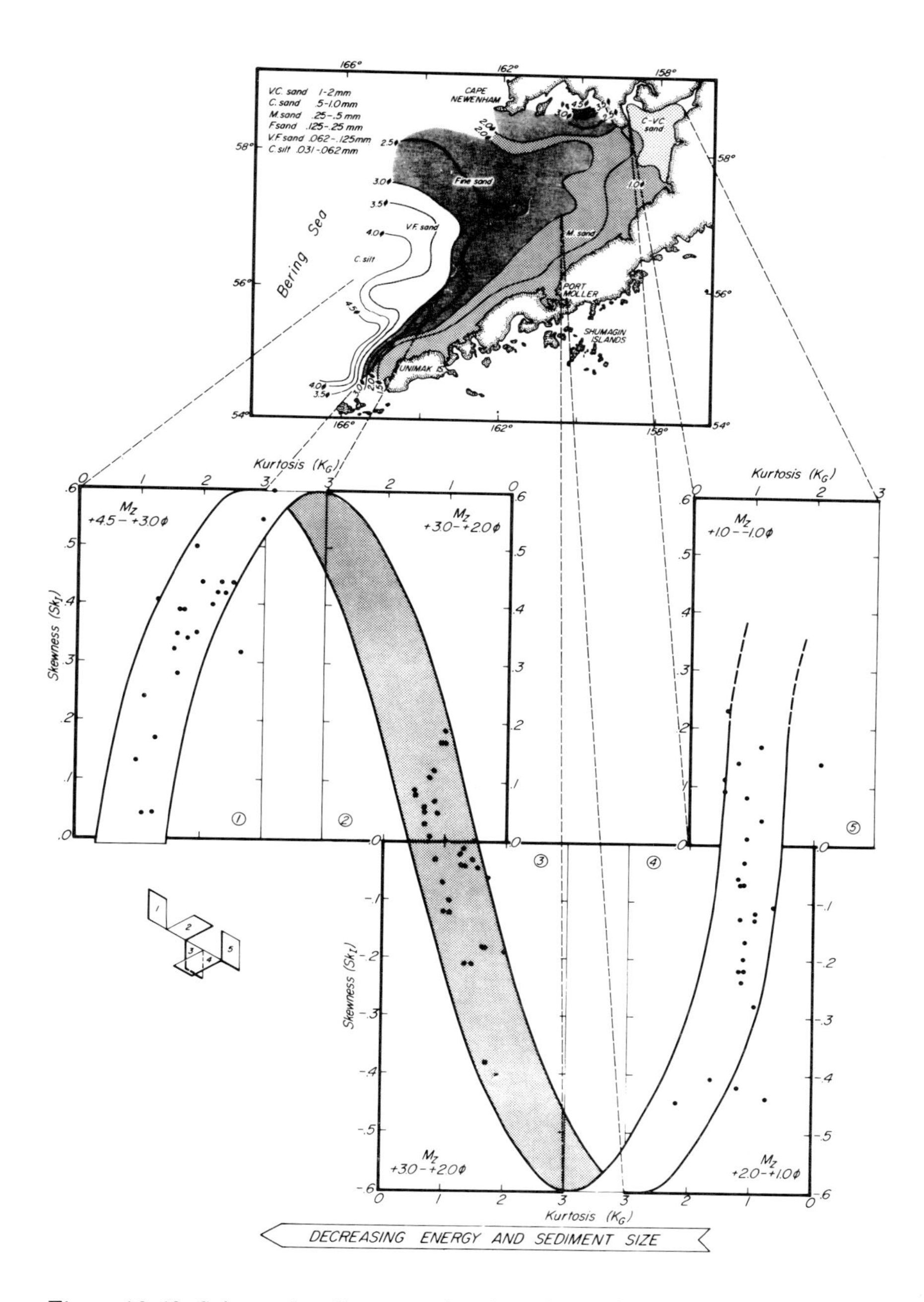

Figure 10-69. Schematic diagram showing interrelation between sediment parameters and water energy in Bristol Bay.

These sediments are deposited on shelf areas with water depths >60 m. The episodic severe storms, however, are more intense, and waves generated by them should be able to resuspend sediments from greater depths. It is suggested that the severe storm waves can effectively roil fine sand in bottom sediments at a depth of 75 m. The wide variation in sediment mean size distribution observed between the 60- and 75-m isobaths is the result of the mixing of sediments transported by both mild and severe storms. Mixing of different sediment modes deposited by mild and severe storms is well manifested in the sediment sorting. Sorting in this zone deteriorates rapidly with increasing water depth; sorting and depth isopleths are parallel (Figs. 10-2 and 10-16). Offshore (>75 m), sediment deposition occurs in the usual graded suspension; these sediments contain increasing silt and clay fractions with increasing water depth.

Locally, the effects of tidal currents are superimposed on the sediment distribution brought forth by wind and storm waves. Average tidal currents measured by Hebard (1961) in Bristol Bay varied from 5 to 85 cm/sec. These currents are competent to transport sand particles and are particularly effective along the northern shores of the Alaska Peninsula and near the head of the bay.

Bottom sediments collected from Bristol Bay during July 1968 contained about 1 to 2% silt and about 1 to 2% clay throughout the shallow (<60 m) region. The sediment sampling followed a period of relatively calm weather. It is suggested that the presence of minor amounts of silt and clay fractions in otherwise sandy sediments is the normal result of usual sedimentation on the shelf during calm weather. Clay and silt deposited on the shallow shelf may be resuspended by occasional storms and carried offshore.

Sediment transport on the southern shelf is achieved in two stages. Initially, under normal atmospheric conditions, the sediments (including silt and clay) are deposited throughout the shelf. The periodic storm and wind waves resuspend bottom sediment and simultaneously destroy the water stratification. Such perturbations of the water structure, particularly in shallow regions, result in a homogeneous water column and resuspended sediment is entrained in the column. The maximum sediment concentrations occur in the near-bottom waters. Continued storm activity and destruction of the water density structure finally result in downslope flushing of the shelf water. The flushing of water transports sediments, both entrained in suspension and as bed load along the bottom.

Coastal and river sediments originating from the Alaska mainland between Nunivak Island and the Yukon Delta are partly deposited along the coast and partly swept away by the permanent northward bound currents. The near-shore deposits are coarse silt, which grade offshore into sand. Offshore, the central Bering Shelf between St. Matthew and St. Lawrence islands is mantled with sediments which have their dominant modes in the fine sand, very fine sand, coarse silt, medium silt, medium clay, and gravel size range. Fine sands are very well sorted and are deposited on the shallow shelf south of St. Lawrence Island, as well as east and north of Nunivak Island. Offshore, and adjacent to the fine sand, the sediment mode becomes moderately sorted very fine sand. The CM pattern of sediments from the central shelf suggests that these sands are transported in uniform suspension (Knebel, 1972). Episodic summer storm and permanent currents may enhance bed load transport and sorting in sand from these regions.

The sediments in Norton Sound are silt, sand, and gravel. The gravel deposits are relict in origin. The distribution of contemporary sediments in Norton Sound is primarily controlled by northward and eastward moving Alaskan coastal water. The near-bottom current speed of Alaskan coastal water during the ice-free season has been reported to vary between 30 and 40 cm/sec. Therefore, the near-shore waters are capable of transporting clay and silt in suspension and sand as bed load. The distribution of bottom sediment texture in Norton Sound conforms well with the movement of near-shore water discussed earlier.

The general net movement of waters in the northern Bering Shelf is toward the north because of the strong southerly winds, which pile water along the southern coast of Seward Peninsula, where sea level may rise as much as 4 m. The Yukon River input, as well as the wind-stressed water, are carried north through the Bering Strait. Local and ephemeral winds, however, can change the current direction and even reverse the northward currents. The secondary effects of wind are the destruction of water structure and wave drift, both of which can cause sediment movement to an average depth of about 20 m. Ultimate flushing of water may occur as a result of vertical mixing. Extensive water movement appears to preclude fine sediment deposition in Chirikov Basin. The sediments in the basin are mostly sand and gravel.

Measurements of sediments in surface water and the distribution of surface plumes as observed in ERTS-1 imagery for the northern shelf suggest that sediments are partly carried to the north toward the Bering Strait and partly to the east into Norton Sound. The waters moving northward through Shpanberg Strait carry part of the Yukon River discharge along their eastern periphery. The northward sediment transport of the Yukon River material is demonstrated by the surface sediment plume, which extends from the Yukon River delta northward toward Nome, and by the high suspended sediment concentration in bottom waters. The eastward sediment transport by the Alaskan coastal water into the Norton Sound is also evident from the bottom sediment texture distribution and the sediment dispersal in the water column.

During winter, the central and northern Bering shelves are covered with ice, which significantly dampens the severe winter storm waves passing underneath. Furthermore, the ice-covered water column is generally isothermal and isohaline and is usually in equilibrium with adjacent waters. Therefore, vertical mixing by waves does not cause lateral water movement. During the ice-free season, however, the moderate storm waves cause mixing of waters in the upper layers. Thus, in the central and northern shelves, sediment transport by storms and wind waves is restricted to summers and to shallow areas.

Distribution of surface sediment plumes and suspended sediment suggest that major rivers and the coast are the main sources of sediment for the Bering Shelf. The river's input during the summer runoff is impressive; river sediments dominate the inner shelf. Estimates of sediment contributed by the Yukon River alone vary between 88 and 100 million metric tons per year. Besides the obvious sources, sediments on the shelf are also contributed by various other sources (biogenic and volcanic) and their contributions locally could be quite significant. For example, the distribution of diatoms in sediments from the south-central Bering Shelf has been described by Oshite and Sharma (1974).

In the Bristol-Kuskokwim Bay region, the Al_2O_3/SiO_2 versus percent SiO_2 plot shows a steady linear decrease in silica content (Fig. 10-70). The sediment samples with high silica content in the lower right-hand corner are from the Kuskokwim Bay region; the sediments with Al_2O_3/SiO_2 between 0.15 and 0.19 are from the head and center of Bristol Bay; and those with values <0.19 are from the near-shore region of the Alaska Peninsula and Togiak Bay. The Al_2O_3/SiO_2 (maturity index) and its variations suggest the Kuskokwim Bay as a major source of sediment in this region. The near-shore sediments from the Alaska Peninsula show a high Al_2O_3/SiO_2 ratio and therefore do not suggest a source in the adjacent peninsula. The heavy mineral distribution in the bay, discussed earlier, suggests that most sediments deposited offshore along the Alaska Peninsula have their origin in the adjacent landmass. The high ratio in these sediments is the result of mineral separation caused by the permanent and the tidal currents, which carry lighter mineral fractions offshore as well as toward the head of the bay. A lag deposit rich in heavy minerals is formed by removal of the light mineral form. The high concentration of heavy mineral, in turn, increases the Al_2O_3/SiO_2 ratio in these sediments. Contrary to the normal increase with distance from source, the high Al_2O_3/SiO_2 in sediments from this region reflects the source and the progressive decrease indicates sediment migration. Therefore, the Al_2O_3/SiO_2 distributions suggest that peninsular sediments are carried northeast toward the bay by permanent water currents and offshore by tidal currents. Similar sediment source and transport mechanisms are indicated by the suspended load distribution and ERTS-1 imagery analysis. The bottom sediment textural distribution also lends support to this scheme.

The triangular diagram with Al_2O_3, $SiO_2 + TiO_2$, and Rest (sum of other elements) as end members also shows a similar regional distribution pattern (Fig. 10-71). The offshore transport of sediments generally results in particle size and elemental differentiation and, therefore, in an increase of aluminum and other elements in sediments. The Kuskokwim Bay sediments are high in silica and low in aluminum and other elements. The sediment locations and their elemental distribution are similar to that discussed for Al_2O_3 and SiO_2 and suggest that sources for sediments in this region are the Kuskokwim Bay and the Alaska Peninsula.

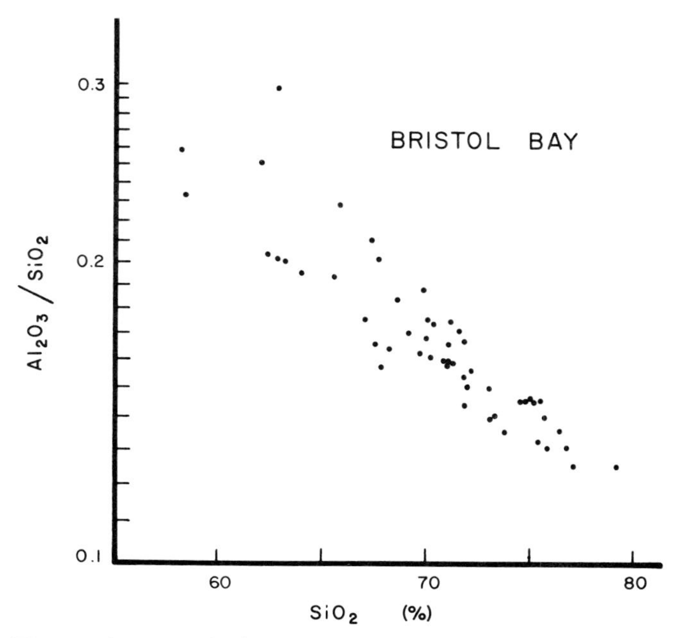

Figure 10-70. Al_2O_3/SiO_2 versus percent SiO_2 plot, Bristol Bay.

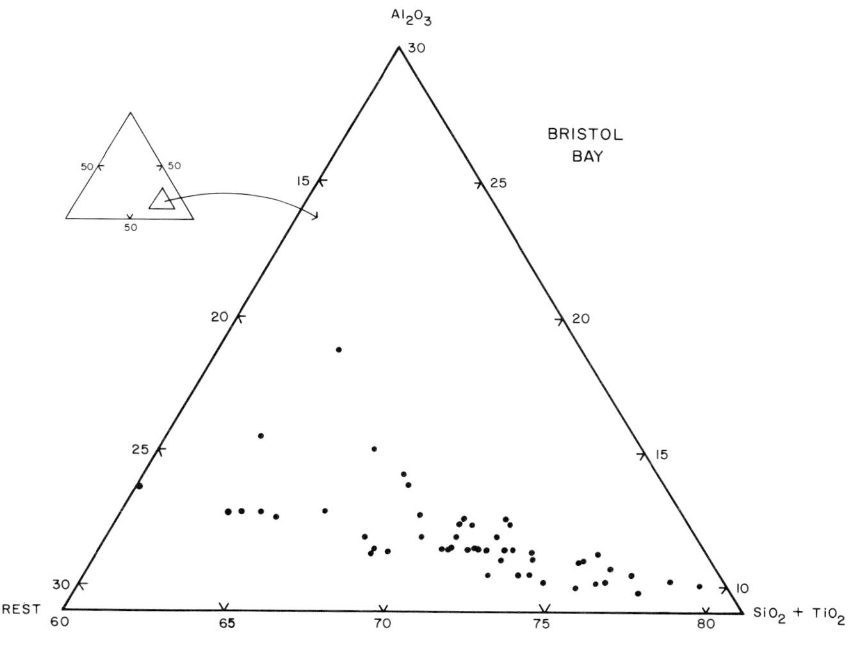

Figure 10-71. Triangular plot with percents Al_2O_3, SiO_2 + TiO_2, and Rest as end members, Bristol Bay.

The Al_2O_3 and SiO_2 relationships of sediments from the open Bering Shelf, exclusive of the semi-enclosed Bristol Bay and Norton Sound, show a linear distribution (Fig. 10-72). The sediments with higher SiO_2 content representing their source are mostly from the southern shores of the Seward Peninsula and a broad region extending from northern St. Lawrence Island, Shpanberg Strait, and Nunivak Island. Offshore, the sediments progressively decrease in their silica content. The triangular diagram with elements as end members also display similar regional distribution (Fig. 10-73).

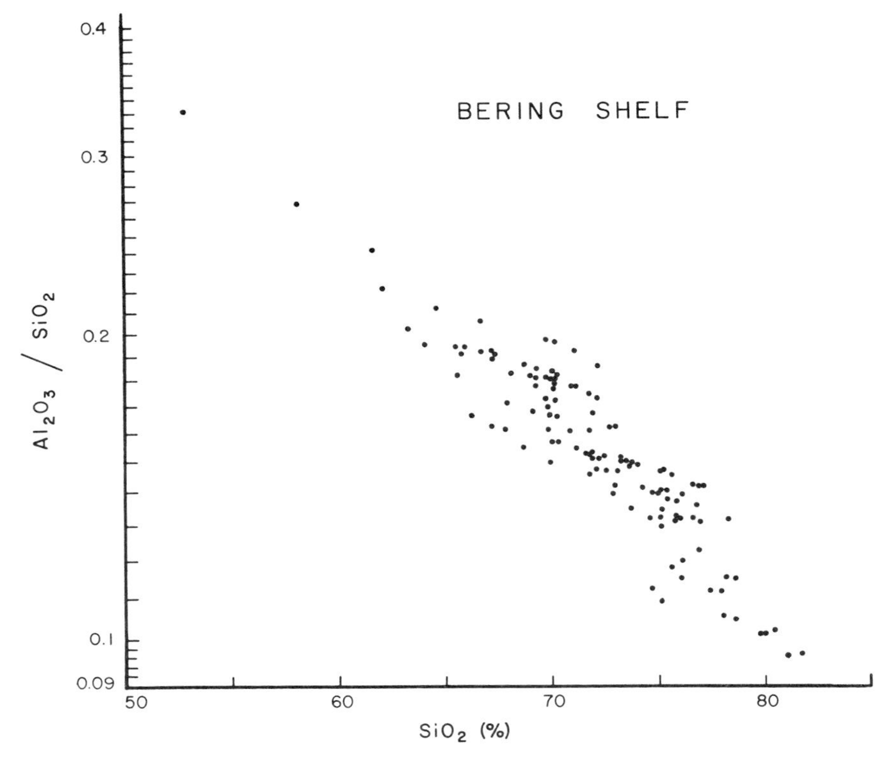

Figure 10-72. Al_2O_3/SiO_2 versus percent SiO_2 plot, Bering Shelf.

The coastal regions of the Seward Peninsula and St. Lawrence Island are covered with relict glacial deposits. Glacial erosion and transport do not cause significant alterations in chemical composition, so the elemental distribution in glaciated sediments is similar to that of the source rock. The high silica content in sediments south of the Seward Peninsula and north of St. Lawrence Island is attributed to glacial gravelly deposits.

The north-south oriented broad region between Nunivak and St. Lawrence islands, adjacent to the Alaska mainland, is also covered with higher silica content sediments. The sediments in this region are continually winnowed by the

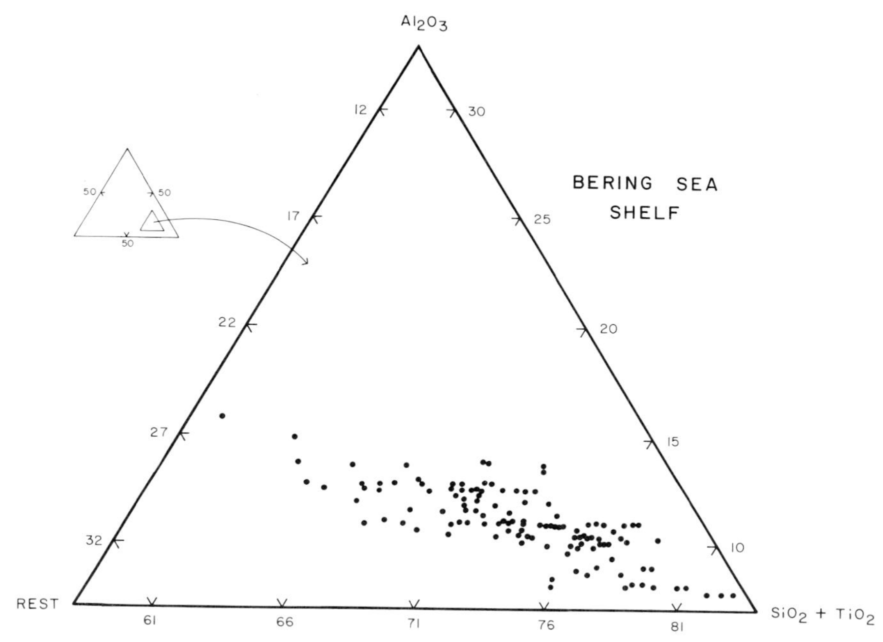

Figure 10-73. Triangular plot with percents Al_2O_3, SiO_2 + TiO_2, and Rest as end members, Bering Shelf.

northward moving permanent currents. The current inhibits deposition of clay and silt. The sand, which covers the area, is also subjected to sorting by the current. Sorting in sand leads to removal of the less resistant feldspathic minerals and a relative increase in quartz grains. The removal of clay and feldspathic sand results in increased silica in sediments from this region. These sediments apparently have their source in the Yukon River to the east, as suggested by the elemental variations in sediments observed on the triangular diagram (Fig. 10-73). Westward in offshore waters, a decrease in silica is complemented by other major elements and aluminum.

In Norton Sound two major sources of sediments, the Yukon River and the Seward Peninsula, are suggested by the Al_2O_3/SiO_2 distribution. Sediment samples with higher silica contents, in the lower right-hand corner of the diagram (Fig. 10-74), are the near-shore deposits from the vicinity of Nome. Some of these sediments are of glacial origin and therefore have a relatively higher silica content. The sediments deposited near the Yukon River delta contain about 73% silica and their Al_2O_3/SiO_2 ratio is only 0.14. Eastward into Norton Sound, the silica in sediments decreases progressively. This decrease implies movement of Yukon River sediments into Norton Sound. The elemental relationship on triangular coordinates also suggests that Yukon River sediments dominate the southern and eastern Norton Sound, while the Seward Peninsula serves as a major source of sediments in the north (Fig. 10-75); Venkatarathnam's (1969) heavy mineral studies point to the same conclusion.

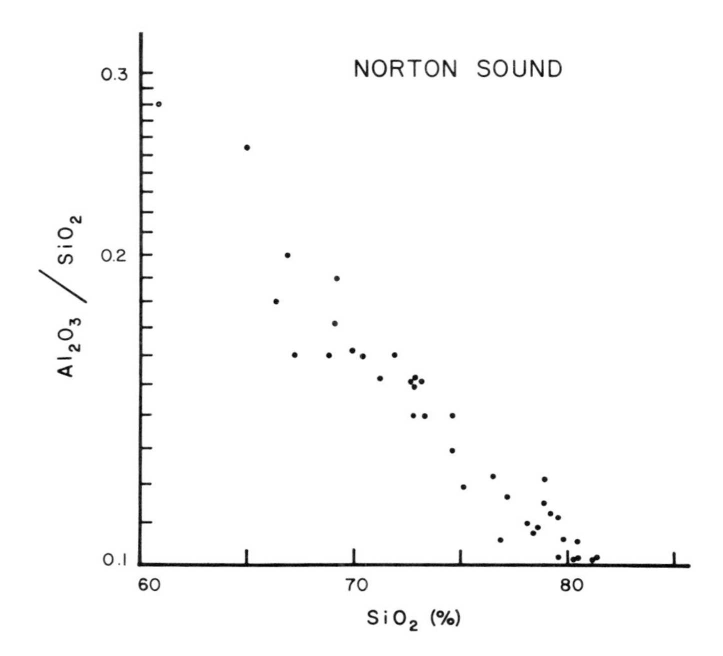

Figure 10-74. Al_2O_3/SiO_2 versus percent SiO_2 plot, Norton Sound.

The sources of sediments and their dispersal on the Bering Shelf are further delineated by elemental ratio plots. The Al_2O_3/TiO_2 ratio is generally low in sediments deposited near the source and increases with increasing distance of sediment transport. Sand deposited in southern Bristol Bay apparently originates in the Alaska Peninsula. The distribution of heavy minerals and their source in Bristol Bay have been studied by Sharma *et al.* (1972). They observed the presence of at least small amounts of reddish-brown pleochroic hornblende in most samples, particularly in samples from north of Unimak Island. This hornblende is characteristic of basalts or hornblende andesite, which are known to outcrop extensively in the Alaska Peninsula. Therefore, the composition and distribution of heavy minerals deposited in the bay suggest their origin in the Alaska Peninsula. The low Al_2O_3/TiO_2 ratio in near-shore sediments and its increase farther offshore indicate that the peninsula is the source of bay sediments (Fig. 10-76). The sediment contribution by the Kvichak River, near the head of the bay, is reflected by the 14 and 16 isopleths. The westward moving sediments, which probably originate in Nushagak and Kvichak bays to the east, are partly diverted into Togiak Bay and are deposited therein. Because the indented bay is protected from storm and tidal currents, these sediments are not flushed out; Togiak Bay apparently does not receive much detritus from land.

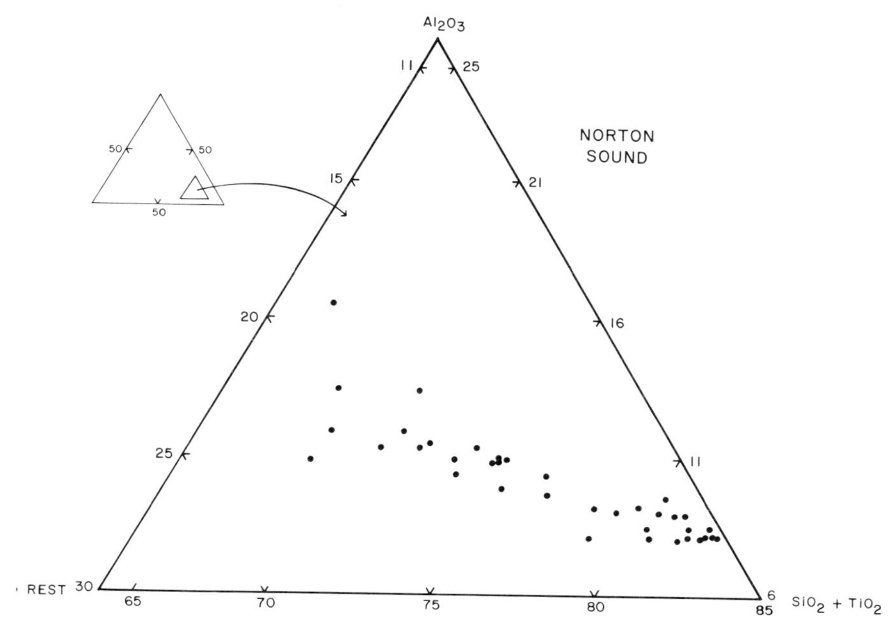

Figure 10-75. Triangular plot with percents Al_2O_3, $SiO_2 + TiO_2$, and Rest as end members, Norton Sound.

The decreasing Al_2O_3/TiO_2 isopleths toward the head of Kuskokwim Bay indicate that Kuskokwim River sediments move out of and along the northwestern shores of the bay. Such a movement of sediment had been observed earlier in the ERTS-1 imagery (Fig. 10-42). The ERTS-1 imagery also displayed a northward intrusion of the westward moving Kvichak-Nushagak surface plume into Kuskokwim Bay along the southeastern shore. The distribution of the Al_2O_3/TiO_2 ratio confirms this northward movement of sediments into the bay, although the Kuskokwim River sediments, it appears, are carried southwest into Bristol Bay and northwest into Etolin Strait. Lisitsyn (1966) postulated that the Kuskokwim River sediments are carried offshore north of St. Matthew Island. The low Al_2O_3/TiO_2 isopleth west of Nunivak Island, however, does not indicate westward migration of these sediments. The southwest and northwest movement of sediments emerging from the bay, on the other hand, is well supported by the suspended load distribution described earlier.

The significance of the Yukon River as a major source for sediment is quite apparent. River sediments are deposited along the Alaska mainland and extend from Cape Romanzof in the south to the Norton Sound in the north and east. Diffusion of sediments into offshore waters to the west and to the northwest

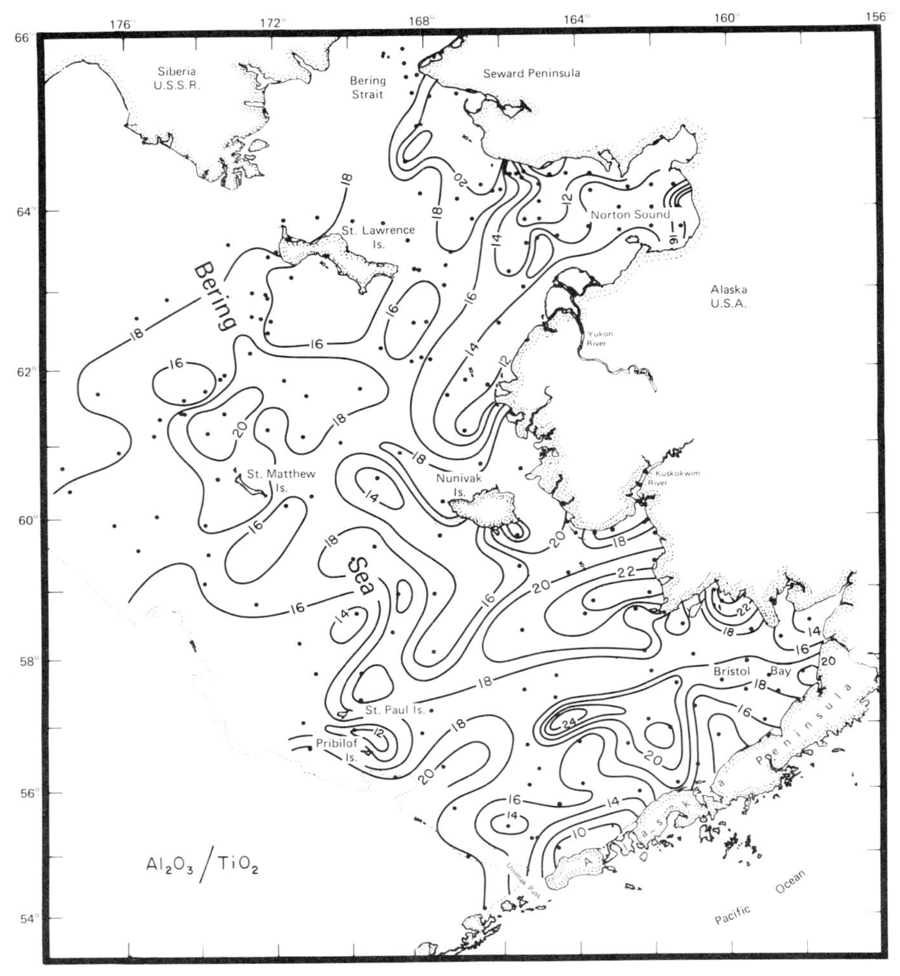

Figure 10-76. Al_2O_3/TiO_2 variations in sediments, Bering Shelf.

toward Nome also occurs. It is interesting to note that the Yukon River sediment dispersal interpreted from Al_2O_3/TiO_2 ratios is almost identical to that observed in ERTS-1 imagery (Fig. 10-48). The dispersal pattern also relates well with bottom sediment texture and suspended load distributions.

The southern shore of the Seward Peninsula provides sediments deposited to the south, but the presence of a high anomaly along the southwestern shore is somewhat puzzling. The high values of Al_2O_3/TiO_2 in this region may result from aggregation of feldspathic sand and relatively lesser amounts of sands with titanium.

St. Lawrence, St. Matthew, Nunivak, and the Pribilof islands provide sediments to the shallow waters that surround them. High Al_2O_3/TiO_2 values in sediment deposited in the region northwest of St. Lawrence Island are probably caused by the provenance. The island has basaltic and granitic plutons which have contributed sediments to the north.

Variations of other ratios, Al_2O_3/Na_2O, K_2O/Na_2O, and CaO/MgO, in sediments are also plotted to delineate the sources and the migratory paths of the sediments (Figs. 10-77 to 10-79). The ratios are higher near the source and decrease with increasing distance of transport from the source. The Al_2O_3/Na_2O ratio distribution in sediments of the Bering Shelf further delineates the sources described earlier. The near-shore sediments deposited along the Alaska Peninsula have their origin in the adjacent land mass to the south. These sediments in part are carried perpendicular to the shore, and partly northeast along the shore. The 4.0 isopleth which extends west and south from the Kuskokwim Bay suggests the movement and distribution of the river sediments.

The Al_2O_3/Na_2O ratios on the central and northern shelf further confirm the Yukon River as a major sediment source to this region. The submerged ridge in Shpanberg Strait (midway between St. Lawrence Island and the Alaska mainland), northwestern St. Lawrence Island, and the southwestern Seward Peninsula also contribute sediments to the central and northern shelves.

A high anomaly midway between St. Matthew and St. Lawrence islands is difficult to explain and the source of sediments for this region cannot be easily traced. The K_2O/Na_2O and CaO/MgO plots (Figs. 10-78 and 10-79), however, tend to relate the sediment of this region to the north, on the southern shores of St. Lawrence Island. The sediment composition in the offshore region west of St. Lawrence and St. Matthew islands appears to be influenced by a source to the west, probably the Gulf of Anadyr.

Prominent sources for the sediments discussed above can also be inferred from the K_2O/Na_2O and CaO/MgO variations (Figs. 10-78 and 10-79). The K_2O/Na_2O should normally decrease with increasing sediment transport distance from the source. In Bristol Bay along the Alaska Peninsula and in eastern Norton Sound, however, the ratio increases. The reverse trend along the Alaska Peninsula is primarily a result of the relatively high concentration of heavy minerals. In the eastern Norton Sound, the high anomalies are caused by increased clay content in sediment. The K_2O/Na_2O distribution also tends to define better the source of sediment as the St. Lawrence Island bank.

The CaO/MgO distribution, besides defining the major sources of sediments, also suggests a general decrease of CaO/MgO in sediments with increasing latitude. The eastward movement of Yukon sediment into Norton Sound is well illustrated in this diagram. An abnormally high value of CaO/MgO in marine sediments from northern St. Lawrence Island suggests that either these sediments are glacial relict deposits or their chemistry is signficantly affected by the provenance. Similar geochemical anomalies in this area were observed in other element ratios.

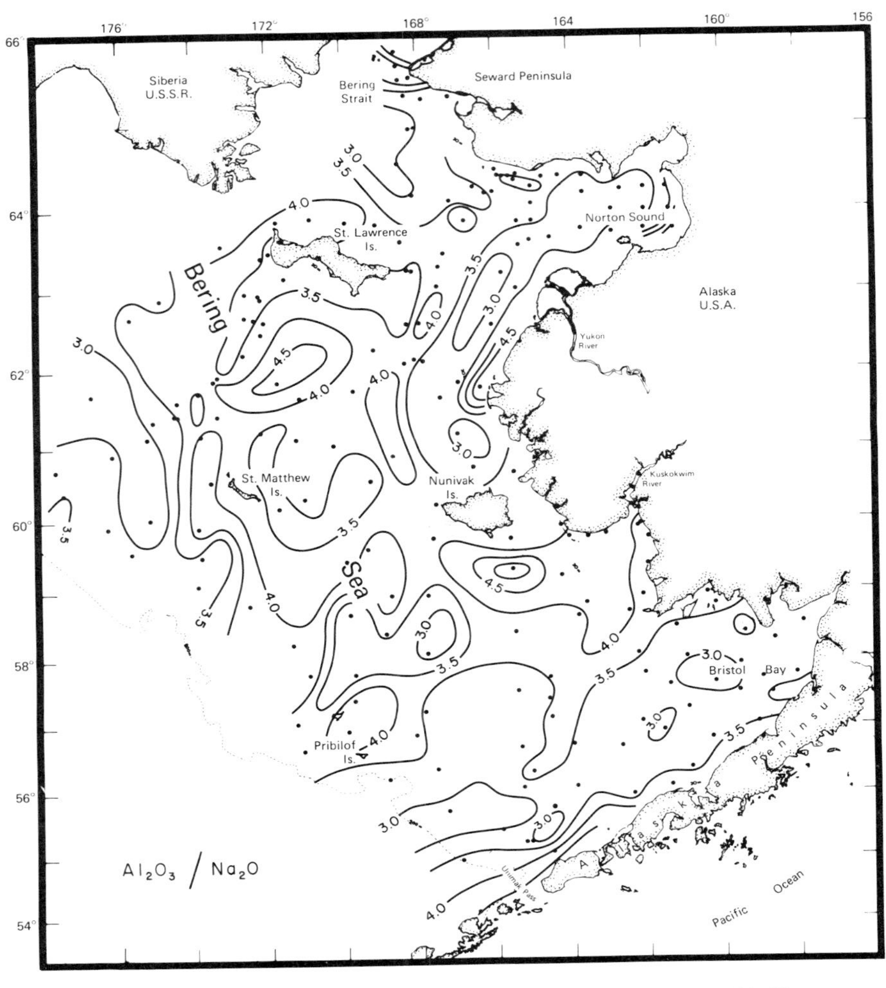

Figure 10-77. Al_2O_3/Na_2O variations in sediments, Bering Shelf.

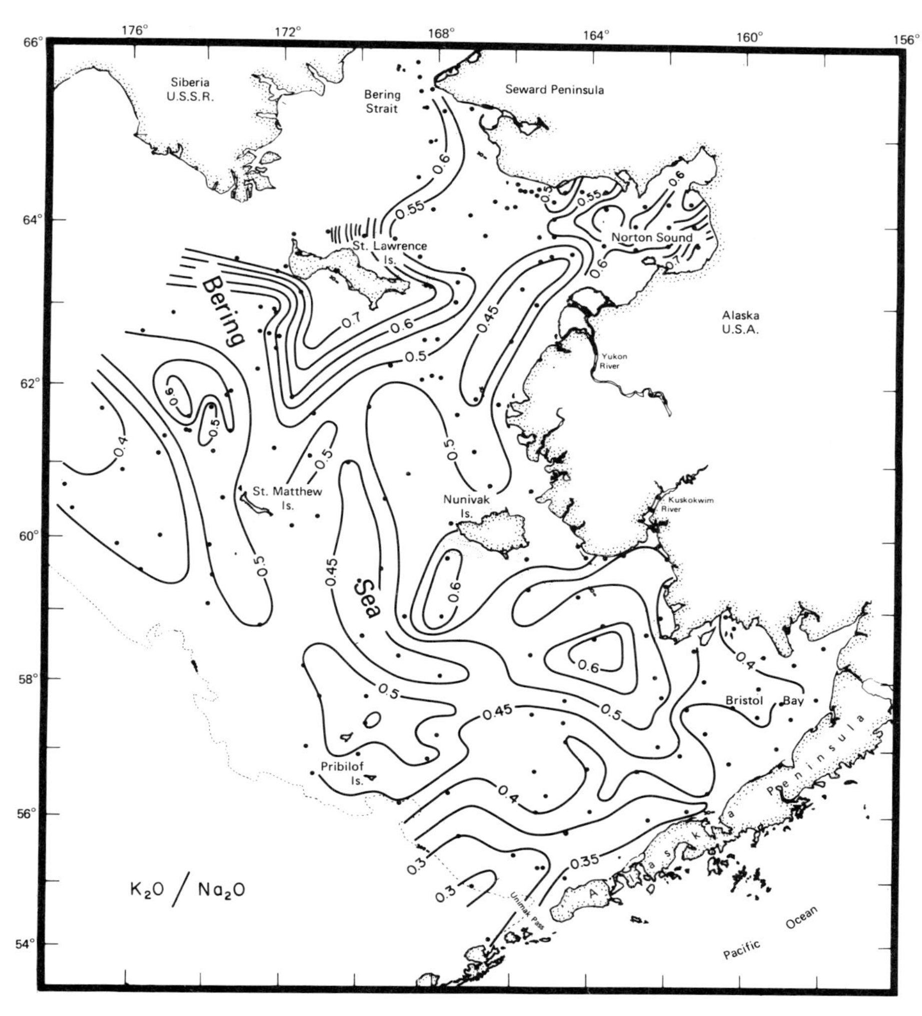

Figure 10-78. K_2O/Na_2O variations in sediments, Bering Shelf.

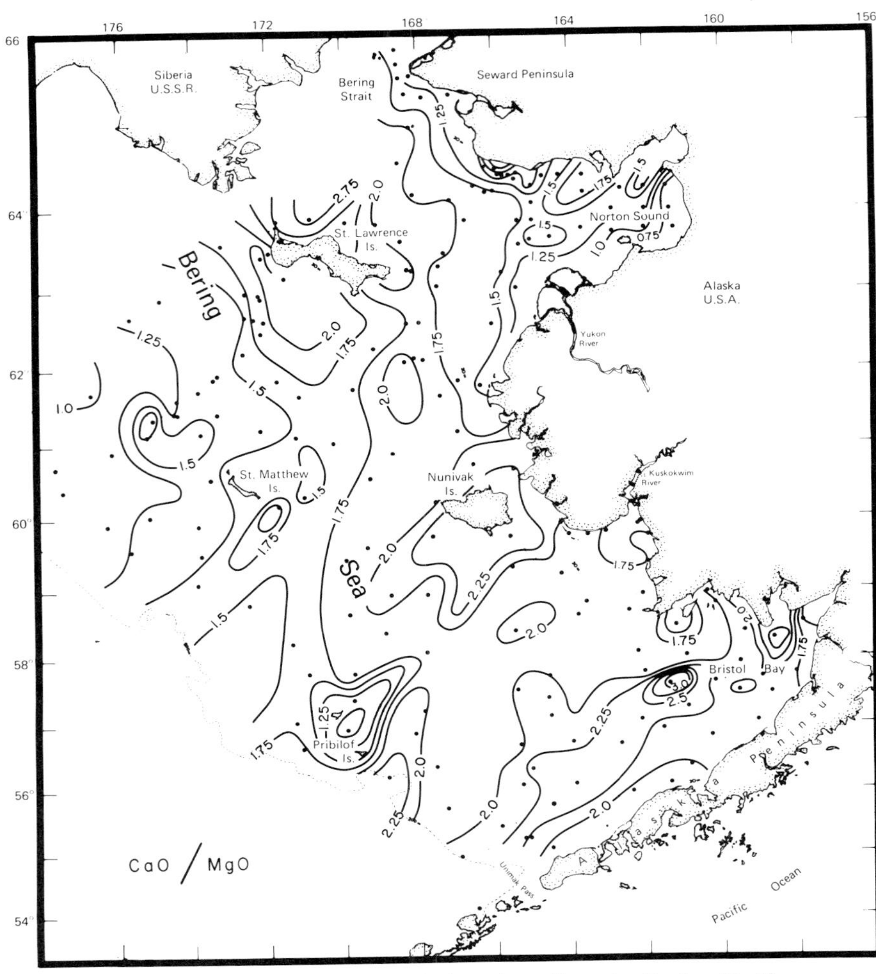

Figure 10-79. CaO/MgO variations in sediments, Bering Shelf.

REFERENCES

Arsen'ev, V. S. (1967). *The current and water masses of the Bering Sea.* Moscow, Izdatel'stvo "Nauka," p. 135.

Askren, D. R. (1972). Holocene Stratigraphic Framework–Southern Bering Sea Continental Shelf. Masters Thesis, Univ. of Washington, Seattle, p. 104.

Askren, D. R., J. S. Creager, and R. J. Echols. (In preparation). Modern sediments and benthic formanifera of the southeastern Bering Sea continental shelf.

Barnes, C. A., and T. G. Thompson. (1938). Physical and chemical investigation in Bering Sea and portions of the North Pacific Ocean. *Univ. Washington Publ. Oceanography* 3(2):35-79.

Berryhill, R. V. (1963). Reconnaissance of beach sands, Bristol Bay, Alaska. U.S. Dept. of the Interior, Bureau of Mines, Washington, D.C. Rept. of Investigations 6214, p. 48.

Churkin, M. (1972). Western boundary of the North American Continental Plate in Asia. *Geol. Soc. Amer. Bull.* 83:1027-1036.

Coonrad, W. L. (1957). Geologic reconnaissance in the Yukon-Kuskokwim delta region, Alaska. U.S. Geol. Survey, Washington, D.C. Survey Map.

Coulter, H. W. (1962). Extent of glaciations in Alaska. U.S. Geol. Survey Misc. Geol. Invest. Map I-415.

Creager, J. S., and D. A. McManus. (1967). Geology of the floor of Bering and Chukchi Seas—American studies. In: *The Bering Land Bridge.* D. M. Hopkins (Ed.), Stanford Univ. Press, Stanford, Calif., pp. 7-31.

Dobrovol'skii, A. D., and V. S. Arsen'ev. (1959). The Bering Sea currents. *Problemy Severa* 3:3-9.

Dodimead, A. J., F. Favorite, and T. Hirano. (1963). Salmon of the North Pacific Ocean. Part II: Review of oceanography of the subarctic Pacific region. *Intl. North Pacific Fisheries Comm. Bull.* No. 13, 195 p.

Environmental Data Service. (1968). *Climatography of the United States.* Nos. 60-49. U.S. Dept. of Commerce, Washington, D.C.

Favorite, F. (1974). Flow into the Bering Sea through Aleutian island passes. In: *Oceanography of the Bering Sea.* D. W. Hood and E. J. Kelley (Eds.), Occ. Publ. No. 2, Inst. of Mar. Sci., Univ. of Alaska, Fairbanks, pp. 3-37.

Favorite, F., and G. Pederson. (1959). Bristol Bay Oceanography August-September, 1938. U.S. Fish and Wildlife Service, Washington, D.C., Special Scientific Rept.—Fisheries, No. 311.

Favorite, F., J. W. Schantz, and C. R. Hebard. (1961). Oceanographic Observations in Bristol Bay and the Bering Sea, 1939-41 (USCGT Redwing). U.S. Fish and Wildlife Service, Washington, D.C., Special Scientific Rept.—Fisheries, No. 381, 323 p.

Fernald, A. T. (1960). Geomorphology of the Upper Kuskokwim River, Alaska. U.S. Geol. Survey Prof. Paper 1071-G, pp. 191-275.

Fujii, T. (1973). Hydrographical conditions of the Bering Strait in the summer of 1972. *Hokkaido Univ., Fac. Fish. Bull.* 24(1):42-48.

Gates, G. D., and G. Gryc. (1963). Structure and tectonic history of Alaska. In: *Backbone of the Americas,* O. E. Childs and B. W. Beebe (Eds.), Amer. Assoc. Petrol. Geol. Mem. 2, Tulsa, Oklahoma, pp. 264-277.

Gershanovich, D. E. (1963). Bottom relief of the main fishing grounds (Shelf and Continental Slope) and some aspects of the geomorphology of the Bering Sea. In: *Soviet Fisheries Investigations in the Northeast Pacific. Part I.* All-Union Scientific Research Institute of Marine Fisheries and Oceanography (VNIRO) Trudy v. 48. P. A. Moiseev (Ed.), Transl. by Israel Program for Scientific Translations, Jerusalem, 1968, pp. 9-78.

Goodman, J. R. J. H. Lincoln, T. G. Thompson, and F. A. Zeusler. (1942). Physical and chemical investigations: Bering Sea, Bering Strait, Chukchi Sea, during the summers of 1937 and 1938. *Univ. of Washington Publ. Oceanography* 3(4):105-169.

Grantz, A., S. L. Wolf, L. Breslav, T. C. Johnson, and W. F. Hanna. (1970a). Chukchi Sea, seismic reflection and magnetic profiles-1969, between northern Alaska and International Date Line. U.S. Geol. Survey, Open-File Rept., p. 26.

Grantz, A., S. L. Wolf, L. Breslav, T. C. Johnson, and W. F. Hanna. (1970b). Reconnaissance geology of the Chukchi Sea as determined by acoustic and magnetic profiling. In: *Proceedings of Geologic Seminar on North Slope of Alaska.* W. L. Adkinson and W. W. Brosge (Eds.), Amer. Assoc. Petrol. Geol., Pac. Section, Los Angeles, Calif., pp. F1-F28.

Grim, M. S., and D. A. McManus. (1970). A shallow seismic-profiling survey of the northern Bering Sea. *Marine Geol.* 8:293-320.

Hatten, C. W. (1971). Petroleum potential of Bristol Bay Basin, Alaska. In: Future petroleum provinces of the United States. *Amer. Assoc. Petrol. Geol. Mem.* 15(1):105-108.

Hebard, J. F. (1961). Currents in Southeastern Bering Sea. *Intl. North Pacific Fisheries Comm. Bull.* 5:9-16.

Hopkins, D. M., and D. W. Scholl. (1970). Tectonic development of Beringia, late Mesozoic to Holocene. *Amer. Assoc. Petrol. Geol. Bull.* 54(12):2486 (abstract).

Inman, D. L., and C. E. Nordstrom. (1971). On the tectonic and morphological classification of coasts. *J. Geol.* 79:1-21.

Karlstrom, T. N. V. (1957). Tentative correlation of Alaskan glacial sequences, 1956. *Science* 125(3237):73-74.

Karlstrom, T. N. V. (1964). Quaternary Geology of the Kenai Lowlands and Glacial History of the Cook Inlet Region, Alaska. U.S.G.S. Prof. Paper 443, U.S. Geol. Survey, U.S. Govt. Prtg. Off., Washington, D.C., 69 p.

King, P. B., Jr. (1969). Tectonic Map of North America. U.S. Geol. Survey Map.

Kitano, K. (1970). A note on the thermal structure of the eastern Bering Sea. *J. Geophys. Res.* 75(6):1110-1115.

Knebel, H. J. (1972). Holocene sedimentary framework of the east-central Bering Sea Continental Shelf. Ph.D. Thesis, Univ. of Washington, Seattle, 187 p.

Knebel, H. J., and J. S. Creager. (1973). Yukon River: Evidence for extensive migration during the Holocene Transgression. *Science* 79:1230-1231.

Knebel, H. J., and J. S. Creager. (1974). Heavy minerals on the east-central Bering Sea Continental Shelf. *J. Sed. Petrology* 44:553-561.

Konishi, R., and M. Saito. (1974). The relationship between ice and weather conditions in the eastern Bering Sea. In: *Oceanography of the Bering Sea.* D. W. Hood and E. J. Kelley (Eds.), Occ. Publ. No. 2, Inst. of Mar. Sci., Univ. of Alaska, Fairbanks, pp. 425-450.

Kummer, J. T., and J. S. Creager. (1971). Marine geology and Cenozoic history of the Gulf of Anadyr. *Marine Geol.* 10:257-280.

LaFond, E. C., and D. W. Pritchard. (1952). Physical and oceanographic investigations in the eastern Bering and Chukchi seas during the summer of 1947. *J. Mar. Res.* 11:69-86.

Lisitsyn, A. P. (1966). *Recent Sedimentation in the Bering Sea.* Transl. by Israel Program for Scientific Translation, Jerusalem. U.S. Dept. of Commerce, Natl Sci. Foundation, Washington, D.C., 614 p.

McManus, D. A. (In preparation). Distribution of bottom sediments on the continental shelf, northern Bering Sea.

McManus, D. A., and C. S. Smyth. (1970). Turbid bottom water on the continental shelf of the Northern Bering Sea. *J. Sed. Petrology* 40:869-873.

McManus, D. A., J. C. Kelley, and J. S. Creager. (1969). Continental shelf sedimentation in an Arctic environment. *Geol. Soc. Amer. Bull.* 80:1961-1984.

McManus, D. A., K. Venkatarathnam, D. M. Hopkins, and C. H. Nelson. (1974). Yukon River sediment on the northernmost Bering Sea Shelf. *J. Sed. Petrology* 44(4):1052-1060.

Menard, H. W. (1967). Transitional types of crust under small ocean basins. *J. Geophys. Res.* 72(12):3061-3073.

Moore, D. G. (1964). Acoustic-reflection reconnaissance of continental shelves: eastern Bering and Chukchi seas. In: *Papers in marine geology: Shepard commemorative volume.* R. L. Miller (Ed.), The Macmillan Co., New York, pp. 319-362.

Muller, E. H. (1952). Glacial history of the Makuek district, Alaska peninsula, Alaska. *Geol. Soc. Amer. Bull.* 63(12):1284.

Natarov, V. A., and N. P. Novikov. (1970). Oceanographic conditions in the southeastern Bering Sea and certain features of the distribution of halibut. In: *Soviet Fisheries Investigations in the Northeastern Pacific, Part V.* P. A. Moiseev (Ed.), pp. 292-303. Transl. from Russian by Israel Program for Scientific Translation, Jerusalem, 1972. U.S. Dept. of Commerce, National Tech. Int. Center, Springfield, Va.

Neiman, A. A. (1961). Certain regularities in the quantitative distribution of the benthos in the Bering Sea (in Russian). *Okeanologia* 1:294-304.

Nelson, C. H. (1970). Late Cenozoic history of deposition of northern Bering shelf. *Amer. Assoc. Petrol. Geol.* 54(12):2498 (abstract).

Nelson, C. H., and D. M. Hopkins. (1969). Sedimentary processes and distribution of particulate gold in northern Bering Sea. U.S. Geol. Survey Open-File Rept., 50 p.

Nelson, C. H., and D. M. Hopkins. (1972). Sedimentary processes and distribution of particulate gold in the northern Bering Sea. U.S. Geol. Survey Prof. Paper 689, 27 p.

Nelson, C. H., D. M. Hopkins, and D. W. Scholl. (1974). Tectonic setting and Cenozoic sedimentary history of the Bering Sea. In: *Oceanography of the Bering Sea.* D. W. Hood and E. J. Kelley (Eds.), Occ. Publ. No. 2, Inst. of Mar. Sci., Univ. of Alaska, Fairbanks, pp. 485-516.

Ohtani, K. (1969). On the oceanographic structure and the ice formation on the continental shelf in the eastern Bering Sea. *Hokkaido Univ. Fac. Fish. Bull.* 20:94-117.

Oshite, K., and G. D. Sharma. (1974). Distribution of recent diatoms on the eastern Bering Shelf. In: *Oceanography of the Bering Sea.* D. W. Hood and E. J. Kelley (Eds.), Occ. Publ. No. 2, Inst. of Marine Sci., Univ of Alaska, Fairbanks, pp. 541-552.

Péwé, T. L. (1953). Multiple glaciation in Alaska. U.S. Geol. Survey Circular 289, pp. 5-19.

Porter, S. C. (1967). Glaciation of Chagvan Bay area, Southwestern Alaska. *Arctic* 20(4):227-246.

Ratmanof, G. E. (1937). Explorations of the seas of Russia. Publications of the Hydrological Institute, Leningrad, n. 25, p. 175.

Roden, G. I. (1967). On river discharge into the northeastern Pacific Ocean and the Bering Sea. *J. Geophys. Res.* 72(22):5613-5629.

Saur, J. F. T., R. M. Lesser, A. J. Carsola, and W. M. Cameron. (1952). Oceanographic cruise to the Bering and Chukchi Seas, summer 1949. Part 3: Physical observations and sound velocity in the deep Bering Sea. U.S. Navy Electronics Lab. Rept. 298, San Diego, Calif., 38 p.

Saur, J. F. T., J. P. Tully, and E. C. LaFond. (1954). Oceanographic cruise to the Bering and Chukchi Seas, summer 1949. Part 4: Physical oceanographic studies; U.S. Navy Electronics Lab. Rept. 416, San Diego, Calif., 31 p.

Scholl, D. W., and E. C. Buffington. (1970). Structure evolution of Bering Continental Margin: Cretaceous to Eocene. *Amer. Assoc. Petrol. Geol. Bull.* 54(12): 2503 (abstract).

Scholl, D. W., and D. M. Hopkins. (1969). Newly discovered Cenozoic basins, Bering Shelf, Alaska. *Amer. Assoc. Petrol. Geol. Bull.* 53:2067-2078.

Scholl, D. W., and M. S. Marlow. (1970). Diapirlike structures in southeastern Bering Sea. *Amer. Assoc. Petrol. Geol. Bull.* 54(9):1644-1650.

Scholl, D. W., E. C. Buffington, and D. M. Hopkins. (1966). Exposure of basement rock on the continental slope of Bering Sea. *Science* 153:992-994.

Scholl, D. W., E. C. Buffington, and D. M. Hopkins. (1968). Geologic history of the continental margin of North America in the Bering Sea. *Marine Geol.* 6:297-330.

Scholl, D. W., E. C. Buffington, and M. S. Marlow. (1975). Plate tectonics and the structural evolution of Aleutian-Bering Sea region. In: *Contribution to the geology of the Bering Sea Basin and adjacent regions.* R. B. Forbes (Ed.), Geol. Soc. Amer. Special Paper 151, pp. 1-31.

Scholl. D. W., H. C. Green, and M. S. Marlow. (1970). The Eocene age of the Adak "Paleozoic?" locality, Aleutian Islands, Alaska. *Geol. Soc. Amer. Bull.* 81:3583-3592.

Sharma, G. D. (1972). Graded sedimentation on Bering Shelf. In: *Proc. 24th Intl. Geological Congress,* Section 8, Ottawa, Canada. Harpell's Press Cooperative, Gardenvale, Quebec, pp. 262-271.

Sharma, G. D. (1974a). Contemporary depositional environment of the eastern Bering Sea. In: *Oceanograpby of the Bering Sea.* D. W. Hood and E. J. Kelley (Eds.), Occ. Publ. No. 2, Inst. of Mar. Sci., Univ. of Alaska, Fairbanks, pp. 517-552.

Sharma, G. D. (1974b). Geological oceanography of the Bering Shelf. Chapter 5, in: *Marine Geology and Oceanography of the Arctic Seas.* Y. Herman (Ed.), Springer-Verlag, New York, pp. 141-156.

Sharma, G. D. (1974c). Geological oceanography near Nome. Chap. 7, in: *Environmental Study of the Marine Environment Near Nome, Alaska.* Inst. of Marine Sci., Univ. of Alaska, Fairbanks, Rept. R-74-3, pp. 111-142.

Sharma, G. D., J. D. Kreitner, and D. W. Hood. (1971). Sea ice characteristics in Bering Sea. *Proc. 1st Intl. Conf. on Port and Ocean Engineering under Arctic Conditions,* Tech. Univ. of Norway, Trondheim, Norway, Vol. I, pp. 211-220.

Sharma, G. D., A. S. Naidu, and D. W. Hood. (1972). Bristol Bay: A model contemporary graded shelf. *Amer. Assoc. Petrol. Geol. Bull.* 56(10):2000-2012.

Sharma, G. D., F. F. Wright, J. J. Burns, and D. C. Burbank. (1974). Sea surface circulation, sediment transport, and marine mammal distribution, Alaska Continental Shelf. ERTS Project 110-7, Final Rept. to NASA, Inst. of Mar. Sci., Univ. of Alaska, Fairbanks.

Shepard, F. P. (1963). *Submarine Geology.* Harper and Row Publishers, New York, 557 p.

Takenouti, A. Y., and K. Ohtani. (1974). Currents and water masses in the Bering Sea: A review of Japanese work. In: *Oceanography of the Bering Sea.* D. W. Hood and E. J. Kelley (Eds.), Occ. Publ. No. 2, Inst. of Mar. Sci., Univ. of Alaska, Fairbanks, pp. 39-57.

U.S. Coast and Geodetic Survey. (1964). United States Coast Pilot 9, Alaska Cape Spencer to Beaufort Sea. 7th ed. Washington, D.C., U.S. Govt. Prtg. Off., 348 p.

U.S. National Ocean Survey. (1973a). Tide Tables 1974, West Coast of North and South America. U.S. Dept. of Commerce, Washington, D.C.

U.S. National Ocean Survey. (1973b). Tidal Current Tables 1974, Pacific Coast of North America and Asia. U.S. Dept. of Commerce, Washington, D.C.

U.S. Navy Hydrographic Office. (1961). Climatological and Oceanographic Atlas for Mariners. Vol. II, North Pacific Ocean, Washington, D.C.

Venkatarathnam, K. (1969). Clastic Sediments of the Continental Shelf of the Northern Bering Sea. Univ. Washington, Seattle, Spec. Rept. No. 41, pp. 40-61.

Venkatarathnam, K. (1971). Sediments on the Continental Shelf of Northern Bering Sea. U.S. Geol. Survey Open-File Rept., 93 p.

Verba, M. L., G. I. Gaponenko, A. N. Orlov, V. I. Timofeev, and I. E. Chernenkov. (1971). Deep structure and oil and gas prospects of the northwestern part of Bering Sea (in Russian). Nanchuo-Jssled; Just. Geol. Arktiki (NIIGA), Geofiz. Metody Razvedki v. Arktike, No. 6, pp. 70-74. (English translation available from Natl. Transl. Center, Chicago, Ill.).

CHAPTER 11
Chukchi Sea Shelf

INTRODUCTION

The Alaskan Chukchi Sea Shelf lies between 65°40′and 73°00′N and 156°00′ and 171°00′W and covers an area of 580,000 km^2 (Fig. 11-1). To the south the Chukchi Shelf is separated from the Bering Sea by a narrow constriction, the Bering Strait, and in the north it is bordered by an abrupt escarpment, the Beaufort Scarp, that descends to the floor of the Arctic Ocean. The 200-m isobath is about 50 km offshore of Point Barrow. The Alaskan Chukchi Sea has an unusual shoreline–the entire coast consists of a narrow coastal plain with the exception of precipitous cliffs near Cape Lisburne and Cape Thompson. Because it is poorly drained small streams and lakes are numerous. From Cape Lisburne northward, the shelf abuts the low Arctic Coastal Plains and the shoreline features an almost continuous chain of barrier islands with lagoons.

GEOLOGY

Comparison of major structural trends in western Alaska, eastern Siberia, and Wrangel Island suggests that the Chukchi Sea is a part of a large continent that has recently submerged. The bedrock stratigraphy and structural lineation exposed on Wrangel Island, for example can be extended southwestward to Cape Lisburne, where similar stratigraphy and structure are observed. The recent test wells and the geophysical data suggest that the pre-Tertiary rocks exposed in the Brooks Range underlie the Tertiary Basins of the northern Chukchi Shelf. For this reason, it is believed that the geologic history of the Chukchi Sea is similar to the onshore region of western Alaska. Buffington *et al.* (1950), LaFond and Pritchard (1952), Carsola (1954), and Hopkins (1959) conducted the early investigations. In recent years the discovery of oil on the North Slope contributed subsurface data that provided additional information to reconstruct the geology of the shelf region (Brosgé and Tailleur, 1971; Brosgé and Dutro, 1973; Churkin, 1973). The layered rocks of the Brooks Range, North Slope, Northern Chukchi Shelf, and Beaufort Shelf can be grouped into five sequences that reflect major stages of tectonic evolution in Arctic Alaska (Alaska Geological Society, North Slope Stratigraphic Section, 1971 and 1973).

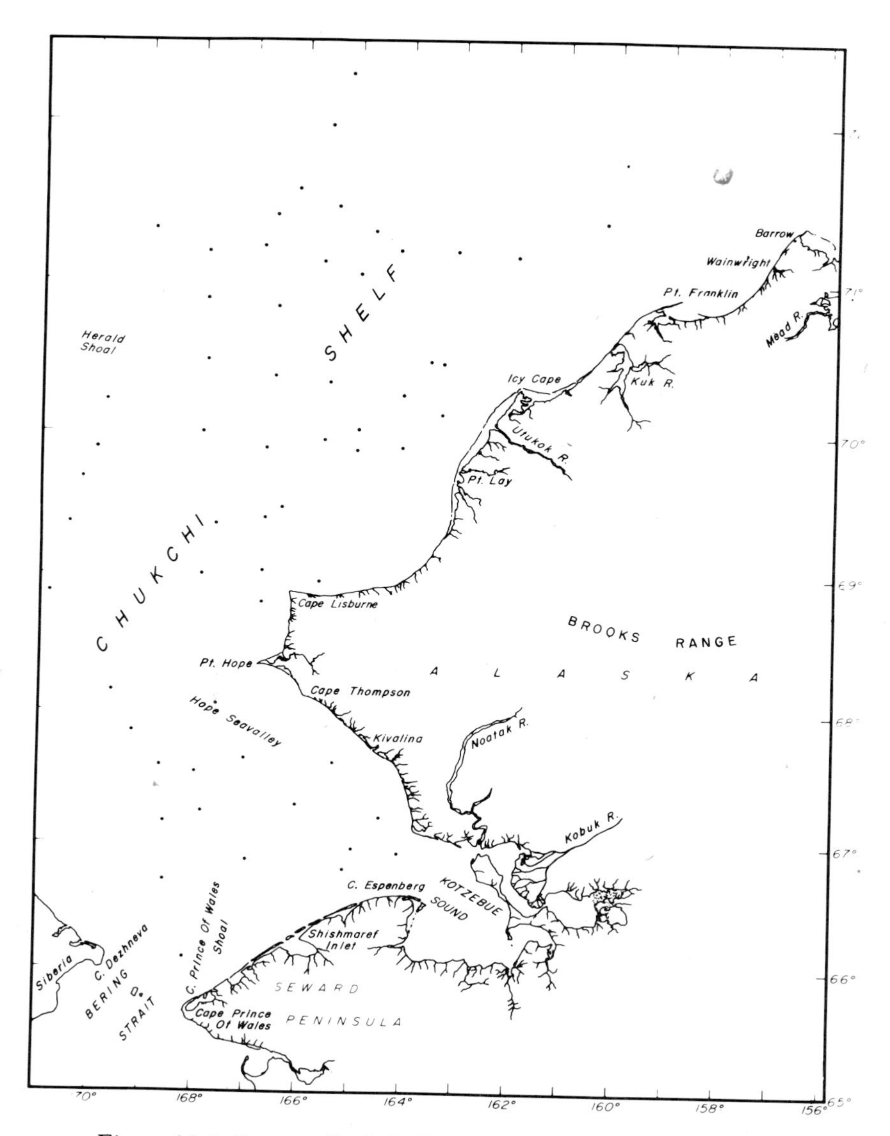

Figure 11-1. Eastern Chukchi Sea. Dots indicate station locations.

The earliest known Paleozoic feature of the Chukchi Sea is the east-west oriented Colville Geosyncline in the north. Between Cambrian and Devonian times, the geosyncline was filled with the first sequence of rocks, both eu- and miogeosynclinal facies, which are metamorphosed. Sometime during the Devonian the region was uplifted and a broad, stable Arctic platform extended along northern Alaska. Marine sedimentation continued from Late Devonian or

Early Mississippian through Early Cretaceous times and created the second sequence of rocks. Clastic and carbonate sediments derived from a northern source area were deposited on the platform. Near Prudhoe Bay these rocks contain giant oil and gas accumulations. At the close of the Jurassic or the beginning of Cretaceous time, an intense tectonic activity compressed the southern part of the platform and thrust Paleozoic to Cretaceous rocks northward onto the central part of the platform. The compression resulted in the uplifting of the Brooks Range-Herald Arch, which subsequently supplied sediment detritus for the third sequence. Cretaceous sedimentary rocks of the Colville Geosyncline underlie most of the northern Chukchi Shelf between Herald and Barrow arches.

The major orogenic episode occurred during the Cretaceous; however, continued uplift in the Brooks Range supplied detritus northward during the Tertiary. In the southern Chukchi Sea, the postorogenic crustal extension formed the Hope Basin. Deformed rocks of the Brooks Range can be followed from coastal outcrops in Alaska and the Chukotka Peninsula under the Hope Basin.

The geologic events in the Chukchi Sea during the Tertiary have been described by Hopkins (1959). He concluded that from the Middle Eocene until Middle Pliocene time the Chukchi Sea floor was above sea level. The area may have submerged briefly as a result of crustal warping during the Late Miocene, which resulted in deposition of the fourth sequence. During the Tertiary a stream-sculptured topography similar to that observed in western Alaska probably developed on the platform. At the beginning of the Pleistocene epoch crustal warping resulted in marine transgression and the final sequence of sediments was deposited on the Chukchi Shelf. The sediments deposited during the Pleistocene have covered the Tertiary topography and formed a monotonously flat platform. The only remnant of the Tertiary landscape in the Chukchi Sea are the Hope Valley, the Barrow Canyon and the Herald Shoal.

The Pleistocene epoch connotes a repeated ice cover during glacial stages and the waning of glaciers during interglacial stages. Each glacial cycle was followed by changes of sea level in the Chukchi Sea. The highest sea level attained during the interglacial stage is estimated at about 30 m higher than present sea level. During intense glacial intervals the sea level fell about 100 m below the present strandline.

On the Alaska mainland along the eastern Chukchi Sea the unconsolidated Quaternary sediments overlie Tertiary or Cretaceous sediments. The Quaternary sediments are mostly fine-grained sand and silt and occasionally clay and gravel, which are the product of frosting and aeolian deposition. The entire onshore region is underlain by permafrost, which acts as a cement for these sediments. However, annual freezing and thawing in the upper layer forms loose sediments, which are modified by alluvial, lacustrine, and aeolian reworking and tundra soil development. Along the shore, thermal, fluvial, and marine processes erode the tundra coast and rework the near-shore sediments to produce sandy and gravelly barrier islands and beaches. Clay and silts are dominant in the coastal lagoons.

BATHYMETRY

The shelf is monotonously flat and almost featureless. The average depth of this wide platform is 50 m and regional gradients range from two minutes to unmeasurably gentle slopes (Creager and McManus, 1966). The major topographic features of the Chukchi Shelf are Herald Shoal, Cape Prince of Wales Shoal, Hope Sea Valley, and Barrow Canyon (Fig. 11-2). Herald Shoal lies in the central Chukchi Shelf and is less than 14 m deep. It is part of a northwest-southeast stratigraphic and structural trend that extends from Cape Lisburne on the Alaskan mainland northwest to the Herald-Wrangel Islands (Hopkins, 1959).

Another conspicuous topographic feature, Cape Prince of Wales Shoal, extends from the eastern margin of the Bering Strait northward for about 120 km. The lobate feature, with 2×10^{10} cubic meters of sediments, has been described in detail by McManus and Creager (1963). The shoal is narrow and less than 10 m below sea level near the Bering Strait but broadens rapidly northward, attaining a width of approximately 50 km. The broad distal end of the shoal lies under 50 m of water.

The prominent depression, S-shaped, east-west oriented Hope Valley, lies south of Point Hope. The submarine valley originates in the vicinity of Kivalina and Cape Thompson and extends northwest and west. The deepest part on the southern shelf lies along the Hope Valley. The Barrow Canyon lies along the slope.

HYDROLOGY AND HYDROGRAPHY

Oceanographic studies in the Chukchi Sea have been conducted by Saur *et al.* (1954), Aagaard (1964), Creager and McManus (1966), Fleming and Heggarty (1966), Coachman and Aagaard (1966, 1974), and Ingham and Rutland (1972). Water characteristics of the Chukchi Sea are dominated by three factors: (1) winter ice cover, (2) influx of Bering Sea water, and (3) the coastal surface runoff. Most of the year the waters of Chukchi Sea are covered by winter ice and polar pack ice. Ice begins to form in early October and its southward growth proceeds rapidly. By late October or early November ice clogs the Bering Strait. Break-up occurs about mid-June in the southern Chukchi Sea and ice begins to recede northward. The coastal regions are covered by shorefast ice for about eight months. Generally, August and September are months with the least sea ice. The extent of open water along the Alaskan coast during summer varies seasonally and is dependent on the wind field and winter ice cover. Easterly and southerly winds keep the ice at some distance from the coast.

The formation of yearly ice and the extension of polar pack ice into the Chukchi Sea form water masses typical of arctic regions. Its ice cover keeps the water temperature of the near-surface layers close to the freezing point, because of its salinity, and extrudes salt from the ice to underlying waters. Water masses

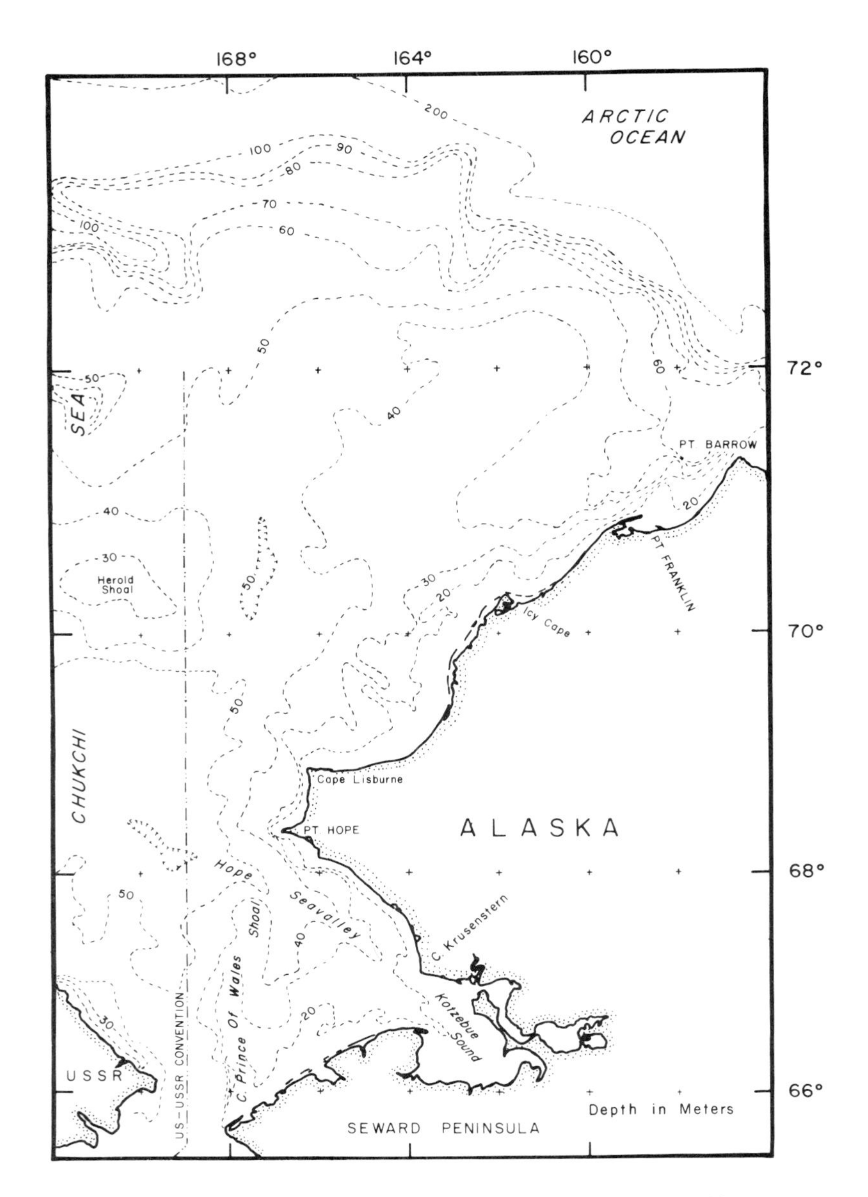

Figure 11-2. Bathymetric chart, eastern Chukchi Shelf.

formed as a result of the ice cover are continually modified by the inflow of Bering Sea water.

Coachman and Tripp (1970) reported a northward flow of water through the Bering Strait of the order of $1 \times 10^6\, m^3/sec$, flowing under the impetus of a surface slope. The Bering Sea water transport through the strait varies considerably and appears to be dependent on the wind regime. A variability of as much as a factor of 2 in transport may occur during one week (Coachman and Aagaard, 1974). These investigators also observed occasional net southward transport through the Bering Strait.

The amount of surface flow contributed by the adjacent Alaska mainland is low. The total river discharge to the Chukchi Sea is estimated at $2.5 \pm 1 \times 10^3 m^3/sec$. The average annual precipitations at Kotzebue, Cape Lisburne, and Barrow are 200, 180, and 125 mm, respectively. No measurements of evaporation in the Chukchi Sea have been reported. Estimates of evaporation over the Arctic Basin reported range between 400 mm/yr (Mosby, 1962) and 300 mm/yr (Fletcher, 1966).

Water masses formed in the Chukchi Sea vary seasonally and the variations are the result of the influence of the ice cover. Bering Sea water influx, and surface water runoff. Saur *et al.* (1954) described the water masses observed in the Chukchi Sea during the summer of 1949. They found water having a temperature $>6.6°C$ and a salinity $<30.5°/_{oo}$ along the coast of Alaska (Alaskan coastal water) and offshore an intermediate water having a temperature between 4 and 6.3°C and a salinity between 30.6 and $32.2°/_{oo}$. They also observed that the Alaskan coastal water extended vertically to the bottom. The temperature and salinity measurements by Aagaard (1964) during October 1962 also indicated the presence of two water masses of slightly different water properties than those observed by Saur *et al.* (1954) . Aagaard reported Alaskan coastal water having a temperature $>1°C$ and a salinity $<31°/_{oo}$, underlain by intermediate water having a temperature $>2°C$ and of salinity between 31.5 and $32.5°/_{oo}$. Fleming and Heggarty (1966), during August 1960, observed similar water masses with a distinct temperature-salinity boundary that extended from the surface to the sea floor. The warmer and less saline water, the Alaska coastal water, was found along the Alaska mainland between the Bering Strait and Icy Cape. They also observed a southwestward surface intrusion of warm and saline water (7°-10°C, $>32°/_{oo}$) in the near-shore region lying between Cape Lisburne and Icy Cape.

The temperature and salinity measurements from the Cape Lisburne-Icy Cape region obtained by Ingham and Rutland (1972) during September-October 1970 did not correspond closely with water mass properties defined by Saur *et al.* (1954) and Aagaard (1964). It appears that the lack of agreement between the recently observed values and previously defined water masses may have been the result of seasonal variations and Bering Sea water influx, which greatly influence the water properties of the coastal waters. Ingham and Rutland (1972) concluded that the distributions of temperature and salinity in the eastern Chukchi Sea are influenced by Alaskan coastal runoff, melting of sea ice, freezing of sea ice, and bottom water influx from the central Bering Strait.

The water circulation in the eastern Chukchi Sea is dominated by the Bering Sea water influx, which sets up an almost permanent northward current, and the local wind regime. Near-surface and bottom currents in the southeastern Chukchi Sea during August 1959 and 1960 have been described by Creager (1963). Water currents in the Bering Strait and north-northeast of Bering Strait during summers and winters have been obtained by Coachman and Tripp (1970) and Coachman and Aagaard (1974). The average current in the Bering Strait varied from 15 to 35 cm/sec (Creager, 1963), and these current measurements were fairly uniform throughout the water column. Data revealed a general northward flow from the Bering Strait which approximately paralleled the coast. Once past the Bering Strait, the current bifurcates and flows toward the north and northeast. The northeast flow setting proceeds along the northern coast of Seward Peninsula and, upon arriving near the mouth of Kotzebue Sound, is deflected toward Point Hope. Near Point Hope, the flow gains speed (50 cm/sec) and merges with the northward flowing component. After leaving Point Hope the combined current again bifurcates. A branch of the northward flow continues on the west side of Herald Shoal, while the main coastal branch flows northward and eastward along the Alaskan coast and enters the Arctic Ocean near Point Barrow. Coachman and Aagaard (1974) reported that during July 1972 the northward transport through the Cape Lisburne section was $1.3 \times 10^6 m^3/sec$ with approximately one-third moving northwestward toward Herald Shoal and two-thirds northeastward toward Point Barrow.

Tides are small in the Chukchi Sea and the tidal range along the eastern coast, on an average, is less than 30 cm. The tides are of the semi-diurnal type (Creager, 1963; Wiseman *et al.*, 1973).

Currents, particularly near the surface, in the northern Chukchi Sea are influenced more by regional winds than by northerly currents originating in the Bering Strait. The water movement in the near-shore region, especially during the open-water months, appears to be predominantly controlled by atmospheric conditions, primarily wind stress and solar heating. Wind-driven currents cause variations in sea level far in excess of those produced by tides (up to 3 m). These sea level changes strongly influence the water mass properties in the near-shore areas and undermine the beach by subjecting it to wave action. The combined effect of wind and wave then sets up the local current system.

The prevailing winds and resulting wave influences, to some extent, also control the coastal morphology in the northern Chukchi Sea, thus forming large cape systems (Point Hope, Cape Lisburne, Icy Cape, Pt. Franklin, and Pt. Barrow). Northerly winds generate waves that dominate the shoreline northeast of each cape, whereas waves approaching from the southwest are generated by westerly winds and these waves dominate the southwestern sections of the capes. The combined effect of winds and waves results in a convergence of near-shore transport systems, thereby causing accretion of land at the capes or points and divergence in the central part of the system, where erosion of the shoreline is common. The general configuration of the coast is also influenced by the structural lineaments in this region.

Deflections of currents by protruding landforms (capes and points) generally causes separation of the current and formation of eddies past the cape. Evidence of an eddy northeast of Cape Prince of Wales has been reported by McManus and Creager (1963). Formation of clockwise eddies in the regions of capes (Cape Lisburne, Icy Cape) have been observed by various investigators. Southward flowing coastal currents between Icy Cape and Cape Lisburne were recorded by Fleming and Heggerty (1966). Similar currents accompanying northerly winds at Point Lay were observed by Wiseman *et al.* (1973). Periodically the current system of these eddies is augmented by the prevailing wind.

SEDIMENTS

The eastern Chukchi Sea floor is mostly covered with gravel, sand, and silt. Gravel occurs as long narrow belts along the shore and as a few isolated patches in offshore regions. Gravel deposits also form beaches along the sea cliffs and adjacent areas. Sand predominates in the near-shore area and in the proximities of the major sediment input, while silts with clay are deposited offshore through settling. The clay content in offshore sediments is minor and varies between 5 and 35%.

Sediment distribution in the southeastern Chukchi Sea has been described in detail by Creager (1963). The results of sediment analysis of over 475 samples from the entire Chukchi Sea have been discussed by Creager and McManus (1966 and 1967) and McManus *et al.* (1969). These papers provide a sediment description based on a good station coverage. The sediment distribution presented here (Figs. 11-3 to 11-7) is essentially that published by these investigators.

Sediments in the Chukchi Sea grade offshore from sandy gravel to sandy clayey silt (Fig. 11-6). The Bering Strait, the southern extremity of the Chukchi Sea, is covered with sand and some patches of gravel. Northward and northeastward of the strait, the sea floor is covered mostly with moderately to poorly sorted sand. The sand forms a north-south oriented lobate feature; the Cape Prince of Wales Shoal (McManus and Creager, 1963). To the northeast, the sand extends to the mouth of Kotzebue Sound and continues northwest along the shore to Kivalina. A narrow belt of gravel covers the coast and the near-shore area between Kivalina and Cape Lisburne. Between Cape Lisburne and Pt. Barrow, the entire near-shore area consists of sand with gravelly offshore bars separating numerous lagoons from open waters.

Coarse sediments with mostly sand and gravel are also found on and around Herald Shoal in the central Chukchi Shelf. The northwest oriented shoal lies between Cape Lisburne and Wrangel Island and under water less than 40 m deep. Herald Shoal is bordered on the east and west by narrow channels mantled with clayey silt sand.

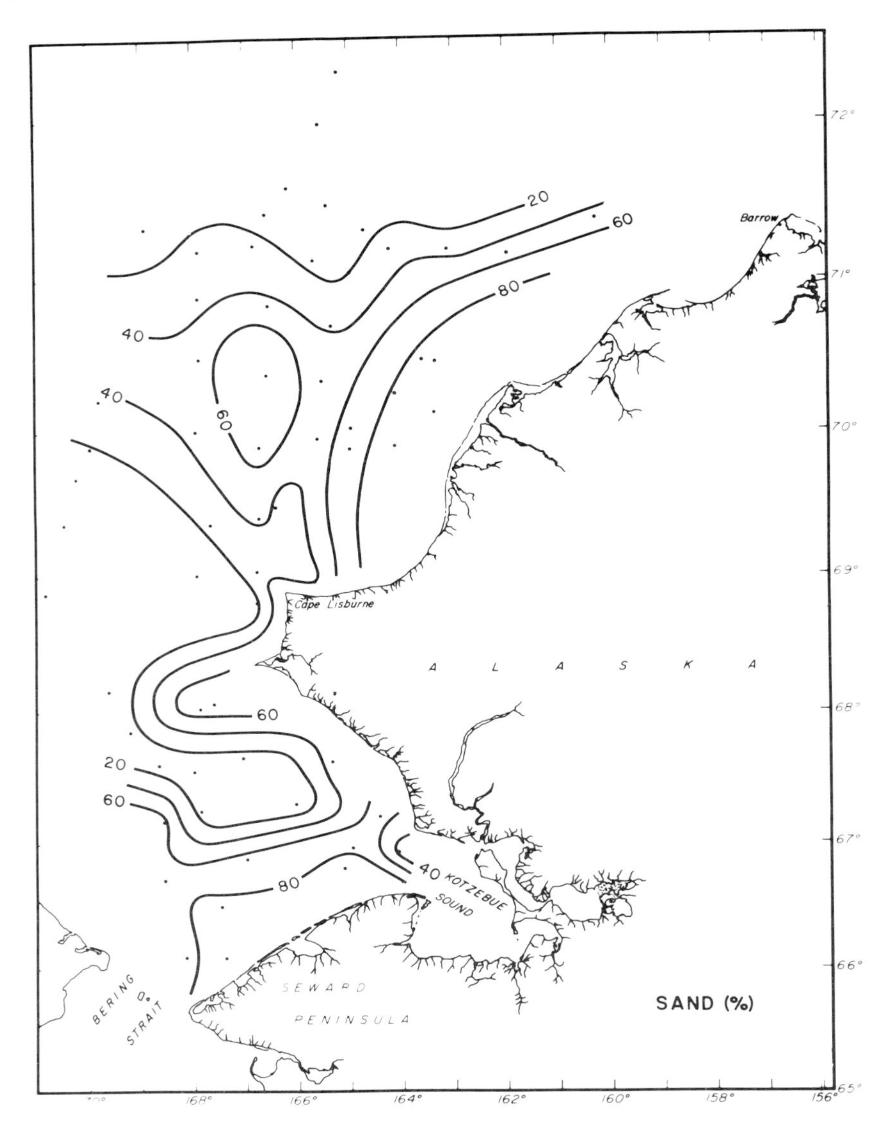

Figure 11-3. Weight percent in sand in sediments, Chukchi Shelf.

Seaward, with the exception of a few irregularities, the sediments generally become progressively finer, consisting of mostly clayey sand silt. The sediments with a dominant silt fraction cover large offshore areas was of Point Hope and northwest of Point Barrow. At first glance the plots showing distributions of sand, silt, and clay do not appear to relate the grain size to water depth (Figs. 11-3 to 11-6), but a careful inspection reveals that sandy, clayey silts are mostly

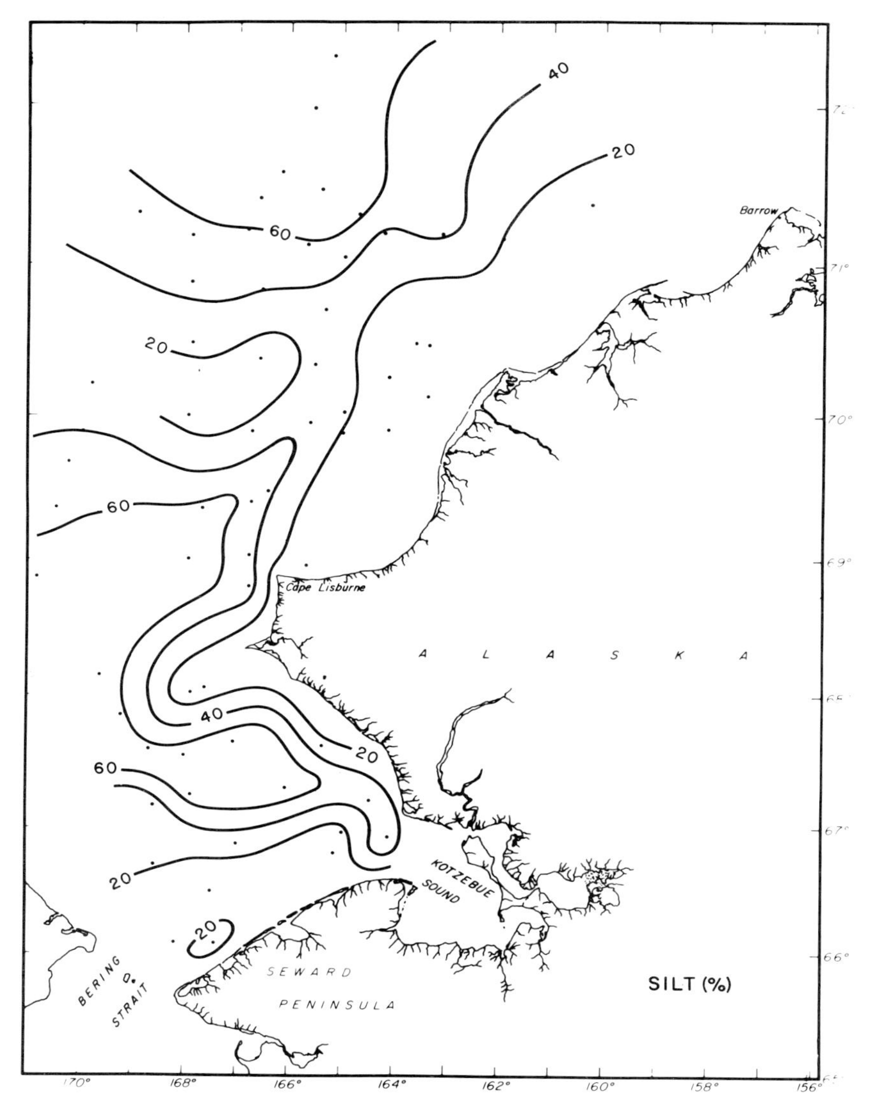

Figure 11-4. Weight percent silt in sediments, Chukchi Shelf.

deposited at water depths >50 m. In shallower regions silty sand is commonly deposited.

The sorting in sediments is related mostly to water energy. In areas of intense current and wave action the sands are moderately well sorted, while in relatively quieter environments sands are poorly sorted. Gravelly deposits near the shore and on Herald Shoal are poorly to very poorly sorted. Sandy, clayey silts, also

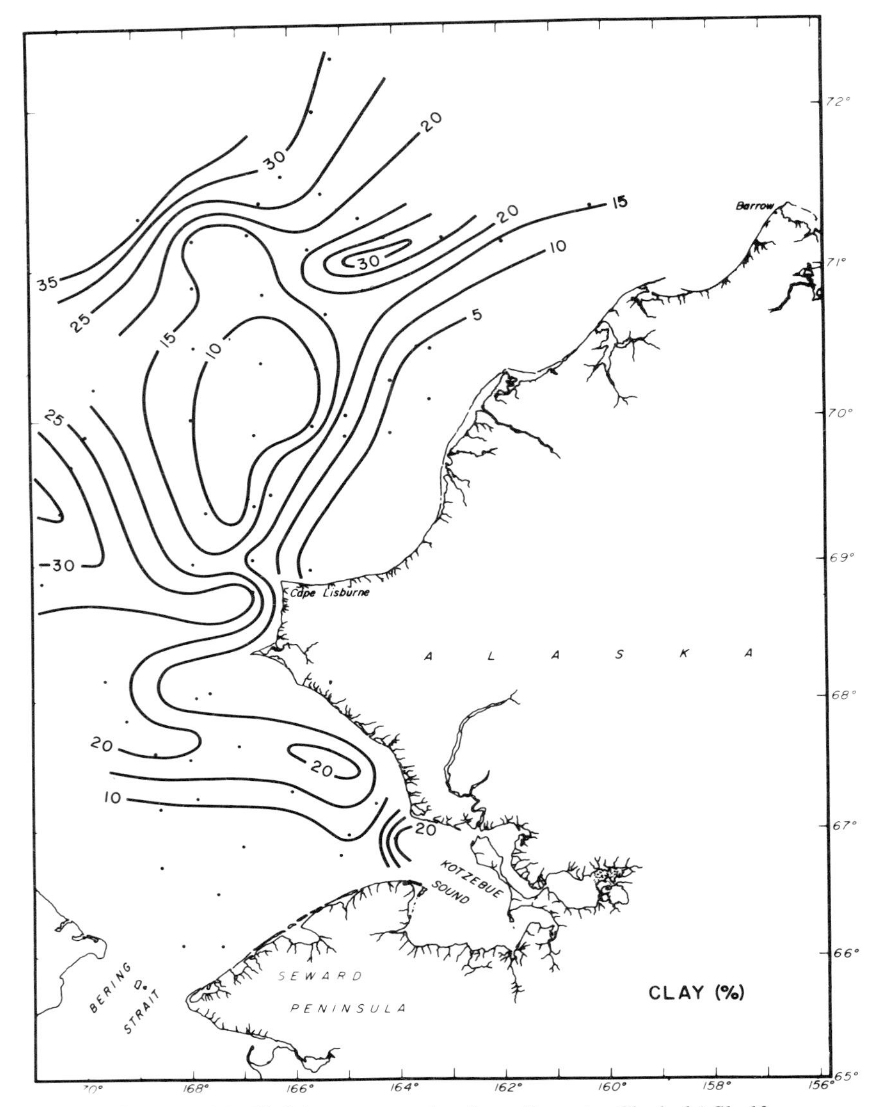

Figure 11-5. Weight percent clay in sediments, Chukchi Shelf.

poorly sorted, are deposited offshore in relatively quiescent environments. Ten granulometric variables from over 400 bottom sediments were subjected to factor analysis by McManus *et al.* (1969) to delineate the sedimentary environments in the Chukchi Sea. Three factors, representing mud, sand and gravel, provide some insight concerning processes of sedimentation on the Chukchi Shelf. McManus *et al.* (1969) suggested that sands deposited along the northern

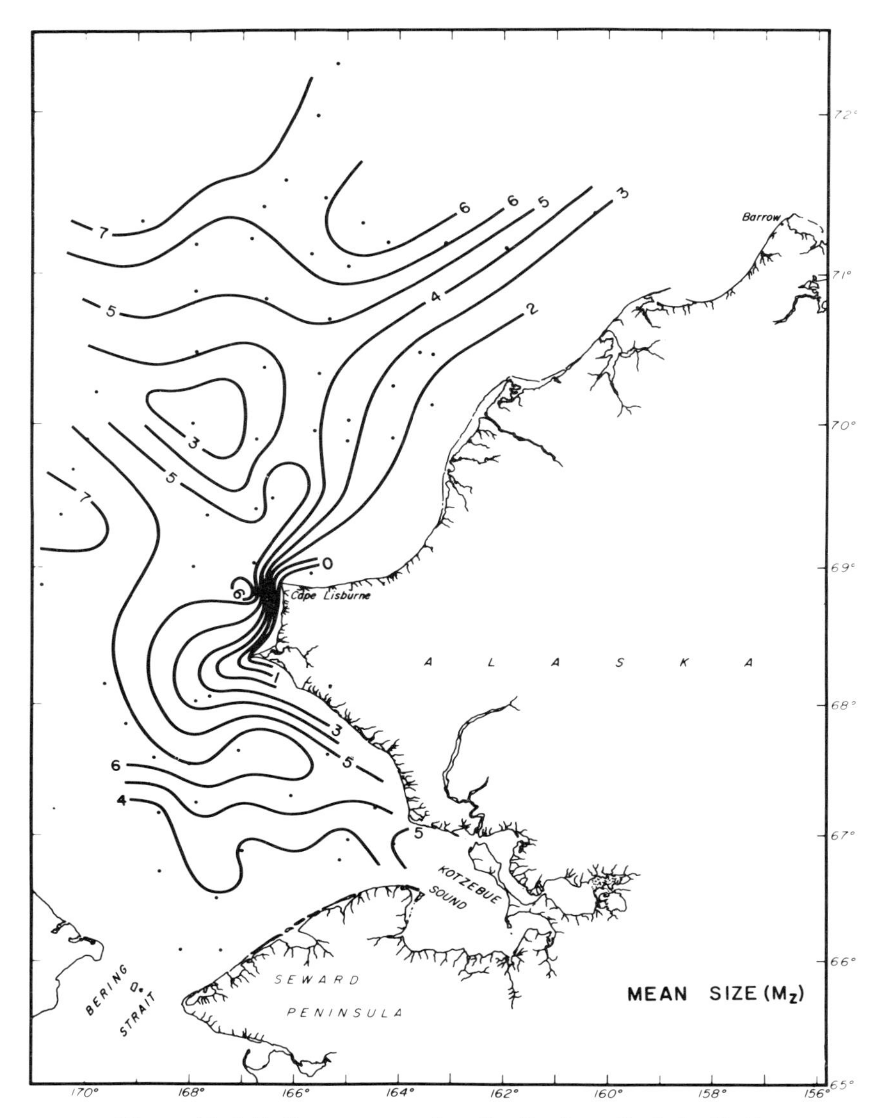

Figure 11-6. Sediment mean size distribution, Chukchi Shelf.

shores of the Seward Peninsula are the result of wave sorting, whereas sands deposited near the mouth of Kotzebue Sound are influenced by tidal currents. Sand transported by currents mantles the near-shore region between Kotzebue Sound and Point Hope. The coarse sand and gravel observed along the northern shores of Cape Lisburne and offshore on and around Herald Shoal are considered as relict and residual sediments. Most of the offshore region is covered with

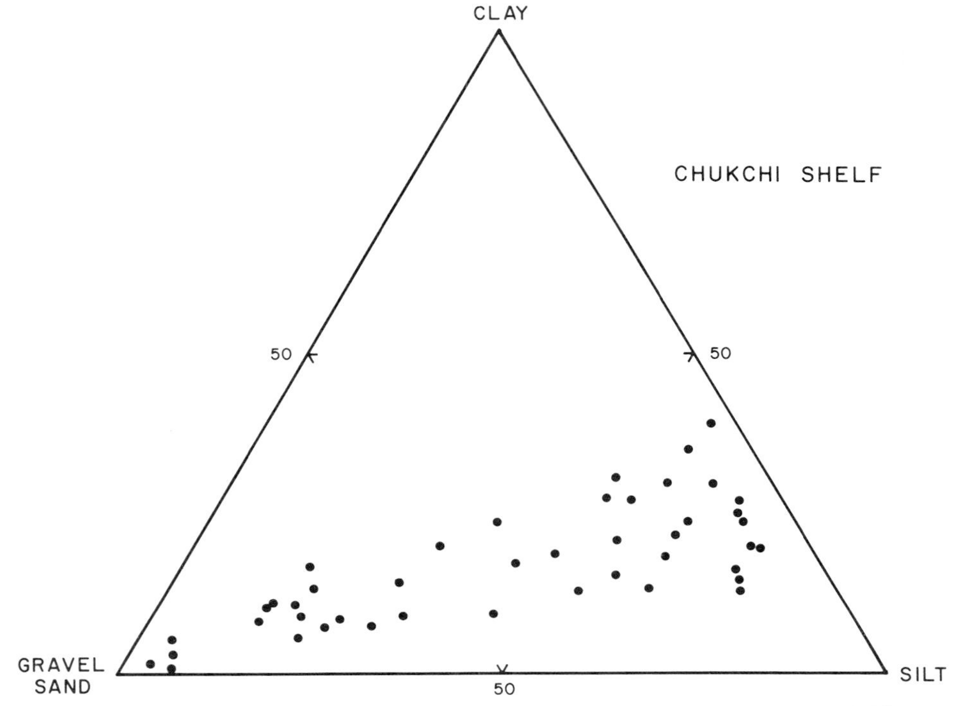

Figure 11-7. Percents gravel-sand, silt, and clay in sediments, Chukchi Shelf.

modern silt and clayey silt. These fine sediments, according to McManus *et al.* (1969), are deposited through particle settling from the wash load of the shelf surface and bottom turbid waters.

A local reconnaissance study of clay minerals and their distribution in the <2 μm fraction of bottom sediments has been reported by Naidu and Sharma (1972). The region studied included the near-shore environment between Point Hope and Icy Cape. The dominant mineral in this fraction was illite, constituting between 50 and 63.3 percent of the total sample. The next two minerals in the order of abundance were chlorite and kaolinite. Smectite was found in minor amounts.

GEOCHEMISTRY

The geochemical parameters of sediments from the Chukchi Sea, like texture, vary considerably. The major element concentration distributions in sediments reflect various geochemical anomalies, which may reflect distinct geochemical environments. These environments may be complementary to the sedimentary environments described earlier.

Aluminum content in sediments from the Chukchi Shelf varies over a wide range (1.5-7.0%). The abnormally low aluminum content is observed in coarse sediments deposited along the shore between Cape Lisburne and Wainwright (Fig. 11-8). Gravel and sand deposited in the Bering Strait, on Herald Shoal, and along the shore between Kivalina generally contain between 3.0 and 4.5% aluminum. The percent aluminum increases rapidly in sediment deposited offshore.

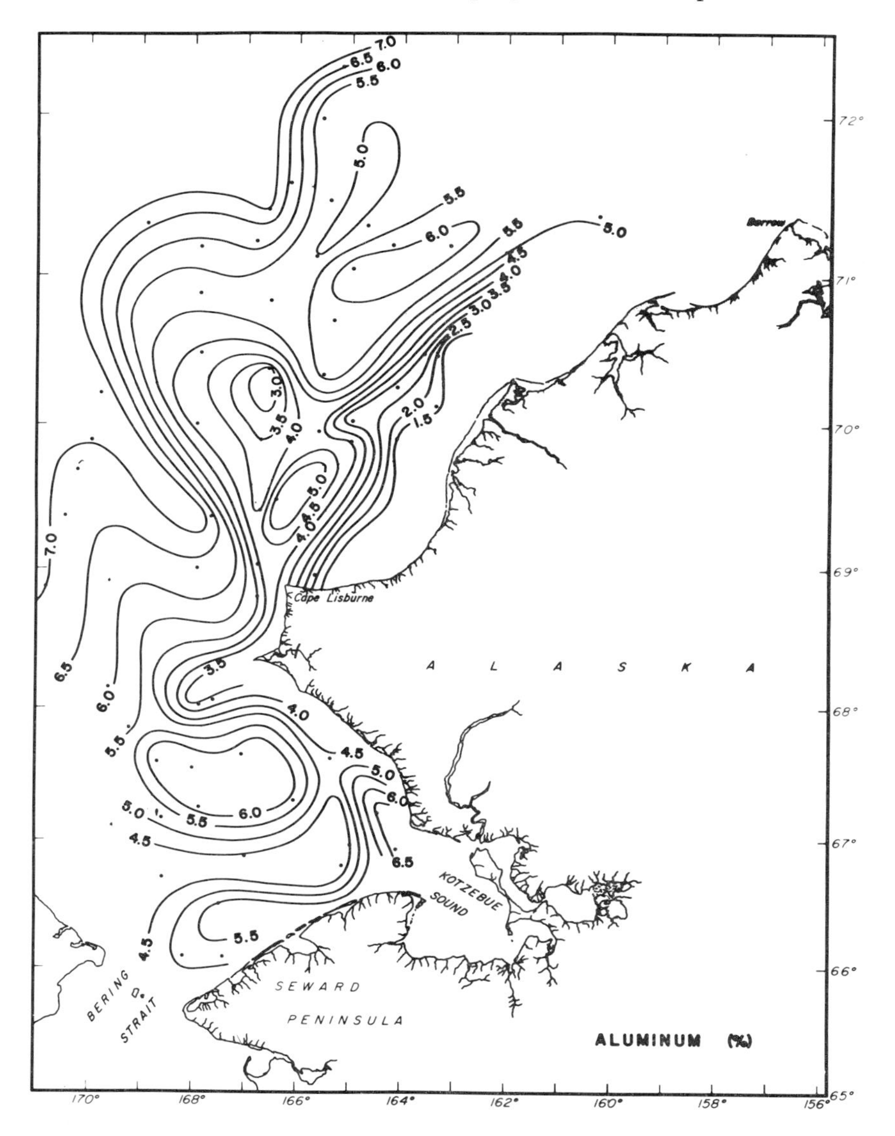

Figure 11-8. Percent aluminum in sediments, Chukchi Shelf.

The fine sediments, sandy and clayey silts, contain the most aluminum and they are found near the mouth of Kotzebue Sound, a circular area lying between the Bering Strait and Point Hope, and in the offshore areas northwest of Point Hope and Point Barrow.

The distribution of iron in sediments is somewhat similar to that of aluminum (Fig. 11-9). Iron content in sediments varies between 1.5 and 4.0%. Low iron

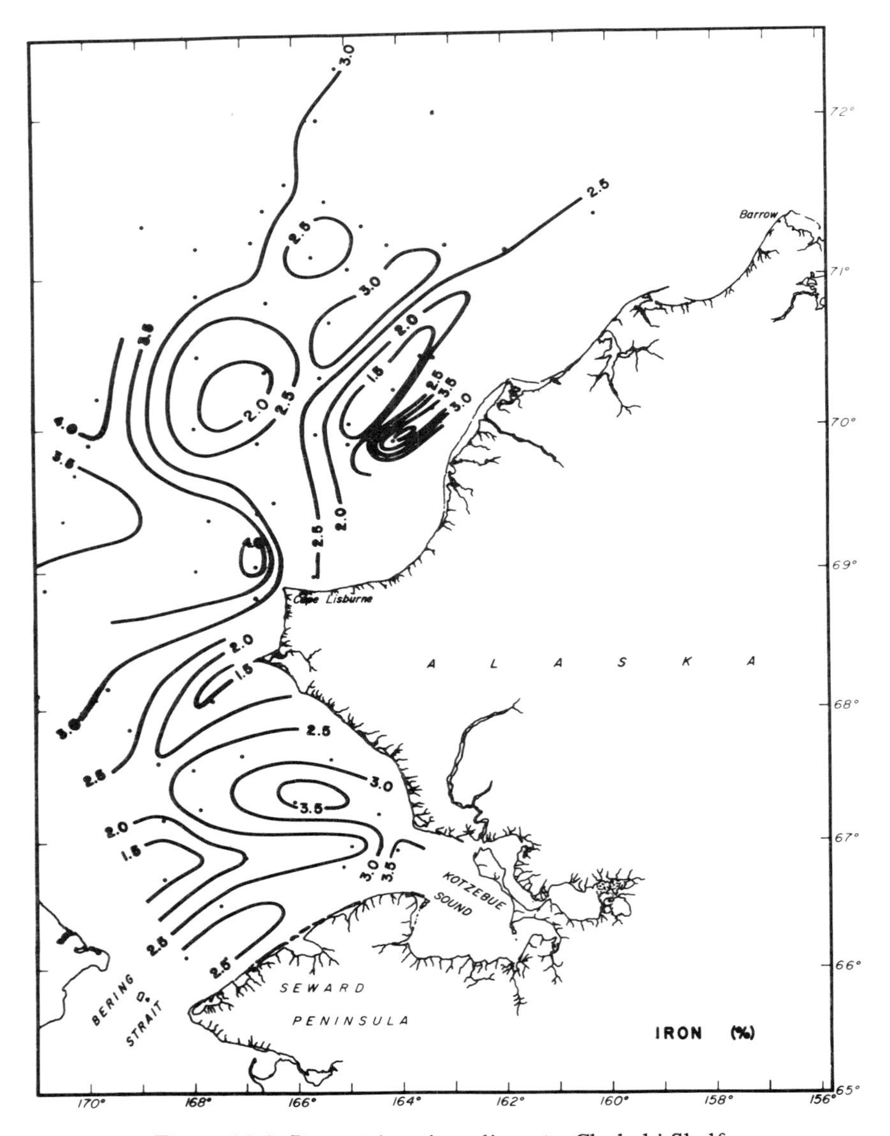

Figure 11-9. Percent iron in sediments, Chukchi Shelf.

content occurs in coarse sediments deposited north of the Bering Strait, southwest of Point Hope, and in an elongated region northwest of Icy Cape. Locally, sediments west of Icy Cape, although coarse in texture, have a relatively higher content of iron. Fine sediments, sandy, clayey silt deposited offshore, contain the maximum percentage of iron.

The calcium distribution in Chukchi Shelf sediments indicates two anomalous regions of relatively high calcium content north of the Bering Strait (Fig. 11-10).

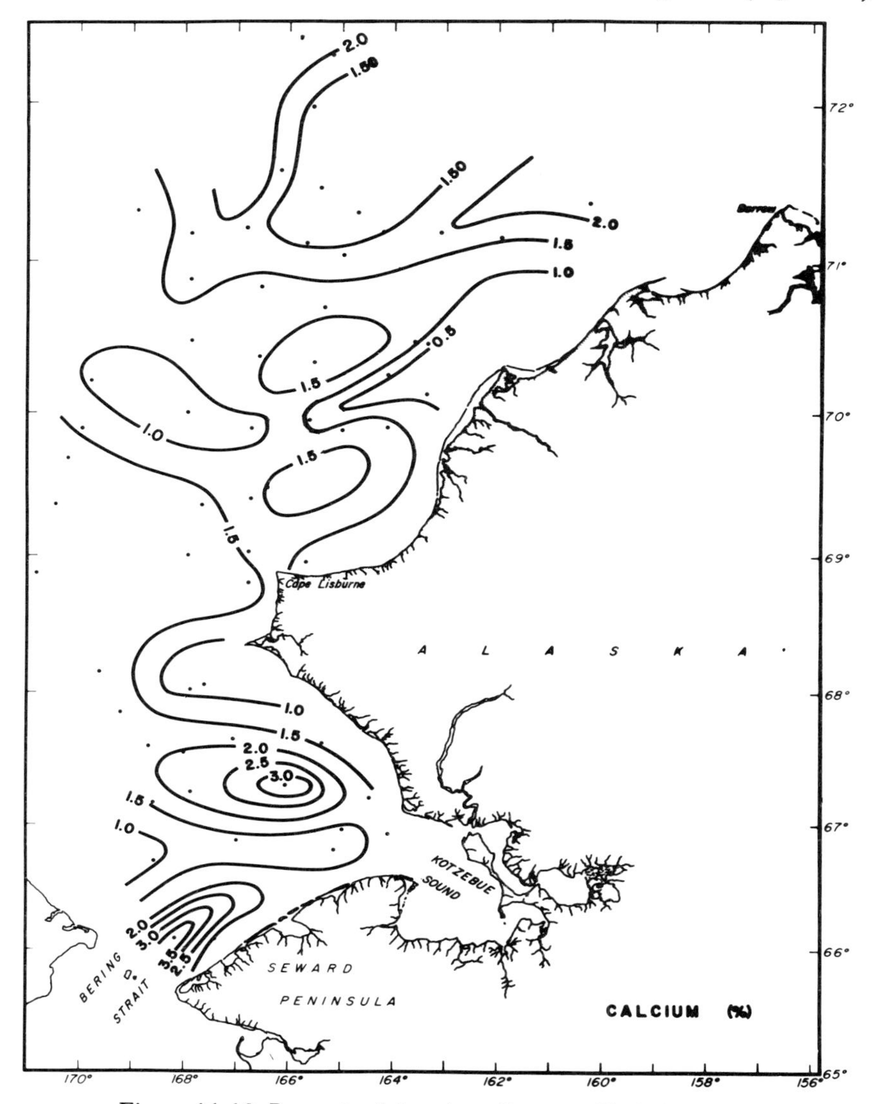

Figure 11-10. Percent calcium in sediments, Chukchi Shelf.

Generally, the calcium content of sediments increases seaward; however, the seaward increase is not related to the sediment texture. For example, the coarse sediments from the Bering Strait have a relatively high calcium content; but sediments of similar texture from the vicinity of Point Hope and Herald Shoal contain little calcium. Therefore, the distribution of calcium in sediments suggests that the high calcium content in anomalous regions may be the result of increased concentration of calcareous skeletons.

The magnesium content in sediments from the Chukchi Shelf varies between 0.25 and 1.5% (Fig. 11-11). This is slightly less than the percentage of magnesium found in sediments from the lower latitude Alaskan Shelf. The distribution of magnesium appears to be strongly related to the sediment texture. The coarse sediments from the Bering Strait, the near-shore region between Kivalina and Wainwright, and Herald Shoal contain between 0.25 and 0.75% magnesium. Medium-grained sediments consisting mostly of sandy clayey silt have a significantly higher content of magnesium. The distribution of magnesium is analogous to the distribution of aluminum. This similarity in distribution, increase in magnesium and aluminum with decrease in sediment mean size, suggests that the primary sources for magnesium and aluminum in sediments are the silicate minerals.

Sodium and potassium in Chukchi Shelf sediments have similar distributions (Figs. 11-12 and 11-13). The concentrations of sodium and potassium in sediments generally appear to be related to the sediment texture. Coarse sediments, mostly gravel, in the Bering Strait and along the coast have a low content of sodium (0.5-1.0%). Seaward with decreasing particle size, the sediments become richer in sodium (1.5%-1.5%). The percentage of potassium in sediments varies between 0.5 and 1.5% and is, on an average, lower than that observed in sediments from other Alaskan shelves. Sediments having a low potassium content are deposited along the northeastern coast between Point Hope and Wainwright. The sand and gravel from the northern Bering Strait contain 1.5% potassium, which is comparable to that found in silts deposited near the entrance of Kotzebue Sound and offshore regions of the shelf. It is not known why this unusually high concentration of potassium in coarse sediments occurs.

The manganese content in sediments from the Chukchi Shelf varies over a wide range. Interestingly, no measurable manganese was detected in the coarse sediments deposited near Icy Cape (Fig. 11-14). Because manganese always has been present in sediments from other Alaskan shelves, the absence of manganese in these coarse sediments is somewhat puzzling. Moreover, manganese is generally associated with iron in sediments. In spite of the abundance of iron in these sediments, they do not contain manganese. Overall, the manganese content of sediments increases offshore and appears to be related to the sediment mean size; fine sediments contain higher amounts of manganese than coarse sediments.

The titanium content, which has a similar distribution as manganese, varies within a narrow range (0.15-0.5%). Titanium in sediments relates well to sediment mean size: near-shore coarse sediments contain less titanium, whereas offshore fine sediments have a relatively higher titanium content (Fig. 11-15).

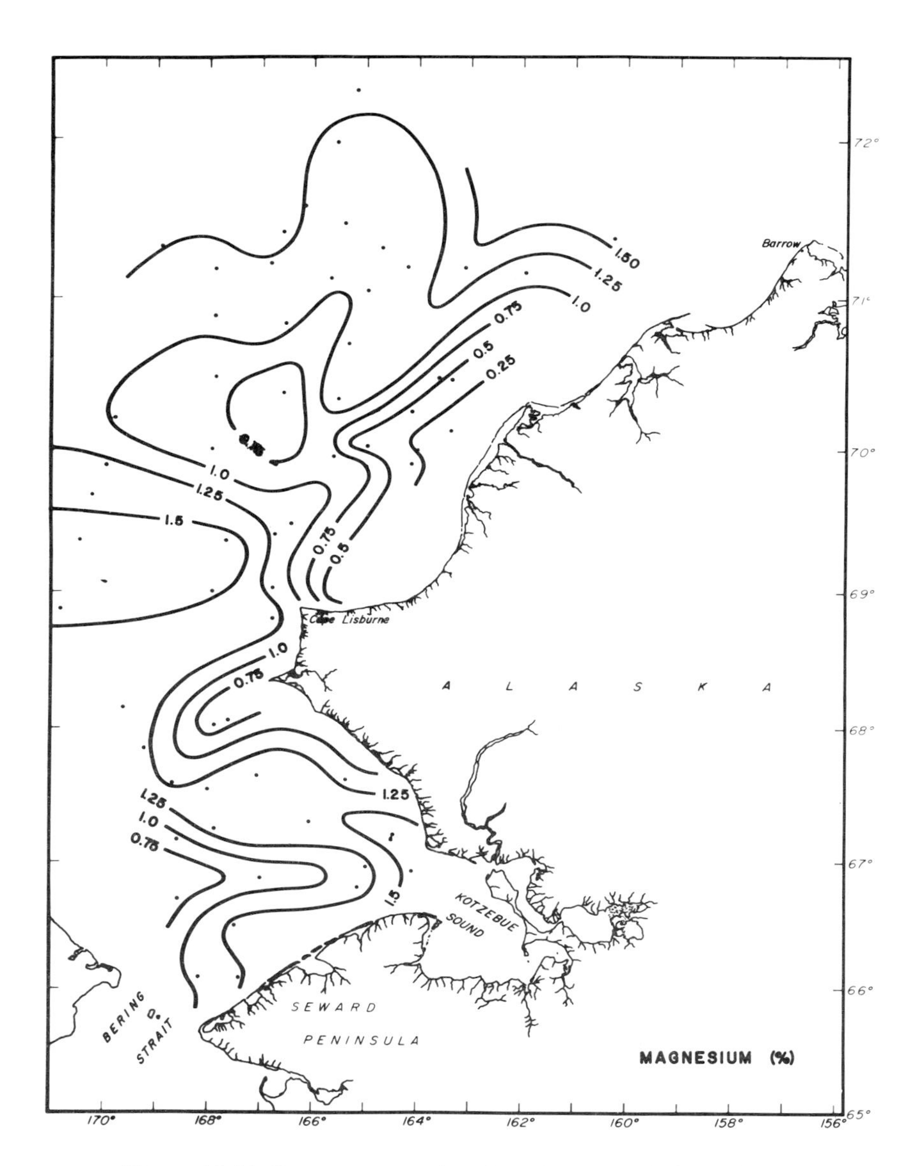

Figure 11-11. Percent magnesium in sediments, Chukchi Shelf.

The distribution of titanium is quite similar to the distributions of other major elements.

The variation of minor elements in sediments from the Chukchi Shelf is complex. In some cases, the minor element concentrations are obviously related to sediment texture, in others the provenance appears to control the distributions.

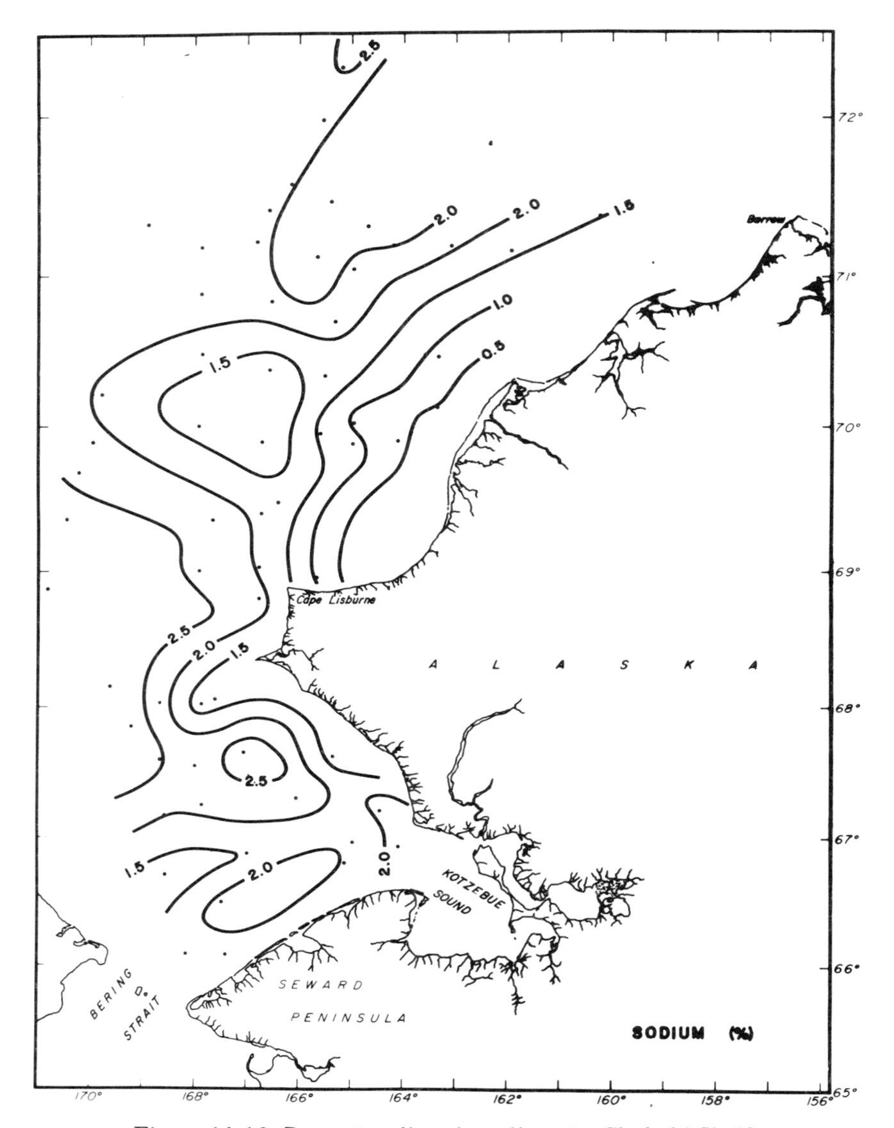

Figure 11-12. Percent sodium in sediments, Chukchi Shelf.

Therefore, the minor element distribution often does not correspond to the distribution of major elements.

Sand from Prince of Wales Shoal near the Bering Strait contains about 300 ppm barium, which decreases northeastward (Fig. 11-16). This trend is similar to that of calcium described earlier. Close to the entrance of Kotzebue Sound, the barium content in sediments increases. The coarse sand and gravel deposited

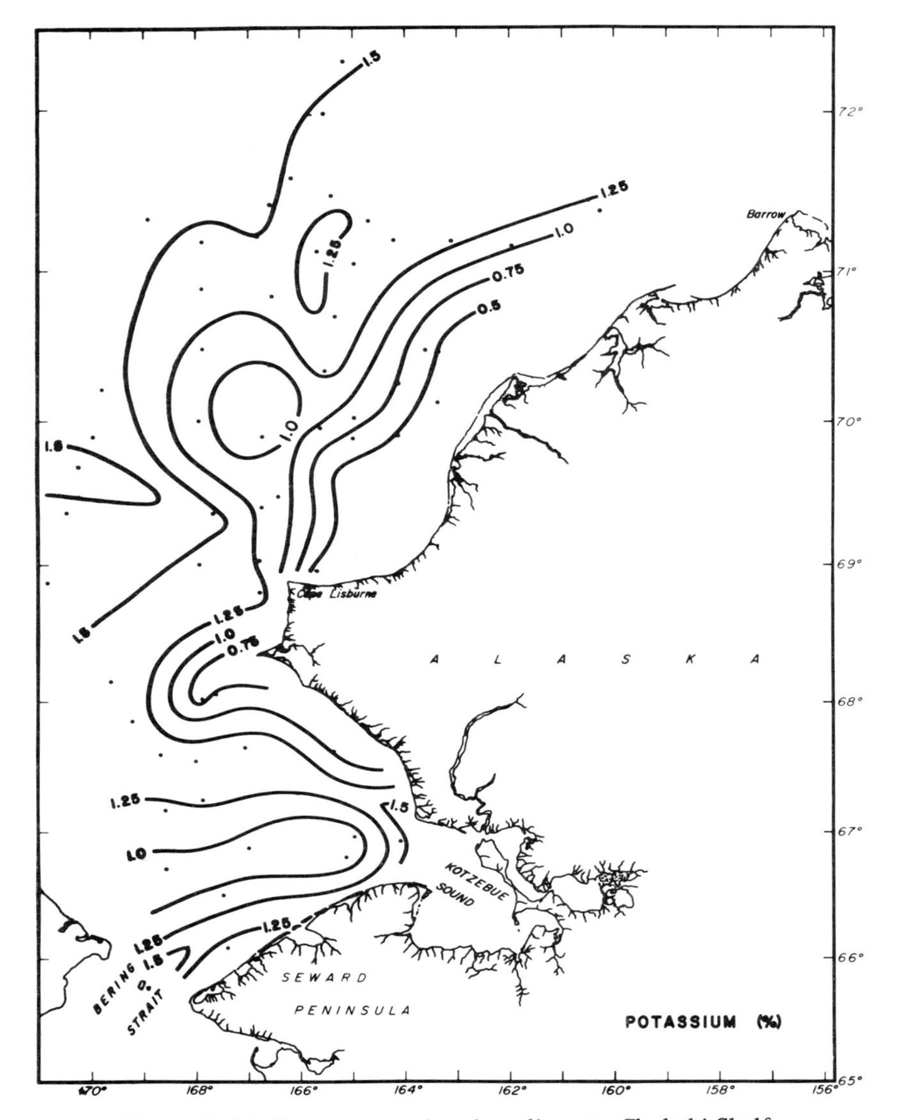

Figure 11-13. Percent potassium in sediments, Chukchi Shelf.

between Kotzebue Sound and Point Hope, however, contain the maximum amount of barium. It is difficult to determine whether this high anomaly in coarse sediments is a result of increased biogenic contributions or is because the source rock provides barium-rich sediments to this region.

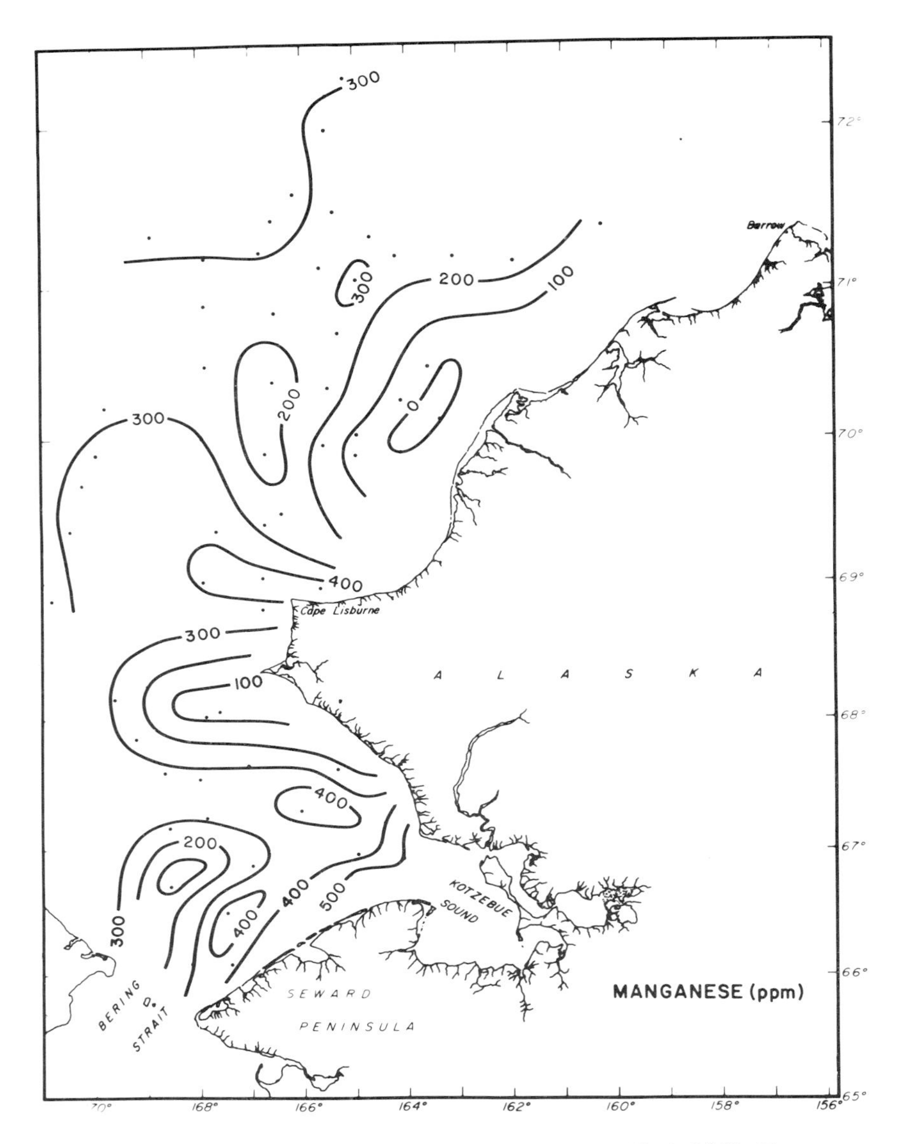

Figure 11-14. Manganese (ppm) in sediments, Chukchi Shelf.

Northward, along the shore the barium in sediments is relatively low and gradually increases seaward. The offshore sediments generally contain between 300 and 400 ppm barium. Interestingly, sediments from Herald Shoal do not exhibit the decrease in barium content normally associated with the substantial increase in grain size.

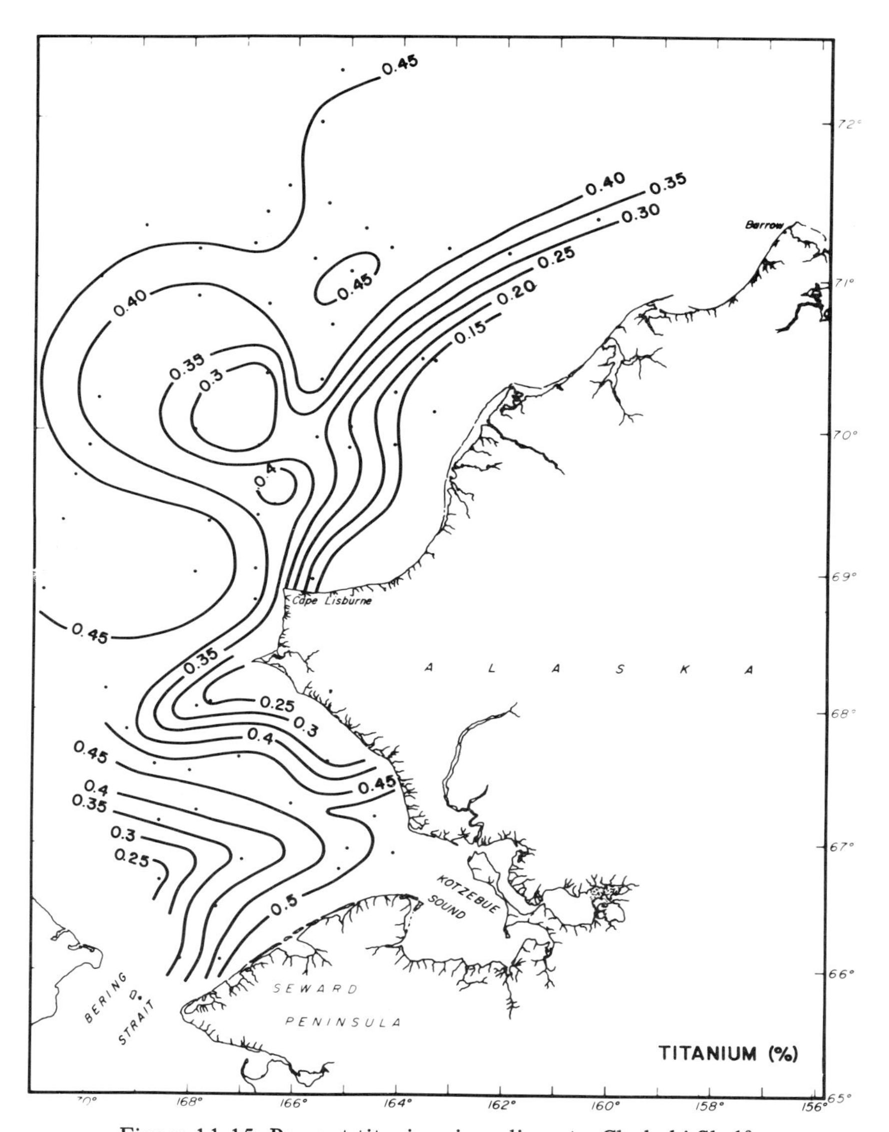

Figure 11-15. Percent titanium in sediments, Chukchi Shelf.

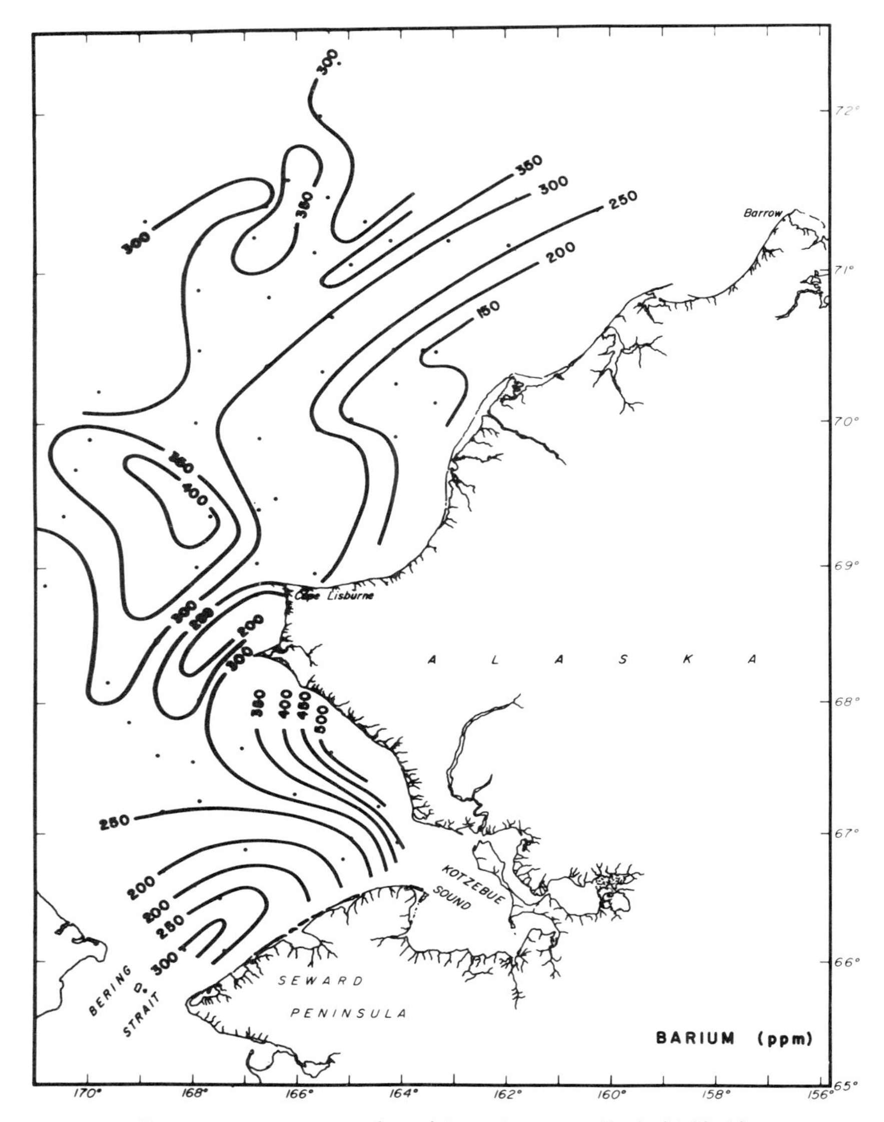

Figure 11-16. Barium (ppm) in sediments, Chukchi Shelf.

The distribution of chromium in general relates well to the sediment texture (Fig. 11-17). Gravel deposits, north of the Bering Strait, the near-shore region, and along the coast generally contain less than 10 ppm chromium. The coarse sediments from Herald Shoal are also relatively poor in chromium and contain between 30 and 40 ppm; the slightly higher chomium content anomaly near Kivalina is perhaps again caused by the source rock.

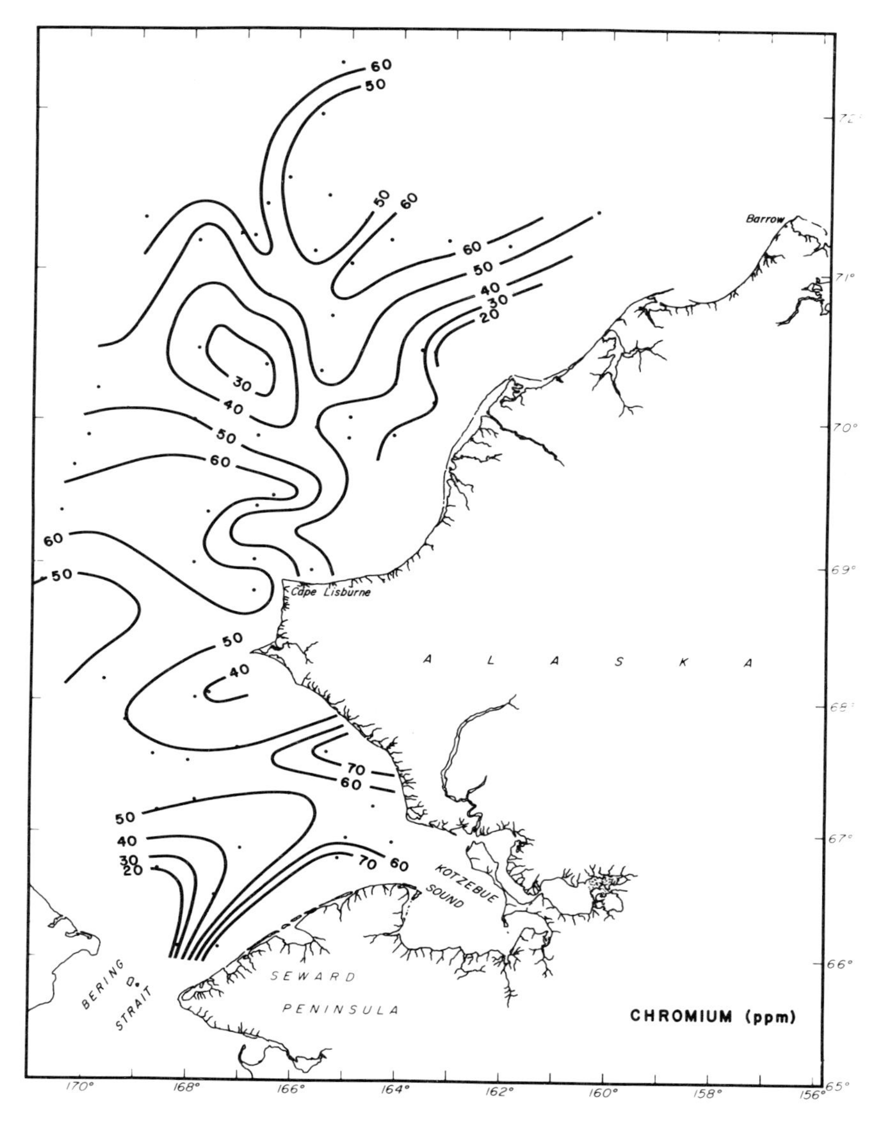

Figure 11-17. Chromium (ppm) in sediments, Chukchi Shelf.

Offshore, sand and silt contain between 40 and 65 ppm chromium. However, the sediments with maximum chromium content lie along the northern shore of the Prince of Wales Shoal and the Kotzebue Sound entrance.

Cobalt content in Chukchi Sea sediments is rather low (8-24 ppm) and varies within a narrow range (Fig. 11-18). Low concentrations of cobalt are observed in near-shore sediments from north of the Bering Strait, Seward Peninsula, and Cape Lisburne. Low values were also found in coarse sediments from Herald

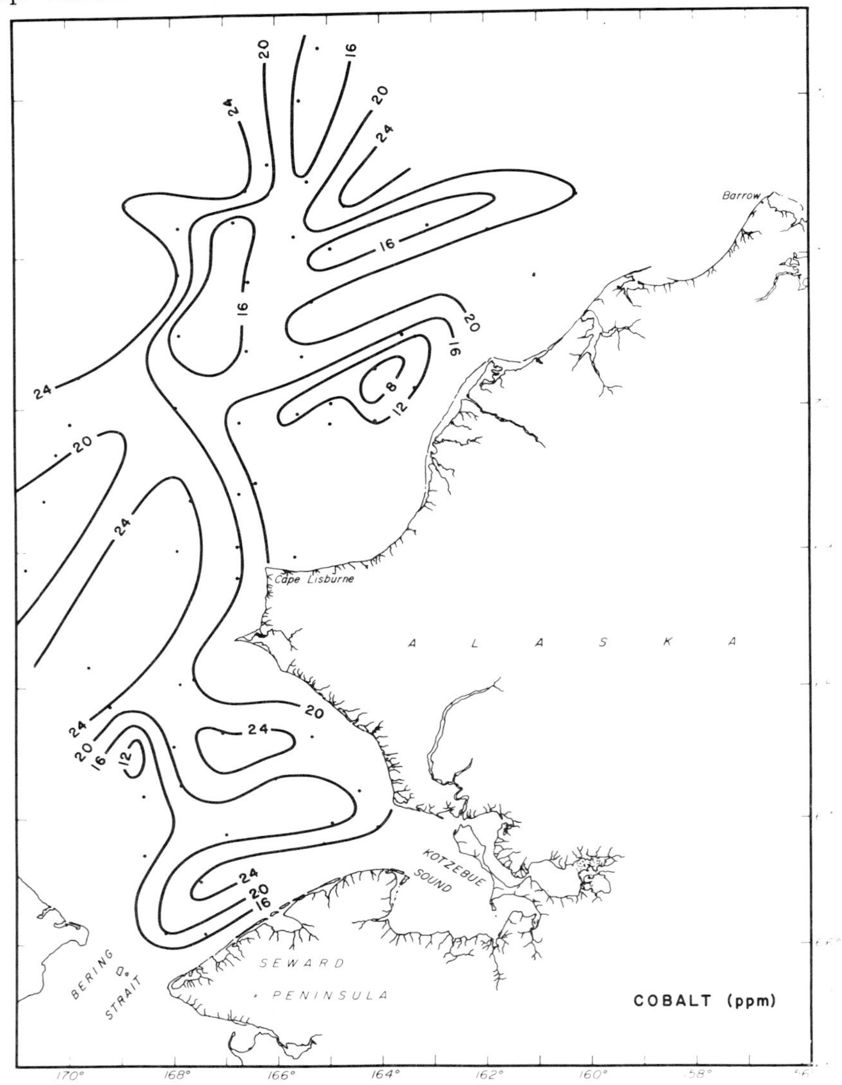

Figure 11-18. Cobalt (ppm) in sediments, Chukchi Shelf.

Shoal, northwest of the entrance to the Kotzebue Sound, contain relatively more cobalt. The distribution of cobalt in sediments appears to be related to the sediment texture.

Variations in the distribution of copper in sediments are rather small: the range lies between 25 and 35 ppm. The variations in copper content do not correspond to the changes in the textural parameters of the sediments. A relatively higher content of copper occurs in sediments that appear to originate in Kotzebue Sound and extend northeastward. Sediments with a high copper content also mantle a broad region seaward of Cape Lisburne (Fig. 11-19). Isolated regions of sediments containing 35 ppm copper are also observed on the northern Chukchi Shelf.

The distribution of nickel in sediments from the Chukchi Shelf is somewhat complex. Two, rather small but conspicuous, sedimentary regions with low nickel content lie north of the Bering Strait and off Icy Cape (Fig. 11-20). Sediments deposited near the entrance of Kotzebue Sound contain maximum concentrations of nickel but the concentration in sediments decreases rapidly to the northwest. Farther offshore, however, the nickel content of sediments increases and reaches 32 ppm. The seaward distribution of nickel in sediments is rather unique; on the inner shelf it decreases seaward, but on the middle and outer shelves it increases with increased distance from shore. This suggests that the nickel content of sediments is not related to the sediment texture.

The strontium content of Chukchi Shelf sediments varies significantly. The sediments off Icy Cape contain 50 ppm or less strontium, sediments from Prince of Wales Shoal and a small region midway between the Bering Strait and Point Hope contain >225 ppm strontium (Fig. 11-21). The distributions of strontium and calcium in sediments from the Chukchi Shelf are similar, which suggests that the concentrations of both these elements may have been influenced by a common factor, possibly the biogenic contribution.

The zinc content of sediments shows a systematic distribution and is primarily related to the sediment grain size. Coarse sediments from the northern Bering Strait, the near-shore deposits between Kivalina-Point Hope-Cape Lisburne and Wainwright, and the shallow deposits on Herald Shoal contain <50 ppm zinc (Fig. 11-22). Offshore, finer sediments contain >50 ppm zinc. Sediments from a small region in the north central Chukchi Shelf contain the maximum amount of zinc (125 ppm). Although the sediments from this region are fine grained, the reason for the anomalously high zinc content remains obscure.

SEDIMENT SOURCE AND TRANSPORT

The Chukchi Shelf displays a sediment grading that ranges from gravelly sand to clayey silt (Fig. 11-6). The textural distribution brings forth two important points. First, most of the shelf is covered with sand and silt and there is a paucity of the clay-sized fraction in the sediments. Clay content in sediments, on an

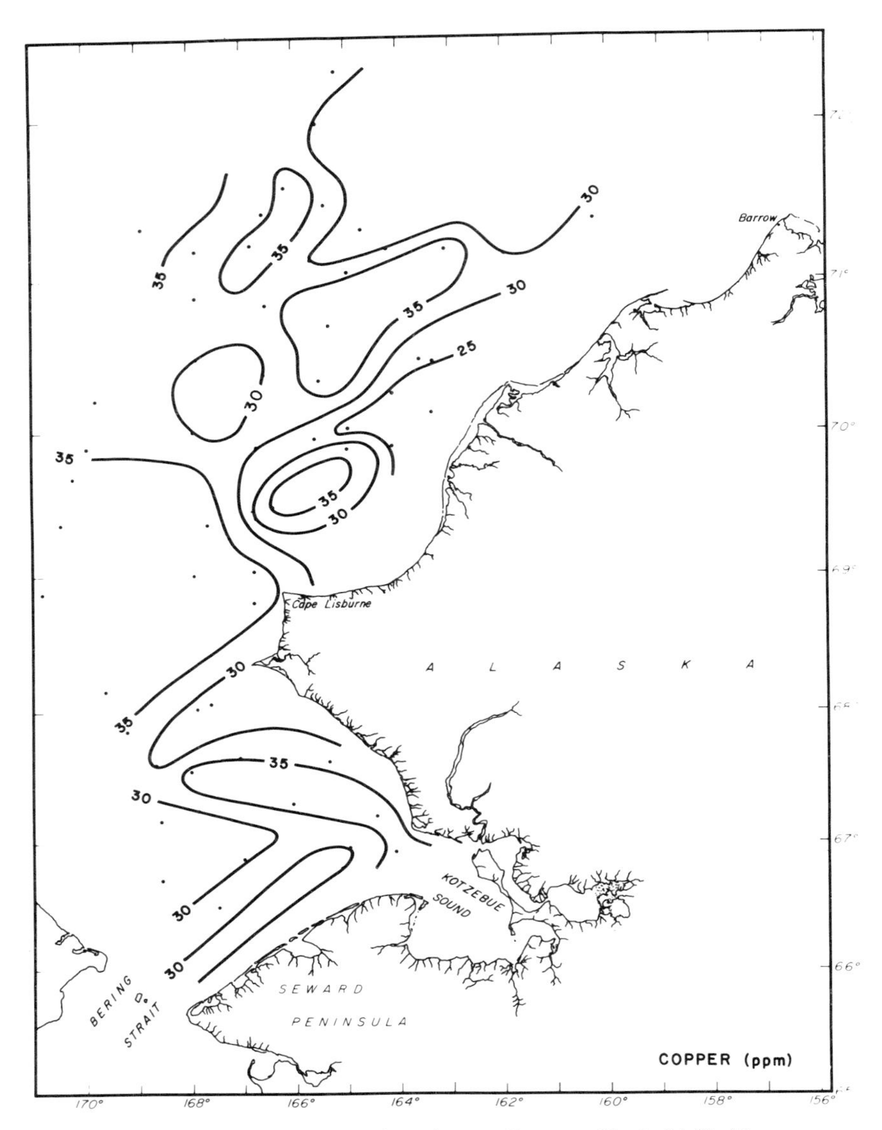

Figure 11-19. Copper (ppm) in sediments, Chukchi Shelf.

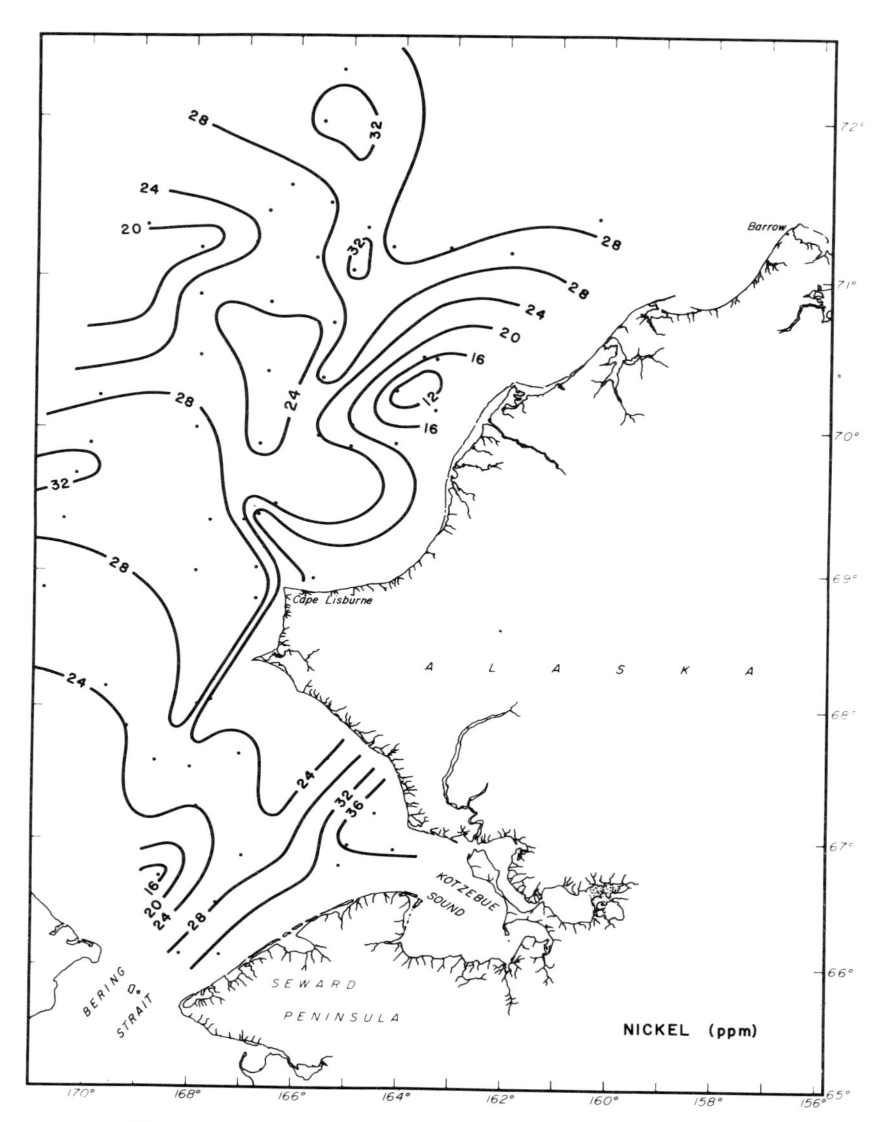

Figure 11-20. Nickel (ppm) in sediments, Chukchi Shelf.

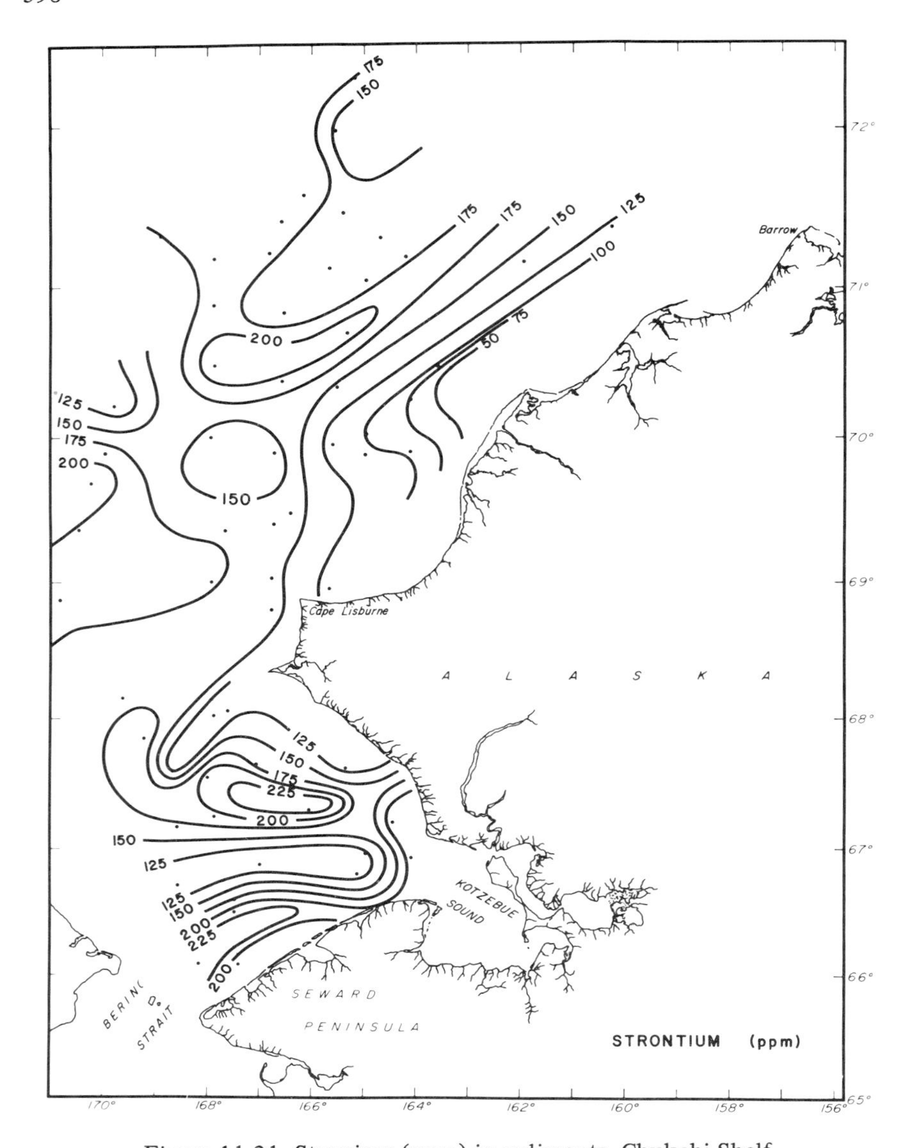

Figure 11-21. Stronium (ppm) in sediments, Chukchi Shelf.

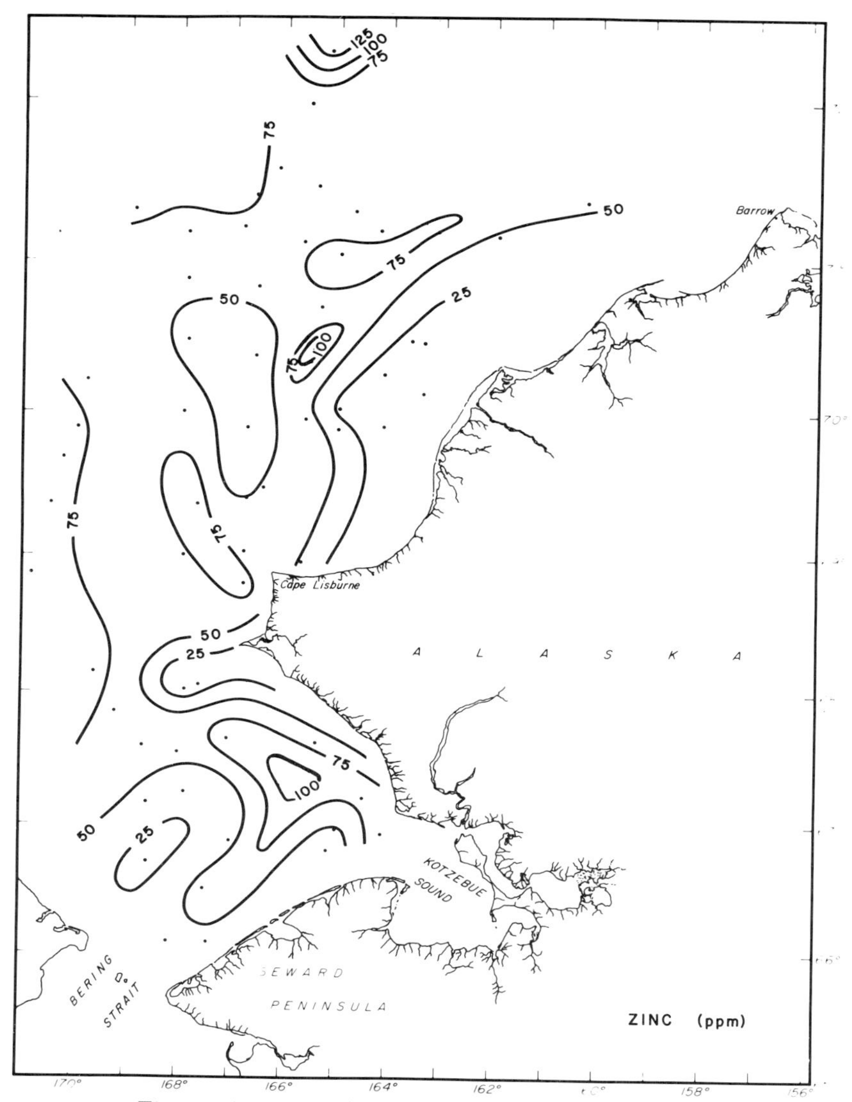

Figure 11-22. Zinc (ppm) in sediments, Chukchi Shelf.

average, does not exceed 20%. The ubiquitous paucity of clay in sediments reflects a high-energy environment on the shelf, or a lack of clay-sized particles in the source sediments, or both. Second, the sediment mean size is generally related to the bathymetry of the shelf. Coarse sediments deposited in shallow regions grade seaward into the sandy clayey silt of deeper water. The sediment size and water depth relation suggests that wave action on the shelf could be an important factor for sediment transport on the shelf.

Although the bottom sediments from the Chukchi Shelf have been studied in detail, a few measurements of sediments in suspension have been made. A reconnaissance survey of the distribution of the suspended matter throughout the Chukchi Sea has been reported by Loder (1971). The suspended load collected during July 27-August 2, 1968, included measurements above the shallow thermocline (10-15 m), as well as below it, and the load varied from 0.55 to 9.85 mg/ℓ. High concentrations were observed mostly in the southern Chukchi Shelf water. In the Bering Strait and its vicinity, the suspended load ranged between 1.25 and 4.10 mg/ℓ.

South of the Bering Strait, the suspended load decreased with increased water depth, and as the water moved northward through the strait the distribution became almost uniform (homogeneous). North and northeast of the strait the suspended load in surface water as well as in the subsurface water increased significantly, particularly in the northeast, where approximately a four-fold increase of the load in the subsurface (9.85 mg/ℓ) was observed. Northward, the suspended load decreased to 1.65 at the surface. Near Point Hope, however, the suspended load in surface water reached 5.35 while at depth it was only 1.10 mg/ℓ.

In the northern Chukchi Sea, the suspended load in the surface water is generally low (<1.0 mg/ℓ) and invariably increases with depth. The near-shore regions contain a higher load in suspension than the offshore regions.

Reconnaissance measurements of suspended loads in the surface waters of the Bering Strait and Chukchi Sea obtained from two separate cruises during 1973 are shown in Figure 11-23. The distribution of the load in suspension is based on rather sparse station locations; however, the distribution clearly delineates the regions of high and low suspended loads. A higher load in suspension is observed in the Bering Strait and over Herald Shoal. The regions with low loads occur north of the Bering Strait, the offshore area west of Icy Cape, and east of Wrangel Island.

The local and sporadic ERTS-1 images that are available provide additional information concerning the sediment source and transport in the surface waters of the eastern Chukchi Sea. The relative distribution of the suspended load in Figure 11-24 (bottom image) displays a typical divergence. The southeastward moving water along the Siberian coast merges with the western part of the north and northeastward moving Bering Sea water. A mixing of Siberian coastal and Bering Sea waters north of the strait is also supported by the suspended load distribution shown in Figure 11-23. The strong northward flow of water through the Bering Strait deflects the coastal southeastward moving plume to the north near Point Netan. The western boundary of the plume emerging from the strait and its confluence with the southeastward moving plume is clearly visible in the ERTS-1 imagery. It appears that the region shown is the mixing ground for the different water masses.

Eastward, the distribution of the surface suspended load in the southeastern Chukchi Sea embayment is somewhat complex (Fig. 11-25). In the lower right

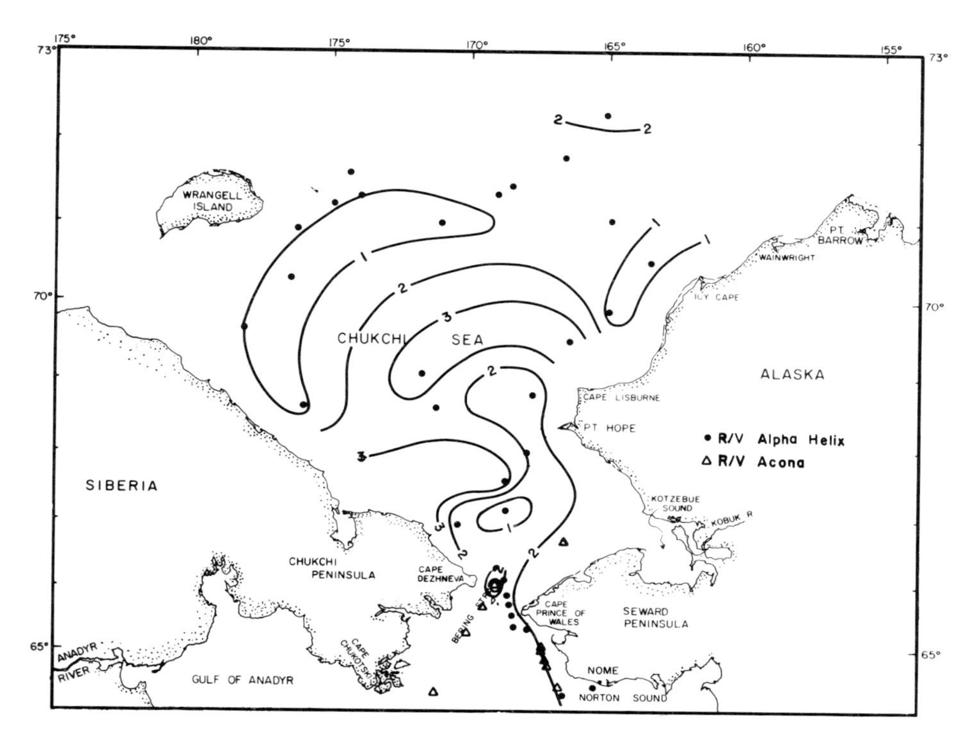

Figure 11-23. Distribution of the surface suspended load, Chukchi Shelf.

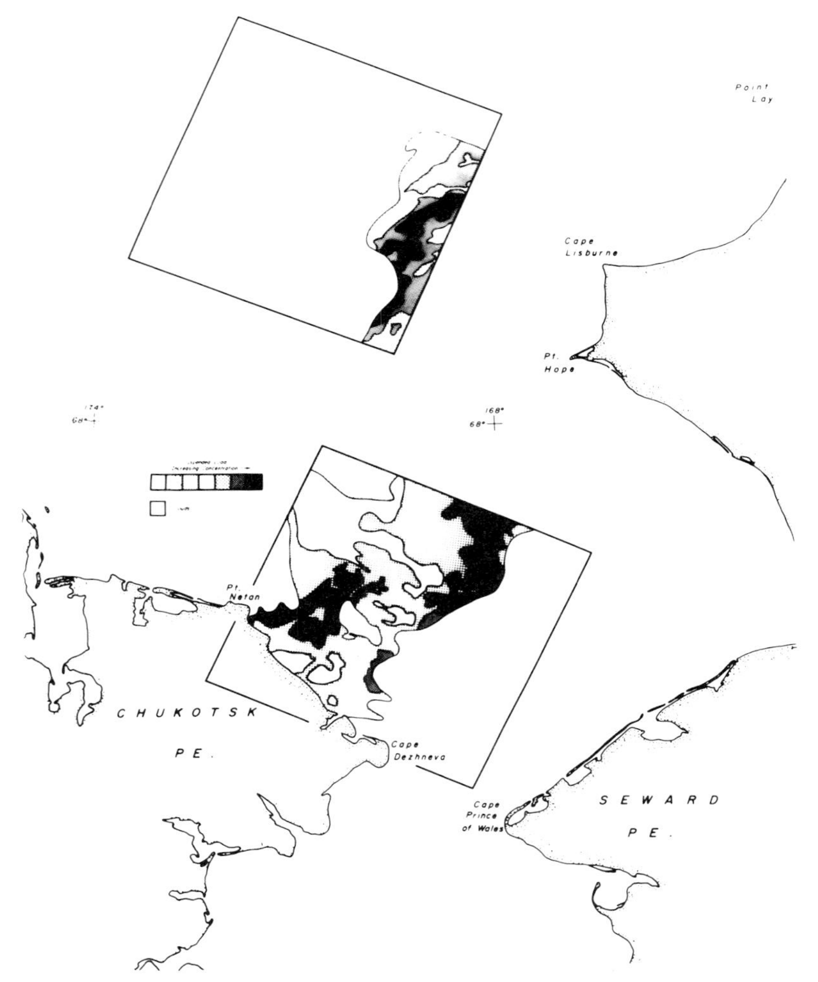

Figure 11-24. Isodensity distribution of reflectance in satellite imagery showing relative suspended load in near-surface waters off Chukotsk Peninsula and central Chukchi Shelf on 2 August, 1973 (top image) and on 22 September, 1973 (bottom image).

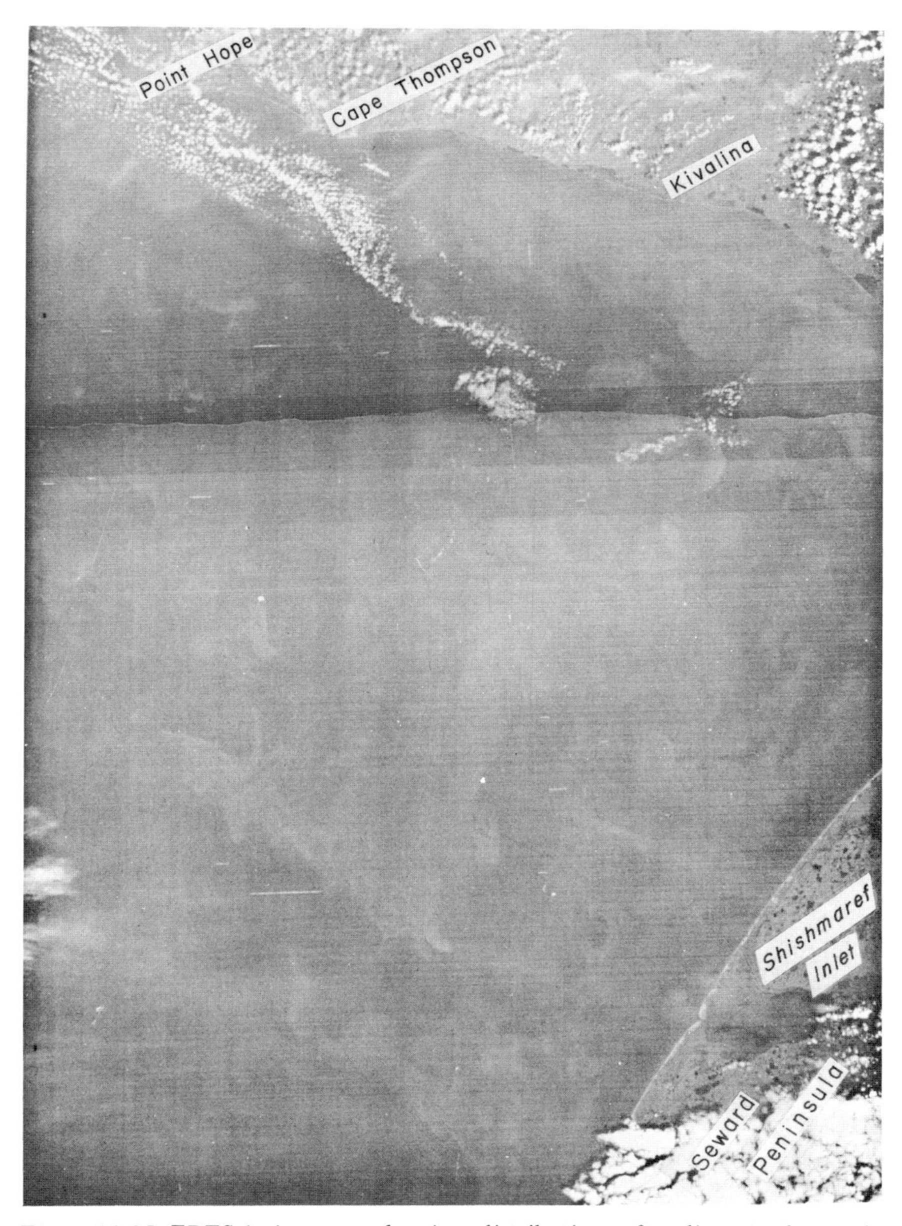

Figure 11-25. ERTS-1 imagery showing distribution of sediment plumes in waters off northwestern shore of Seward Peninsula.

corner of the image lies a part of the northwestern shore of the Seward Peninsula, with Shishmaref Inlet in the center. Part of the Point Hope cuspate is visible in the top left corner of the figure. The image suggests that considerable sediments in the surface waters which leave the Bering Strait are carried northward and northeastward. Along the northern slopes of the Seward Peninsula the sediments appear to drift northeastward along the coast as well as diffuse northward. A well-defined near-shore plume observed between Kivalina and Point Hope indicates that sediments are mostly retained within the coastal waters and are carried northwestward.

The density analyses of several ERTS-1 images from the Kotzebue Sound region (Figs. 11-26 to 11-28) reveal that northwestward moving Bering Sea water from the Bering Strait does not carry its load into the Kotzebue Sound. The Bering Sea water is mostly deflected to the north and moves northwestward along the shore. However, northwestward divergence of the main current near the entrance of Kotzebue Sound forms a small clockwise gyre near Cape Espenberg. As a result of the reduction in current speed, the sediments rapidly settle out, depositing a spit (Cape Espenberg) near the south shore of the sound entrance and a chain of islands to the south.

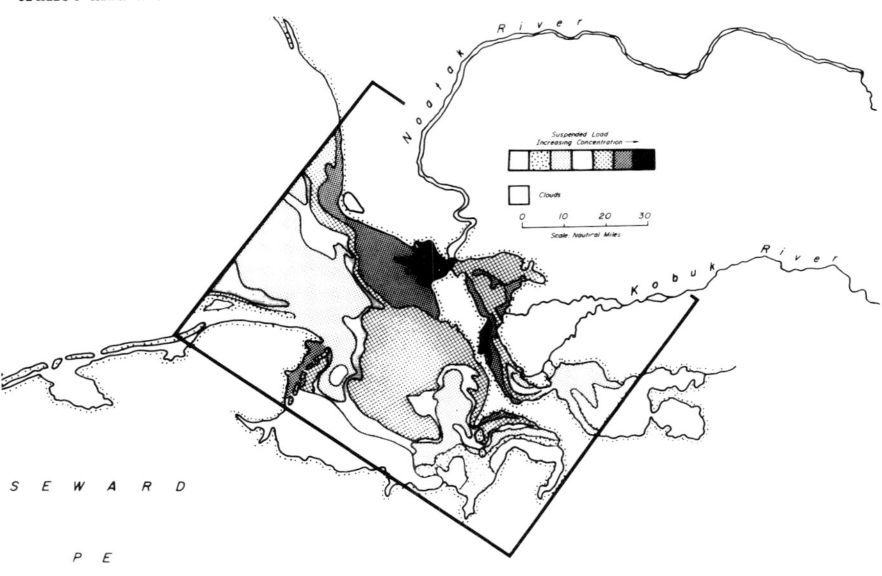

Figure 11-26. Isodensity distribution of reflectance in satellite imagery showing relative suspended load in near-surface water in Kotzebue Sound on 30 August, 1973.

In Kotzebue Sound dense sediment plumes are invariably present near the mouths of the Kobuk and Noatak rivers. Both rivers evidently contribute significant amount of sediment to the sound. The sediment brought by the Kobuk

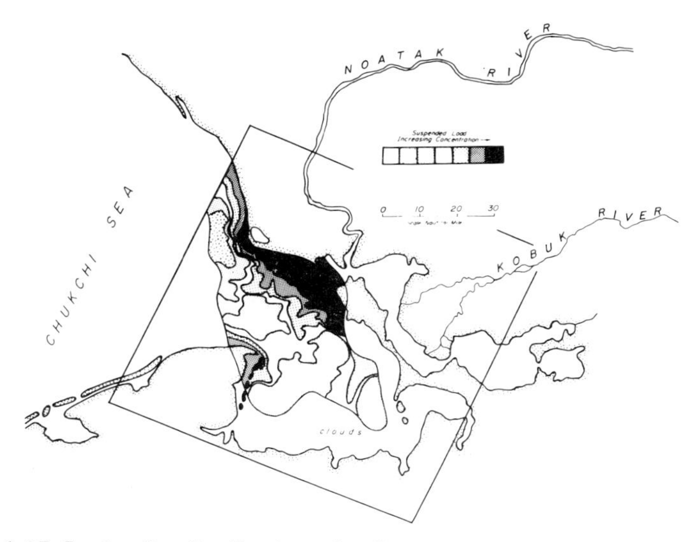

Figure 11-27. Isodensity distribution of reflectance in satellite imagery showing relative suspended load in near-surface water in Kotzebue Sound on 5 October, 1973.

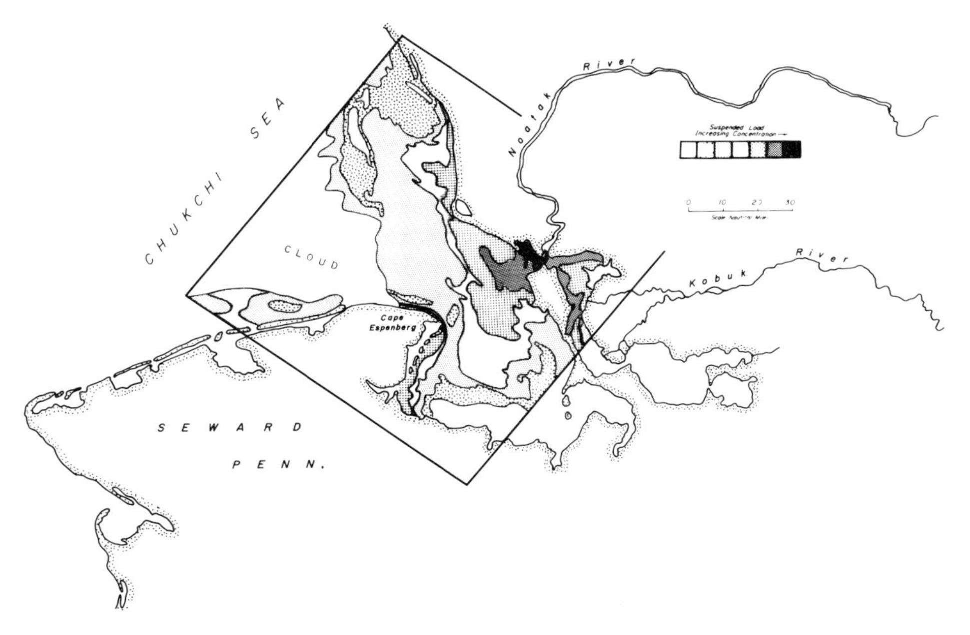

Figure 11-28. Isodensity distribution of reflectance in satellite imagery showing relative suspended load in near-surface water in Kotzebue Sound on 6 October, 1973.

River appears to settle out in the eastern sound, Hotham Inlet. The Noatak River sediments, on the other hand, are carried mostly southward. Some of these sediments, however, are carried eastward with the flood tide and the rest westward and out of the sound during the ebb tide. Once out of the sound, the sediments plume merges with the general coastal water extending northwestward toward Point Hope.

A few available ERTS-1 images of the region north of Point Hope show a well-defined near-shore plume extending from Cape Lisburne to Point Barrow. The entire coast is bordered by lagoons and inlets, which retain large quantities of sediments in suspension. These sediments are generally carried into the near-shore zone by tides or simply are spilled over the narrow breaks in offshore bars. The high sediment concentration in suspension is mostly restricted to near-shore waters and rapidly decreases seaward. It appears that most coastal sediments are carried northward along the shore (Fig. 11-29). In the embayment north of Cape Lisburne (Fig. 11-30) the distribution of the surface suspended load suggests the presence of a clockwise gyre.

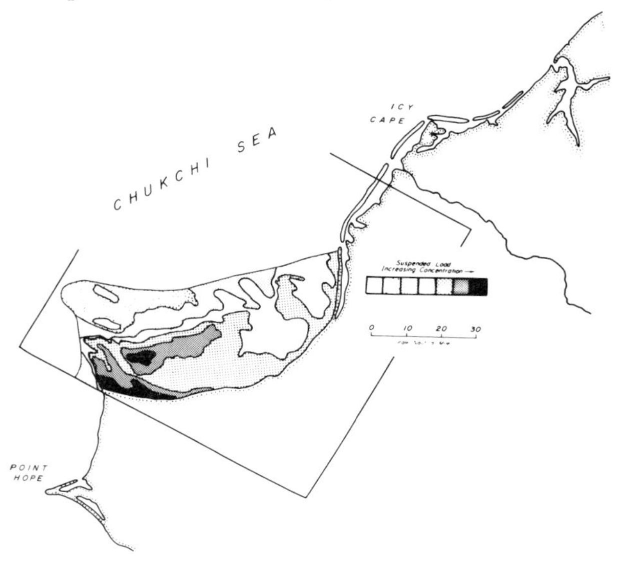

Figure 11-29. Isodensity distribution of reflectance in satellite imagery showing relative suspended load in near-surface waters between Cape Lisburne and Point Lay, Chukchi Shelf on 8 September, 1973.

It is interesting to note that the regions of high suspended load in Chukchi Sea are invariably mantled with sand and gravel. These regions lie in and north of the Bering Strait, along the Alaskan coast, and on Herald Shoal. The clay and silt size particles in these regions are apparently kept in suspension by the currents and, therefore, identify the regions of high-energy environment.

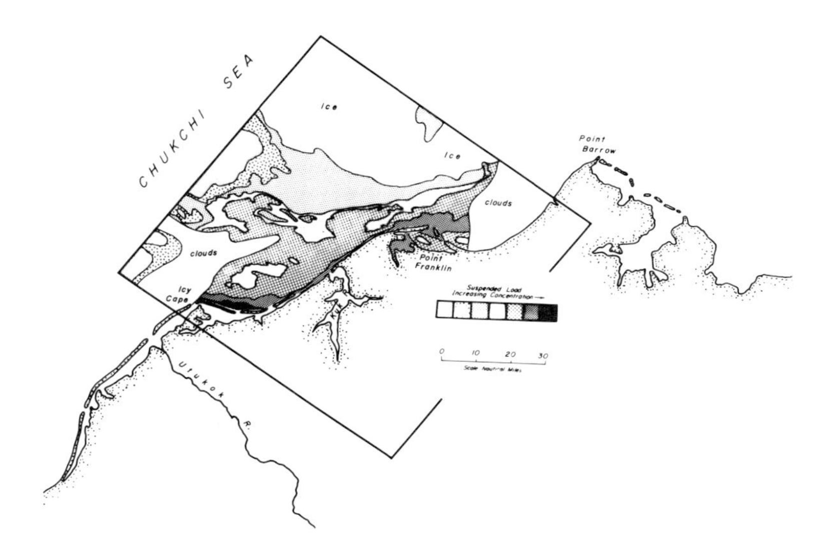

Figure 11-30. Isodensity distribution of reflectance in satellite imagery showing relative suspended load in near-surface waters between Icy Cape and Point Franklin, Chukchi Shelf on 1 September, 1973.

As noted in earlier passages, the strong currents in the Bering Strait region are caused by the northward flowing stream transports water on the order of $1 \times 10^6 m^3/sec$ which flows under the impetus of a surface slope (Coachman and Tripp, 1970). The near-shore region along the Alaskan mainland is dominated by the warm, less saline Alaskan coastal water. This water consists partly of northward flowing Bering Sea water and partly of runoff water from the coastal streams and rivers. In the southeastern Chukchi Sea a pronounced vertical discontinuity between coastal and offshore waters suggests that lateral mixing is quite small, and most of the fresh water is confined to the near-shore zone. Depending upon the amount of input from the Bering Strait and coastal region, the coastal water, on an average, achieves current speeds up to 72 cm/sec (Creager and McManus, 1966). The current intensity is generally uniform throughout the water column. The magnitude of the current clearly indicates that silt and clay size particles in the coastal waters can be retained in suspension and transported within the water column.

Storms and associated surges along the coast, however, frequently destroy the structure of the coastal water and the receding mixed waters carry some sediments offshore. The coastal sediment carried offshore eventually diffuses into the northward bound Bering Sea water.

The shallow region of Herald Shoal has two north-south depressions along its east and west flanks. Much of the northward moving Bering Sea water moves along these channels, which, therefore, serve as major arteries for the northward flow. Some water movement occurs over the shoal itself, however, and the coarse sediment cover on the shoal is probably a result of the winnowing of fine sediments during storms.

The textural and suspended load distributions indicate that the sediment movement over the Chukchi Shelf is primarily controlled by the general northward flowing coastal and offshore currents. The elemental ratio plots and the distribution of sediment maturity index further support this inference. The distribution of aluminum-titanium ratios (Fig. 11-31) suggests offshore and northeastward movement of sediment along the coast of the northwestern Seward Peninsula. Sediment originating from Kotzebue Sound and from the coast between the sound and Point Hope are carried northwestward along the coast. Near Point Hope, the sediments appear to move northward through channels along the east and west of the shoal. A somewhat high anomaly in the ratio, observed off Icy Cape, appears to be caused by the influx of sediment from the northwest. The source for this sediment may be either the westward moving Beaufort Sea water or sediments carried southward by Arctic sea ice.

Although the increase in Al_2O_3/TiO_2 isopleths indicates the movement of sediments, the decreasing Al_2O_3/Na_2O isopleths also imply pathways for sediments. The distribution of the Al_2O_3/Na_2O suggests sediment movement similar to that observed earlier (Fig. 11-32). The aluminum-sodium ratios, however, further signify that the sources for sediments in the northern Chukchi Sea may lie to the northeast and the northwest. The sediment input from the northeast may be the result of the intrusion of the Arctic water into the Chukchi Sea. The sediments from the northwest may have been brought by southward moving winter ice as well as by waters from the Siberian Arctic Shelf. The Al_2O_3/SiO_2 versus SiO_2 (Fig. 11-33) suggests a progressive decrease in aluminum compensated by a related increase in silica on the Chukchi Shelf. The linear function, in general, is caused by elemental differentiation during transport so that the maturity of sediments increases with increased transport (Fig. 11-33). A closer analysis of the station location and maturity index of sediments, however, indicates that near-shore sediments contain less aluminum and more silica (low Al_2O_3/SiO_2). Therefore, near-shore sediments are more mature than those deposited offshore (Fig. 11-34). The higher maturity of the near-shore sediments is caused by swift currents, which prevail in this region. The abrasion and diminution of the particle size of light minerals, particularly feldspars, and removal of fines by currents in these regions cause an increase of siliceous sands in sediments. This process results in a relative enrichment of silica over aluminum in the sediment.

The rapid decrease in K_2O/Na_2O north of the Bering Strait clearly signifies northward and northeastward movement of sediments (Fig. 11-35). Sediment movement along the shore and offshore is also apparent. In the northeast, the region surrounding Herald Shoal is high in potash-soda ratio, but the ratio decreases eastward. The distribution of isopleths suggests that the sediments carried by the east and southeast moving water are deflected to the north by the Bering Sea water described earlier.

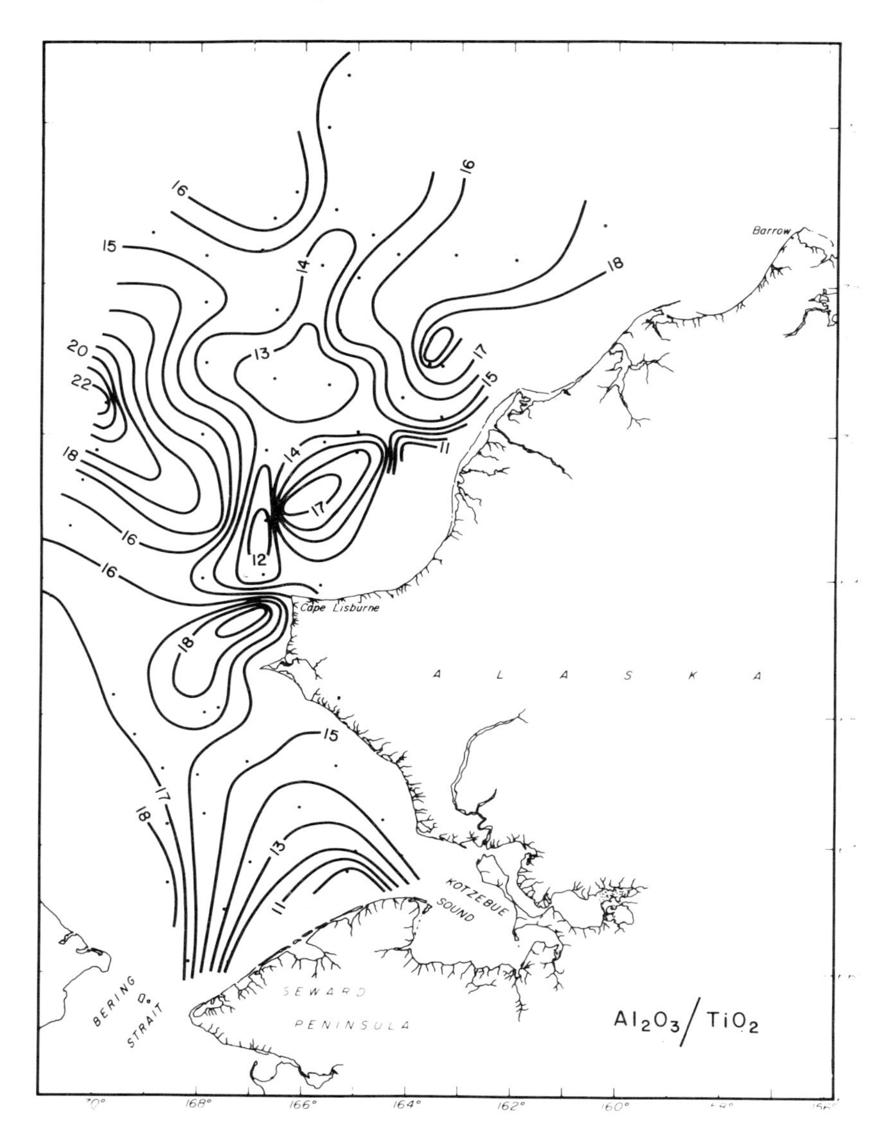

Figure 11-31. Al_2O_3/TiO_2 variations in sediments, Chukchi Shelf.

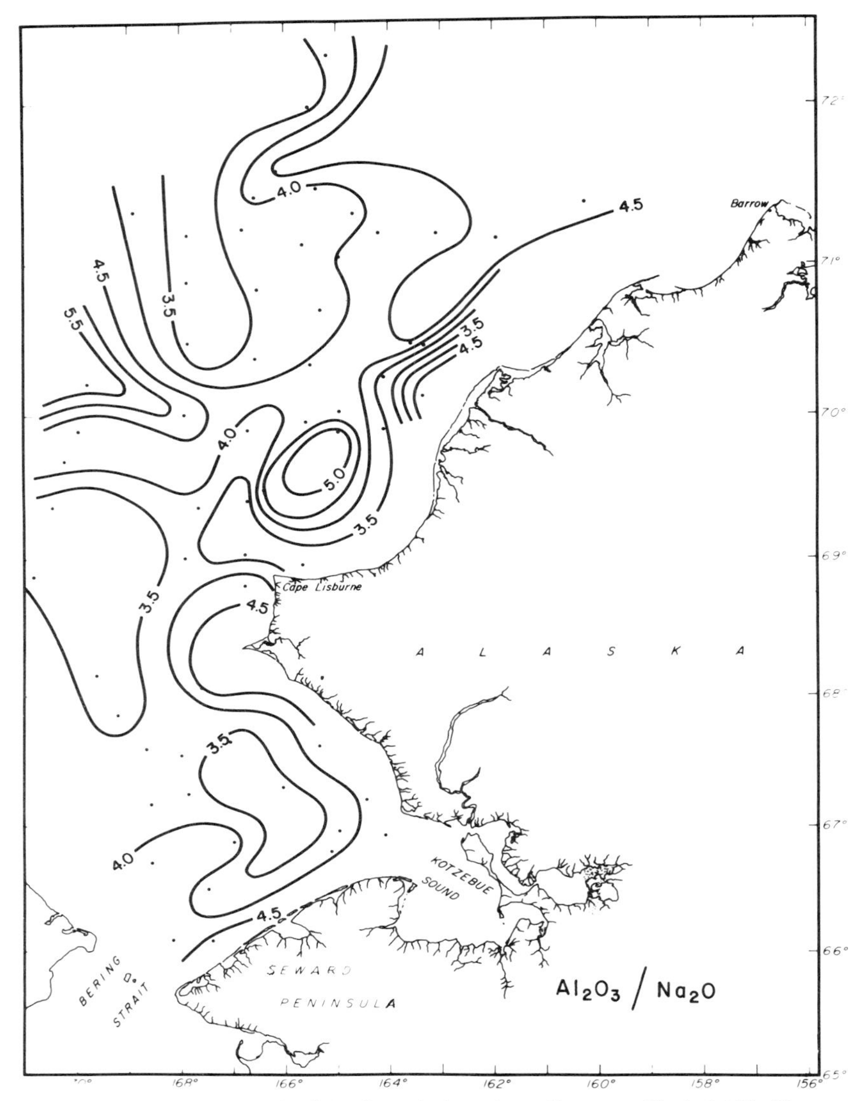

Figure 11-32. Al_2O_3/Na_2O variations in sediments, Chukchi Shelf.

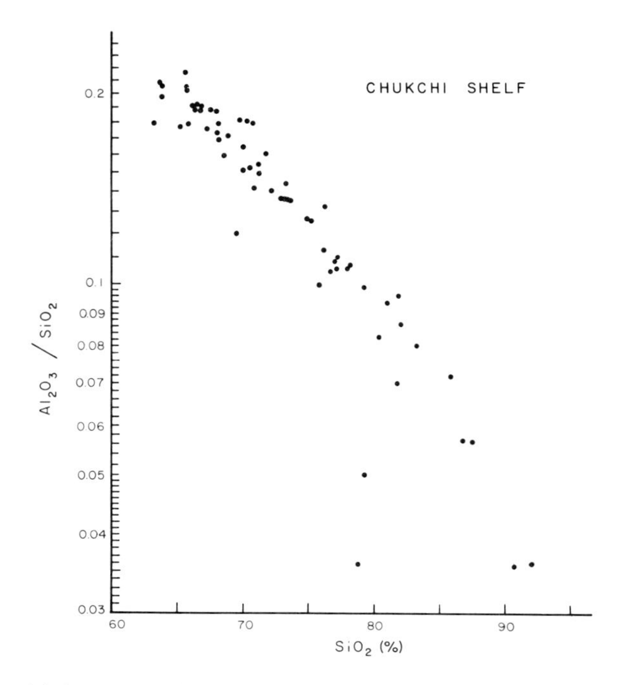

Figure 11-33. Al_2O_3/SiO_2 versus percent SiO_2 plot, Chukchi Shelf.

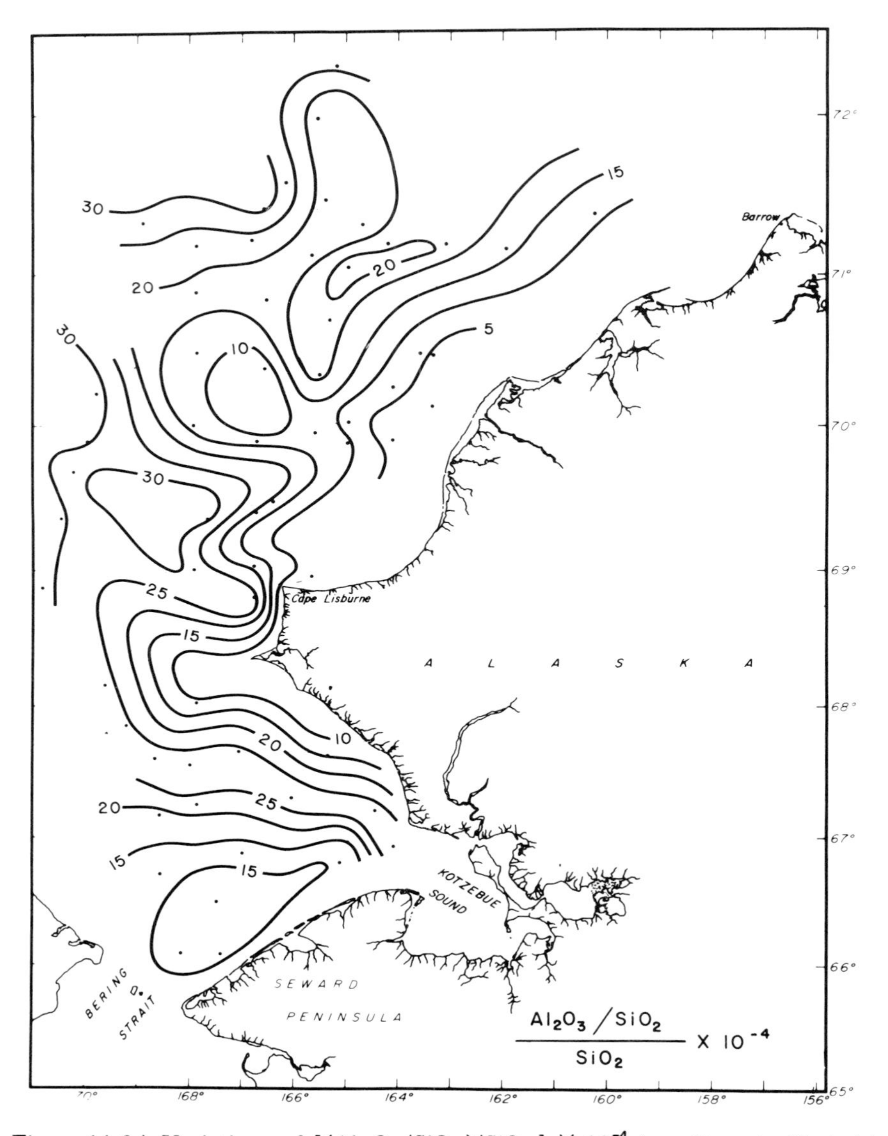

Figure 11-34. Variations of $[(Al_2O_3/SiO_2)/SiO_2] \times 10^{-4}$ in sediments, Chukchi Shelf.

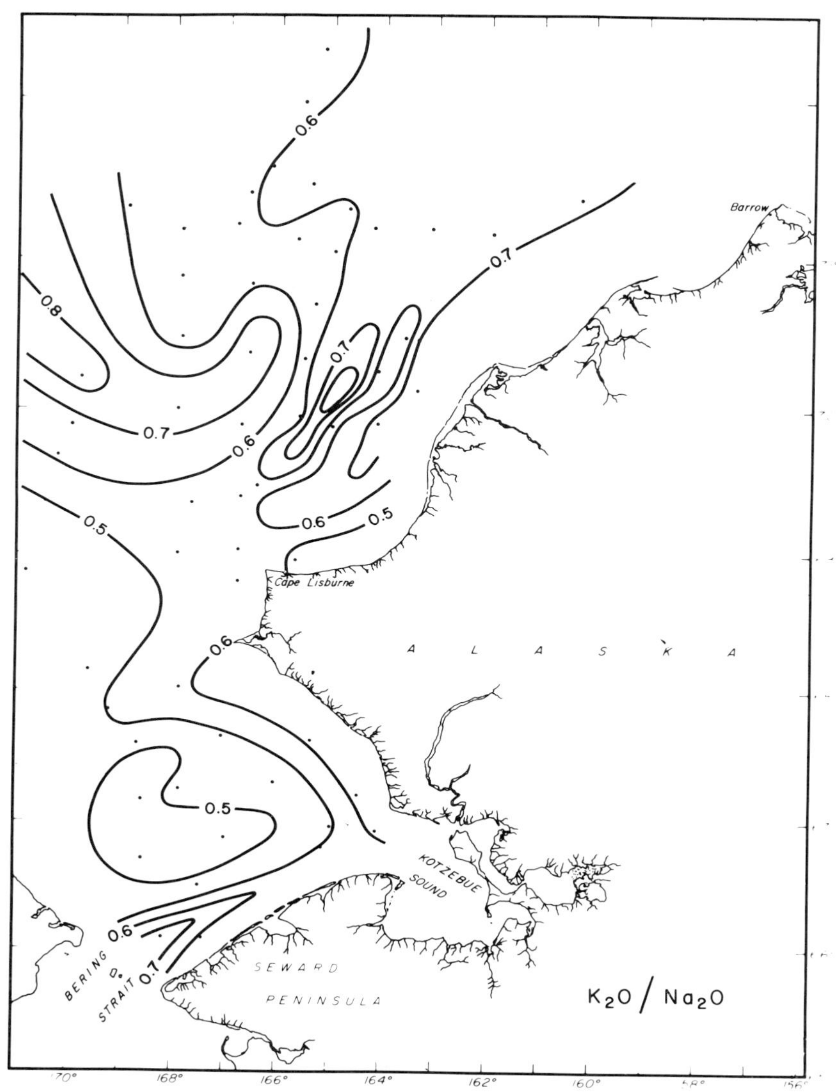

Figure 11-35. K_2O/Na_2O variations in sediments, Chukchi Shelf.

The triangular plot, with alumina, silica plus titania, and the Rest as end members, shows increasing maturity in sediments with increased transport (Fig. 11-36). Compared to other Alaskan shelves the sediments from the Chukchi Shelf appear to be more matured. The higher maturity in sediment is the result of mixing of the Yukon River sediments carried into the Chukchi Sea by the northward flowing Bering Sea water. Continual current action during the northward transport of Yukon River detritus into the Chukchi Sea increases

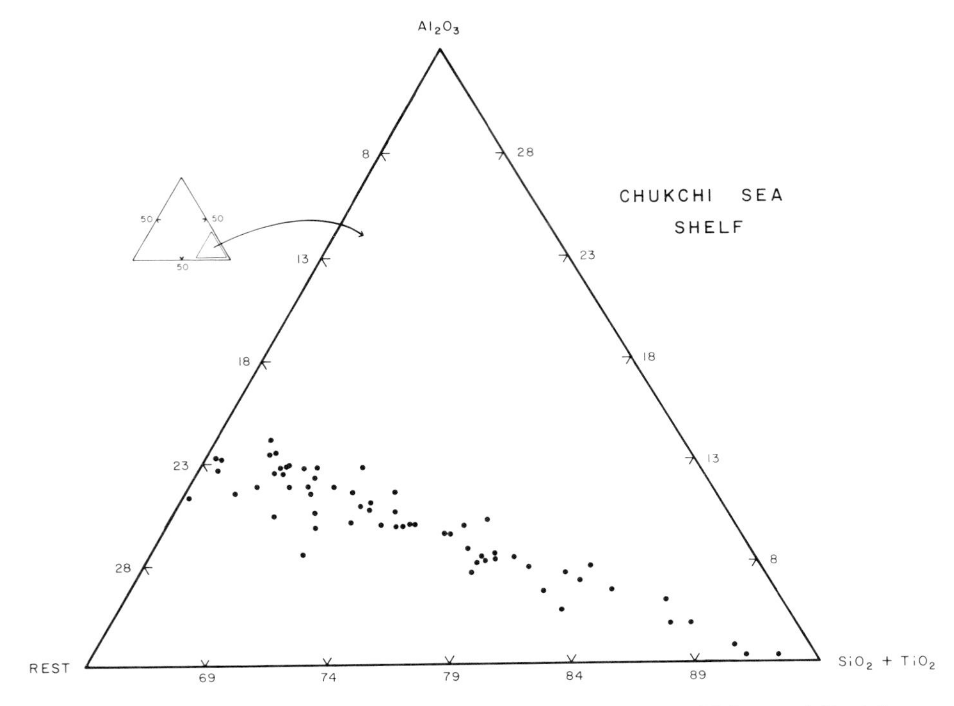

Figure 11-36. Triangular plot with percents Al_2O_3, $SiO_2 + TiO_2$, and Rest in sediments as end members, Chukchi Shelf.

the maturity of the river sediments. Thus, the mixing and deposition of the Yukon River sediments on the Chukchi Shelf results in an overall increase in the maturity of the shelf sediments.

The geochemical and textural parameters of sediments from the Chukchi Shelf suggest two major northward sediment pathways. The coastal water generally channels the terrigenous sediments contributed by the Alaska mainland. Offshore, the impetus of the Bering Sea water transports the sediments to the north from several sources, including the Bering Sea to the south and the Russian Arctic Shelf to the west.

REFERENCES

Aagaard, K. (1964). Features of the Physical Oceanography of the Chukchi Sea in the Autumn. M.S. Thesis, Univ. of Washington, Seattle.

Alaska Geological Society, North Slope Stratigraphic Section. (1971). Barrow Gas Field to Prudhoe Bay Oil Field to outcrops in Sadlerochit Mountain Area. Arctic North Slope, Alaska. Alaska Geol. Soc., Anchorage, Alaska.

Alaska Geological Society, North Slope Stratigraphic Section. (1973). Prudhoe Bay Oil Field to outcrops at Ignek Valley, Arctic North Slope, Alaska. Alaska Geol. Soc., Anchorage, Alaska.

Brosgé, W. P., and I. L. Tailleur. (1971). Northern Alaska Petroleum Province. In: Future Petroleum Provinces of the United States–Their Geology. I. H. Cram (Ed.). *Amer. Assoc. Geol. Mem.* .15:68-99.

Brosgé, W. P., and J. T. Dutro. (1973). Paleozoic Rocks of Northern and Central Alaska. *Amer. Assoc. Petrol. Geol. Mem.* 19:361-375.

Buffington, E. C., A. J. Carsola, and R. S. Dietz. (1950). Oceanographic cruise to the Bering and Chukchi Seas, summer 1949. Part I: Sea floor studies. U.S. Navy Electronics Lab. Rept. No. 204, p. 26.

Carsola, A. J. (1954). Recent marine sediments from Alaskan and northwest Canadian Arctic. *Amer. Assoc. Petrol. Geol. Bull.* 38(7):1552-1586.

Churkin, M. (1973). Paleozoic and Precambrian Rocks of Alaska and Their Role in its Structural Evolution. U.S. Geol. Survey Prof. Paper 740, 64 p.

Coachman, L. K., and K. Aagaard. (1966). On the water exchange through Bering Strait. *Limnol. Oceanogr.* 11:44-49.

Coachman, L. K., and K. Aagaard. (1974). Physical oceanography of Arctic and Subarctic Seas. In: *Marine Geology and Oceanography of the Arctic Seas.* Y. Herman (Ed.). Springer-Verlag, New York, pp. 1-73.

Coachman, L. K., and R. B. Tripp. (1970). Currents north of Bering Strait in winter. *Limnol. Oceanogr.* 15(4):625-632.

Creager, J. S. (1963). Sedimentation in a high energy, embayed, continental shelf environment. *J. Sed. Petrol.* 33(4):815-830.

Creager, J. S., and D. A. McManus. (1966). Geology of the southeastern Chukchi Sea. In: *Environment of Cape Thompson Region, Alaska.* N. J. Wilimovsky and J. M. Wolfe (Eds.), U.S. Atomic Energy Comm., Oak Ridge, Tenn., pp. 755-786.

Creager, J. S., and D. A. McManus. (1967). Geology of the floor of Bering and Chukchi Seas—American studies. In: *The Bering Land Bridge.* D. M. Hopkins (Ed.), Stanford Univ. Press, Stanford, Calif., pp. 7-31.

Fleming, R. H., and D. Heggarty. (1966). Oceanography of the southeastern Chukchi Sea. In: *Environments of the Cape Thompson Region, Alaska.* N. J. Wilimovsky and J. M. Wolfe (Eds.), U.S. Atomic Energy Comm., Oak Ridge, Tenn., pp. 697-754.

Fletcher, J. O. (1966). The Arctic heat budget and atmospheric circulation. Proceedings of the Symposium on the Arctic Heat Budget and Atmospheric Circulation, Lake Arrowhead, Calif., Jan 31-Feb 4, 1966. The Rand Corporation, Memoir RM-5233-NSF, pp. 25-43.

Hopkins, D. M. (1959). Cenozoic history of the Bering Land Bridge. *Science* 129:1519-1528.

Ingham, M. C., and B. A. Rutland. (1972). Physical oceanography of the eastern Chukchi Sea off Cape Lisburne-Icy Cape. In: *WEBSEC-70, An ecological survey in the eastern Chukchi Sea, September-October 1970.* U.S. Coast Guard Oceanographic Rept. 50, CG 373-50, pp. 1-86.

Lafond, E. C., and D. W. Pritchard. (1952). Physical Oceanographic investigations in the eastern Bering and Chukchi seas during summer of 1947. *J. Marine Res.* 11:69-86.

Loder, T. C. (1971). Distribution of dissolved and particulate organic carbon in Alaskan polar, sub-polar and estuarine waters. Unpublished Ph.D. Dissertation, Univ. of Alaska, Fairbanks.

McManus, D. A., and J. S. Creager. (1963). Physical and sedimentary environments on a large spitlike shoal. *J. Geol.* 71:498-512.

McManus, D. A., J. C. Kelley, and J. S. Creager. (1969). Continental shelf sedimentation in an arctic environment. *Geol. Soc. Amer. Bull.* 80:1961-1984.

Mosby, H. (1962). Water, salt and heat balance of the North Polar Sea and of the Norwegian Sea. *Geofys. Publ.* 24(11):289-313.

Naidu, A. S., and G. D. Sharma. (1972). Texture, mineralogy and chemistry of Arctic Ocean Sediments. Inst. of Mar. Sci., Univ. of Alaska, Fairbanks. Rept. R72-12. p. 31.

Saur, J. F. T., J. P. Tully, and E. C. Lafond. (1954). Oceanographic cruise to the Bering and Chukchi seas, summer 1949, Part LV. Physical Oceanographic Studies. U.S. Navy Electronics Lab. Research Rept. 416, v. I.

Wiseman, W. J., Jr., J. M. Coleman, A. Gregory, S. A. Hsu, A. D. Short, J. N. Suhayda, C. D. Walters, Jr., and L. D. Wright. (1973). Alaskan arctic coastal processes and morphology. Louisiana State Univ. Coastal Studies Inst., Baton Rouge, Technical Rept. 149, p. 171.

CHAPTER 12
Beaufort Sea Shelf

INTRODUCTION

The Beaufort Shelf among other Alaskan shelves is currently being systematically investigated. Recent developments related to petroleum exploration activities have drawn the attention of both the scientific and industrial communities to this area. Particularly, the increasing awareness of the importance of the fragile arctic ecology has initiated a variety of environmental investigations. The understanding of the delicate arctic environment and its processes, it is hoped, will help preserve the natural state of the region with minimum perturbations during the development and utilization of the resources from the shelf.

The earliest known attempt to study the sediments and bathymetry of the Beaufort Shelf was conducted by the Canadian Arctic Expedition during 1913 from the *Karluk.* Unfortunately this effort was short lived and ended with the loss of the vessel, data, and several scientists (Stefansson, 1944; Cushman, 1920). A summary of the sediment characteristics and topography of the Arctic Basin was described by Emery (1949). The first comprehensive geologic investigation of the Beaufort Sea was carried out by Carsola (1954a and b). Geologic investigations conducted on the shelf during the early 1970s were presented at the Symposium on Beaufort Sea Coast and Shelf Research held in San Francisco on January 7-9, 1974. This chapter is a summary of many investigations, details of most of which have been published in *The Coast and Shelf of the Beaufort Sea* (Reed and Sater, 1974).

GEOLOGY

The geologic evolution of the Beaufort Shelf and the region north of the Brooks Range has been discussed by various investigators and the current understanding of the geology and the related events of the region has been described by Pitcher (1973). In essence the evolution of the Beaufort Shelf conforms closely with the northern Chukchi Shelf described in Chapter 11. The major geologic feature of the shelf is a subsurface broad regional structure, the Barrow Arch. The axis of this arch follows the Beaufort Coast from the foothills of the Brooks Range to Point Barrow and continues onto the Chukchi Shelf. The arch developed during Late Jurassic or Early Cretaceous time on the northern provenance of the Arctic Platform. Consequently, a southward-sloping platform south of the arch and a northward-sloping continental margin north of the arch evolved. Cretaceous and Tertiary sedimentation north of the arch prograded the Beaufort continental terrace north from the arch across the presently formed continental margin.

Morphologically, the entire coastal belt between Point Barrow and the Alaska-Canada border can be divided into three groups: barrier islands, tundra bluffs, and river deltas (Short *et al.*, 1974). Four irregular, discontinuous barrier island chains occupy 52% of the coast and are composed of coarse sand and gravel. The tundra bluffs with gravelly beaches form 32% of the coast. The deltas take up 16% of the coast and consist of mixtures of sand, silt and clay.

Landward, the coast is adjoined by an extensive plain, with tundra soils, patterned ground, lakes, ponds, streams and braided rivers. The plain is underlain by the Quaternary Gubik Formation, which consists of an unconsolidated mixture of silt and fine-grained sand with clay and gravel. The Gubik Formation overlies Cretaceous or Tertiary sediments (Black, 1964).

BATHYMETRY

The detailed bathymetry of the central and outer shelves of the Beaufort Shelf has been described by Carsola (1954a and b). The floor of the Beaufort Shelf is characterized by three main physiographic features: (1) the inner shelf, which grades gently from the shoreline to the 30-m isobath; (2) the outer shelf, with a slightly steeper gradient, which lies between the 30-m and 70-m isobaths; and (3) the continental slope, which falls fairly steeply from the shelf edge. In comparison with other Alaskan shelves Beaufort Shelf is unusually shallow. The shelf break occurs between the 50- and 70-m isobaths and the average depth of the shelf is 37 m. The width of the shelf ranges from 55 km in the east to 110 km in the west, and the average width is 72 km (Figs. 12-1 and 12-2).

The upper reaches of the innershelf are interrupted by low deltaic mudflats where prominent rivers meet the sea. Much of the near-shore region is marked by low, narrow, sand and gravel barrier islands. Several east-west shoals have been observed and presumably are the submerged extensions of the barrier island chain. Numerous spits and bars associated with headlands are also present.

The outer shelf is essentially flat with some micro-relief, most of which results from the physical interaction of grounded ice with the shelf. Other morphologic features are the submarine hills, so-called submarine pingos (Shearer *et al.*, 1971). The peaks of some of these pingos lie within 10 m of sea level. These topographic highs are ice-cased conical mounds up to 300 m in diameter at their base and rise 20 to 50 m to form sharp peaks, which are generally breached.

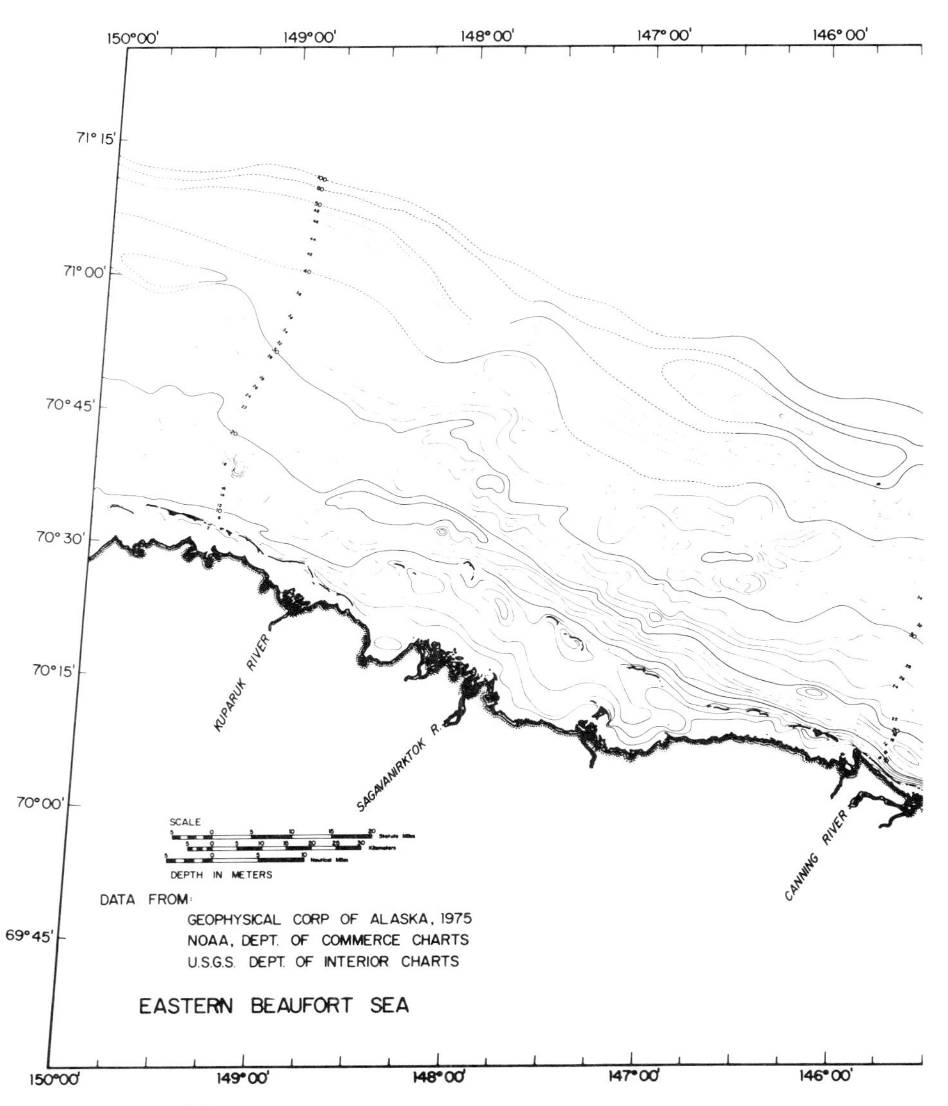

Figure 12-1. Bathymetric chart, Eastern Beaufort Sea.

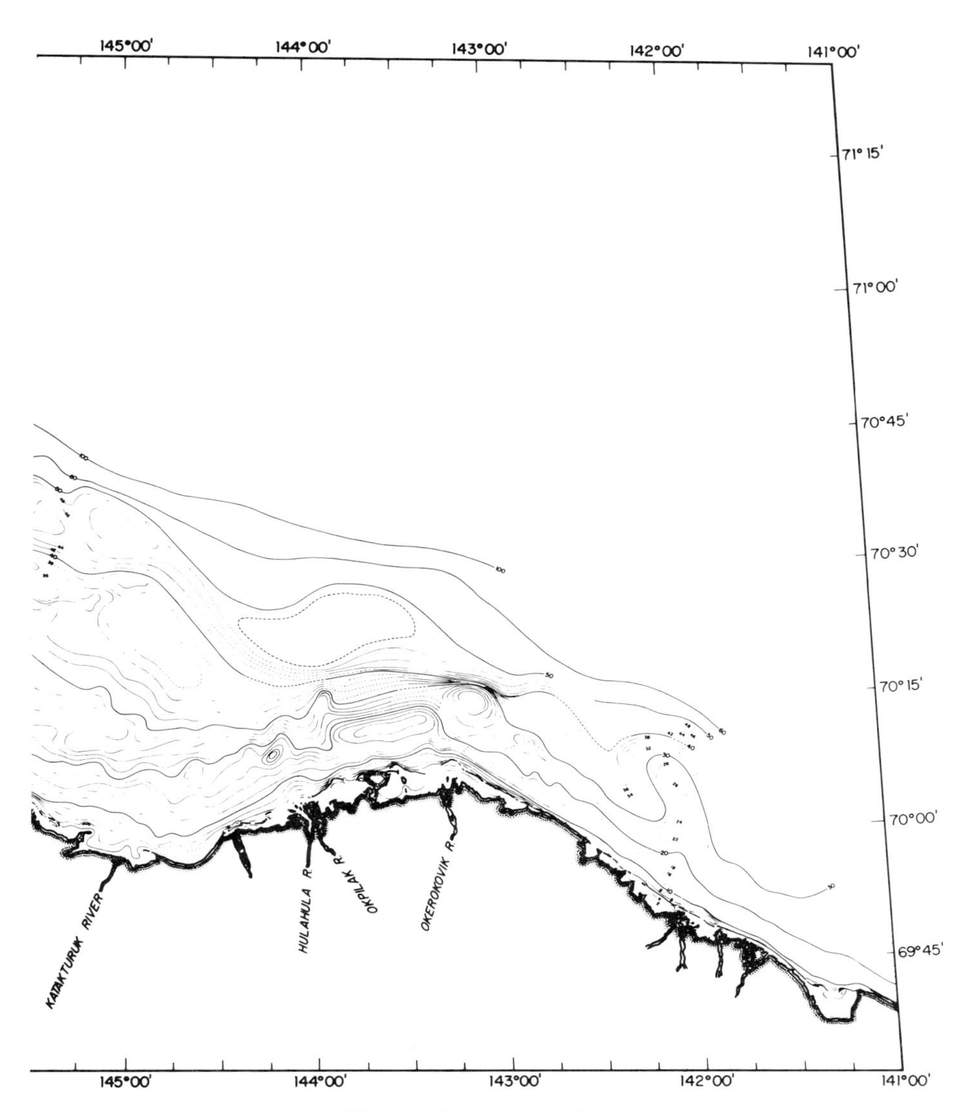

Figure 12-1 continued.

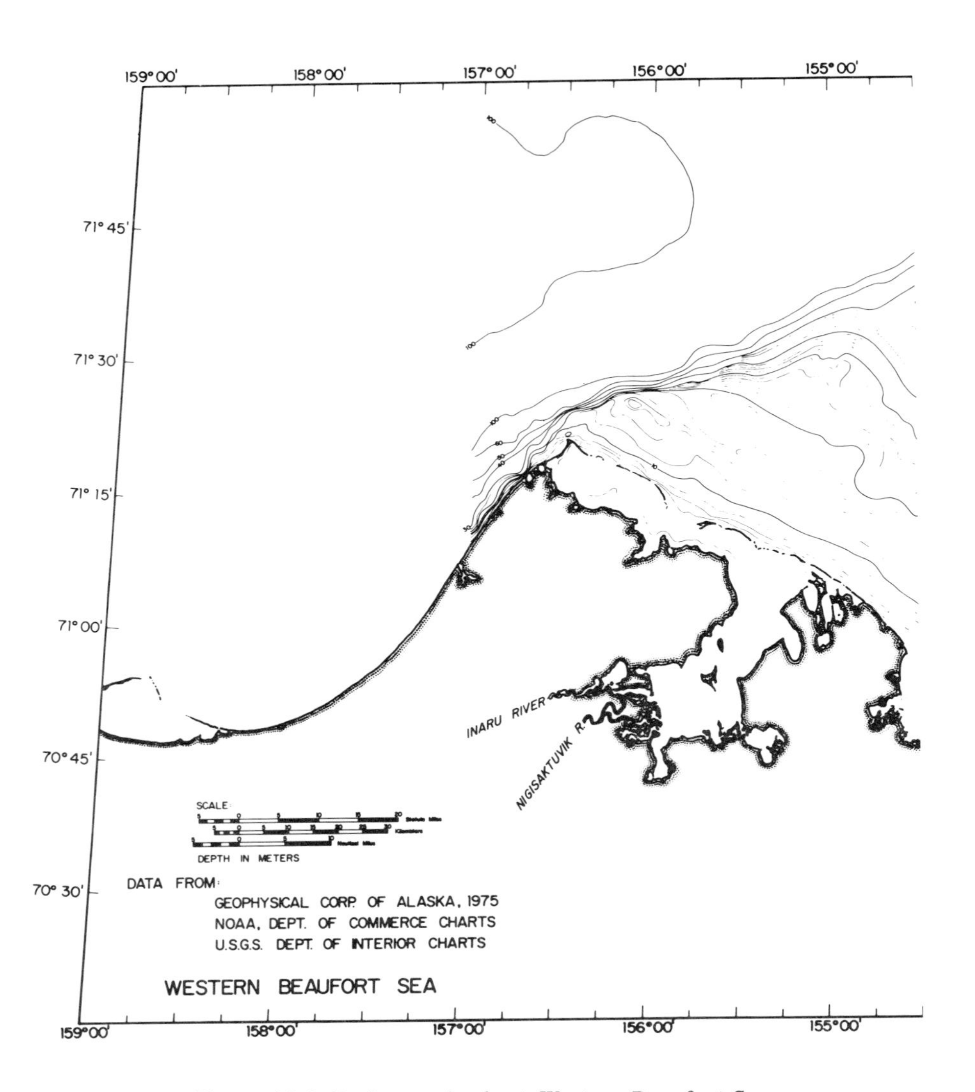

Figure 12-2. Bathymetric chart, Western Beaufort Sea.

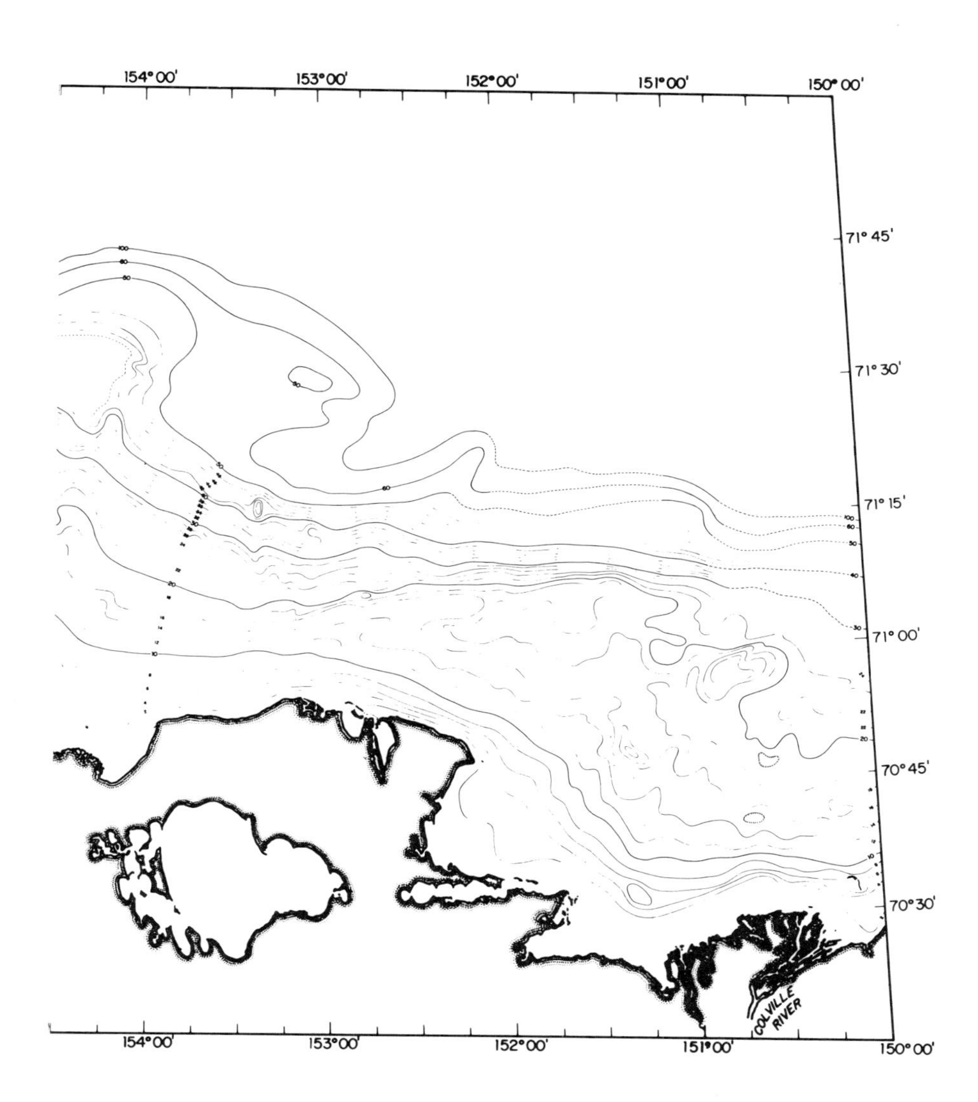

Figure 12-2 continued.

Subsequently ice gouging forms grooves or furrows, resulting in mixing of sediments and complex shelf bathymetry.

The ice-gouged features observed on the sea floor range from V-shaped to U-shaped depressions with sharp-crested ridges (Reimnitz and Barnes, 1974). The largest gouge surveyed by these authors was about 10 m wide and nearly 2 m deep; however, they added that average gouge ranged to 1 m deep and 2 to 3 m wide.

Evidence for the presence of offshore and sub-seabottom permafrost in the Beaufort Sea has been reported by various investigators (MacKay, 1972; Lewellen, 1974). Molochuskin (1973) reported the presence of frozen ground beneath the Laptev Sea of Siberia. He observed the presence of sub-bottom permafrost in water ranging from 4 to 900 m. The data collected by various investigators suggest that sub-seabottom permafrost exists over much of the Beaufort Shelf. The mean annual temperature of the near-shore bottom water is -1.3°C and it has been suggested that permafrost is aggrading under the Holocene bars and shoals and along some shorelines (Lewellen, 1974). Sediment temperatures in the shallow region of Beaufort Shelf (<2-m isobath) ranged between -2.2°C and -4.6°C (Reimnitz and Barnes, 1974) and are indicative of subbottom permafrost.

The distribution of ice-bonded permafrost, however, is probably dependent upon the sediment texture. From limited observations it appears that ice-bonded permafrost is mostly confined to the areas of coarse-grained sub-seabottom material where abundant fresh water has been available to form inter-granular ice. Some of the seabottom morphologic features can be related to the ice-bonded permafrost.

The continental slope is steep and its gradient generally varies between 5 and 10°, with local steeper gradients. The slope intermittently is covered by large slumps.

HYDROLOGY AND HYDROGRAPHY

Because of logistic problems in winters oceanographic parameters gathered so far on the Beaufort Sea pertain mostly to the spring and summer seasons. The shelf water characteristics are an integral part of the water masses of the Arctic Ocean but are highly influenced by the seasonal processes. During winter, for about nine months, the shelf is dominated by ice. In July the heavy polar pack ice begins to recede northward and the warmer temperature melts the shorefast ice, making the shelf of the western Beaufort Sea generally ice free from late July through September.

Yearly freezeup and breakup times along the Beaufort Sea coast may vary widely. For example, freezeup in the west at Point Barrow may occur any time between September 1 and December 15, and in the east at Sachs Harbor (Banks Island, Canada) between late September and late November (Kovacs and Mellor,

1974). Similarly, ice breakup has occurred at Point Barrow as early as mid-June and as late as the end of August, while at Sachs Harbor it occurs from late June to late July. The sporadic occurrence of northerly winds in summer can cause the arctic pack ice cover to move toward the shore and cover the inner shelf.

According to Kovacs and Mellor (1974), the winter ice cover on the southern shelf can be divided into three broad zones: (1) the shorefast ice zone, (2) the seasonal pack ice zone, and (3) the polar pack ice zone. The quasi-immobile shorefast ice zone consists of seasonal ice. The ice is about 2 m thick and generally extends seaward up to the 20-m isobath. At the seaward edge of the shorefast ice the seasonal pack ice zone begins with a distinct narrow shear zone of brecciated ice and extends seaward to between 100 and 200 km. The undeformed seasonal pack ice is also about 2 m thick and has a salinity of $3°/_{oo}$ to $14°/_{oo}$. The polar pack ice zone of the Arctic Ocean is characterized by thick multi-year ice floes. The old multi-year floes vary in thickness from 2.1 to 4.5 m (Bushuev, 1964) and their salinity ranges between $0°/_{oo}$ and $6°/_{oo}$.

The distribution of three types of ice on the shelf is, of course, highly dependent on meteorologic conditions, such as air temperature; cloud cover; wind speed; snow fall; oceanographic conditions, including salinity, water depth, currents, and sea state; and finally the surface runoff. The ice conditions are also significantly altered by seasonal intense storms. These storms move polar pack ice shoreward and build pressure ridges. The onshore movement of polar pack ice naturally causes extensive deformation along the shear zone of the seasonal pack ice.

The summer water masses on the shelf are controlled by the three major Arctic Ocean water masses and the surface outflow. The Arctic Ocean consists of (1) the Arctic surface water (0-200 m), which includes Bering Sea water; (2) Atlantic Ocean water (200-900 m); and (3) the Arctic bottom water (below 900 m) (Coachman, 1969). Normally, on the shallow shelf, the Arctic surface water exerts most influence, except when upwelling along the slope causes intrusion of intermediate Atlantic water. The Arctic surface water on the shelf and slope is interlayered with Bering Sea water. Hufford (1974) observed Bering Sea water at three different depths: near the surface (with temperatures above 0°C because of summer heating); at about 75 m, where it is characterized by a subsurface temperature maximum (when overlain by cold, less saline local surface water); and at 125 m (identified by a temperature minimum that represents Bering-Chukchi Sea winter water).

The coastal effluent, particularly from large rivers such as the Colville River, also affects near-shore water mass characteristics. The most striking feature of the arctic drainage is the intense annual discharge during a relatively short period. For example, in 1962 the Colville River discharged 43% of the 16×10^9 m^3 annual discharge and 73% of the 5.8×10^6 tons annual load in suspension during a three-week period around the spring breakup (Arnborg *et al.*, 1967). Although the contribution of fresh water by the surface runoff of the Colville River (the largest river draining into the western region) has been calculated over a few years, the freshwater contributions made by other streams remain unknown. The yearly runoff into the Alaskan Beaufort Sea is estimated at about 400 km^3 (Antonov, 1958). Most of this discharge (approximately 80% of the total) occurs in June and is carried by the Colville, Kuparuk, Sagavanirktok, and Canning rivers. The water characteristics described below are a summary of the data collected by various agencies and investigators.

The inner shelf (0-30 m isobaths) shows the largest variations in both temperatures and salinity during summer as well as winter. The variability is the result of its proximity to shoreline, inflow of fresh water, and surface interface phenomena. The temperature during summer ranges from -1.5 to 14°C, and salinity ranges from 0.7‰ to 31.6‰. The winter temperature is generally below 0°C, varying from -0.5°C to 0.5°C. Winter salinity is somewhat lower than summer salinity, with a range of 21.8-31.0‰.

The ranges of summer temperature and salinity on the outer shelf are -1.8°C to 5.0°C and 11.5‰ to 33.0‰, respectively. Temperatures and salinities in winter are quite uniform, with ranges of -1.5°C to -1.1°C and 32.0‰ to 33.0‰, respectively. The mean annual bottom temperature for the Beaufort Sea is -1.3°C (Lewellen, 1974).

The pattern of water circulation on the shelf is not well defined. A few current measurements have been recorded on the shelf waters, and most of these were obtained during the relatively short summer months. Because the surface currents are largely controlled by local seasonal winds, the net movements of water masses at the surface as well as at shallow depths during different seasons cannot be assessed from short-term summer measurements. Summer measurements obtained during 1971 and 1972 by Hufford (1974) suggest that surface currents near the shore vary from 0 to 60 cm/sec and are primarily caused by easterly winds. The surface currents, however, can reverse their directions for short periods due to abrupt changes in wind directions. In the subsurface, below the strong pycnocline (>20-m depth), the currents are independent of the wind direction and are predominantly eastward at a speed of <15 cm/sec. Offshore, the circulation is dominated by the clockwise Beaufort Sea Gyre (Coachman, 1969).

SEDIMENTS

Surface sediments from the Beaufort Shelf have been described by Carsola (1954a). Naidu and Sharma (1972), Naidu and Mowatt (1974), and Barnes and Hopkins (1978). Although much of the narrow shelf is relatively smooth, the textural distribution of sediments is complex. In particular, there is a lack of seaward coarse- to fine-grained sediment grading across the shelf. Sediments predominantly consist of poorly sorted gravelly silty sand, well-sorted clayey silt, and moderately to well-sorted silt and sand.

Sediment texture on the shelf shows three broad regional distributions. The shelf break and the shelf slope are covered with poorly sorted gravelly silt-clay. These bimodal sediments have their primary mode in coarse sand and secondary mode in fine clayey silt. The shelves west of the Colville River and east of Prudhoe Bay are mantled with well-sorted silt. Finally, the shelf between the Colville River and Prudhoe Bay is floored with poorly sorted, gravelly silty sands, which often extend to the shelf break. These sands, on an average, contain between 2 and 10% gravel, but locally the gravel content may exceed 60%.

The heavy mineral composition of the sand fraction (0.062-2.00 mm) of sediments from this region has been investigated in detail by Leupke (1975). She reported that heavy minerals constituted only 5% or less of the sand fraction. Twenty-four heavy minerals were identified. Higher concentrations generally were observed offshore, but not significant association between percent of heavy minerals and either mean grain size or shelf depth of sample was apparent. Occurrence of large numbers of species and lack of mineral separation suggest that the sediments are not transported over an extended distance. The mineral composition also suggests a heterogeneous source for the sediments.

The distributions of iron-stained aggregates and garnet in sediments and the adjacent coastal plain deposits provide some clues as to the sources for these components. The iron-stained aggregates, with their high concentrations in near-shore sediments, apparently originate in the adjacent sea cliffs. Higher concentrations of garnet west of Cape Halkett reflect a higher percent of garnet in the Gubik Formation extensively exposed west of Colville River.

The clay mineralogy of the fine fraction ($<2\ \mu m$) has been studied by Naidu and Mowatt (1974). The clay mineralogy of the fine fraction of sediments consists predominantly of illite (56 to 69%), with subordinate amounts of chlorite (14 to 30%) and kaolinite (5 to 10%), and small amounts of smectite (trace to 9%). Based on the clay mineral distributions of the local rivers, it appears that the fine sediment derived from the river drainage on the coastal plains is largely deposited between Point Barrow and Barter Island.

GEOCHEMISTRY

The distributions of some selected elements in sediments, interstitial waters and gravel encrustations have been studied by Naidu and Mowatt (1974). These investigators have documented a relative deficiency of various major and minor elements in the northern high-latitude shelf sediments, as compared to shelf sediments of the lower latitudes. A study of the distribution patterns of these elements in shelf sediments was also included in these investigations.

SEDIMENT SOURCE AND TRANSPORT

The vast Alaskan drainage area providing sediment to the Beaufort Sea includes portions of the Brooks Range, Arctic Foothill and Arctic Coastal Plain provinces. Presently glacial ice mostly erodes the upper reaches of the Brooks Range. Abundant glacial deposits along the Brooks Range suggest that part of the coastal plains may have been glaciated during the recent past. These glacial deposits are currently being reworked and partly carried northward by rivers and streams. The discharge mostly flows over low-lying plains with tundra soils, vegetations, patterned ground, lakes, ponds, and numerous streams.

Sediment input and its subsequent distribution on the shelf is seasonal. In the coastal zone (<20-m isobath) the sediment influx is related to spring and summer events: river breakup, ice breakup, open-water conditions and beach thaw. Offshore, the sediment movement is affected by the winter pack ice and sub-ice conditions.

The river flooding and breakup in late May or early June marks the beginning of summer conditions. The breakup melts the fast ice, which eventually becomes thin and rises off the bottom, and open water begins to form near river mouths. As the fast ice continues to melt, the open water extends offshore. By July, the remaining fast ice is broken up by wind and waves and the melting of fast ice is symbolic of the open-water season, which usually lasts until late September.

The bluffs undergo thermoerosion during summer. The slow melting of permafrost ice on and in the beach causes subsidence of overlying beds, which provide sediments to the near-shore zone.

Terrigenous sediments, mostly consisting of sand, silt, and clay, upon entering the sea are reworked and partly redistributed on the shelf and slope. The reworking of sediments by grounded ice at depth <20 m is somewhat limited. Reworking of bottom sediments by ice gouging on the shelf at depth >20 m has been reported by Reimnitz and Barnes (1974).

Relatively warm river floodwater that enters the Beaufort Shelf carries organic and inorganic matter. Some of this detritus is deposited on the sea ice, while much of it is carried to the sea. River detritus brought to the sea generally increases with increased flow. Walker (1974) estimated just under 300,000 tons of suspended sediments in the seaward edge off the Colville River during early stages of the river breakup. River sediments carried either over or under the ice are deposited in the near-shore region. Besides the river detritus, the near-shore region also receives sediments from bluff. During summer, considerable thawing of ice in the bluff causes slumping of bluffs and sediments are introduced to the surf zone.

Open-water conditions prevailing during summer cause sediments to be transported more readily by normal marine processes. The wind-induced currents in the near-shore region (depth <20 m) may reach speeds up to 100 cm/sec (Reimnitz and Barnes, 1974). Sedimentation on the shallow shelf is therefore dominated by wave- and current-related processes. Occasional severe storms also play a significant but cataclysmic role in overall sediment transport in the area (Hume *et al.*, 1972). The net transport of water to the west in the near-shore regions causes a westward movement of sediments. The westward drift of sediments is indicated by the deflection and westward extension of plumes emerging from major rivers and westward migration of barrier islands, spits and near-shore bars. Westward sediment migration on the Alaskan Beaufort Shelf has been also suggested by Barnes and Reimnitz (1974). They observed westward fining and improved sorting in offshore sediments and the progressive widening of the shelf. A net westward transport of sediments in the near-shore zone has been proposed by Wiseman *et al.* (1973) and Short *et al.* (1974). A similar pathway in the far offshore region has been suggested by Campbell (1965).

Ice and ice-related sedimentary processes dominate the Beaufort Shelf during winter. Shoreward of the 20-m isobath the near-shore region is covered by shorefast ice for 8-10 months annually. Sediment transport during ice cover, however, is scant. Small quantities of sediments are transported by adfreezing to the bottom of ice and by ice rafting. Seawards of the 10-20 m isobaths, in the area of shear zone, an extensive interplay between ice- and water-related processes contributes to the complex sediment distribution. Ice-gouging, current, and wave ripple marks are common in this region (Reimnitz and Barnes, 1974). Seaward of the shear line, the central shelf is criscrossed by the deep-keeled pressure ridges and the grounded ice islands, both of which gouge the shelf bottom. Ice gouging causes extensive mixing of bottom sediments. Current scours, commonly observed around grounded ice, remove fine fractions of sediments from the bottom.

Offshore, the outer shelf is covered by the pack ice. Movement of pack ice and subsurface currents mostly controls the sediment distribution on the outer shelf.

The source for sediments deposited near-shore lies along the coastal bluffs and the coastal plain to the south. The bluffs undergo thermo-erosion during summer. The slow melting of permafrost ice causes subsidence and provides sediments to the near-shore zone. The low-lying coastal plain and the northern slope of the Brooks Range are drained to the north. This large drainage area provides much detritus to the shelf, most of which is discharged into the near-shore region.

The sources for coarse material, in particular the gravel, deposited on the middle and outer shelves remains obscured. The striated angular character of the gravel suggests their glacial ice transport. Scrutiny of shorefast ice and ice ridges indicates that presently ice does not carry gravel offshore. Lack of conclusive evidence for glaciations on the coastal plain during the recent past further shrouds the possibility of an earlier source of the gravels found on the shelf. On the basis of the lithology of the gravel, Rodeick (1974) suggested the southern part of the Canadian Archipelago as their source.

The interaction between open-water hydrodynamic and ice-related processes results in a highly complex distribution of sediments. This complex distribution, nevertheless, can be explained provided that the events are followed in the proper sequence. During the recent low sea level stand it is safe to assume that the present middle and outer shelves have been under the influence of the Arctic gyre. The east-west trajectory of peripheral ice of the clockwise moving gyre would normally be grounded in these parts of the shelf. Thus, ice carried sediments imbedded in it and released them on the shelf. The distribution of gravel conforms quite well with the movement of ice. The glacially derived sediments were probably ice gouged and some finer fractions were elutriated.

During the subsequent transgression the ice-rafted sediments were reworked by wave action and typical gravel lag deposits were formed. The rise in sea level also resulted in an influx of sediments from the coastal plain and the coast. These sediments normally would cover the lag gravel were it not for intense mixing by the ice gouging. The combined action of ice and water ultimately reworked clay and silt fractions.

Hydrographic data (Bornhold, 1975) provide some insight into the present sedimentary regime. North of the Mackenzie River, water from depth rises southward along the canyon and, as it reaches the shelf, it bifurcates into east and west streams. This intrusion of water may be the combined effect of Mackenzie flow to the east and the wind-related net westward water movement on the near-shore shelf. As a result of the upwelling and wind-induced circulation, an elongated clockwise gyre with its westward flow along the near shore and its eastward flow below the pack ice and at depth along the shelf edge develops on the Alaskan Beaufort Shelf. The coarse sediments in the near-shore area are generally under the westward moving portion of the gyre, while finer fractions, which reach offshore, fall under the influence of eastward flowing water. This scheme explains well the eastward migration of clayey sediments along the outer shelf. Clay mineral distribution of sediments has been studied by Naidu and Mowatt (1974) and they proposed an easterly drift for these minerals.

The distribution of clay minerals in the <2-μm sediment fraction, studied by Naidu and Mowatt (1974), however, suggests that at least some eastward transport of sediments in the central and outer shelves occurs. These investigators also pointed out an anomalously low thickness of contemporary sediments on the eastern Alaskan Shelf of the Beaufort Sea and suggested that fine-grained sediment deposition is probably being bypassed in this shelf area and that the sediments are carried offshore by ice.

In view of the available hydrographic data, it appears that on the shallow shelf (<20 m water depth) the dominant water flow is indeed to the west. Sediments in the near-shore region are continually reworked by waves, currents, and ice and, therefore, most silt- and clay-sized sediments are retained in suspension, part of which is ultimately transported offshore by tidal currents. In the near-shore region, however, the coarser sediments (coarse silt, sand and gravel) continue to move westward as observed by various investigators.

Offshore, with decreased turbulence in the water column, the fine-grained sediments begin to settle and seaward grading of sediments begins. During settling, however, the particles encounter the eastward moving subsurface current observed by Hufford (1974) and Mountain (1974). It appears that the eastward moving subsurface current may be the prime carrier for fine-grained sediments. Based on clay mineralogy Naidu and Mowatt (1974) have suggested an eastward transport of the sediments on the central and outer shelves.

The anomalous presence of gravel on the eastern Alaskan Shelf has been discussed by Naidu and Mowatt (1974). The petrology and the texture of the cobbles and pebbles indicate that the bulk of the gravel in this region was ice rafted on the continental ice breakup during the Late Wisconsin. Probably the slow rate of sedimentation on this shelf region has precluded the burial of the paleo-gravel deposits.

REFERENCES

Antonov, V. S. (1958). The role of continental runoff in the current regime of the Arctic Ocean. *Problemy Severa.* 1:52-64 (title translated).

Arnborg, L., H. J. Walker, and J. Peippo. (1967). Suspended load in the Colville River, Alaska, 1962. *Geografiska Analer* 49:131-144.

Barnes, P. W., and D. M. Hopkins. (1975). Geological Science. In: *Interim Synthesis Beaufort/Chukchi, Environmental Assessment of the Alaskan Continental Shelf.* G. E. Weller (Ed.). Dept. of Commerce, NOAA, Boulder, Colorado, pp. 101-133.

Black, R. F. (1964). Gubik Formation of Quaternary Age in Northern Alaska. U.S. Geol. Survey Prof. Paper 302-C, pp. 59-91.

Bornhold, B. D. (1975). Suspended matter in the southern Beaufort Sea. Beaufort Sea Project Tech. Rept. #25b. Dept. of Environment, Victoria, B.C., Canada.

Bushuev, A. V. (1964). Plasticity and isostatic equilibrium of an ice cover. *Trudy, Arktiki i Antarktiki Institute* 267:105-109.

Campbell, W. J. (1965). The wind-driven circulation of ice and water in a polar ocean. *J. Geophysical Research* 70:3279-3301.

Carsola, A. J. (1954a). Recent marine sediments from Alaskan and northwest Canadian Arctic. *Amer. Assoc. Petrol. Geol. Bull.* 38:1552-1586.

Carsola, A. J. (1954b). Microrelief on arctic sea floor. *Amer. Assoc. Petrol. Geol. Bull.* 38:1587-1601.

Coachman, L. K. (1969). Physical oceanography in the Arctic Ocean: 1968. *Arctic* 22(3):214-224.

Cushman, J. A. (1920). Foraminifera. In: *Canadian Arctic Expedition, 1913-1918.* Rept. v. 9: Annelids, parasitic wurms, protozoans, etc., Pt. M. Kings Printer, Ottawa, 13 p.

Emery, K. O. (1949). Topography and sediments of the Arctic Basin. *J. Geol.* 57:512-521.

Hufford, G. L. (1974). Dissolved oxygen and nutrients along the north Alaskan shelf. In: *The Coast and Shelf of the Beaufort Sea.* J. C. Reed and J. E. Sater (Eds.), Arctic Inst., Arlington, Virginia, pp. 567-568.

Hume, J. D., M. Schalk, and P. W. Hume. (1972). Short term climate changes and coastal erosion, Barrow, Alaska. *Arctic* 25:272-78.

Kovac, A., and M. Mellor. (1974). Sea ice morphology and ice as a geologic agent in the southern Beaufort Sea. In: *The Coast and Shelf of the Beaufort Sea.* J. C. Reed and J. E. Sater (Eds.), Arctic Inst., Arlington, Virginia, pp. 113-161.

Lewellen, R. I. (1974). Offshore permafrost, Beaufort Sea, Alaska. In: *The Coast and Shelf of the Beaufort Sea.* J. C. Reed and J. E. Sater (Eds.), Arctic Inst. North America, Arlington, Virginia, pp. 417-461.

Luepke, G. (1975). Heavy-mineral trend in the Beaufort Sea. U.S. Geol. Surv. Open File Rept. 75-667, p. 29.

MacKay, J. R. (1972). Offshore permafrost and ground ice, southern Beaufort Sea. *Can. J. Earth Sci.* 9:1550-1561.

Molochuskin, E. N. (1973). The effect of thermal abrasion on the temperature of permafrost rocks in the coastal zone of the Laptev Sea. *2nd Intl. Proc. Permafrost Conf., Moscow* 2:52-58.

Mountain, D. G. (1974). Preliminary analysis of Beaufort Shelf circulation in summer. In: *The Coast and Shelf of the Beaufort Sea.* J. C. Reed and J. E. Sater (Eds.), Arctic Inst. North America, Arlington, Virginia, pp. 27-42.

Naidu, A. S., and T. C. Mowatt. (1974). Clay mineralogy and geochemistry of continental shelf sediments of the Beaufort Sea. In: *The Coast and Shelf of the Beaufort Sea.* J. C. Reed and J. E. Sater (Eds.), Arctic Inst. North America, Arlington, Virginia, pp. 493-510.

Naidu, A. S., and G. D. Sharma. (1972). Geological, biological and chemical oceanography of the southeastern Chukchi Sea. In: *WEBSEC-70, An ecological survey in the eastern Chukchi Sea,* September-October. 1970. U.S. Coast Guard Oceanographic Rept. 50, CG373-50, pp. 173-195.

Pitcher, M. G. (Ed.) (1973). Arctic geology. *Amer. Assoc. Petrol. Geol. Mem.* 19:8.

Reed, J. C., and J. E. Sater (Eds.) (1974). *The Coast and Shelf of the Beaufort Sea.* Arctic Inst. North America, Arlington, Virginia, 750 p.

Reimnitz, E., and P. W. Barnes. (1974). Sea ice as a geologic agent on the Beaufort Sea Shelf of Alaska. In: *The Coast and Shelf of the Beaufort Sea.* J. C. Reed and J. E. Sater (Eds.), Arctic Inst. North America, Arlington, Virginia, pp. 301-354.

Rodeick, C. A. (1974). Marine gravel deposits of the Beaufort Sea Shelf. In: *The Coast and Shelf of the Beaufort Sea.* J. C. Reed and J. E. Sater (Eds.), Arctic Inst. North America, Arlington, Virginia, p. 511.

Shearer, J. M., R. F. Macnab, B. R. Pelletier, and T. B. Smith. (1971). Submarine pingos in the Beaufort Sea. *Science* 174:186-818.

Short, A. D., J. M. Coleman, and L. D. Wright. (1974). Beach dynamics and nearshore morphology of the Beaufort Sea coast, Alaska. In: *The Coast and Shelf of the Beaufort Sea.* J. C. Reed and J. E. Sater (Eds.), Arctic Inst. North America, Arlington, Virginia, pp. 477-488.

Stefansson, V. (1944). *The Friendly Arctic.* Macmillan, New York, p. 812.

Walker, H. J. (1974). The Colville River and the Beaufort Sea: Some interactions. In: *The Coast and Shelf of the Beaufort Sea.* J. C. Reed and J. E. Sater (Eds.), Arctic Inst. North America, Arlington, Virginia, pp. 513-540.

Weiseman, W. J., Jr., J. M. Coleman, A. Gregory, A. S. A. Hsu, A. D. Short, J. N. Suhayda, C. D. Walters, Jr., and L. D. Wright. (1973). Alaskan Arctic Coastal Processes and Morphology. Louisiana State University, Coastal Studies Institute, Tech. Rept. No. 149, p. 171.

CHAPTER 13
Textural and Geochemical Evolution

TEXTURAL EVOLUTION

The Alaskan Shelf provides an excellent opportunity to evaluate the changes in the sediment texture evolved as a result of diverse weathering (temperate to polar) at the source. The latitudinal changes in the terrestrial weathering and transport agencies may significantly influence the character of detritus deposited on the shelf. It must be pointed out that the initial character of the mineral detritus is, to an extent, altered by the marine processes operating on the shelf. In spite of the differentiation in detritus during marine transport, remnants of the original texture character formed near the source are expected to be retained on the shelf. One of the prime purposes of describing the textural distribution of shelf sediments will be to relate them to their source. This, however, would require some understanding of major transport processes on the shelf.

Besides the usual weathering process, there are two processes in Alaska that are prevalent at the source area; glaciation and permafrost. In southern Alaska, the solifluction in the adjacent mountains, glaciation and readjustment of landscape in areas that are now ice free but were glaciated 15,000 years ago are the main processes of weathering. The most important features of inshore sediment processes are related to talus fall, landslides, and avalanches. Sorting is extremely poor in these deposits. The permafrost occurs in regions with low mean annual temperatures and meager precipitation, and sediments from these regions generally contain a high percentage of silt. Besides the weathering, the nature of the source rock also influences the texture of detritus and, depending on the climatic condition, the source rock in relation to its lithology may provide a dominant grain size mode or several modes.

Once the detritus enters the marine environment its movement is mostly controlled by the water mass gradient, wave action, tidal and permanent currents, and occasional storm surges and slides. The coastal water masses generally carry finer fractions of detritus offshore as suspension load, characteristically as large plumes. The coarse fractions are generally transported near the bottom as bed load. Sediment distribution in northern latitudes (Beaufort, Chukchi and Bering seas), to some extent, is also influenced by ice gouging and rafting.

On the shallow shelf, waves by far provide the most energy for sediment transport. Depending on the wave approach to the coast most waves will cause onshore-offshore sediment movement. The incident wave reaching shore at an angle may generate a longshore current. Its generation is best explained as resulting from the longshore component of the momentum flux associated with the waves (the radiation stress). The longshore current, interacting with wave surf, in turn produces a sand transport along the beach. Under normal conditions, as the current picks up sediments for transport, it must then deposit a corresponding amount. The longshore current, therefore, may erode sediments and deposit them further up current. The transport and deposition along the shore, however, is dependent on the increase or decrease in cumulative wave energy and on the configuration of the shore. Excellent examples of longshore currents and the movement of sediments can be observed along the northeastern Chukchi Sea coast. Between capes sediments are transported to the northeast by longshore currents generated by winds from the west and to the southwest by longshore currents developed by the winds from the north. In either case, the sediments are eroded from the central part of the crescent-shaped shore and are carried to the extremities of the crescent to form prominent capes. In some parts of the Alaskan Shelf, longshore currents are the primary agents for sediment transport.

Generally, tidal currents exercise a great influence on the sediment transport on the central and outer shelves. The tides over the most part of the Alaskan Shelf are small in magnitude. Overall, the tidal range progressively decreases northward. It is a few meters in the southeast Alaska and decreases to a few decimeters in the Beaufort Sea. In some areas, however, tidal amplitude is magnified and the resulting currents may solely control the sediment transport in that region. Exceptionally high tidal currents occur in Cook Inlet, and these currents, to a large extent, control sedimentation in the inlet. Tidal currents in Bristol Bay, however, also influence sediment distribution in the bay.

Numerous seasonal and permanent currents have been observed on the Alaskan shelves. Along the northeastern Pacific rim, the omnipresent strong oceanic counterclockwise Alaska Current sweeps past the edge of the shelf margin and produces large eddies on the shelf. The current primarily sets the pace and direction for the inflowing fresh water as well as sediment transport on the shelf. As an effect of this current, the Copper River forms an unusual delta near the mouth. Instead of the normal offshore accretion, an abnormally elongated delta extends laterally along the shore, part of which spills over into Prince William Sound.

The northward flow of the Pacific water into the Bering Sea influences sediment distribution in parts of the Bering and Chukchi shelves. The presence of coarse sediments beneath the paths of currents and the occurrence of finer sediments in the adjacent zones is the direct result of intense movement of water. The coarsening of sediments on elevated segments of the shelf floor (Tarr Bank in the central Gulf of Alaska Shelf), as compared with lower areas, is an inevitable consequence of more intense movement of water on the topographic high.

It is difficult to discern the sole effects of either tides or currents on the sediments because superimposed on these effects are the influence of waves. Winter and summer surface waves are of sufficient energy to resuspended sediments at depths of 40-50 meters. Internal waves, on the other hand, propagate shoreward along density interfaces in the thermocline. The summer thermocline on the shelf is at shallower depths (30 m) than winter thermoclines are (60 m). The effect of these waves on sediment transport is therefore seasonal.

Most regions of the Alaskan Shelf are subjected to moderately severe to severe storms. These storms in some regions are quite frequent and, therefore, markedly affect sedimentation on the shallow shelf. Each storm produces waves that are competent to resuspend some of the sediments deposited during previous quiet periods. The energy imparted during the storm may not only texturally grade the shelf sediments but simultaneously alter the slope of the shelf. Apparently marked changes in the sediment mean size and the sediment textural parameters along somewhat steeper slopes in Bristol Bay suggest that the mean size-slope relationship may be controlled by storm waves. The sedimentation in Bristol Bay suggests that the slope of the shelf is a function of both wave energy and sediment mean size. For example, the wave effectively removes particles smaller than 3 ϕ from the shallow shelf with less than 60-m water depth. Beyond that depth the wave energy imparted at the bottom diminishes rapidly. The sediments removed from the shallow shelf and carried offshore beyond the 60-m isobath rapidly settle out because of loss of wave energy. Settling of sediments having a wide spectrum of particle size (less than 3 ϕ carried as the traction and more than 3 ϕ in suspension) along a narrow zone causes steepening of the slope.

Finally, the influx of sediment into marine environment may affect the sediment texture on the shelf. The rate of detritus input may sometimes overwhelm the capacity of the marine agents to effectively sort sediments. Typical of the excessive rate of sedimentation and the resulting textural distribution are the near-shore deposits along the eastern Gulf of Alaska and along the distal end of the Yukon River delta.

GEOCHEMICAL EVOLUTION

The source of sediments on the Alaskan Shelf mostly lies in the rugged tectonic belt which for the most part forms the coastline. Because the amount of detritus formed at the source tends to increase exponentially with increasing source elevation the rugged coastline should provide large amounts of detritus to the shelf. Primarily, the detritus is carried to the shelf by rivers, glaciers, and streams. Coastal sediments eroded by ice, waves, thawing, and slumping are also found on the shelf. The detritus, on encountering the shelf environment, undergoes further mechanical as well as chemical differentiation, and the transformation of the detritus into marine sediments results in pronounced textural and elemental changes. The extent of the transformation, of course, depends on various factors, such as climate, agents of weathering and transport, and rates of deposition.

The high-latitude provenance is mostly eroded by ice, snow, and snowmelt. Rocks in the hinterland disintegrate by mechanical processes, such as glacial abrasion and frost action, and chemical weathering is minimal. They are transported by glaciers, rivers, and mass wasting processes (along the most rugged coast) directly into the sea. Soil-forming processes are minimal compared to more temperate latitudes. Because of the small-scale chemical breakdown at the source, the detritus, on entering the shelf, actively reacts with marine water and undergoes significant chemical readjustments. Because large amounts of water for chemical reaction are available in which the sediment particles are constantly moved (stirring effect), it is to be expected that the chemical weathering of silicates, especially those that have not undergone the process of soil formation, takes place to a far greater extent on the shelf than on the land. This applies particularly to the finely grained material because of its relatively large surface area. These particles are strongly reactive. In addition, these particles can remain in suspension for longer periods because of their low settling velocities. Coarse particles (sand), which settle out rapidly, are subjected to mechanical crushing by wave action and undergo successive physico-chemical adjustments. The textural and chemical changes in sediments are, of course, primarily dependent on the transient time during which sediments are retained in a certain physicochemical environment.

It is apparent that significant chemical differentiation, accompanied by textural changes, occurs on the shelf. Such differentiation causes elements to migrate in solution as well as in sediment particles. Some elements are partly carried in clay minerals, where they may substitute for Al, and partly in detrital minerals, either as single grains or incorporated in rock fragments. There are, therefore, much more erratic variations of these elements in the sand fraction than in clays. The elemental distribution in sand mostly depends on the sorting processes, which tend to concentrate, with varying degree of efficiency, the detrital mineral, leading to a marked dispersion of the element in sand. In spite of the scatter in element distribution, the dispersal of detritus eventually develops an overall lateral geochemical gradient on the shelf.

One of the methods to trace the migratory paths for detritus on the shelf is to discover characteristic trends of chemical variations for the sediments, especially trends based on rather immobile elements (e.g., Al and Ti). By such "fingerprinting" the lateral geochemical gradients, such as textural gradients, should delineate source and migration of sediments. Furthermore, close scrutiny of these changes should provide insight into the transformation of land detritus into marine sediments and the elemental migration in the marine environment.

CHEMICAL TRANSFORMATION

The evolution of sedimentary rocks, in general, is the reverse of the Bowen series of mineral formation in igneous rocks. The progressive degradation of mafic minerals is apatite-iron oxides-olivine-pyroxene-amphiboles-biotite and in felsic minerals it is Ca-plagioclase-(Ca-Na)-plagioclase-(Na-Ca)-plagioclase-Na-plagioclase-K-feldspar muscovite-quartz. The distribution of silica and alumina, along with other elements, as in igneous rocks, can be good indicators of geochemical differentiation and sediment weathering.

The evolution of sedimentary rocks on the shelf, besides the normal trend, is concurrently influenced by the marine physical environment, which influences both the textural and the elemental differentiation. The trend on the shelf follows the development leading to sandstone, silt, shale, and carbonate rocks. The distribution of carbonate rocks on the Alaskan Shelf is rather restricted; this discussion, therefore, is mainly directed toward the evolution of sandstone and shales.

Sandstones display several transient stages prior to attaining the final matured stage as quartz sandstones (end member). Initially, the composition of sandstones is primarily dependent on the provenance, tectonic setting, and climatic conditions at the source. The least weathered and most complex, chemically as well as mineralogically, are the graywackes. Their composition is similar to glacial tills. Graywackes are sedimentary deposits that usually contain some undecomposed fragments of igneous and other rocks from erosion of orogenic belts. Typically, graywacke consists of Ca-feldspar. Plagioclase is readily decomposed during the early stages of weathering and the presence of plagioclase in graywackes reflects the residual primary rock material typical of relatively mild weathering and a nearby source rock.

Goldberg and Griffin (1964) showed that the plagioclase to quartz ratio is strictly controlled by climate. The ratio is maximum in the ice and cold tundra part of humid zones as well as in the arid zones of the northern and southern hemispheres. Minimum plagioclase to quartz ratios occur in the equatorial and warm parts of the humid zones, where the feldspar weathering rate is greatest. The relatively weak actions of weathering also results in the predominance of Na-feldspar over K-feldspar and therefore low K_2O/Na_2O ratios.

Another common variety of sandstone, generally evolved during the early stages of weathering, is arkose. Arkoses are relatively rich in K_2O and Na_2O and, in general, K_2O dominates Na_2O. These sandstones contain lower amounts of ferrous iron and magnesium than do graywackes. Typically the arkosic sandstones are polymictic and characteristically have a low content of quartz and a high content of feldspar, micas, and unstable weathered pyroxenes, amphiboles, and epidote. These features suggest initial weathering stages that pertain only to partial removal of sodium and, to a lesser degree, of potassium and involves a very slight alteration in alumino-silicate and silicates, including even the most unstable ones.

The content of quartz in sediments is an index of chemical maturity and the final matured sediment is pure quartz sand. Quartz grains, compared to other mineral grains, resist physical and chemical breakdown during marine weathering and transport. The feldspars relative to quartz, on the other hand, break down more readily in the marine environment and, therefore, decrease with increased transport and wave action. The degree of maturity resulting from such mineral differentiation in sandstone can be expressed chemically. The immature sandstones are rich in alumina, soda, and potash, and matured sandstones are rich in silica.

While the residual sand generally becomes richer in quartz, the finer sediment transport offshore becomes enriched in alumina. In general, the increase in alumina is accompanied by a slight increase in MgO, Fe_2O_3, and TiO_2. These elemental differentiations are primarily due to potassium hydrous aluminum silicates and iron-rich magnesium-bearing chlorites, which tend to aggregate in silt and finer fractions. This leads to the development of silt and shale.

The chemical composition of sediments, as discussed above, is closely related to sediment grain size. In general, this is to be expected in marine environments, because chemical maturity is rarely attained without corresponding textural differentiation. Both the chemical and the textural maturity evolve through the differential movements of various oxides. Such oxides as SiO_2 are more inert than others (Na_2O, K_2O, CaO, MgO, and TiO_2) and tend to remain in the residue (sandstones), whereas the more mobile oxides tend to move offshore (silt and clay).

The interrelationship between the chemical composition and the texture of the sediments can be readily seen in Figure 13-1. The plot clearly shows development of two end members, e.g., clays and quartz sands. It is also interesting to note that sediments from southeastern Alaska, the Gulf of Alaska, Prince William Sound, and the Aleutian Shelf are mostly silt and clay, whereas the sediments on the Bering Chukchi, and Beaufort shelves are silt and sand, with some approaching quartzic sands. This distinct differentiation in sediments may have been caused by the climatic differences. Toward the circumpolar zones chemical weathering is drastically reduced because of the lower annual temperature and is gradually replaced by physical weathering.

The southern regions, southeast Alaska, the Gulf of Alaska, Prince William Sound, and the Peninsular (Aleutian) Shelf, lie in the maritime climate and are rich in clay. The northern shelves, which lie in transitional and arctic climatic zones, distinctly lack clay fractions in the sediments. This lack of clayey sediments may be due to the high-latitude weathering, which inhibits formation of clay at the source. Conversely, the paucity of clay size sediments on the shallow shelves may have been the result of the high-energy environment, which prevents deposition of fine fractions. Because no apparent correlation between sediment parameters and the latitude has been observed it may be deduced that the textural and to some extent the chemical anomalies between the two regions are related to environments of deposition.

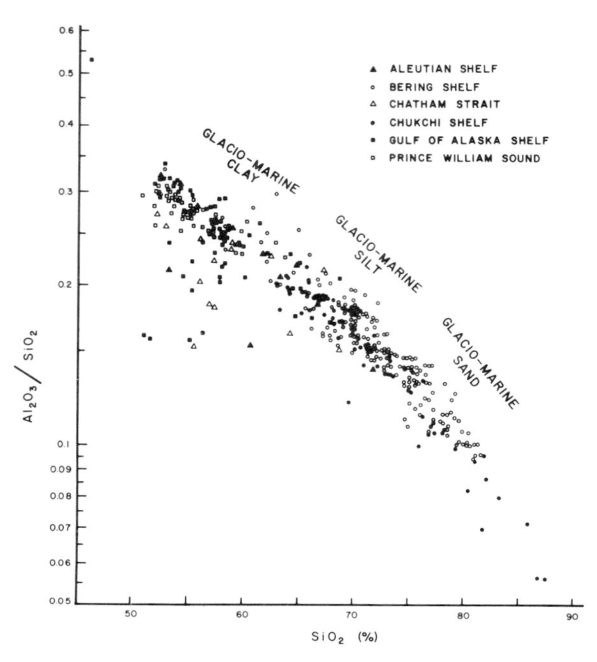

Figure 13-1. Al_2O_3/SiO_2 versus percent SiO_2 plot for Alaskan Shelf sediments.

Finally, the differences in sediment chemistry between southern and northern regions of the Alaskan shelves may have been due to the differences in relief at the source. Although the intensity of decay process is related to climate, the duration of time needed to decompose is controlled by relief. The provenance along the southern region has a high relief and undergoes rapid erosion, so that feldspar escapes destruction. This leads to deposition of silicate minerals. In northern regions the relief is low and the rate of erosion is slow. This results in, relatively to a greater extent, decomposition of feldspar, providing quartz-rich sediments to the northern shelves. Quartzitic detritus, when brought to a high-energy shallow shelf, generally leads to deposition of quartz sands.

The triangular compositional diagram (Fig. 13-2) obtained by plotting as end members percentages of Al_2O_3, SiO_2 + TiO_2, and Rest (combined percentages of other major elemental oxides) shows the development of the Alaskan Shelf sediments. The average content of silica and titania in igneous rocks is about 60% (Clarke, 1924; Goldschmidt, 1954; Brotzen, 1966). The source rock, on weathering and marine transport, evidently follows two distinct trends; one leading to end member quartz sand (lag deposit of resistant minerals) and the other leading to clays (mechanical and chemical breakdown of minerals). It is suggested that those sediments having combined silica and titania percentages of about 60% represent detritus that has undergone the least mineralogic-elemental differentiation. These include sediments from the northern coastal areas, in the vicinity of the Copper River, as well as the glacial sediments from the shelf break. Consequently, the sediments with significant departures in composition from the average igneous rock indicate an appreciable degree of mineralogic-elemental differentiation.

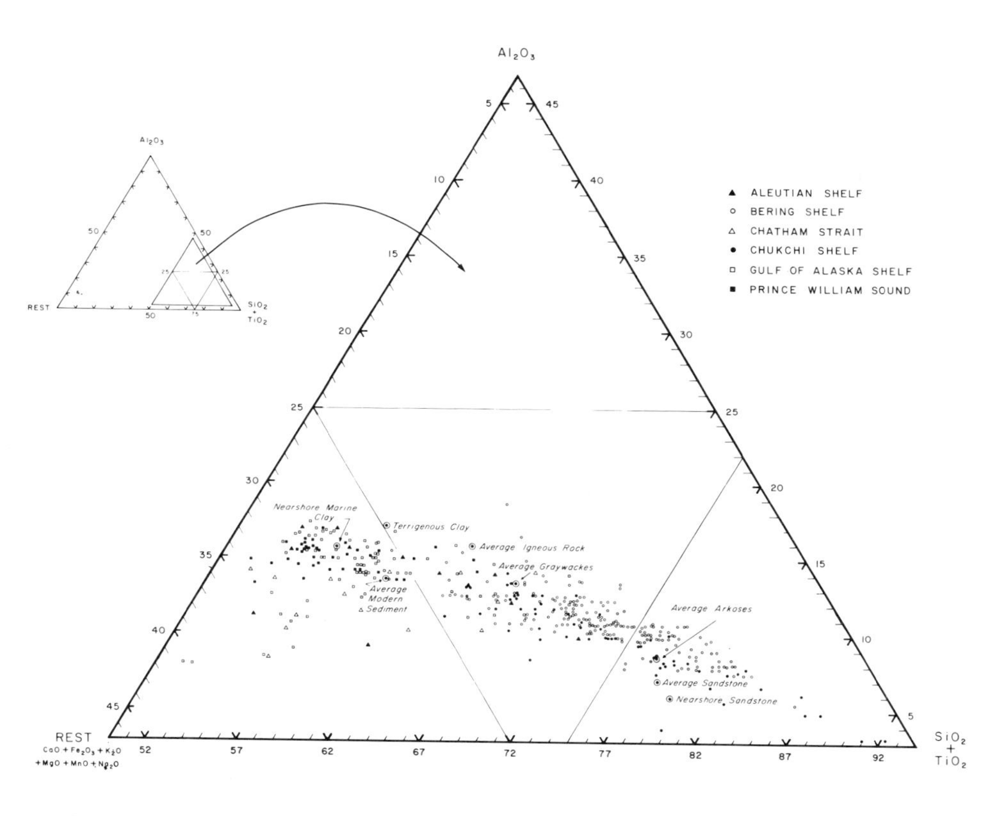

Figure 13-2. Triangular plot with percents Al_2O_3, SiO_2 + TiO_2, and Rest in sediments as end members, Alaskan Shelf.

The major trends of the evolution of sediments from various sources and their degree of maturity can best be observed by plotting log (SiO/Al_2O_3) versus log [(CaO + Na_2O)/K_2O] (Fig. 13-3). This plot indicates that sources for the Alaskan Shelf sediments mostly lie in the rock composition that varies between andesite and grandiorite. Composition of some sediments from the Aleutian Shelf lies close to andesite-basalt. Outcrops of andesite-basalt along the Alaska Peninsula are well known and apparently these sediments have their sources to adjacent exposures. Few samples from the Gulf of Alaska and the Aleutian shelves fall closer to the carbonate regime. As described earlier these sediments indeed contain calcareous shells and contain up to 50% or more calcium carbonate. The sources for shelf sediments suggested by the plot, therefore, appear to be well founded.

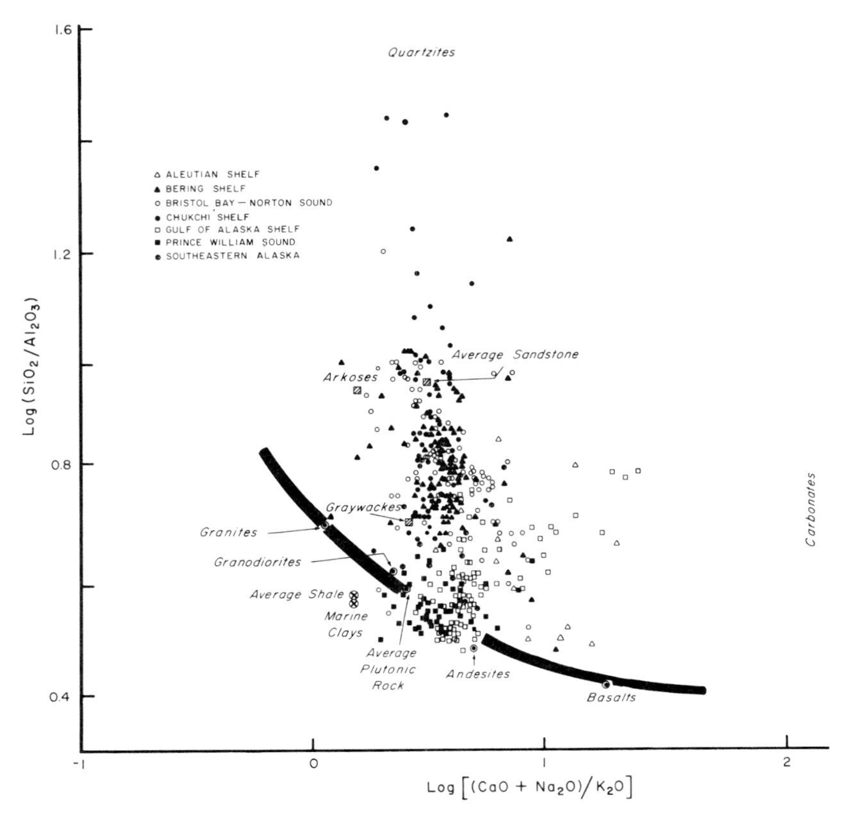

Figure 13-3. Log (SiO_2/Al_2O_3) versus log [($CaO + Na_2O)/K_2O$] plot, Alaskan Shelf sediments.

The major development trend observed in the plot, once again, consists of graywackes-arkose-quartz sand and clay. It may also be inferred that as the sands become matured, they become more quartzitic. The development of clay is obviously in the opposite direction, where silica-alumina ratios slightly decrease and relatively more CaO and Na_2O than K_2O are leached out with increasing weathering and transport.

The development of sand and the clay is further illustrated by the K_2O/Al_2O_3 plot. In reference to the composition of average igneous rocks the successive maturity in sands results in both loss of Al_2O_3 and only slight loss in K_2O (Fig. 13-4). This is to be expected because K_2O is relatively more mobile than Al_2O_3.

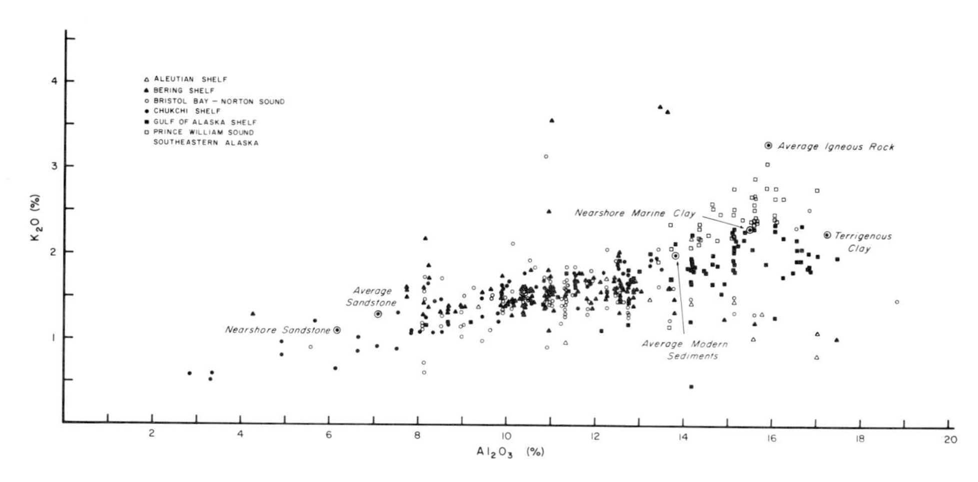

Figure 13-4. Percent K_2O versus percent Al_2O_3 plot, Alaskan Shelf sediments.

Ratios of pairs of common and minor elements are thus capable of providing a variety of geochemical information concerning sediment source and transport (maturity) on the shelf. Nanz (1953), for example, studied the chemical composition of clays associated with various types of sandstones to determine relationships between maturity of sandstones and that of associated clays. The clays associated with the more matured sand are themselves more mature than those associated with the immature sands. The average Al_2O_3/Na_2O ratio of two clays associated with quartz sandstoned was 125, whereas the corresponding average of two clays of graywacke was only 11. Alumina is the most inert oxide, whereas soda is mobile; the Al_2O_3/Na_2O, therefore, is a good maturity index for determining the transport of the fine-grained sediment on the shelf.

Other elemental ratios that reflect sediment movement on the shelf include Al_2O_3/TiO_2, which is designed to reflect the mineralogic-elemental separation resulting from hydraulic action. The K_2O/Na_2O index is based on the relative rates of potash and soda feldspar weathering and also enrichment of K in clays. Last, the CaO/MgO reflects the mobility of these two elements in relation to sediment movement on the shelf.

The chemical composition of sediments, in relation to the texture, is therefore suggestive of the extent of weathering (maturity) during or after deposition, The alumina-silica versus silica (percent) ratio is generally a good index of the grain size on the shelf sediments (Fig. 13-1.).

Once the major evolutionary trends for sediments have been defined, it will be equally important to relate these trends to physical environments. The sands are mostly deposited in four environments: very high energy, high energy, moderate energy, and low energy (Fig. 13-5). Each environment has specific processes or a set of processes that cause a characteristic geochemical (elemental-mineralogic) and textural differentiation.

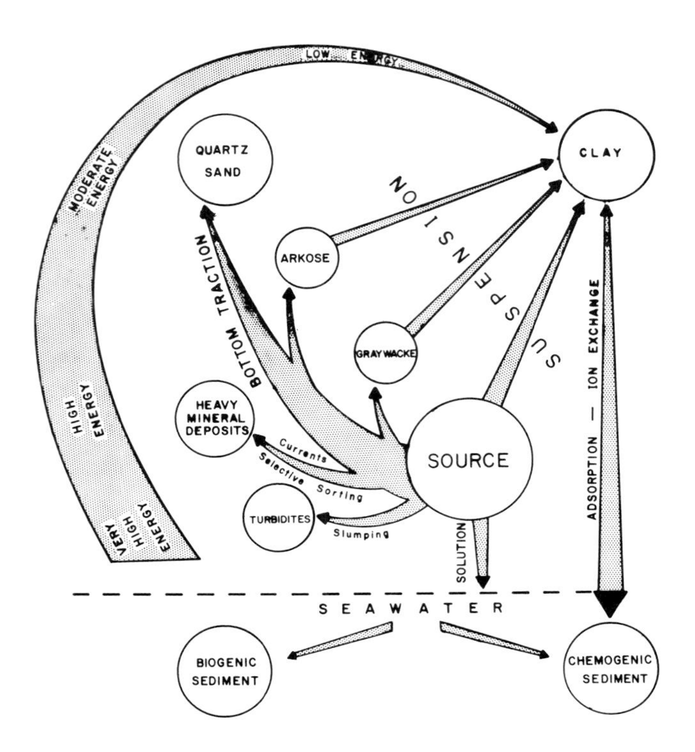

Figure 13-5. Relationship between sedimentary processes and sedimentary facies.

The very high energy environments are mostly episodal. Nevertheless, because of extremely large-scale mass movement they are important agents of sediment transport and differentiation. In the southern region of the Alaskan Shelf, frequent earthquakes as well as soil creeps along the steep slopes cause slumping. As a result large amounts of sediments are introduced into the marine environment. Instantaneous movement of voluminous sediments results in turbidity currents, which cause profound vertical and to a lesser extent lateral textural differentiation.

The near-shore area, particularly the surf zone, is a region of wave action. Some near-shore areas are also swept by the longshore currents. The wave action and the current both lead to considerable textural and mineral sorting. In regions of high energy, particles of low specific gravity minerals (light minerals) are preferentially removed, leaving a lag deposit rich in minerals with higher specific gravity, the so-called heavy minerals. The degree of such selective sorting in sands, of course, varies significantly regionally.

The differences in efficiency of sorting in sand on the shelf result in an accumulation of mixtures of infinitely different proportions of minerals in sands. It is the variance in proportion of heavy and light minerals in sand that causes significantly wider scatter in the distribution of some elements in sand- than in clay-sized sediments.

MAJOR ELEMENT MIGRATION

Introduction

The chemical composition of shelf sediments is primarily dependent on the migration characteristics of individual elements in mineral components. A few elements are strongly bound as minerals and undergo marine cycling, as a part of a lattice. Most elements, however, as a result of marine processing, attain elemental interrelations that are significantly different from the parent rocks. In other words, the distribution of these elements in sediment does not closely correspond to the provenance.

During marine weathering, a rapid decrease of sodium, calcium and magnesium; the slower loss of potassium and silicon; and still slower loss of aluminum and iron occurs. After the cations are set free, the Al-Si-O frameworks of the original silicate minerals are in part decomposed and in part reconstituted in to the frameworks of clay minerals, so that only a part of the silicon and very little of the aluminum and iron find their way to solution. Because chemical analysis provides only the relative amounts of various elements in sediments at a particular stage of weathering, it is difficult to determine how much of any one element has been lost.

The relative loss or gain of elements in sediments during weathering can be evaluated by comparing their content with that of alumina. This procedure is primarily based on the assumption that alumina in sediments does not change appreciably during weathering. This is a reasonable assumption because relative to all other rock constituents alumina shows the greatest increase in weathered sediments and the least abundance in sea water. The distribution of elements in relation to aluminum in various sediment size fractions generally reflects various modes in which an individual element migrates during marine transport. In particular, such relationships indicate the extent of migration of an element as a part of a crystal lattice and in solution. In view of the complex behavior of elements during marine transport, it is pertinent that the migration of each element be discussed separately.

In view of the large number of sediment parameters obtained, the interrelation between various parameters is best determined using a correlation matrix (Table 13-1). Parameters with significant coefficient correlation (>0.321) are included in the chart. The correlation matrix chart readily provides the degree of the interrelationship between various sediment parameters. The interelemental association, to some extent, suggests migration characteristics of various elements on the shelf. Elements with high correlation coefficients appear either to migrate in

TABLE 13-1

Correlation Matrix of Sediment Parameters

Variable Number	Mean Size	Sorting	Skewness	Kurtosis	Gravel (%)	Sand (%)	Silt (%)	Clay (%)	Al_2O_3 (%)	CaO (%)	MgO (%)	Fe_2O_3 (%)	K_2O (%)	Na_2O (%)
Mean Size	1.000	-	-	-	-0.576	-0.685	0.677	0.854	0.581	-	0.568	0.536	0.439	0.522
Sorting		1.000	-	-0.324	0.485	0.569	-	0.371	-	-	0.316	-	-	-
Skewness			1.000	0.335	-	-	-	-	-0.323	-	-0.356	-0.408	-	-
Kurtosis				1.000	-	0.395	-	-0.404	-	-	-0.408	-0.384	-	-
Gravel (%)					1.000	-	-0.334	-	-	-	-	-	-	-0.352
Sand (%)						1.000	-0.733	-0.804	-0.492	-	-0.585	-0.537	-0.510	-0.337
Silt (%)							1.000	0.435	0.324	-	-	-	-	-
Clay (%)								1.000	0.636	-	0.754	0.732	0.591	0.484
Al_2O_3 (%)									1.000	-	0.717	0.732	0.426	0.706
CaO (%)										1.000	0.361	-	-	-
MgO (%)											1.000	0.824	0.400	0.487
Fe_2O_3 (%)												1.000	0.390	0.570
K_2O (%)													1.000	0.425
Na_2O (%)														1.000
MnO (%)														
TiO_2 (%)														
SiO_2 (%)														
Ba (ppm)														
Co (ppm)														
Cr (ppm)														
Cu (ppm)														
Ni (ppm)														
Sr (ppm)														
Zn (ppm)														
Organic Carbon (%)														
Weight Loss (%)														
Latitude														
Longitude														

MnO (%)	TiO_2 (%)	SiO_2 (%)	Ba (ppm)	Co (ppm)	Cr (ppm)	Cu (ppm)	Ni (ppm)	Sr (ppm)	Zn (ppm)	Org. Carb. (%)	Weight Loss (%)	Latitude	Longitude	Variable Number
0.344	0.560	-0.579	0.374	-	-	0.604	0.570	-	0.553	-	0.705	-	0.378	Mean Size
-	-	-0.334	-	-	-	-	-0.480	-	-	-	0.354	-	-0.351	Sorting
-0.446	-	0.336	-	-	-	-	-	-	-	-	-	0.331	0.389	Skewness
-	-	0.397	-	-	-	-	-0.347	-	-0.363	-	-0.358	-	-	Kurtosis
-	-	-	-	-	-	-	-	-	-	-	-	-	-	Gravel (%)
-	-0.524	0.587	-	-	-	-0.617	-0.631	-	-0.589	-	-0.764	-	0.381	Sand (%)
-	0.356	-	-	-	-	0.410	-	-	0.363	-	0.595	-	-	Silt (%)
0.524	0.626	-0.732	0.419	-	-	0.666	0.779	-	0.644	0.327	0.747	-	-0.621	Clay (%)
0.661	0.721	-0.833	0.438	-	-	0.489	0.556	-	0.579	-	0.494	-0.441	-0.560	Al_2O_3 (%)
0.404	-	-0.562	-	0.374	-	-	-	0.825	-	-	-	-0.368	-0.440	CaO (%)
0.709	0.726	-0.866	-	-	0.328	0.559	0.705	-	0.586	-	0.563	-	-0.718	MgO (%)
0.790	0.780	-0.868	-	-	0.346	0.523	0.687	-	0.574	-	0.526	-0.342	-0.735	Fe_2O_3 (%)
-	-	-0.419	0.483	-	-	0.450	0.434	-	0.413	-	0.429	-	-0.324	K_2O (%)
0.487	0.564	-0.667	0.369	-	-	0.334	-	-	0.397	-	0.438	-0.467	-0.387	Na_2O (%)
1.000	0.731	-0.770	-	-	-	0.342	0.545	0.383	0.442	-	-	-0.490	-0.668	MnO (%)
	1.000	-0.761	-	-	-	0.453	0.602	-	0.533	-	0.457	-	-0.582	TiO_2 (%)
		1.000	-0.327	-	-	-0.510	-0.694	-0.496	-0.603	-	-0.649	0.428	0.736	SiO_2 (%)
			1.000	-	-	0.313	0.419	-	0.373	-	-	-	-	Ba (ppm)
				1.000	-	-	-	0.457	-	-	-	-	-	Co (ppm)
					1.000	-	0.394	-	-	0.324	-	-	-0.406	Cr (ppm)
						1.000	0.518	-	0.525	-	0.533	-	-0.336	Cu (ppm)
							1.000	-	0.582	0.333	0.530	-	-0.697	Ni (ppm)
								1.000	-	-	-	-0.472	-0.431	Sr (ppm)
									1.000	-	0.527	-	-0.351	Zn (ppm)
										1.000	-	-	-	Organic Carbon (%)
											1.000	-	-0.356	Weight Loss (%)
												1.000	-	Latitude
													1.000	Longitude

association or to have a similar migration characteristics. In the following discussion the correlation coefficient matrix has been exclusively used to describe the migration of elements.

Aluminum

Aluminum, after silicon, is the most abundant element in detrital sediments. It is also one of the least mobile elements and its distribution to a large extent is not affected by complex marine processes, including biologic processes. The aluminum content in sediments, therefore, would be a good indicator for the estimate for the terrigenous fraction in marine sediments. This concept is further extended to determining the relative accumulation rate of an element as detritus and as a chemically bound aggregate. The rates are simply obtained by comparing the ratios of distributions of an element and aluminum in the sediments and in the parent rocks. This provides a good approximation. The change in element to aluminum ratios in relation to sediment texture generally delineates the various modes of migration and accumulation of the element in sand, silt, and clay fractions.

The distribution of aluminum in Alaskan Shelf sediments is primarily related to the sediment texture (Fig. 13-6). The coarse sediments mostly contain lesser amounts of aluminum than fine sediments. The wide scatter in the distribution of alumina in the range of coarse sand and gravel (Fig. 13-6) is primarily due to

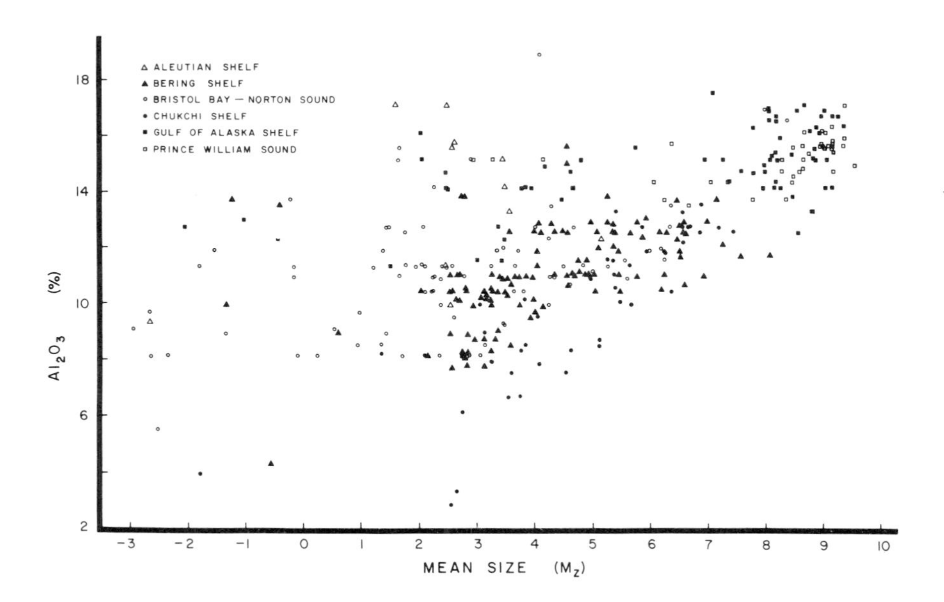

Figure 13-6. Percent Al_2O_3 versus mean grain size plot, Alaskan Shelf sediments.

the mixing of sediments from various origins. Sediments in these size classes are generally a mixture of relict glacial, biogeneous, and contemporary sediments. The mean grain size composition of relict glacial sediments, although similar to the source rock, is not compatible with those evolved through normal marine transport. The second component, biogenous sediments, mostly consists of coarse shell fragments, and mean grain size increases disproportionately. Furthermore, these shell fragments (mostly calcareous) are $CaCO_3$ in composition and are devoid of aluminum. These factors, therefore, cause large variations of aluminum content in gravel and coarse sand-sized sediments.

The larger scatter in the range of fine sand and coarse silt is primarily caused by the variance of mineralogic and chemical differentiation to which these sediments are subjected. The arkosic and graywacke sands close to a source are relatively weakly weathered and contain higher amounts of aluminum than the quartzitic sand. Accumulation of monomineralic grains, mostly quartz, is typical in areas of high energy, where continual wave action generally causes chemical and mechanical weathering of the feldspars. These are relatively lower in alumina.

The regional distribution of aluminum in sediments is of significant interest (Table 13-2). South of the Alaska Peninsula, the average content of aluminum in sediments decreases from Prince William Sound-Gulf of Alaska to the Aleutian Shelf. This successive decrease follows the progressive increase in sediment grain size and decreasing latitude. In contrast, north of the peninsula, on the average, the aluminum decreases with decreasing sediment grain size. Furthermore, the decrease occurs in spite of a progressive increase in percent clay in the sediment. Normally, aluminum content is directly related to percent clay in the sediments.

The unusual distribution of aluminum in relation to the sediment texture on the Bering and Chukchi shelves is primarily due to high-energy environment that prevails over most of the shelves. The wave action on the shallow shelves actively breaks down and removes feldspar and thus forms lag deposits rich in more resistant quartzitic sand and silt. The increase of mineral quartz is well reflected in the progressive increase in silica. The aluminum-silica relationship in terms of sediment texture is well displayed in Figure 13-2.

Iron

The iron content in sediments described throughout this book was measured as total iron (Fe). No separate analyses were made for the determination of Fe^{2+} and Fe^{3+}. The sediment color and the limited Eh and pH measurements (for details, see Chapter 7, "Port Valdez") suggest that part of the soluble iron has precipitated as Fe^{3+}. It is commonly known that most iron in sedimentary rocks occurs as ferric oxide. Consequently, it was considered appropriate to report the oxide of iron as Fe_2O_3.

Obviously, a major part of the iron in shelf sediment is carried in mineral lattices. The influence of detritus minerals on the regional distribution of iron in sediments is quite apparent. An appreciably higher concentration of iron occurs

TABLE 13-2

Regional Averages of the Sediment Parameters

Sediment Parameters		Northeastern Pacific Shelf Regime				Bering – Chukchi Shelf Regime		
		Prince William Sound (Avg 45 samples)	Central Gulf of Alaska (Avg 68 samples)	Northwestern Gulf of Alaska (Avg 12 samples)	Bristol Bay (Avg 53 samples)	Bering Shelf (Avg 123 samples)	Norton Sound (Avg 52 samples)	Chukchi Shelf (Avg 64 samples)
Mean Size		8.08	5.58	2.57	2.08	3.94	3.00	4.75
Sorting		2.75	3.41	1.33	0.99	1.75	2.56	2.46
Skewness		0.03	0.17	0.25	0.04	0.38	0.37	0.46
Kurtosis		0.93	1.01	1.90	1.43	1.61	1.85	1.38
Gravel	(%)	3.25	9.30	7.08	1.42	3.27	14.94	3.88
Sand	(%)	3.50	23.25	79.19	87.59	56.06	47.91	38.22
Silt	(%)	36.57	30.75	9.06	8.75	30.84	26.08	42.00
Clay	(%)	56.67	36.70	4.67	2.24	9.83	11.07	15.90
Al_2O_3	(%)	15.16	14.22	13.75	12.10	11.06	10.14	9.76
Fe_2O_3	(%)	8.25	7.39	6.41	5.17	3.73	4.14	4.21
CaO	(%)	4.03	6.84	7.88	3.79	2.83	1.95	2.15
MgO	(%)	4.14	3.59	2.51	1.88	1.72	1.69	1.84
K_2O	(%)	2.42	1.78	1.36	1.42	1.57	1.55	1.55
Na_2O	(%)	4.41	3.33	4.06	3.56	3.07	2.73	2.57
MnO	(%)	0.147	0.136	0.132	0.101	0.057	0.069	0.035
TiO_2	(%)	0.977	0.937	0.904	0.685	0.631	0.767	0.630
SiO_2	(%)	55.27	57.56	59.57	69.11	72.31	74.62	73.03

TABLE 13-2 (Continued)

Sediment Parameters		Northeastern Pacific Shelf Regime				Bering – Chukchi Shelf Regime		
		Prince William Sound (Avg 45 samples)	Central Gulf of Alaska (Avg 68 samples)	Northwestern Gulf of Alaska (Avg 12 samples)	Bristol Bay (Avg 53 samples)	Bering Shelf (Avg 123 samples)	Norton Sound (Avg 52 samples)	Chukchi Shelf (Avg 64 samples)
Ba	(ppm)	511	389	352	353	369	342	287
Co	(ppm)	44	43	31	18	11	19	18
Cr	(ppm)	107	133	31	27	39	45	50
Cu	(ppm)	52	37	23	12	31	27	32
Ni	(ppm)	50	48	31	20	24	28	26
Sr	(ppm)	313	408	422	324	252	197	155
Zn	(ppm)	103	91	63	56	66	56	58

in sediments deposited adjacent to sources with basic igneous rocks, some of which are exposed along the Alaska Peninsula.

Soluble iron added to the sea also, to some extent, influence the iron content in sediments. Solubility of iron in fresh water (pH 6) is much higher (10^5 times) than in sea water (pH 8.5). Continental drainage, therefore, carries a significantly higher content of soluble iron than that found in sea water. Because of its polyvalency and its easy bonding with organic matter, soluble iron has many modes of occurrence. It can occur in four forms: in the free state as pigment; in the carbonate state as siderite; in the sulfide state as colloidal pyrite, such as hydrotroilite; and in the silicate state. The precipitation of soluble iron as $Fe_2(OH)_3$ may be accompanied by coprecipitation of several other major and minor elements. Such a process may contribute to a higher concentration of iron and probably of other elements in estuarine sediments.

Marine clays generally undergo some compositional variation and, therefore, are susceptible to elemental substitution. The increasing iron content with increasing clay content observed in sediments may be caused in part by such a substitution, particularly the substitution of soluble Fe^{2+}. Carroll (1959) stated that between pH 4.5 and 8.0, Fe^{3+} does not enter exchange reaction because within this range of pH, Fe^{3+} would form soluble hydroxides or oxides. In contrast, iron oxide in the colloidal state can be absorbed on the surfaces of clay particles because of clay minerals.

More importantly the iron and clay content association in sediments may be solely because of the commonality of their environment of deposition. The colloidal iron-like clay particles are carried in suspension and deposited in relatively quiescent deep waters.

The distribution of iron in the shelf sediments is related to the sediment mean size (Fig. 13-7). Iron content in sand and silt is generally low, with a gradual increase with a decreasing grain size; but in clay fractions it increases dramatically. The large scatter of iron content in sand, as observed in Figure 13-7 is a result of differences in provenance mineralogy and selective sorting through hydraulic action. Sediments along the Alaska Peninsula have their origin in the volcanic rocks and these sediments are continually subjected to wave action and longshore drift, leading to hydraulic separation of light and heavy minerals in sands. The lag sand deposited in the near-shore area is generally rich in ferro-magnesium minerals and so has a high iron content. Offshore, on the other hand, the sand from regions under major currents tends to be quartzitic and so poor in iron. Depending upon the source and the hydraulic action, the iron content in sand can vary over a wide range.

The extreme variations of iron in gravelly sediments are also related to the source and the sediment transport. The gravelly sediments are mostly biogenic or glacial relict. The coarse biogenic sediments consist predominantly of calcareous shells ($CaCO_3$) and, therefore, contain very little iron. The glacial relict gravel, in contrast, contains a strikingly high content of iron, sometimes even higher than contemporary marine sediments. The higher concentration of iron

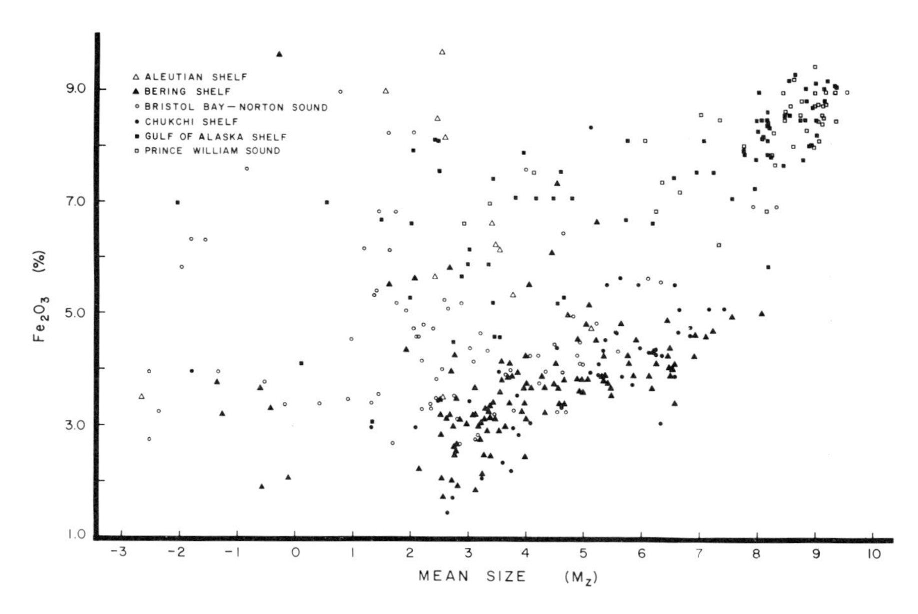

Figure 13-7. Percent Fe_2O_3 versus mean grain size plot, Alaskan Shelf sediments.

may be attributed to the lack of chemical weathering and elemental differentiation, which leads to higher stability of iron in glacial sediments.

The distribution of iron generally covaries with that of aluminum, and the Al_2O_3/Fe_2O_3 in average sediments and average sandstone are 2.4 and 3.5, respectively. The alumina to ferric oxide ratios in shelf sediments vary over a wide range (1.5 to 3.8). It is assumed that the detritus generally carries alumina and ferric oxide within a narrow range and the wide fluctuations observed in the ferric oxide versus alumina plot (Fig. 13-8) are caused by iron transported by various other modes. The higher amounts of iron in relation to aluminum in coarser sediments is primarily observed in glacial relict as well as in lag deposits rich in heavy minerals. The Al_2O_3/Fe_2O_3 in these sediments varies from 1.5 to 2.6. In contemporary sand and silt deposited under normal marine environments, the ratio varies between 2.4 and 3.6, and both elements are mostly carried in crystal lattices of various minerals.

The percent clay and iron distribution in sediments shows a strong direct relationship (Table 13-1 and Fig. 13-8). In marine environments, the hydroxides of iron as small particles and colloids are carried in suspension and, therefore, tend to aggregate in very fine fractions of sediments. Iron carried in suspension along with clays is deposited in relatively quiet environments, such as bays, deep inlets and offshore regions. These areas serve as settling tanks and sediment traps, have a high rate of sedimentation and generally contain fine sediment. Colloidal iron, after deposition, becomes part of the fine sediments. Depressions filled with clay,

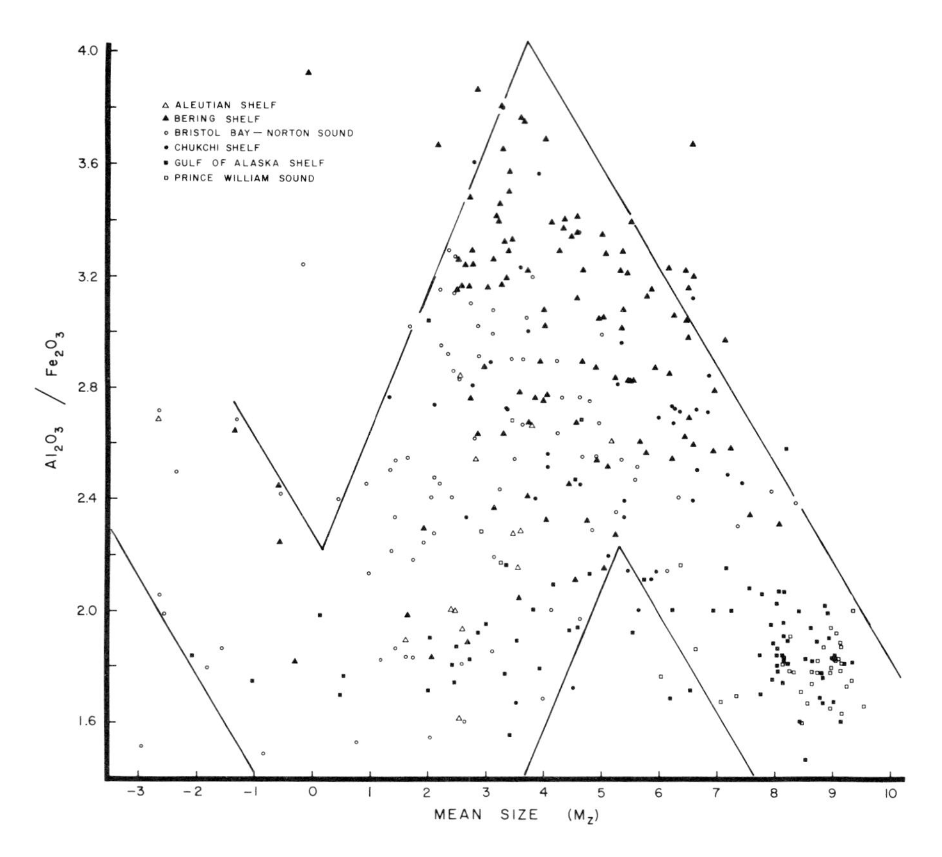

Figure 13-8. Al_2O_3/Fe_2O_3 versus mean grain size plot, Alaskan Shelf sediments.

therefore, generally have high iron content. On the other hand, the leveled shelves of high energy have a low rate of iron accumulation. When the alumina-ferric oxide ratio in fine sediments increase considerably, *i.e.*, where it is more than one would expect in shales (2.4), then the excess of iron may have been present as ferric hydroxide.

In some areas of the shelf, particularly in semi-enclosed basins, an oxidized surface layer, somewhat rich in iron, was observed. This apparently occurs as a result of the upward migration of mobile ferrous forms in the sediments. The upward movement results in the conversion of ferrous to ferric forms. The ferric forms are less mobile and, therefore, tend to concentrate in the surface layer. Some of the Fe_2O_3 observed in surfaces layers may have precipitated from overlying sea water.

The influence of climate on the distribution of iron in sediments appears to be absent or negligible. The distribution of iron in relation to latitude does not show any variation (Table 13-1).

Calcium

The detritus component of calcium in sediments is brought in as feldspar, mainly as (Na_2-Ca)-plagioclase, and varies within a narrow range. The anomalous distribution of calcium in sediments is largely controlled by the calcareous organic remains.

Because of the varying proportions of biogenic contributions the calcium in sediments varies over a large range (1.2-24.5%). High anomalies occur locally and biogenic calcium accumulates principally on shoals and in near-shore regions. These regions generally have a pebbly substrate, which provides a proper habitat for biogenic growth. Almost all areas of prolific benthic communities with calcareous shelled organisms and accumulations have high-energy environments, including flushing of large amounts of water associated with tidal and permanent currents. Such environments inhibit deposition of silt and clay, which causes suffocation and burial of benthic calcareous organisms.

In normal marine sediments, *i.e.*, sediments without excessive biogenic calcareous matter, the distribution of calcium lies within a relatively narrow range (2.4-4.0%). The variation in calcium, in part, is related to sediment mean size. In the gravel-sand-silt range there is a slight increase in calcium with decreasing sediment grain size. In clayey sediments, however, there is noticeable increase in percent calcium.

Besides the local anomalous distribution, the calcium in sediments appears to vary regionally. On the average, the calcium in sediments from the Pacific Shelf (Prince William Sound, 4.03%; Gulf of Alaska, 6.84%; Aleutian Shelf, 7.88%; all values are regional average as CaO) is markedly higher than in sediments from the Bering and Chukchi shelves (Bristol Bay, 3.79%; Bering Shelf, 2.82%; Norton Sound, 1.95%; and Chukchi Shelf, 2.15%). These latitudinal variations do not appear to be related to the sediment mean size (Table 13-2).

Magnesium

The migration of magnesium in the ocean is quite complex. The complexity arises from its ability to migrate in solution, as adsorbed ion, and in the crystal lattices of various major minerals. In spite of its various modes of migration, the magnesium in sediments varies within a limited range of 1.25 and 2.8%, and its distribution is strongly related to the sediment texture (Fig. 13-9). The MgO versus sediment mean size plot shows a gradual increase in magnesium with a decrease in sediment mean size in the sand and silt range. In the clay fraction, however, the magnesium content is significantly higher than in the sand-silt fraction. The large scatter in coarser sediments (gravel-sand-silt) is primarily an artifact of varying amounts of clay-sized minerals in the sediments. Percent magnesium varies in direct relation to the clay content such that magnesium increases steadily with increase in percent clay in sediment. It is apparent that magnesium is mostly carried in association with clays.

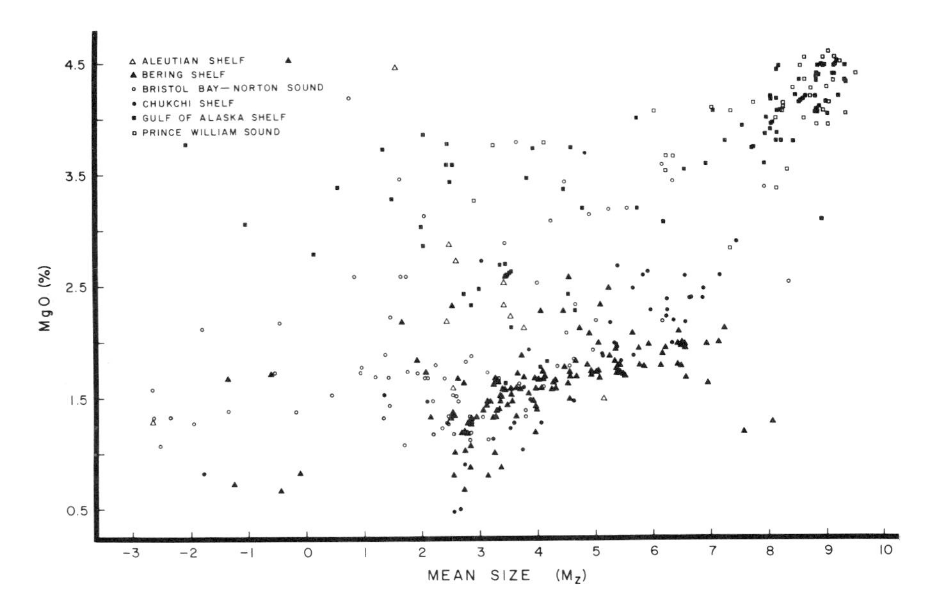

Figure 13-9. Percent MgO versus mean grain size plot, Alaskan Shelf sediments.

In the lower range of the plot (Mg = 2.5% and Al = 14.0%) the increase in MgO relative to Al_2O_3 is slight, with an average MgO/Al_2O_3 of 0.155. These sediments are mostly sand and silt. The increase in ratio (slope) is primarily controlled by the amount of fine sand and silt, which generally contain higher percentages of heavy minerals (Fig. 13-10).

The sediments in the upper range of the plot are mostly from Prince William Sound and the Gulf of Alaska, containing mostly clay fractions. The average MgO/Al_2O_3 value in these sediments is 0.27. The increase in ratio is the result of dominant change in the mineralogy. Feldspar and quartz mostly constitute sand. Aluminum is the major component in feldspar, and magnesium is carried as a minor component. Illite and chlorite, however, are the dominant minerals in clay fractions. Both carry significant amounts of magnesium, particularly chlorite, which often contains MgO and Al_2O_3 in equal amounts. Moreover, x-ray diffraction analysis of clay fractions invariably showed the presence of a fair content of magnesium-rich silicates, mostly hornblende. which should also increase the MgO/Al_2O_3 ratio in fine-grained sediments.

The increase of magnesium in fine sediment as a result of changes in mineralogy is also reflected by the CaO/MgO versus mean size plot (Fig. 13-11). Calcium, mostly found in feldspar and in carbonate minerals, is more abundant in the sand-silt fraction than in the clay fraction. The question remains, however, whether magnesium is totally transported as an integral part of the clay lattice or partly by adsorption/cation-exchange processes.

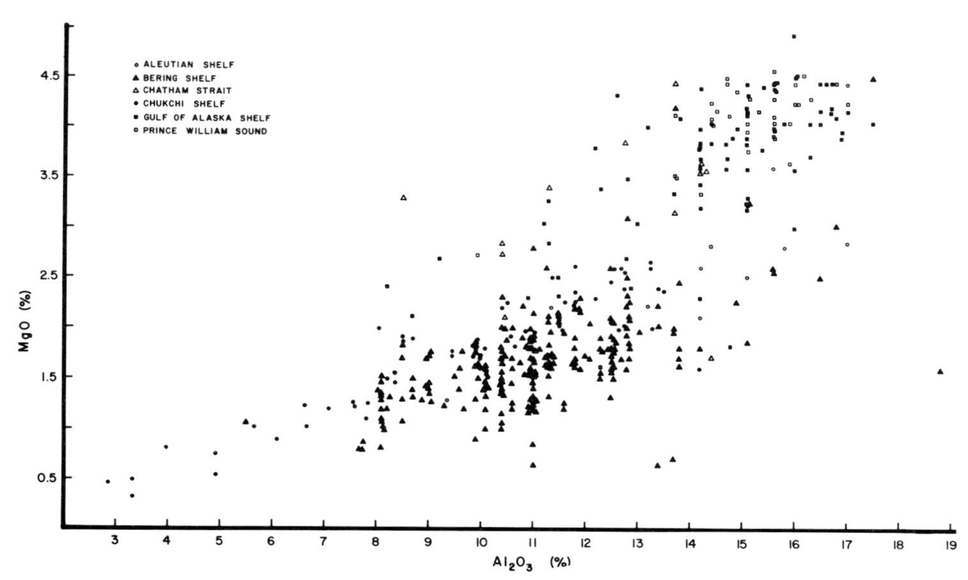

Figure 13-10. Percent MgO versus percent Al_2O_3 plot, Alaskan Shelf sediments.

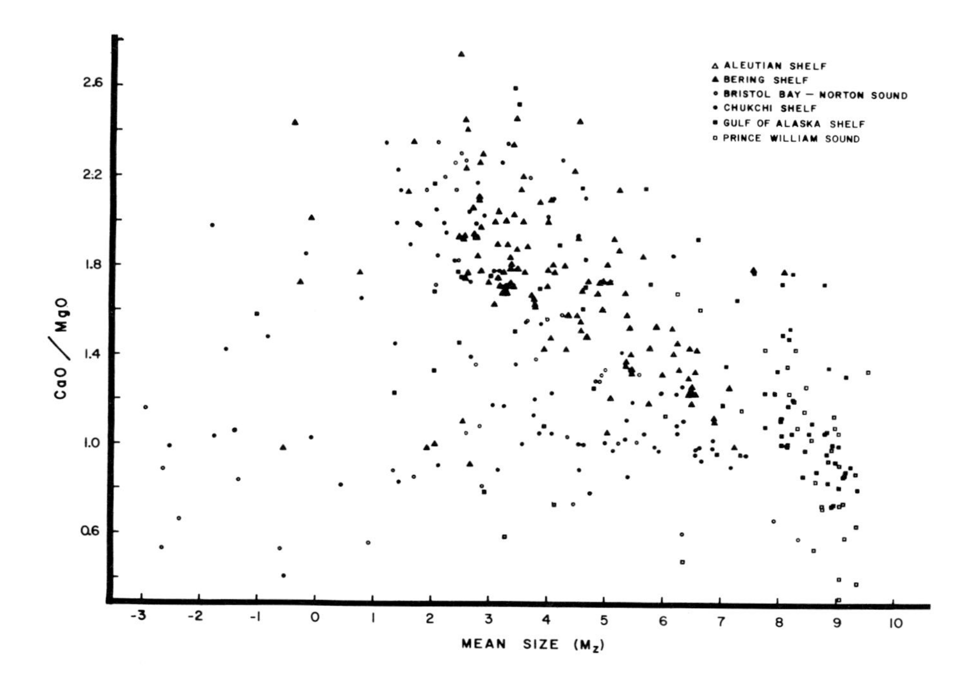

Figure 13-11. CaO/MgO versus mean grain size plot, Alaskan Shelf sediments.

In order to further elucidate the mineralogic association of magnesium in clay minerals, percentages of magnesium were plotted against chlorite-illite values. This plot was restricted to sediments from the Gulf of Alaska and Prince William Sound, which normally contain high clay content. In view of the smaller particle size of chlorite and its association with magnesium, it was anticipated that with the increase in the chlorite-illite ratio there would be a corresponding increase of magnesium in sediments. This, however, was not observed. Interestingly iron also did not show any increase. The lack of association between magnesium and chlorite contents may be due to the varied distribution of Al, Mg and Fe in the tetrahedral, octahedral, and brucite sheets. It appears that in fine-grained sediments magnesium content is not significantly controlled by the relative distribution of chlorite and illite, because both minerals are capable of carrying magnesium in lattices and as adsorbed cation. These processes are of a complex nature.

On a regional basis, the distribution of magnesium is variable. The northeast Pacific Shelf sediment averages suggest an increase in magnesium in response to the increase in clay percent and general decrease in sediments mean size. The distribution of magnesium in the Bering and Chukchi shelves sediments apparently is related to neither clay content nor the mean size (Table 13-2). The gradual northward decrease may be the result of a lessening in intensity of chemical weathering. In relation to calcium, however, the northward decrease in magnesium is markedly less (Fig. 13-12). On the Pacific Shelf the organic sedimentation, which is quite high locally, leads to the predominance of calcium over magnesium. However, in the northern regions, where the clays are reconstituted

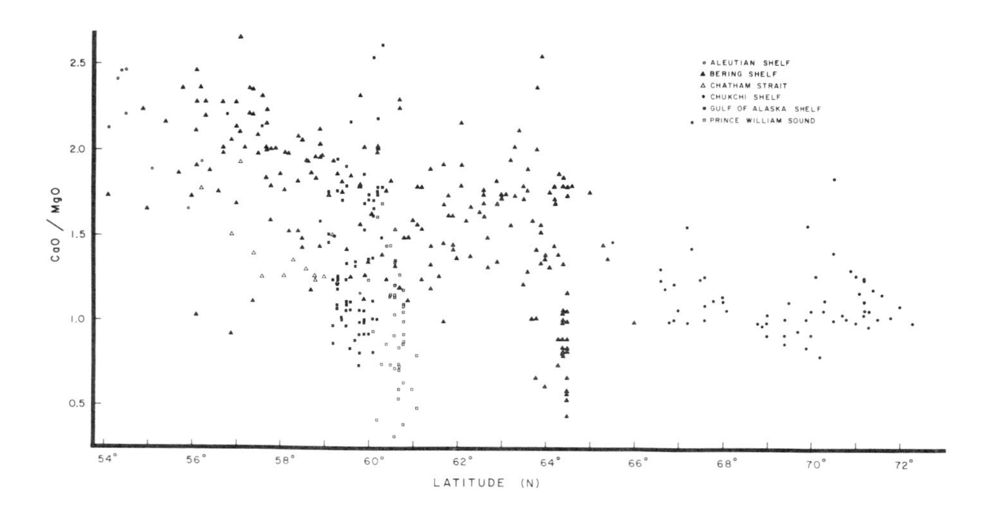

Figure 13-12. CaO/MgO versus latitude plot, Alaskan Shelf sediments.

as open, interstratified, or degraded minerals, magnesium is transferred from sea water to sediments. Chlorite and mixed layers with chloritic layers are magnesium clay minerals and it is well known that illite contains magnesium. Ca does not come into play.

Sodium

Among the major elements sodium is the most mobile. It is carried as a major constituent of various minerals found throughout the entire spectrum of sediment grain size. In sand and silt, sodium is present in feldspars and in clays it occurs as an exchangable cation. Nevertheless, sodium in sediments increases with decreasing grain size (Fig. 13-13). Apparently, the sodium content in the coarse sediment fraction varies over a wide range. These variations are caused by sorting of sand and silt by wave action and water currents during transport, resulting in differentiation of feldspars and quartz. More resistant quartzitic sands are low in sodium but arkosic sands, rich in feldspars, are high in sodium. Such differentiation causes extreme variations in concentrations of sodium in sand and silt.

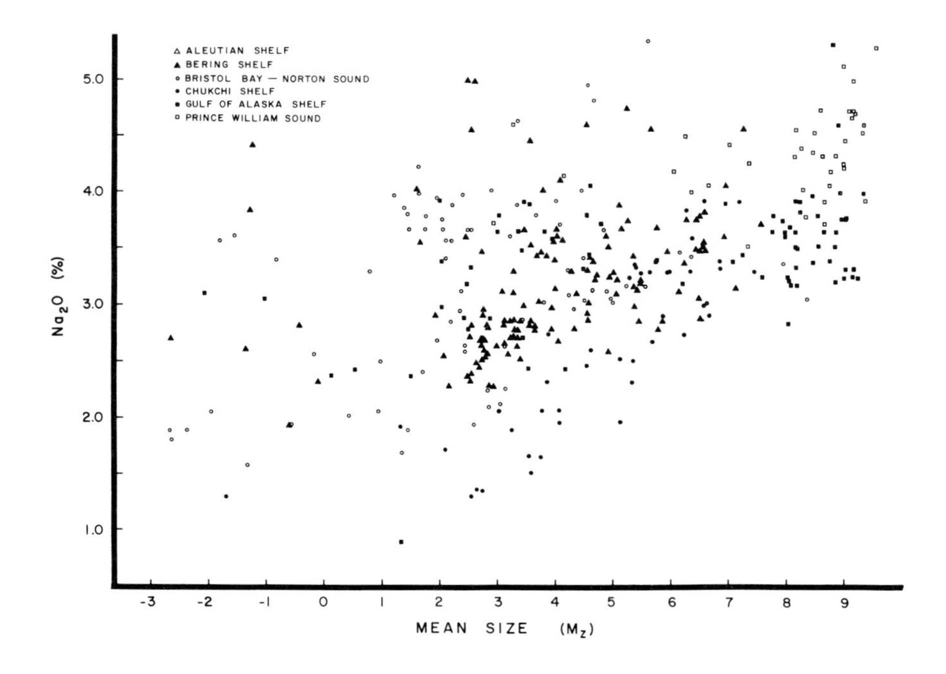

Figure 13-13. Percent Na_2O versus mean grain size plot, Alaskan Shelf sediments.

In comparison with sediments from lower latitudes, the sodium in sediments from the Alaskan Shelf is slightly higher. This probably related to the sediment weathering. The glacially eroded sediments generally contain higher contents of K, Na, Fe, Al, and Ti. This characteristic composition of sands of glacial origin is also reflected by their polymictic mineralogy, which is low in quartz and contains relatively higher contents of feldspar, micas and unstable, weathered pyroxenes, amphiboles, and epidote. The mineralogic and chemical composition of sediment formed in the northern latitudes, therefore, reflect very initial stages of weathering and only partial removal of sodium from sediments. The alumino-silicates and silicates, including even the most unstable minerals, are altered only slightly.

There are significant differences in the regional distribution of sodium in sediments (Table 13-2). On an average basis, the clayey sediments of Prince William Sound contain 4.4% Na_2O, which is slightly higher than the average of 3.6% found in acid igneous rocks. The higher concentration is an artifact of fine clay, which predominates the sediments from this region. Although the sediments from the Gulf of Alaska contain more clay content than those from the Aleutian Shelf, the percent sodium in the latter is higher. The increase in sodium, particularly in sediments from the southern Aleutian Shelf, is the result of the alkali-rich magmatic rocks exposed to the adjacent Alaska Peninsula and serving as the source for these sediments.

Northward, on the Bering and Chukchi shelves, with the exception of some sediments from the near-shore area of the Alaska Peninsula in Bristol Bay, the sodium content decreases with increasing latitude. This decrease is particularly noteworthy because of the successive increase in percent clay content in sediments. The northward decrease of Na_2O in sediment is related to the depositional environment. The northern shelves are shallow and, therefore, are areas of significant wave action. Undoubtedly, wave action will tend to break down feldspar and inevitably lead to enrichment of quartz. Such mineralogic differentiation is indeed reflected by the simultaneous decrease in both alumina and soda with increasing latitude and a complementary northward increase of silica content in sediments (Table 13-1).

The Al_2O_3/Na_2O ratio is an excellent measure of the extent of elemental separation in sediments because aluminum is one of the most stable, whereas sodium is the most mobile major element. The ratio is, therefore, a good index of marine weathering and sediment transport. When plotted, these ratios reflect the major pathways of sediments on the shelf. It is interesting to note that Al_2O_3/Na_2O separation is not pronounced (Fig. 13-14) and that the sediments are weakly weathered. Compared to the sediments from tropical regions the elemental separation has not been extensive.

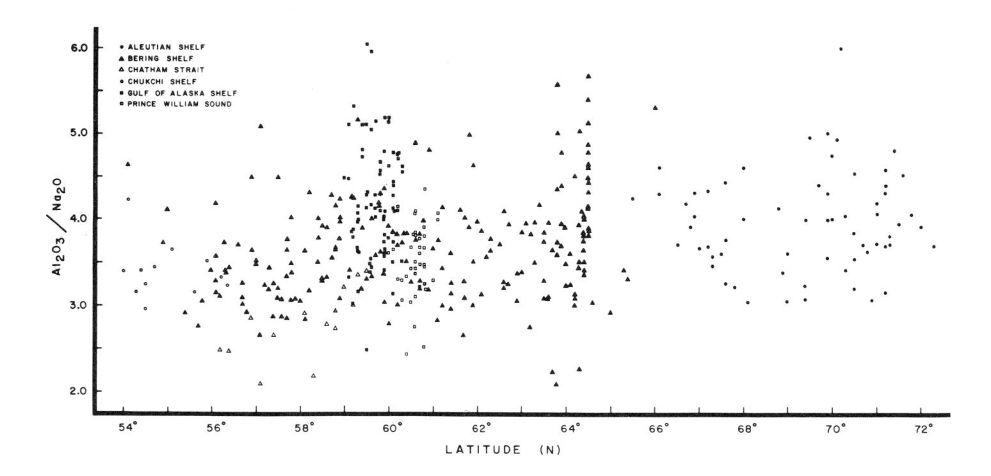

Figure 13-14. Al_2O_3/Na_2O versus latitude plot, Alaskan Shelf sediments.

Potassium

The distribution of potassium is closely related to the sediment mean size (Fig. 13-15). It increases slightly with decreasing particle size in sand and silt fractions, but there is a noticeable increase in the clay size fraction. Percent potassium also increases linearly with percent clay content in sediments (Table 13-1). In sand and silt fractions, the potassium is carried in feldspars and is also included in the micaceous muscovite, and thus arkosic sands are rich in potassium. Because of its association with silicates, the potassium distribution shows a close coherence with the distribution of aluminum (Table 13-1).

The relatively higher accumulation of potassium in clay fraction is primarily a result of an increased content of mineral illite and secondarily due to ion exchange of potassium from sea water.

The distribution and transport of potassium in fine sediments as mineral illite is further suggested by the plot of percent potassium versus chlorite-illite ratios. With increasing chlorite fraction relative to illite mineral the potassium content decreases steadily.

Because of disparity in the concentration, the behavior of potassium in comparison to that of sodium in the hydrosphere has been studied in fair detail. The potassic character of sediments and the sodic character of the sea water is explained on the basis of ionic and chemical properties. Accordingly, it is quite evident that K ion from sea water is preferentially adsorbed by the fine-grained particles of the sediments. It is K^+ and not Na^+ that fulfills the crystallochemical conditions required for the stability of mica and of illite. As the extent of marine weathering increases, more and more sodium is transferred to the sea, potassium is retained in the sediments. Potassium to sodium ratios, therefore,

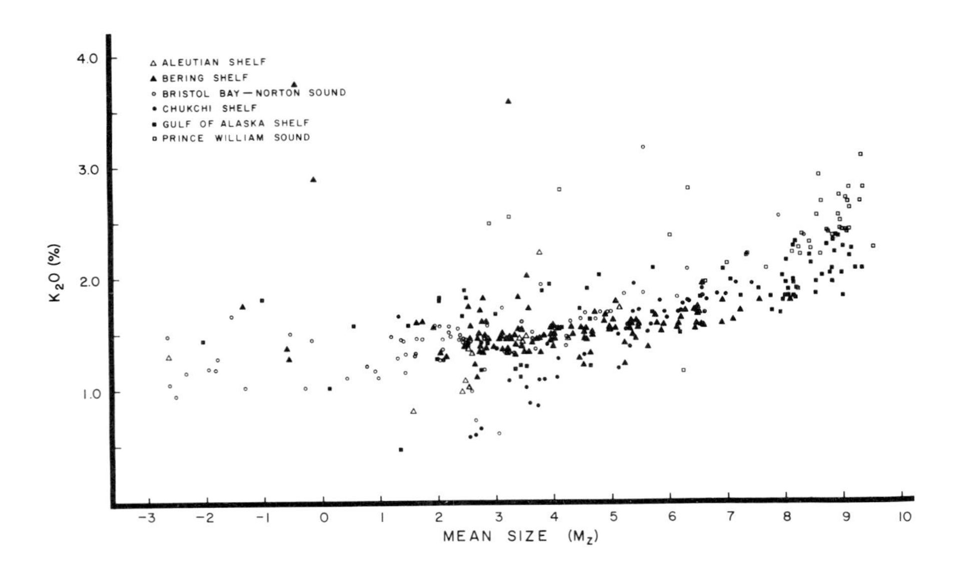

Figure 13-15. Percent K_2O versus mean grain size plot, Alaskan Shelf sediments.

can be a good indicator of degree of marine weathering and distance of transport.

It is interesting to note that potassium relative to sodium in the Alaskan Shelf sediments is consistently lower. This is unusual because the potassium to sodium ratio in average acid igneous rock is about 1.1. The unusual distribution perhaps can be explained by the K_2O/Na_2O versus latitude plot (Fig. 13-16). The sediments from the vicinity of the Alaska Peninsula (open circles), in general, are low in K_2O/Na_2O. The low values originate in the basic rocks that contribute sediments to this region. The average for K_2O/Na_2O in basic igneous rock is only 0.45, which is also the average for sediments deposited between 54° and 60°N latitude. Northward of 60°N, K_2O/Na_2O does not change, reflecting uniformity of weathering in the arctic region.

The lower than usual values of K_2O/Na_2O in the Alaskan sediments may be caused by two factors. First, the lower than usual intensity in weathering in the source region and the marine transport may not permit sufficient differentiation of elements. Second, the deficiency of potassium over sodium may simply be mixing of sediments contributed by basic as well as igneous rocks. In view of the higher content of sodium available in basic than in igneous rocks the latter case is more probable. The low values for K_2O/Na_2O are indeed observed in the Peninsular (Aleutian) Shelf and Bristol Bay regions (Table 13-2).

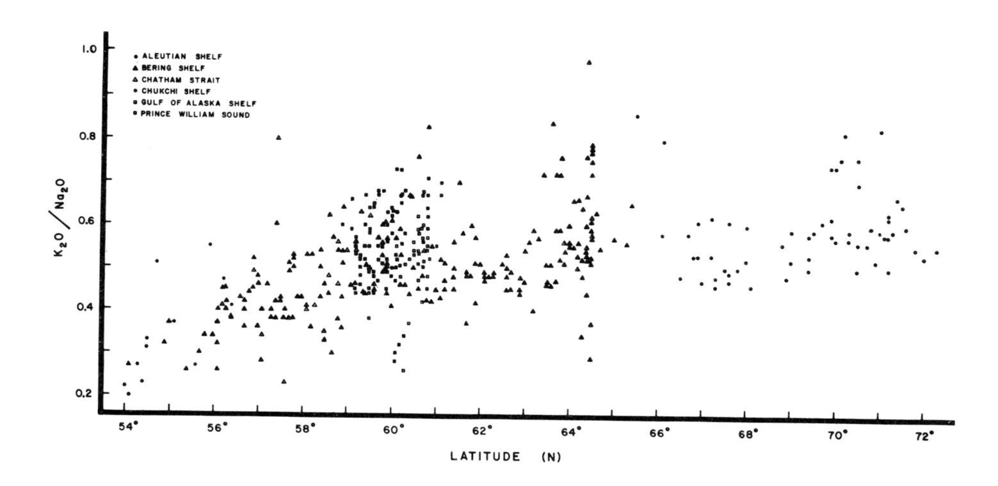

Figure 13-16. K_2O/Na_2O versus latitude plot, Alaskan Shelf sediments.

Titanium

On the Alaskan Shelf, titanium mostly enters as the product of river drainage and coastal abrasion. Like aluminum, titanium is relatively immobile, and in a marine environment it commonly migrates either as a mineral component or as particulates in suspension.

As mineral components, titanium and aluminum are related crystallochemically (Table 13-1) and they are present in the silicates. Although Al enters both feldspars and sheet silicates, Ti enters sheet silicates in substantial amounts but hardly at all into feldspar, so that a characteristic feature of most sediments that contain detrital micas and clay minerals is a marked positive correlation of Ti. This is well reflected by the increasing Ti with decreasing mean grain size. It should be noted that decrease in mean grain size is directly related to the clay content in sediments.

In the mechanically abraded coarse detritus, because of the high specific gravity of the titaniferrous mineral, titanium has its maximum concentration in the littoral sediments. Such near-shore concentrations of titaniferrous minerals generally occur in zones of waves and currents. Seaward, to about mid-shelf, the titanium content decreases slightly, but farther offshore it again increases rapidly in accompaniment with increasing clay on the outer shelf. The vacillating concentrations of titanium on the shelf can be easily understood if it is realized that two factors, equally important, govern its distribution. Hydrodynamics control the distribution of the titaniferrous heavy minerals in the silt and sand of the littoral zone, and accumulation of mica and clay minerals and colloidal titanium enriches the fine sediments of the offshore regions.

The migration of titanium in the form of a dispersed suspension and its eventual deposition with fine sediments is also important. The titanium in suspension originates from the hydrolysis of titaniferrous minerals, resulting in partial decomposition and dispersion of free hydroxide. The decomposition of titanium minerals (ilmenite, titanomagnetite, rutile, and sphene) leads to new intermediate compounds, such as leucoxene and xanthotitane, both of which are colloids. These colloidal titanium hydroxide compounds either are carried offshore and deposited with clay or form minerals, such as anatase, and lose mobility. It should be noted that titanium in the colloidal form has been observed in oceanic sediments, and it is possible that iron plays an important role in transporting titanium by chemical sorption process.

Some titanium is also deposited on the shelf as an isomorphous impurity in other most common rock-forming minerals. In such cases, it is mostly associated with iron minerals, such as magnetite and other various iron silicates.

The distribution of titania varies between 0.2 and 1.2% and its concentration is related to the sediment texture (Fig. 13-17). It increases with decreasing sediment mean size. The large scatter in the concentration of titanium in sand is primarily the result of accumulation of heavy minerals in sands from Bristol Bay. The heavy sand contains titaniferrous minerals and so enriches the sediment with titanium. The overall distribution of titanium in the Alaskan Shelf sediments suggests two distinct provenances. The Gulf of Alaska Shelf contains, on an average, 0.93% titanium oxide compared to only 0.68% in the Bering and Chukchi shelves. These differences cannot be accounted for by the variations in sediment texture and, therefore, are believed to result from the differences in the provenance providing sediments to these regions.

In the near-source and near-shore zones, hydrodynamic sorting segregates titanium as heavy sands, whereas aluminum is carried offshore as feldspar and clays. Such differentiation is the primary cause of variations in the Al_2O_3/TiO_2 of sediments on the shelf, and the increasing ratio suggests the migratory paths of the sediments. It should be noted that the ratios show no marked trend with latitude (Fig. 13-18). The plot suggests that this ratio in sediments is not significantly influenced by climate and related weathering. Because these variations are caused by mechanical (hydrodynamic) separation such patterns can be explained only by surface-controlled processes, such as tidal and permanent currents.

The Al_2O_3/TiO_2 in acid igneous rock is only about 12, but in basic rocks it is approximately 20. The ratios, therefore, can also identify contributions from various provenances. On an average, the ratio in Alaskan sediments is close to 16, suggesting mixing of detritus originating from acid as well as basic igneous provinces. Evidence for significant contributions from basic rocks was also provided by K_2O/Na_2O ratio in the sediments discussed earlier.

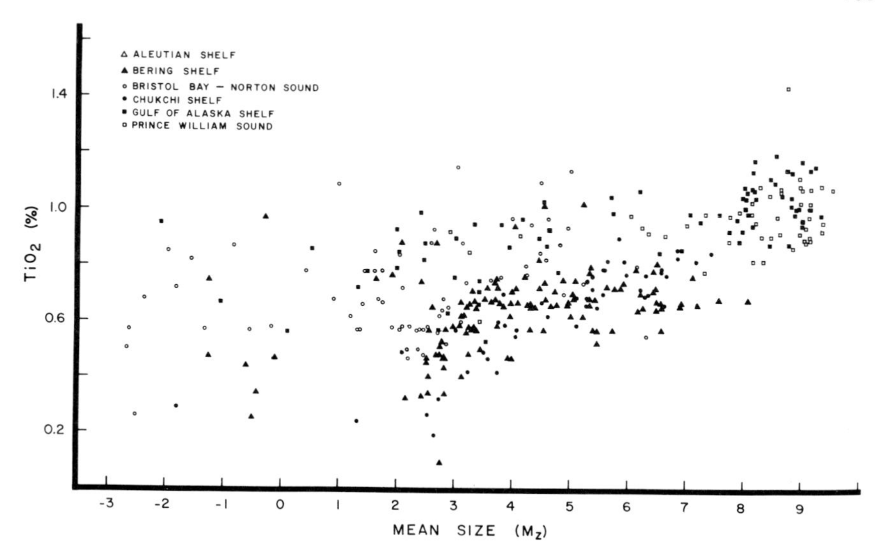

Figure 13-17. Percent TiO_2 versus mean grain size plot, Alaskan Shelf sediments.

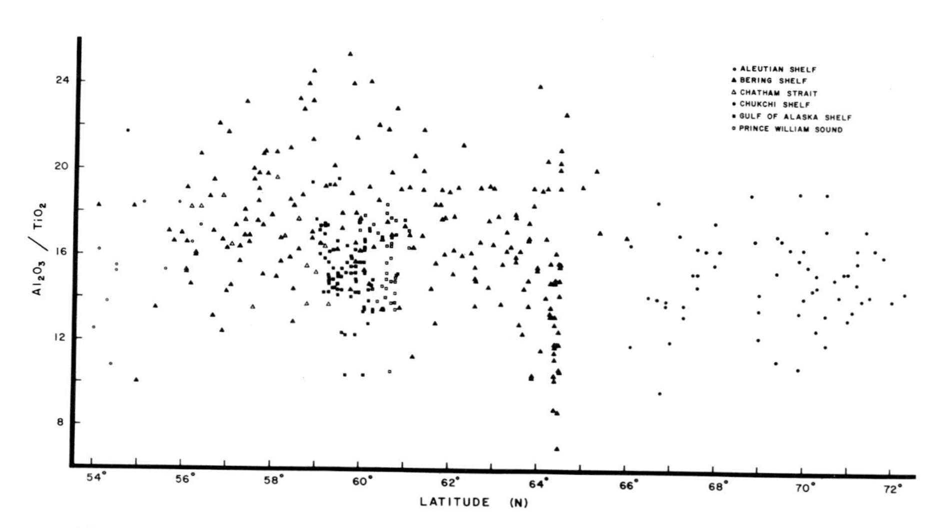

Figure 13-18. Al_2O_3/TiO_2 versus latitude plot, Alaskan Shelf sediments.

Manganese

The migration of manganese in sediments in complex and its distribution is partly controlled by the provenance and partly by its coherence with other elements (Table 13-1). Manganese and texture of sediments are not related, but distribution of manganese, to an extent, is related to the percent clay in sediments. The clay-manganese relationship is even somewhat complex.

When the distribution of manganese in sediments is described on a regional basis, however, it becomes quite apparent that its content is primarily related to the provenance. Sediments from the Gulf of Alaska and Bristol Bay regions contain 0.06% or less (Table 13-2). In general, with the slight exception of the sediments from Norton Sound, there appears to be a slight northward decrease of manganese content in sediments. The slight increase in Norton Sound, primarily, is the result of the Yukon River clayey sediments deposited in this region.

One of the major sources of manganese in solution and in fine suspension is the river runoff. Manganese in river water is about one magnitude higher than that observed in ocean water. A large part of the soluble manganese brought by surface runoff, upon entering the marine environment, must be removed from solution. Manganese and iron behave chemically alike to a first approximation and, therefore, manganese may follow iron in the hydrosphere. This may be the reason for the close association between iron, clay, and manganese in sediments (Table 13-1).

Manganese in sediments closely follows iron, and the MnO/Fe_2O_3 varies mostly between 0.01 and 0.02. In sand and silt it is about 0.01 or slightly higher, and in clayey sediments from Prince William Sound it is about 0.02. These concentrations are similar to those found in modern sediments. The average MnO/Fe_2O_3 in acidic and basic igneous rocks is 0.015 and 0.021, respectively. From the comparison of these ratios it appears that only slight chemical differentiation of iron and manganese in sediments occurs on the shelf. The separation is primarily mechanical and is well manifested by the distributions of iron as well as manganese in the near-source and littoral zones.

TRACE ELEMENT MIGRATION

Introduction

Continental runoff and coastal erosion are the principal sources for trace elements in the shelf sediments. The distribution of a trace element in the water as well as in sediments of the shelf is affected by the chemical properties and the migration mobility of the element, by physiographic conditions, and by the petrologic features of the source region. The trace elements may be carried in solution and as finely divided suspended ditrital. They are also found as absorbed ions on the clayey sediments. Considerable quantities of these elements are retained in crystal lattices of minerals, which remain relatively undecomposed and are deposited on the shelf. This is particularly significant in areas where the shelf is bordered by mountainous belts, and river detritus, generally, carries trace elements as suspended matter. However, lowland rivers (Yukon River) debouching in the Bering and Chukchi shelves may carry trace elements in solution.

Of particular interest in southern Alaska is the direct interaction of rock flour with water and subsequent leaching of trace elements. The catchment areas of the streams and rivers are covered with glacial sediment. These are leached of their trace elements and subsequently carried to the shelf. Even more important, perhaps, is the large amount of glacier flour that is carried to the shelf and, on reaction with sea water, releases a variety of elements.

Another factor that may significantly affect the distribution of trace elements on the Alaskan Shelf is the climate. The onshore climate varies from maritime subtropical in the south to dry arctic (polar) in the north. Because of the changes in weathering it is stipulated that the mobility of elements should decrease northward. The long cold winters and the permafrost in source areas retard the decomposition and separation of elements, and therefore the overall migration of trace elements. Changes in weathering pattern may also retard the formation of clay minerals that serve as carriers for many trace elements in marine environments. Besides weathering, many of the trace elements enter the shelf chiefly in lattice positions within the clay minerals.

The source rocks in southern Alaska are dominantly metamorphic and granitic, with some sedimentary sequences and, therefore, considerable amounts of trace elements will be retained within the lattices of micaceous constituents. This region also contains exposed igneous rocks, which upon weathering release trace elements in solution as a result of breakdown of high-temperature minerals and formation of clay minerals. In such cases, the clays not only carry trace elements in their lattices but adsorb additional quantities from solution during their transfer on the shelf.

Sediments brought to the northern shelves mostly have their sources in sedimentary and metamorphic rocks. The colder climate inhibits extensive weathering and therefore coupled with a lack of primary mineral sources for the trace elements, the weathered sediments chiefly carry trace elements only in their lattices.

Barium

The distribution of barium in Alaskan Shelf sediments mostly varies between 250 and 500 ppm. It has a bimodal distribution in relation to the sediment texture. The coarse sediments (gravel, sand, and silt) on the average contain about 350 ppm, whereas the finer sediments (clays) have a somewhat higher

(500 ppm) barium content (Fig. 13-19). It is quite obvious that the concentration in coarser sediments varies significantly and the cause for such a scatter is difficult to explain.

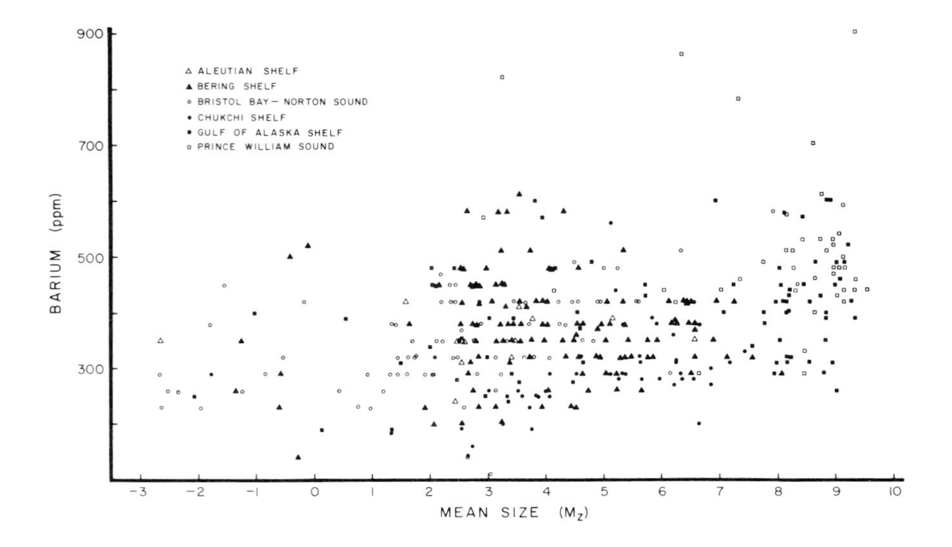

Figure 13-19. Barium (ppm) versus mean grain size plot, Alaskan Shelf sediments.

The higher content of barium in clay is primarily a result of increased illite, which carries barium with potassium. There is, however, no coherence between the distributions of potassium and barium. It should be noted that the barium scarcely follows the distributions of other major or minor elements (Table 13-1).

On the regional average basis, the barium distribution shows a slight but consistent northward decrease in sediments. The average from southeast Alaska and Prince William Sound is about 500 ppm and it successively decreases to 280 ppm in sediments from the Chukchi Shelf (Table 13-2). This decrease is not caused by the colder climate in the northern latitudes (Table 13-1).

Cobalt

Cobalt is a relatively mobile element and, therefore, it readily migrates in detritus as well as in non-detritus phases. Its complex mobility in sediments is well displayed by the poor relationship between cobalt content and the texture of sediments. Sand and silt generally show a large scatter and mostly contain lesser amounts of cobalt that clayey sediments do (Fig. 13-20).

The migration of cobalt in crystal lattices, as well as by other processes, is evident from the variations in Co/Al ratios. The Co/Al ratio is fairly consistent

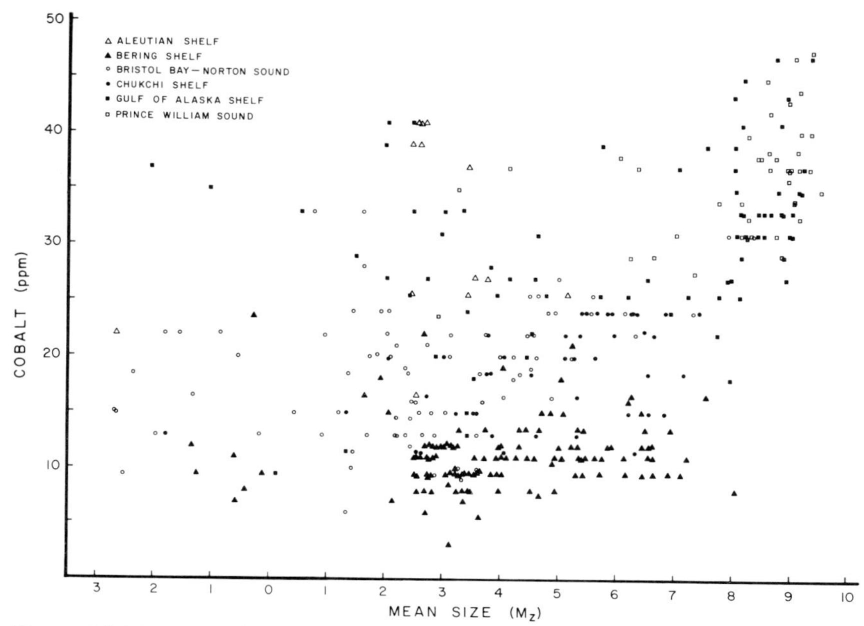

Figure 13-20. Cobalt (ppm) versus mean grain size plot, Alaskan Shelf sediments.

for sediments with 7%-13% Al and with 10-25 ppm Co. With increasing aluminum, however, more than 13%, the cobalt content increases rapidly, suggesting that aluminum-rich sediments (clays) carry cobalt in crystal lattices as well as in association with clays.

Calcium carbonate has been known to scavenge trace elements. It is also known that cobalt (and sometimes nickel and copper) is brought down by calcite depositing organisms. Although the general distribution of calcium is not similar to the cobalt, it appears that coherence of cobalt with calcium is undoubtedly the result of biogenic calcium carbonate in the sediments (Table 13-1).

Distinct regional variations in the distribution of cobalt are obvious (Table 13-1). Pacific Shelf sediments contain twice the content of cobalt than that observed in sediments from the Bering and Chukchi shelves. The regional differences may be the result of the provenance.

Chromium

Chromium in sediments varies over a wide range (10-150 ppm). It varies regionally as well as with texture of sediments (Fig. 13-21 and Table 13-2). On the Gulf of Alaska Shelf, the highest concentration occurs in the central shelf, and slightly lesser quantities are observed in clayey sediments of Prince William

Sound. Southward, along the Aleutian Shelf, the chromium in sediments decreases significantly. In contrast to the central Gulf of Alaska Shelf and Prince William Sound, the sediments from the Bering and Chukchi shelves contain only half as much chromium or less in sediments. Interestingly, however, northward of Bristol Bay the chromium increases and is comparable to the increase in clay content in sediments.

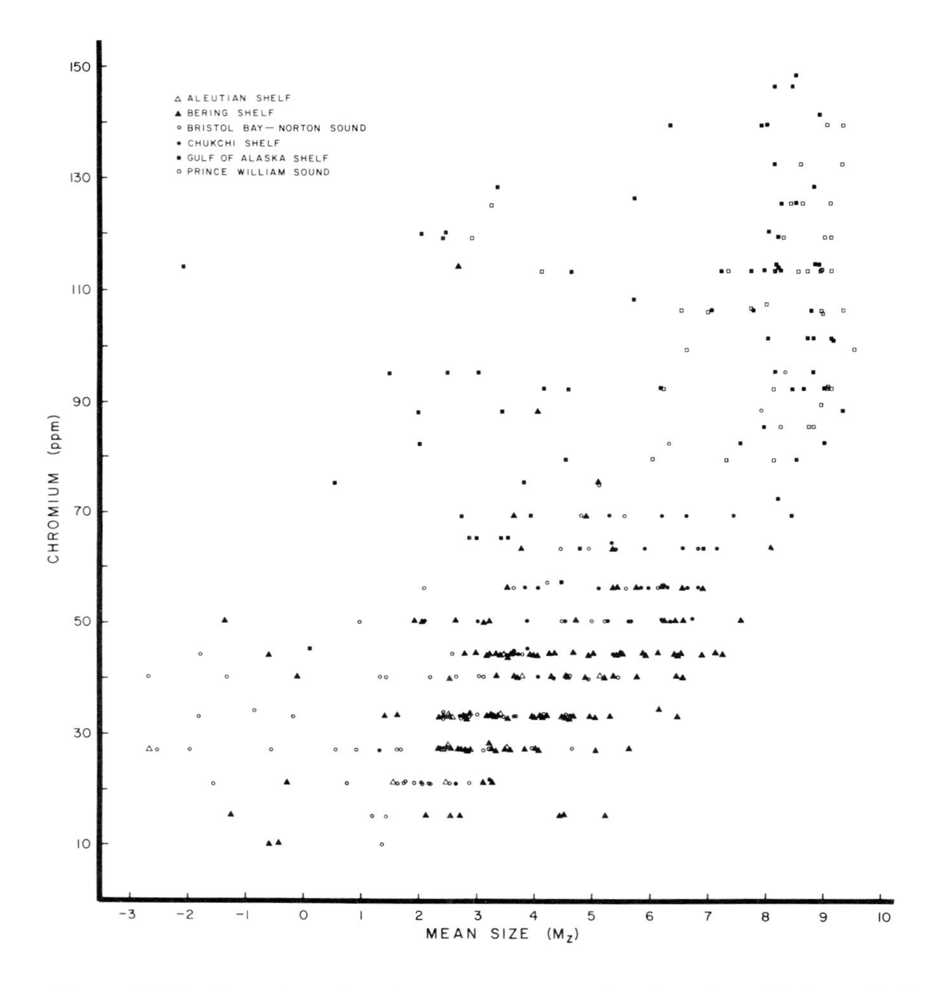

Figure 13-21. Chromium (ppm) versus mean grain size plot, Alaskan Shelf sediments.

The relationship between chromium and sediment texture appears to be somewhat complex. The cobalt versus sediment mean size plot shows a bimodal U-shaped distribution. In gravels, chromium is about 30 ppm and it decreases with decreasing sediment mean size and reaches a minimum of about 15 ppm in coarse sand. With further decrease in sediment mean size, however, the chromium content increases rapidly and reaches a maximum of about 130 ppm in clays (Fig. 13-21).

The complex migration of chromium is also suggested by the chromium and aluminum relationship. Chromium increases moderately with increasing aluminum in sediments but, as in the case of cobalt, the increase in chromium relative to aluminum in sediment with 13% or more alumina is significantly higher. Chromium also displays linear relationships with iron and magnesium percents, which suggests similar modes of migration on the shelf.

The successive northward increase of chromium in sediments on the Bering and Chukchi shelves, with a complementary increase in percent clay and a good linear relationship with magnesium in sediments, suggests that chromium is strongly associated with chlorite and illite. Chromium is perhaps carried in lattice positions of the degraded clays and its association with magnesium indicates that it is carried by both chlorite and illite. The large scatter of chromium concentration in the coarse fraction is the result of the deposition of chromium magnetite-rich sand in near-shore regions, and chromium-poor quartzitic sands in midshelf regions. High-energy environments (wave action and currents) are regions of low sediment accumulation as well as low chromium content. The distribution of chromium in sediments, therefore, may serve as an indicator of sedimentation rate.

Copper

The distribution of copper on the Alaskan Shelf is primarily related to sediment texture and the percent clay in the sediments. The complex relationship between copper and the sediment mean size suggests that copper may be migrating both in association with clay and in mineral crystal lattices. The concentration increases with decreasing grain size (Fig. 13-22). Sand and silt on an average contain about 30 ppm or less copper. In particular, the moderately sorted sand from the Bristol Bay region contains only about 10-15 ppm copper. The paucity of copper in these sediments is probably related to the absence of clays.

The marked increase in concentration of copper in the clay fraction is noteworthy. This increase suggests that mechanical transport of copper is subordinate to other processes. The distribution of copper in relation to aluminum further indicates that copper is very mobile and migrates readily in association with clays. The clay percent and copper content in sediments, of course, show a direct linear relationship; copper increases with increasing clay percent in sediments. This relationship readily accounts for the copper in sediments throughout the Alaskan Shelf (Table 13-1).

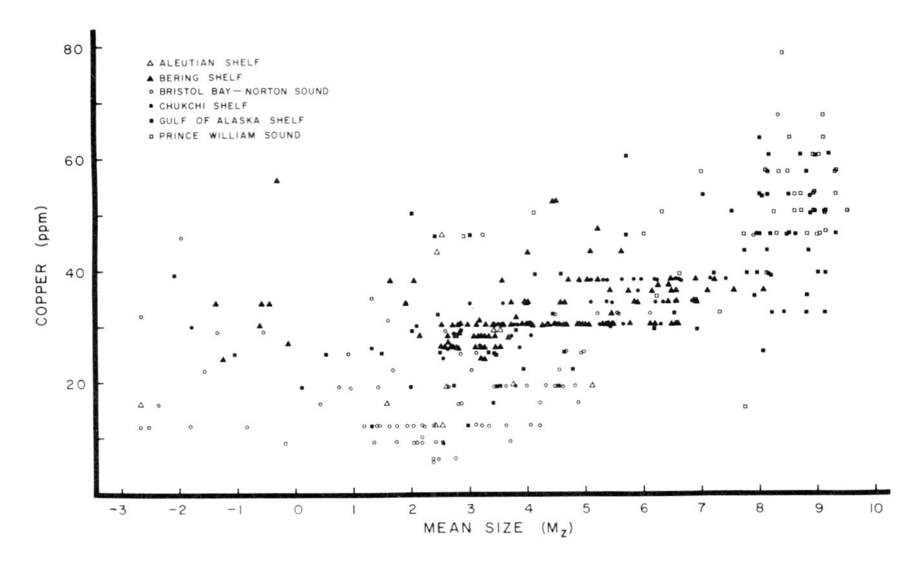

Figure 13-22. Copper (ppm) versus mean grain size plot, Alaskan Shelf sediments.

The relationship between copper and magnesium in sediments appears to be, like that of copper and sediment texture, bimodal and therefore of some interest. For the most part, the copper increases from 7-40 ppm with a complementary increase in magnesium of from 0.7-3.0%. With further increase in magnesium the rate of increase in copper accelerates, however, and the slope of the curve steepens. Magnesium and copper both display strong relationships with clays in sediments. The marked change in slope with increasing magnesium thus may indicate that significant amounts of copper in clays, besides carried in crystal lattice, are transported in the form of highly mobile compounds (absorbed material, soluble salts, etc.).

Nickel

Among the trace elements discussed so far nickel appears to be the most mobile component of the sediments. Its mobility, particularly with the finer fractions of the sediments, is well illustrated by its relation to the sediment texture (Fig. 13-23). Nickel displays a strong relationship with sediment particle size. In gravel-sand-silt fractions it increases moderately with decrease in mean size and in clay fraction it increases rapidly.

Migration of nickel in relation to detritus is noted in the Ni/Al ratio. Sand-silt sediments reveal a significantly lower ratio than clayey sediments. Similar variation in ratio is observed in relation to the distribution of iron and manganese. The magnesium, however, indicated a direct linear relationship with nickel such

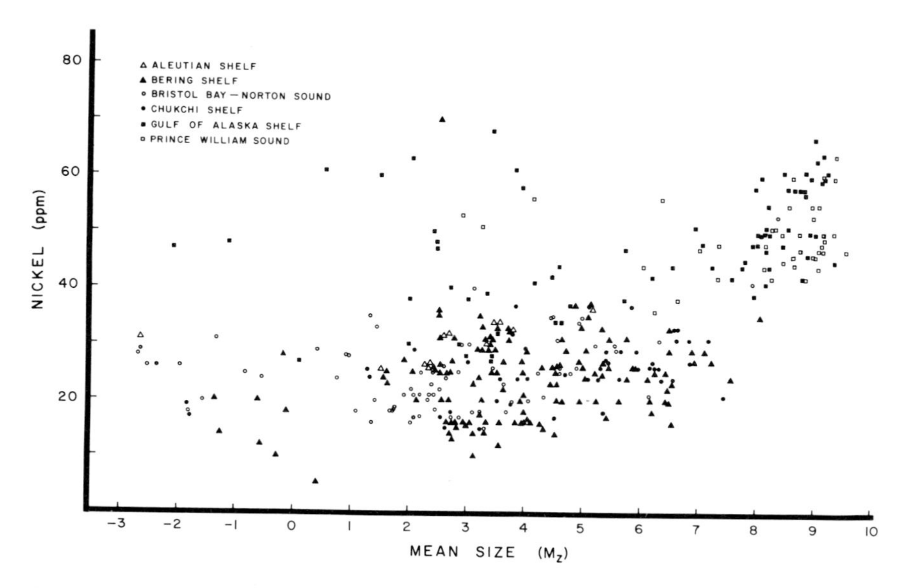

Figure 13-23. Nickel (ppm) versus mean grain size plot, Alaskan Shelf sediments.

that nickel increased with increasing magnesium. Nickel also displays a direct linear relationship with percent clay in sediments (Table 13-1).

The elemental ratios and textural relationship of nickel indicate increasing migration with clay, iron and manganese in sediment. The coherence with iron and manganese is important. The increase with the proportion of iron and manganese may be the result of adsorption and of autocatalytic oxidation processes. Hydroxides of iron and manganese formed on the shelf have high specific surfaces, and the accumulation of these transitional metals along with cobalt may have been caused by conjugate processes.

The regional average distribution of nickel in sediments further exemplifies its association with clays in sediments (Table 13-2). Besides the textural control, the effect of provenance is also superimposed on the distribution of nickel. The exception to this, however, is noted in sediments from the Chukchi Shelf. The low content may be related to colder climate, which should significantly reduce the mobility and thus the supply of nickel to the shelf. The sediments from the Pacific Shelf have markedly more nickel than those deposited on the northern shelves.

Strontium

The distribution of stronrium is complex because its content varies regionally. It varies between 50 and 830 ppm; most values, however, lie between 100 and 400 ppm. Table 13-1 does not show any correlation between strontium concentration and sediment mean size. However, the sediments from the Chukchi

Sea and Norton Sound, when considered separately, distinctly show a slight increase with decreasing mean size in sediments.

In general, strontium does not show coherence with other sediment parameters, with the exception of contents of calcium and cobalt. Distributions of calcium and strontium have a very strong coherence, and cobalt and strontium are related to a lesser extent. The strontium-calcium association is the result of biogenic processes, which contribute strontium as well as calcium carbonate to skeletal material. Turekian and Kulp (1956) stated that strontium in calcite n exceed 610 ppm. A high strontium content in sediments thus appears to be elated to increased content of skeletal constituents in sediments.

Strontium in sediment decreased with increasing latitude. The northward decrease may be the result of both a decrease in biogenic calcium carbonate production and a lessening of chemical weathering.

Zinc

The distribution of zinc in sediments is related to the mean size and even more strongly related to the percent clay in sediments (Fig. 13-24). Both relationships are linear. The large scatter in the very coarse-coarse sand size fraction is somewhat puzzling and cannot be explained. The distribution in the coarser sediments (gravel) is even more complex and does not follow the general trend.

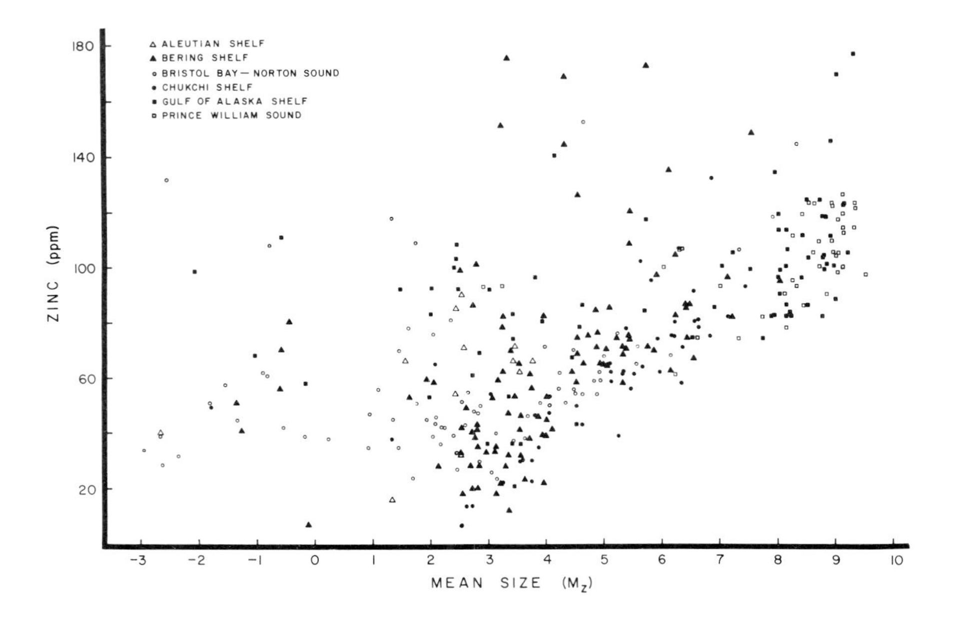

Figure 13-24. Zinc (ppm) versus mean grain size plot, Alaskan Shelf sediments.

Among trace elements zinc tends to follow nickel and copper. Its distribution also indicated coherence with percent aluminum, iron, magnesium, titanium and clay. This coherence may simply be related to its association with clay that has served as a carrier for the others.

The regional distribution of zinc is primarily related to the sediment texture. There is, however, a slight deficiency of zinc in sediments in the northern shelves.

REFERENCES

Brotzen, O. (1966). The average igneous rock and the geochemical balance. *Cosmochim. Acta* 30:863-868.

Carroll, D. (1959). Ion exchange in clays and other minerals. *Geo. Soc. Amer. Bull.* 70:749-780.

Clarke, F. W. (1924). The data of geochemistry. U.S. Geol. Survey Bull. 770 (fifth edition), 841 pp.

Goldberg, E. D., and J. J. Griffin. (1964). Sedimentation rates and mineralogy in the South Atlantic. *J. Geophys. Res.* 69:4293-4309.

Goldschmidt, V. M. (1954). *Geochemistry.* Oxford University Press, Fair Lawn, N.J., 20 pp.

Nanz, R. H. (1953). Chemical composition of pre-Cambrian slates with notes on the geochemical evolution of lutites. *J. Geol.* 61:51-64.

Turekian, K. K., and J. L. Kulp. (1956). The geochemistry of strontium. Geochim. et Cosmochim. Acta 70:245-296.

CHAPTER 14
Shelf Environments

INTRODUCTION

The large shelf bordering Alaska extends across many physiographic and climatic regions. It is, therefore, not simple to classify the entire shelf into a few identifiable environments. The most commonly used parameters for defining the sedimentary environments are the sediment parameters. Lacking thorough knowledge of the hydrodynamic and other related information, the sedimentary parameters fail to fully elucidate the critical aspects of deposition, and especially the characteristic process over a large area. They may, however, fully characterize the sediments on a regional basis, particularly in a region with one or two identifiable detritus sources and with not too complex transporting agents.

The Alaskan Shelf displays a variety of environmental parameters. It has a complex bottom topography with diverse water movements. The detritus texture and mineralogy at the source vary regionally. Strong shelf currents generally carry sediments across the regional boundaries and mixing of sediments further complicates the attempt to classify the shelf.

A variety of environmental criteria can be chosen to classify the shelf. The criterion can be based on the textural, mineralogic, hydrodynamic or geochemical parameters. Regionally, the Alaskan Shelf has been described in terms of textural and geochemical parameters. Limited information on the mineral character of sediments is also included. Fundamentally, the mineralogy of sediments primarily controls the geochemical parameters; therefore, mineralogic character is well represented by the geochemistry of the sediments.

Sparse information and few *in situ* measurements of water flow are available from the Alaskan Shelf. The influence of water movement is mostly reflected by sediment texture and to some extent geochemical differentiation. It is hoped that sediment-hydrodynamic flow interrelationships studied in some detail in a few regions will provide the key to deciphering the water movement patterns in other regions.

FACTOR ANALYSIS

Textural and geochemical parameters probably provide the most comprehensive information; therefore, an attempt can be made to classify the environments of the shelf on the basis of these combined parameters. Such an attempt should, undoubtedly, provide environmental subdivisions based on a wide variety of parameters and, therefore, should be extremely useful. The number of sediment variables studied for each sample is 26 and, in total, approximately 600 samples were processed. In view of the magnitude of the data points, it was deemed necessary to seek the aid of a computer. Recently, many methods for determining multiple associations and relationships between various parameters have been developed. One of the most widely used methods is factor analysis.

Factor analysis has been frequently used to interpret geologic data and has been described by various investigators (Imbrie and van Andel, 1964; McManus *et al.,* 1969). For this study a general factor analysis program (FORMAN) which can be used in two modes—R or Q— is used (Kelley, Dept. of Oceanography, Univ. of Washington, Seattle, personal communication). As a simple example, consider M kinds of tests that have all been given to each of N subjects. To illustrate the R mode first, we might ask whether all the tests look at independent variables or whether there is some redundancy. The $M \times M$ matrix of correlation coefficients between tests offers the answer. The eigenvalues and eigenvectors of this matrix are the basis of the factor analysis. Roughly, the number of eigenvalues greater than one show the number of independent dimensions or factors described by the tests. The eigenvectors govern linear combinations of the original tests that, if constructed, would represent pseudo-tests, or factors that measure the independent dimensions in the testing domain. The factor associated with the largest eigenvalue explains the largest possible part of the variance in the testing space; the next factor explains the largest amount of the remaining variance, etc. In the program these factors are called principal components or unrotated factors. They are a set of orthogonal vectors defining a vector space. Next, the program performs a rotation on the factors so that a new set is produced in which each factor accounts for an approximately equal share of the total variance. These are the rotated factors. Finally, the program produces a set of oblique factors, which are not orthogonal but may be the most interpretable in an intuitive sense. These procedures constitute the R mode analysis.

The Q mode analysis follows all these steps; however, the starting point is the $N \times N$ matrix of correlation coefficients between the N subjects (figured from the tests they have taken). The number of independent factors that are determined among the subjects correspond to clusters of subjects who performed similarly on the tests. The factors themselves show who is in what cluster.

The analysis provided six factors, which accounted on an average for 90 of the variance in each subgroup. Loading values of more than 0.5 provided the index of the affinity of a particular sample and all samples were classified according to their affinity to factors I-VI (Table 14-1). Samples included in each factor were then plotted areally to define the shelf environments (Figs. 14-1 to 14-4).

TABLE 14-1

Loadings of the Variables of Extremal Samples of Each Factor

Variables		Factor I	Factor II	Factor III	Factor IV	Factor V	Factor VI
Mean Size		9.15	2.82	2.76	-2.51	2.43	5.28
Sorting		1.88	2.73	0.45	1.13	0.71	1.93
Skewness		0.04	0.27	-0.21	0.45	-0.23	0.72
Kurtosis		1.06	5.03	1.43	1.31	1.48	1.97
Gravel	(%)	0.00	7.20	0.00	81.70	0.00	0.00
Sand	(%)	1.31	74.94	96.20	16.50	96.67	12.17
Silt	(%)	24.80	11.16	2.20	1.00	1.07	74.65
Clay	(%)	73.89	6.7	1.60	0.80	2.26	13.18
Al_2O_3	(%)	16.07	8.13	10.96	5.48	11.34	11.53
Fe_2O_3	(Total %)	8.59	3.12	3.53	2.76	5.66	4.10
CaO	(%)	2.66	2.03	2.66	1.05	16.52	3.08
MgO	(%)	4.45	1.13	1.33	1.06	2.19	2.16
K_2O	(%)	2.78	1.17	1.42	0.91	0.97	1.54
Na_2O	(%)	4.99	2.23	2.90	0.82	3.58	3.24
MnO	(%)	0.168	0.068	0.059	0.015	0.134	0.041
TiO_2	(%)	0.90	0.68	0.62	0.26	0.74	0.75
SiO_2	(%)	53.06	80.06	75.09	85.93	53.02	70.06
Ig. Loss	(%)	6.33	1.38	1.43			

TABLE 14-1 (Continued)

Variables		Factor I	Factor II	Factor III	Factor IV	Factor V	Factor VI
Ba	(ppm)	590	290	350	260	240	280
Co	(ppm)	44	15	21	95	25	20
Cr	(ppm)	113	33	33	27	27	50
Cu	(ppm)	67	16	6	12	12	30
Ni	(ppm)	60	30	17	26	26	24
Sr	(ppm)	275	220	313	88	605	192
Zn	(ppm)	119	47	48	132	54	39
Org. Carbon	(%)	0.766	0.537	0.190	1.214	2.004	

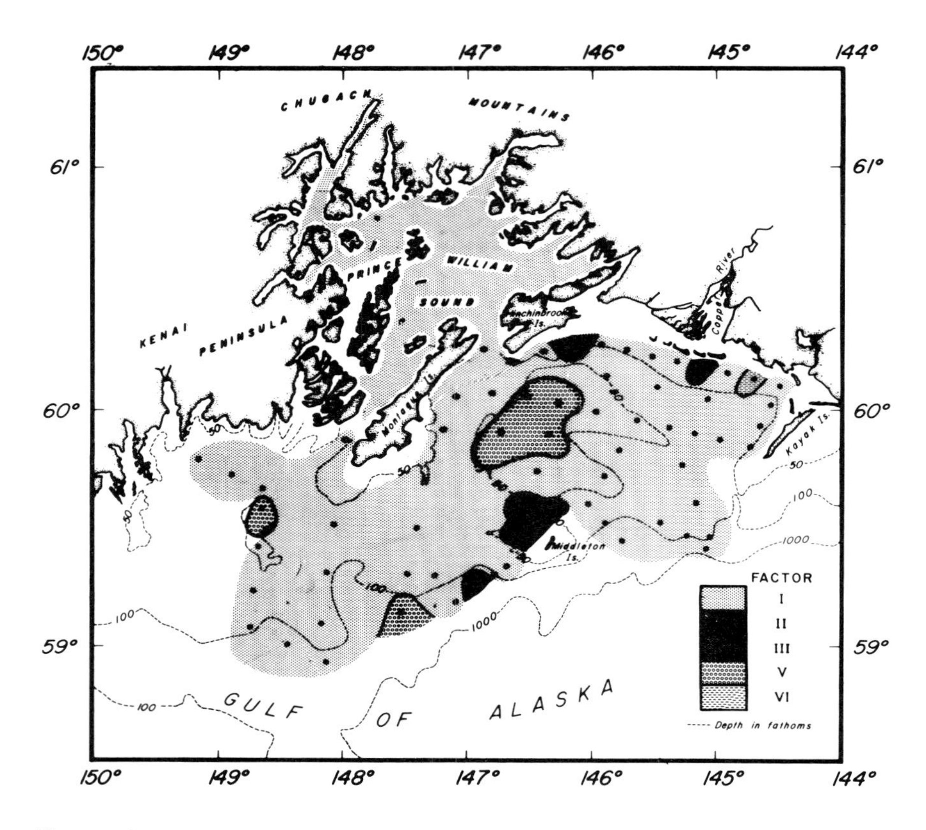

Figure 14-1. Shelf environments on Prince William Sound and central Gulf of Alaska.

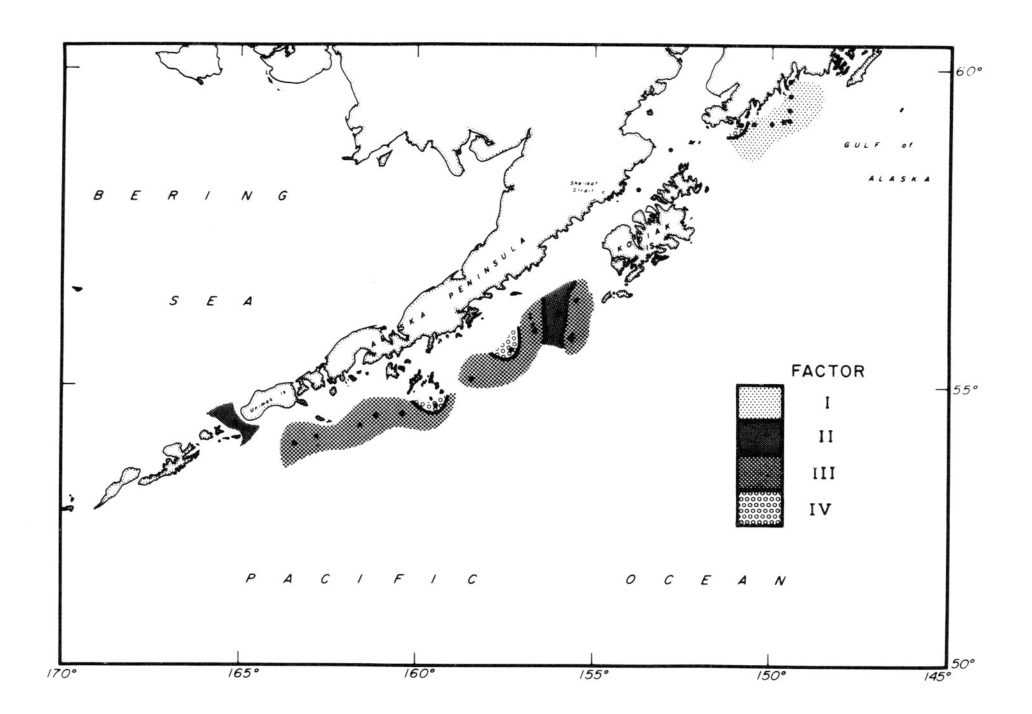

Figure 14-2. Shelf environments on the northwestern Gulf of Alaska.

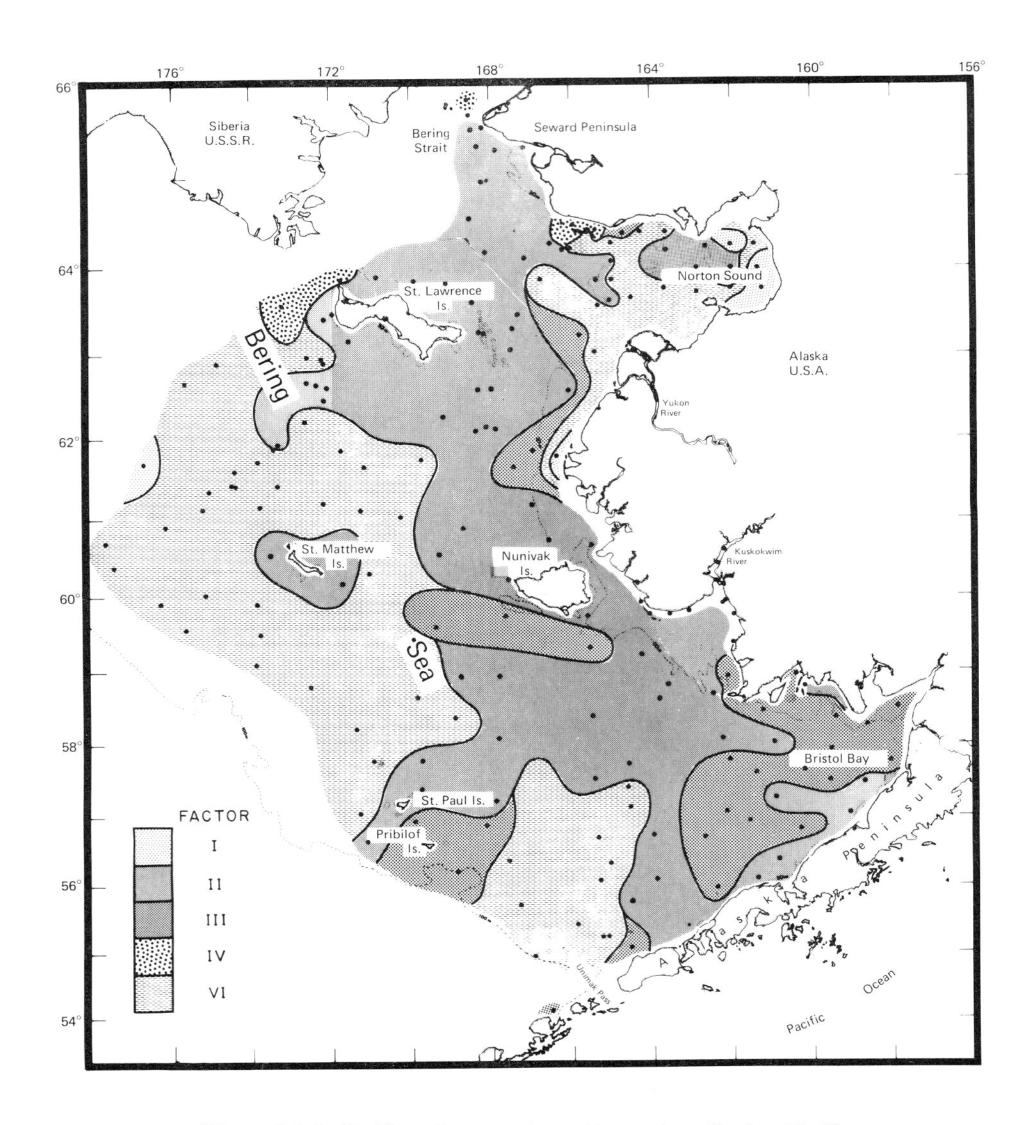

Figure 14-3. Shelf environments on the eastern Bering Shelf.

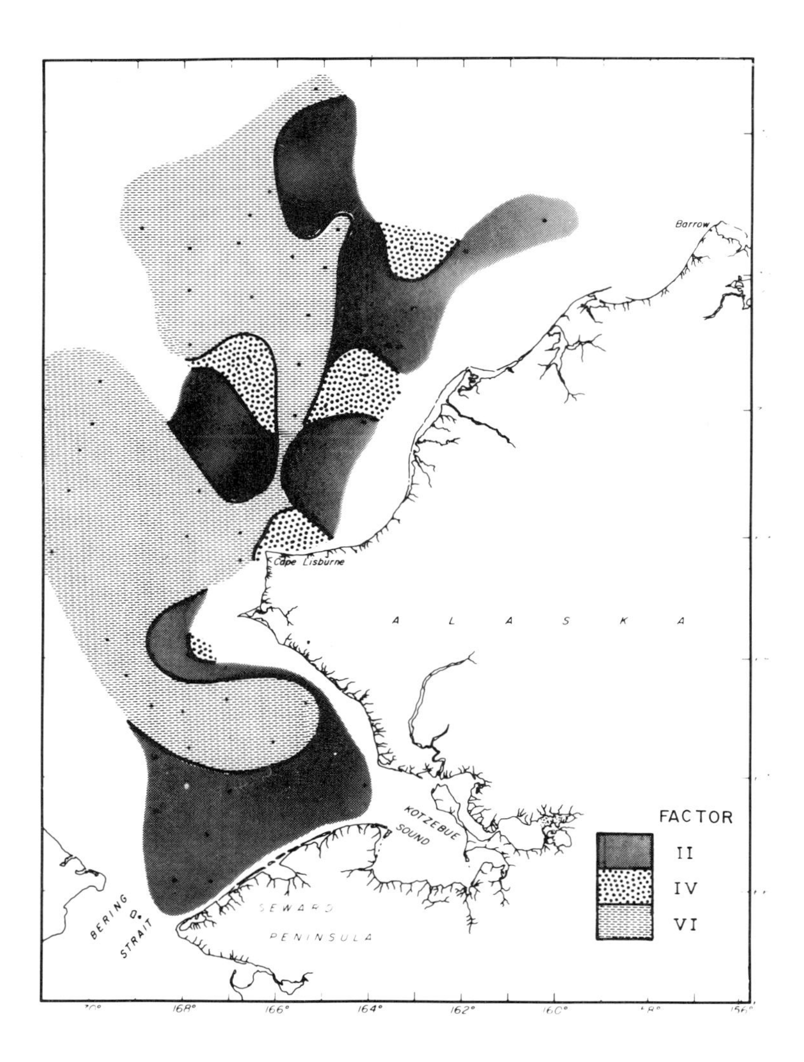

Figure 14-4. Shelf environments on the eastern Chukchi Shelf.

Factor I

The sediments from this factor have a high loading of mean size, clay, aluminum, iron, magnesium, sodium, cobalt, chromium, and copper. Factor I sediments cover the entire Prince William Sound and almost all of the central Gulf of Alaska. These also cover large parts of the northeast Aleutian Shelf. In the Bering Shelf, the distribution of factor I is restricted to areas near Unimak Pass and the near-shore Yukon Delta. In the Chukchi Shelf, the sediments of factor I are absent.

The predominance of clay with some silt in factor I suggests that these sediments are mostly deposited in relatively quiet environments. Such environments permit deposition of sediments carried in suspension. Once deposited it appears these sediments are not subjected to resuspension and elutriation. In most cases the wave action that causes resuspension is inhibited by the water depth. The areas mantled with factor I sediments generally lie under more than 100 m water depth. The near-shore clayey deposits of the shallow Yukon Delta and Copper Delta are the result of extreme influx of the rivers and the inability of the transport agents to carry the clays offshore.

Geochemically, the factor I sediments contain high percentages of aluminum, iron, magnesium and sodium. These sediments also contain high contents of cobalt, chromium and copper. Except for aluminum, all these elements are mobile and their high concentration in sediments is typical of a quiet environment of deep water. These major and minor elements combined form a well-known covariant group. Their association with a fine fraction of the sediments suggests scavenging by clays and possibly with ferric and manganese hydroxides.

Factor II

The major components of this factor are sand and gravel. Geochemical characteristics of these sediments are the dominance of silica and a low content of aluminum, iron and other elements.

The distribution of factor II sediments on the Gulf Shelf is limited and only small patches of these sediments are observed. On the Bering and Chukchi shelves, however, factor II sediments cover significant areas and may extend from near-shore to mid-shelf. These areas lie under water depths of less than 50 m and are regions of swift water currents. The high-energy environment is reflected by the removal of less resistant feldspars and other minerals and higher concentrations of quartzitic sands. In some areas the currents are swift enough to remove even fine and medium sands and thus form gravel lag deposits.

The mineralogic differentiation in the high-energy environment is further reflected by the elemental distributions. Sediments from factor II are extremely rich in quartz and contain about 80% or more silica. This is primarily achieved by intense water movement, which continually breaks down feldspars into smaller grain sizes and eventually carries them down current. This process results in enrichment of more resistant quartz grains in the sediment, leading to high silica and a deficiency in the rest of the elements.

On the Bering Shelf, the areas covered by factor II sediments are known for extensive northward movement of Pacific water. It is, therefore, logical to assume that these sediments signify a high-energy environment.

Factor III

Sediments typical of factor III have a limited distribution and their characteristics are quite unique. The dominant parameters are overwhelming dominance of sand (95%-100%) and good sorting. These sands are typical of shallow regions and generally lie close to the provenance. Although both factors II and III are predominantly sand, the sediments from factor II are distinctly different from those of factor III

In spite of the similarity in texture between factors II and III the chemical characteristics of these factors are distinctly different. While factor II sand represents highly elutriated gravelly to coarse quartzitic sand, the chemical composition of factor III sands is closer to arkosic sand. The sediments are well sorted and a high degree of sorting is the result of wave action. In view of the large area of inner Bristol Bay that is covered by these sediments it appears that sorting in sediments may have been achieved by storm wave action in shallow areas as discussed earlier.

Factor IV

Factor IV sediments are characterized by their coarse grain size distribution and high percentage of gravel. Except for silica, the sediments are extremely deficient in the distribution of other elements. On the Chukchi Shelf the distribution of these sediments is extensive in shallow waters along the shores and on topographic highs. Factor IV sediments also occur along the shelf slope in the Gulf of Alaska and in near-shore areas of Nome and northwest tip of St. Lawrence Island.

The gravelly sediments in the Bering Strait and along the shores of the eastern Chukchi Sea, which also belong to factor IV, appear to be of recent age. The sediment may have been contributed by the shore cliffs and because of swift currents have been robbed of their finer fractions. The areas near and north of Herald Shoal that are covered with factor IV sediments suggest that those sediments may have been carried by ice and after deposition have been reworked by water currents.

The ice transport for factor IV sediments is well established from the detailed work conducted in the Nome area of the Bering Sea. Furthermore, the presence of gravelly sediment along the slope in the Gulf of Alaska at water depth of 1000 m or more is clearly indicative of their ice transport. Distribution of factor IV sediments along the Chukchi shores is somewhat enigmatic. Even though these sediments may be contemporary, their textural distribution may have been influenced by the shorefast ice.

Factor V

The high loading of calcium and strontium in factor V sediments clearly suggests their calcareous nature. Sediments mostly contain calcareous shell fragments and are coarse grained, with an extremely poor sorting coefficient. These sediments are found on the topographic highs and in shallow regions of the Gulf of Alaska and Aleutian shelves, respectively, and their distribution is restricted to isolated patches.

The environment is characterized by shallow depth, restricted detritus input, and extensive flushing of water. The bottom is generally covered with gravelly sediments, which provide firm substrate for the biogenic growth and accumulation of skeletal material. The fine sediments (silt and clay) are generally absent as a result of active currents.

Factor VI

The texture of the factor VI sediments is dominated by the silt fraction, with subordinate fractions of clay and sand. The clay fraction is always in excess of sand. The geochemical characteristics of the sediments, however, are not unique. The distribution of aluminum and iron along with other elements is similar to that observed in factor III. The trace elements, however, show a distinct enrichment in factor VI sediments.

The distribution of factor VI sediments, with the exception of a small area northwest of Kayak Island in the central Gulf of Alaska, is restricted to the Bering and Chukchi shelves. These sediments are deposited in waters of moderate depth and generally cover portions of the outer shelf. Yukon River sediments, characteristically silt of factor VI, are partly deposited as river delta and partly carried offshore into Norton Sound.

Factor VI sediments are mostly found on the Bering and Chukchi shelves. There are two possible reasons for the abundance of factor VI sediments in northern shelves. First, the climatic conditions, *i.e.,* permafrost and weathering, mostly produce gravel, sand, and silt grades, which are distributed on the shelf with seaward decreasing grain size. Second, the factor VI sediments are characteristic of sediments discharged by the Yukon River which are deposited on the Bering as well as on the Chukchi Shelf. The large area north of Unimak Pass is covered with silt and the probable sediment source appears to lie in the catchment areas to the east and northeast drained by the Kvichak and Kuskokwim rivers. The northward moving Pacific waters on the shelf certainly will prohibit movement of Yukon River sediments to Unimak Pass. It therefore appears that factor VI sediments are mostly the product of weak chemical and intense mechanical weathering occurring in a cold climate.

In summary, the major controlling factors for texture and geochemistry of the sediments are sedimentary processes, depth, hydrography and provenance.

REFERENCES

Imbrie, J., and Tj. H. van Andel. (1964). Vector Analysis of heavy-mineral data. *Geol. Soc. Amer. Bull.* 75:1131-1156.

McManus, D. A., J. C. Kelly, and J. S. Creager. (1969). Continental Shelf Sedimentation in an Arctic Environment. *Geol. Soc. Amer. Bull.* 80:1961-1984.

Author Index

Subject Index

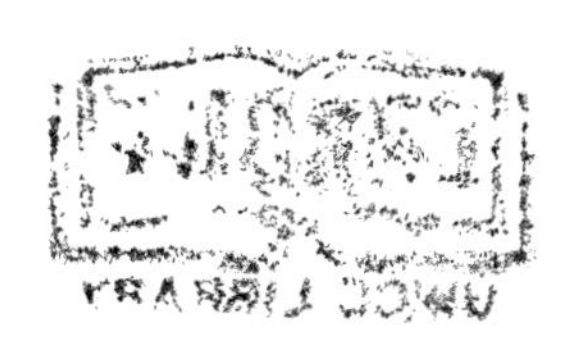